Ungulate Taxonomy

Ungulate Taxonomy

COLIN GROVES and PETER GRUBB

THE JOHNS HOPKINS UNIVERSITY PRESS | BALTIMORE

Printed in the United States of America on acid-free paper
9 8 7 6 5 4 3 2 1

The Johns Hopkins University Press
2715 North Charles Street
Baltimore, Maryland 21218-4363
www.press.jhu.edu

Library of Congress Cataloging-in-Publication Data

Groves, Colin P.
Ungulate taxonomy / Colin Groves and Peter Grubb.
p. cm.
Includes bibliographical references and index.
ISBN-13: 978-1-4214-0093-8 (hardcover : alk. paper)
ISBN-10: 1-4214-0093-6 (hardcover : alk. paper)
1. Ungulates—Classification. I. Grubb, Peter, 1942–2006. II. Title.
QL737.U4G76 2011
569'.6—dc22

2011008168

A catalog record for this book is available from the British Library.

Title-page photo by Bruce Patterson

Special discounts are available for bulk purchases of this book. For more information, please contact Special Sales at 410-516-6936 or specialsales@press.jhu.edu.

The Johns Hopkins University Press uses environmentally friendly book materials, including recycled text paper that is composed of at least 30 percent post-consumer waste, whenever possible.

Contents

Preface

IT HAS NOW been nearly a century since the last major survey of ungulate taxonomy appeared (Lydekker, 1913, 1915a, b; Lydekker & Blaine, 1914a, b). A new treatment would be needed only if, in the meantime, there have been major advances in knowledge and understanding; and indeed there have. This is our first major justification for writing this new book.

In most biota, ungulates are among the most visible fauna on the landscape. In national parks they are readily spotted and much admired by visitors, and it is easy to assume that because these animals are so familiar, everything is known about them, especially how many species there are. Such an assumption would be wrong. Mammalogists expect diversity among small mammals, and seek it out, but they have grown used to accepting that larger ones, because they are thought to be highly mobile (and indeed they may be), are bound to be much less diverse and thus will yield up less information about ecosystems, biogeography, and so on. It may be taken for granted that an antelope of a particular "kind" is, to an extent, expendable: if it becomes endangered in one place, replacements can be found elsewhere and imported in order to top up the local gene pool—such, at any rate, is the assumption, but it needs to be tested: are the local gene pools really exchangeable, or are some of them unique? Taxonomy can go some way toward testing such propositions.

The coming of the molecular revolution has begun to reopen such questions, but morphological research has often not kept up. Because both of us have been traditional skin-and-skull taxonomists, we felt that this was a gap that must be filled, and it was only by going through the different ungulate groups, one by one, that we would find out exactly which gaps remained to be filled, and what research still needed to be done.

We do not intend to present this book as a finished proposition. Time and again, we remark that "more research is necessary"; we want, above all, to stimulate more taxonomic research on ungulates. In some cases we have been able to break new ground, and we hope that these examples will be examined by colleagues and, if found acceptable, used as templates for other studies.

We have both benefited enormously from the assistance of museum curators, discussions with colleagues, and help and advice from families and friends—going back at least 40 years. We would like to take this opportunity to thank (in alphabetical order): Alexei Abramov, Faris Al-Timimi, Renate Angermann, Bruce Banwell, Kurt Benirschke, the late Biswamoy Biswas, Boeadi, Gennady Boeskorov, the late Peter van Bree, Isabella Capellini, Sujit Chakraborty, Lucas Chin, Bill Clark, Cäsar Claude, Juliet Clutton-Brock, Gordon Corbet, Loïc Costeur, Woody Cotterill, the late Peter Crowcroft, Jacques Cuisin, the late Pierre Dandelot, J. C. Daniel, Bijan Dareshuri, Shantini Dawson, Jim Dolan, Keith Dunmall, Feng Zuojiang, Pruthu Fernando, Wolfgang Frey, Val Geist, Alan Gentry, Anthea Gentry, Arnaud Greth, David Happold, Malcolm Harmon, David Harrison, Larry Heaney, Kris Helgen, Mahmoud Reza Hemami, Helmut Hemmer, Daphne Hills, the late John Hill, Kes Hillman, the late Dick Hooijer, Barry Hughes, Paula Jenkins, the late Peter Jewell, Tolga Kankiliç, Mahmoud Karami, Christian Kern, Jonathan Kingdon, Steve Kingswood, Dieter Kock, Katrin Kohmann, Richard Krafft, the late Arlene Kumamoto, Adrian Lister, Ibnu Maryanto, Frieder Mayer, Ji Mazak, the late Vratja Mazák, Erik Meijaard, the late Erna Mohr, Guy Musser, Gertrud Neumann-Denzau, William Oliver, Rohan Pethiyagoda, Francis Petter, Pierre Pfeffer, Roberto Portela Miguez, M. K. Ranjitsinh, Tom Roberts, Jan Robovsky, Kees Rookmaaker, Terri Roth, Klaus Rudloff, Maria Rutzmoser, Jaffar Shikari, Chris Smeenk, Kes Smith, Atanas Tchobanov, Dirk Thys van den Audenaerde, Louise Tomsett, Michel Tranier, Marc Vassart, Wang Sung, Wang Yingxiang, Eileen Westwig, Henning Wiesner, Detlef Willborn, Don Wilson, Roland Wirth, Derek Yalden, Yang Chan Man, Yang Qisen, Zainal-Zahari Zainuddin, Tatjana Zharkikh, and the late Klaus Zimmermann. Above all, we would like to thank Phyll and Eileen, who have been with us and supported us all the time.

WHAT ARE UNGULATES?

"Ungulate" means "having hoofs." This is nowadays taken to imply that part of the terminal phalanx is encased in a sturdy hoof, although Lydekker (1913) required, inferentially, only that the toenail be thickened for it to count as a hoof; hence, his inclusion of elephants and hyraxes in his catalog of ungulates.

This book takes ungulates to be members of the Artiodactyla and Perissodactyla; moreover, not all of the Artiodactyla are ungulates, if the Cetacea should actually be included in the order (as they should be, to make the Artiodactyla monophyletic).

A NOTE ON SYNONYMY

The synonymies given for most of the taxa we list are incomplete. Given that full synonymy lists are readily available in other sources, we have thought it unnecessary to repeat them here. In general we include only those synonyms that are little known or controversial or otherwise contextual. Synonymy entries that we consider doubtful are preceded by a question mark.

IN MEMORIAM: PETER GRUBB

My friend and colleague Peter Grubb died in late December 2006 (Groves, 2006b).

We were first introduced by the late Peter Jewell in 1970, and we repaired to a pub in London to talk about our mutual interests. I knew that Peter had done fieldwork on Soay sheep on the island of Hirta in the Outer Hebrides, and on giant tortoises on Aldabra in the Indian Ocean, so at first I was slightly surprised that he was interested in mammalian taxonomy. It turned out that this was, in fact, a consuming passion for him, one that had overtaken animal behavior. We started to plan strategies, beginning with a research trip around European museums (Berlin, Bonn, Frankfurt, Leiden, Tervuren) in 1973, after which we separated to take up posts overseas, I in Canberra, he in Accra. From these distant centers we would write to each other, gradually work up papers together, and send each other our sole-authored manuscripts for comment. "Groves and Grubb" alternated with "Grubb and Groves" in the literature, and our colleagues got the impression that we were making a takeover bid for the ungulates, perhaps for all the mammals.

The unfortunate thing was that, while my post was a continuing one, and I happily established my teaching and research profile at the Australian National University, Peter's, at the University of Ghana, was a fixed-term appointment. At its termination in the late 1970s, when he returned to the UK, he found a very tight market in the university world. Most zoology departments were not especially interested in taxonomy at the time; he did apply for an advertised post in animal behavior, and was shortlisted, but he was, in effect, not very interested and withdrew. And so it was that he became a schoolteacher.

He was not unhappy teaching biology, but chafed a little under the restrictions of the system. It meant that he had comparatively little time now for his research, but his friends supported him in his applications for leave to make overseas research trips—notably, a very productive one to museums in North America—and his output, plus our joint output, continued, only much more slowly than before. As his reputation grew, he came to be invited to more and more conferences and symposia, though I doubt whether his school ever really realized what a gem they had on their hands. His first and only trip to China, to take part in the cataloging of the rediscovered Heude collection (Braun et al., 2001), was shorter than planned, simply because his employers grumbled at the prospect of him staying away for more than two weeks.

Different professional colleagues will remember him for different things. In ungulate taxonomy, he will long be remembered as the person who replaced assumptions by research and wrote the first sensible things for half a century or more on such groups as African buffaloes (Grubb, 1972), bushbuck (Grubb, 1985a), or wild pigs (Grubb, 1993b); I myself got straightened out about musk-deer in a revealing paper (Grubb, 1982b). Wild pig aficionados think of his astonishing demonstration that warthogs are much, much more interesting than they had envisaged! Meanwhile, primatologists began increasingly to call on his expertise, such as at an important symposium on primate taxonomy in Orlando, Florida, in 2000.

He was always interested in the bigger picture: relationships between forest and savanna-living relatives in Africa, and interrelationships among African forest faunas. His papers on African forest biogeography are by now standard texts: patterns of speciation (Grubb, 1978a), rainforest biogeography (Grubb, 1979, 1982a, 1990b, 1999b, 2001b), and—throwing light on a neglected field—the East African forest biome (Grubb, 1985b). There is no doubt that his expertise, perspicacity, and insights will be greatly missed.

When, after his death from cancer, I finally began to organize a project that we had always had as a future ambition, a book on ungulate taxonomy, I was invited by his wife Eileen, and his old colleague Barry Hughes, to go through his papers. To my astonishment, there were manuscripts almost ready to be sent for publication, and acres and acres of notes and measurements that he had taken in museums over the years but never found time to work up into publishable form. Using his data, and incorporating his manuscripts (in different stages of completion), I decided to make this book the culmination of the "Groves and Grubb" series. My European research trip in 2009 was made thoroughly manageable by the knowledge that a lot of the basic work had already been done by Peter, and I had mainly to fill in gaps.

I will miss my old friend. I will miss our communications by letter and by e-mail, and, even more, I will miss our too-infrequent meetings, usually when I visited London. We would plan to spend some days together at the Natural History Museum, but our time would begin with sitting down together in an Indian restaurant, shaking hands or slapping palms, and at once beginning to talk as if we had never been apart.

Colin Groves

Ungulate Taxonomy

Theory of Ungulate Taxonomy

WHAT IS TAXONOMY?

It is still true that taxonomy is regarded to some extent as the poor relation of ecology and behavior, but it is again becoming more fashionable nowadays, riding, at least partly, on the coattails of the molecular revolution. The general public, too, are becoming more and more aware of biodiversity and get excited about the discovery or identification of new species; taxonomy, of course, is the basis for understanding biodiversity, as was recently emphasized by Cotterill & Foissner (2009) in the context of the importance of museum collections.

At its best, a taxonomic arrangement (above the species level) is a two-dimensional readout of phylogeny: taxa of the same rank are sister groups, and as the rank/time depth nexus becomes accepted, the level at which these sister groups are ranked tells us approximately when they separated in evolutionary time. This is not always possible: there are only so many taxonomic ranks to go around (not a problem if we want to junk ranks altogether and adopt the phylocode, of course—which we do not), splitting may not always be dichotomous, and we sometimes have evidence of past reticulations. Nonetheless, taken cautiously, a taxonomic scheme can perform something of this function.

SPECIES

In essence, a species is "a lineage . . . evolving separately from others and with its own evolutionary role and tendencies" (Simpson, 1961). This is true, in the sense that it says why species are important in the grand scheme of things (as the units of biodiversity), but it is not useful when it comes to actually identifying them and discriminating them from each other. We are in need of an operational definition. One of us has argued extensively (Groves, 2001a, b, 2004) that what is known as the Phylogenetic Species Concept (PSC) fills this need.

In this book, therefore, we define a species (in sexually reproducing organisms) as follows: "A species is the smallest population or aggregation of populations which has fixed heritable differences from other such populations or aggregations" (modified following Eldredge and Cracraft, 1980; Nixon & Wheeler, 1990; and Christofferson, 1995; also see Groves [2001a, b, and especially 2004] for discussions of the implications of this concept).

Note that, under the PSC, species have to be distinguished by fixed character states, but—in most versions of the concept—these character states do not have to be autapomorphic. It is perfectly feasible that, when a parent species splits into two, evolutionary changes may occur, and even accumulate, in just one of the daughter species, the other persisting essentially unchanged from the parent species; this, indeed, is the basis for progressive clines (see Grubb, 2000a).

We must admit that, when looking at the morphological or other phenotypic features (behavioral, ecological) that differentiate species, it is not always possible to demonstrate that the differences are heritable, but it is important to identify the units of biodiversity that are species, so we argue that it is advisable to start by acting as if they are, until it is shown otherwise.

The advantage of the PSC is that it depends entirely on the evidence to hand; there is no extrapolation.

The Biological Species Concept (BSC), which was the dominant concept of (sexually reproducing) species for some half century following on the great evolutionary synthesis of the 1930s, defined species as being "reproductively isolated," meaning that they did not interbreed with one another in a state of nature. Many biologists still use the BSC as their guiding principle, even if only lip service is paid to its primacy. There are, however, two glaring problems with the BSC, and either, by itself, would be sufficient to raise doubts about its universal applicability.

The first problem concerns allopatric populations, that is to say, populations that are completely out of contact with one another, because they are separated by barriers of some kind (water barriers, tracts of unsuitable habitat, mountain ranges). In such cases, the BSC obliges the taxonomist to hypothesize whether interbreeding "might" occur if their ranges were to meet. This results in an impossible situation: how would one know? The distributions of the black-faced impala and the common impala do not overlap; they have no possibility to interbreed in nature, so how should we classify them? Other criteria began to creep in: are they "different enough" to be ranked as distinct species, or not? In short, the BSC offers no guidance when it comes to allopatric populations, and we are reduced to untestable hypotheses. Don't get us wrong; the book by Mayr (1963), premised on the BSC, is a remarkable piece of work, and even today one can learn endlessly from it, but, over time, the species concept it promotes has been shown to have irreparable flaws.

The second problem is, quite simply, that it is wrong. What we have called the molecular revolution has shown that species actually do interbreed in nature; yet they nonetheless persist as discrete entities. White-tail and mule deer are widely sympatric in North America and demonstrably form separate entities in nature, yet they often have each other's mitochondrial DNA (mtDNA)! Evidently they were intermittently interbreeding, undetected by biologists prior to the 1980s. We will say more about this below; but, for the moment, the important point is that to say that different species are "reproductively isolated" can often be shown to be incorrect.

Paterson (1978, 1980) proposed that a species has its own Mate Recognition System, and that this is the overriding criterion for species rank: what he termed the Recognition Species Concept. This puts us back into the realm of hypothesis, though in a rather different way from the BSC. In most living species, behavioral studies are simply not sufficiently advanced to be able to determine, with any degree of certainty, what these recognition features actually are; in the case of fossil species, one can never know. It is easier to have a "good guess" in the case of many ungulates than, for example, in primates or rodents, but that is all it is. There is an intense interest in detecting what sort of interactions seem to matter in a mating context, and how these patterns differ in various species, but one cannot use these observations as the basis for deciding whether they are actually separate species or not, even if one were certain that one did have complete information.

The Phylogenetic Species Concept does not depend on hypothesis, nor on extrapolation; it depends on the evidence to hand. If the evidence before us indicates that, in some character state or other, two operational taxonomic units (OTUs) are discrete (i.e., nonoverlapping), then the two OTUs are to be classified as distinct species. Further data may show that there is, in fact, overlap, so that the decision to class them as distinct species was wrong, but the important point is that the decision always depends on data, never on extrapolation or hypothesis. The only point at which we are forced to extrapolate is in the inference that the OTUs in question are different populations. Species must be populations in any species concept—there is no getting away from this inference, ever—and unfortunately this is just something that we have to live with; once we have made this extrapolation, we can get back to the dataset without making any more hypotheses. If this results in more—perhaps many more—species being recognized than previously, surely it is a small price to pay for exactitude?

This is as close as we can come to putting a finger on the units of biodiversity. The next level— where we start deducing how these different units are related to each other, what ecological roles they perform, and their relationship to geography—is where the excitement begins for many workers, and we can only agree that this is a matter of intense interest (see, for example, the geospecies concept in Grubb, 2006). Likewise, we have to ensure the conservation of these units, and, if there are more of them than we formerly thought, then more care, and probably more finances, will have to be involved. But first we have to determine what the units actually are.

The characteristics of species under the PSC are therefore as follows:

—They are the terminal points on a cladogram; that is, they are the least inclusive phylogenetic units.

—They are discrete units; that is, it is below the species level that relationships are reticulate (and yet some species are undoubtedly the products of hybridization!—see below).

—They are 100% diagnosable; that is, they have fixed heritable differences between them, though these may, in fact, be expressed only in one sex or at one stage in the life cycle.

—They are genetically isolated, though not necessarily reproductively isolated.

The criteria by which species may be recognized may be morphological, or perhaps physiological or behavioral, or they may be base-pair differences in a DNA sequence. We cannot privilege one source of information over another, as long as there is at least a presumption that they are heritable (to be tested, if feasible).

MOLECULES AND MORPHOLOGY

We have found, and recorded in this book, that in many cases the evidences from DNA (the "molecular evidence") and morphology, including morphometrics, are consilient: they give the same results. In other cases, however, they are inconsistent. In former days, when DNA sequencing was its infancy and was very like magic to most nonpractitioners, the demonstration that two hitherto established taxa were not much different in, say, the control region, was held to require that they should be synonymized: DNA was "the answer." We now know much better; we have remembered that phenotypes result from genotypes (plus, potentially, from an input from environmental effects), so these are also, in their own way, telling us about genetics.

There are several reasons why DNA data, especially mtDNA, might not send precisely the same message as morphology. First, we might have the persistence of ancestral polymorphism in DNA, particularly mtDNA, so that two given populations might not have achieved reciprocal monophyly. R. G. Harrison (1998) has calculated that, where *N* is the size of the two populations, and if *N* has remained approximately constant over time, then the probability of reciprocal monophyly in mtDNA sequences is about 35% after *N* generations, 57% after 2*N* generations, and 83% after 4*N* generations. Two daughter populations of 10,000 individuals each—not an inflated figure for many mammalian populations—will therefore need 20,000 generations to have an even chance of reciprocal monophyly; if an average artiodactyl generation can be envisaged as about five years, this translates to 100,000 years. On the other hand, reciprocal monophyly for a nuclear DNA (nDNA) sequence will be, on average, four times that for an mtDNA sequence and, as morphological characters result from nDNA (not from mtDNA), cases where a population will be homogeneous morphologically, but similar to its neighbors in mtDNA, will be uncommon for this reason, but not impossible.

This, however, applies only in the neutral evolution case. If selection is involved, all bets are off. Thus the second reason why DNA sequences and morphology may be offset is if there has been a selective sweep in one or the other. An example of this might be the case of the plains zebra, *Equus quagga*, in which—despite the clear geographical differences in morphology (striping, size, incisor cupping, mane)—there is no such structuring in mtDNA (Lorenzen et al., 2008b). Evidently, a fairly recent advantageous mutation occurred in the mitochondrial genome (remembering that mitochondrial genes code for important functions in metabolism) and moved rapidly through the species.

The third reason, of course, is introgression. Hybridization between two species is frequently asymmetrical. This idea was first put forth more than half a century ago by Flerov, who described hybrids between sika and wapiti in N China and the Russian Far East: "These hybrids are encountered comparatively frequently in the wild state and have long been known to the Chinese. The male wapiti during rutting drives away the weaker spotted deer male and covers his females" (Flerov, 1952:142 [126 of the English translation]). If this superiority of wapiti stags also applied over hybrid stags in F_1 and subsequent generations—as it well might, since the size of the hinds would increase in consecutive backcross generations, and they would become accessible only to wapiti stags—then the proportion of sika nDNA would halve in each successive generation, until we would we end up with populations that were effectively wapiti but with sika mtDNA. This effect, known as nuclear swamping, seems widespread among ruminants, as we will discuss in several places in this book.

SUBSPECIES

Subspecies are, in principle, geographic divisions of a species. There are strong differences in the frequencies of their particular character states, but subspecies are not absolutely different: they overlap. Therefore they are, to an extent, arbitrary and must in no case be reified.

In the past, there were too many subspecies: large numbers of subspecies are listed for most of the species in such key checklists as G. M. Allen (1939), Chasen (1940), Ellerman & Morrison-Scott (1951), Hall & Kelson (1959), and Cabrera (1961). These subspecies may be differentiated into the Good, the Bad, and the Ugly.

The Good subspecies are 100% diagnosable; hence they are actually distinct species masquerading as mere subspecies—victims of the general feeling around the mid-20th century (and certainly under the influence of the BSC) that taxa should be, if possible, relegated to the status of subspecies of the nearest species as long as "they do not occur together" (a form of wording that constantly recurs throughout the checklist of Ellerman & Morrison Scott, 1951). As noted above, the BSC offers no guidance for the allocation of allopatric taxa, so that the classification of these taxa is forever unfalsifiable—it remains purely a matter of opinion.

The Bad subspecies are points along a cline, or are differentiated at very low frequency levels, or simply are based on one or two individuals that seemed outstanding at the time. The samples that they represent may be interesting for population genetics or in some other respect, but they have no taxonomic standing.

The Ugly subspecies are the ones which are left over. Subsequent studies have shown that they can be differentiated from other subspecies (i.e., from other geographic segments of the same species) at high frequencies, but they are not absolutely differentiated or diagnosable. The dilemma is, what to do with them? There does seem to be some advantage in dignifying them with a trinomial, especially for conservation purposes, but as these Ugly subspecies are arbitrary and unfalsifiable, one cannot insist upon it. Nonetheless, this is what we have in mind when we recognize subspecies in this book, unless we specifically state that they are "provisional."

GENERA

Genera are monophyletic groupings of species. If a putative genus is shown to be nonmonophyletic, then it must be either be disbanded (by dividing it up into its monophyletic constituents), or expanded (by including in it the other putative genera whose independent existence rendered it nonmonophyletic in the first place), or narrowed (by taking out all but the nominotypical species and placing the former in the genera to which they are truly related).

Ropiquet & Hassanin (2005) found that the three species of the genus *Hemitragus* (known as tahr) were more closely related to other caprins than to each other: *H. jayakari* to the genus *Ammotragus*; *H. hylocrius* to *Ovis*; and *H. jemlahicus*, the nominotypical species, to *Capra*. They had three choices. First, they could keep all three species in *Hemitragus*, into which they would also sink *Ovis*, *Capra*, and *Ammotragus* (the prior available name would actually be either *Ovis* or *Capra*). Second, they could retain *Hemitragus*, but for *jemlahicus* alone, and place *jayakari* in *Ammotragus* and *hylocrius* in *Ovis*. Third, they could again retain *Hemitragus* to contain *jemlahicus* alone, but erect new genera for the other two species. They chose the third option (putting *jayakari* in a new genus *Arabitragus* and *hylocrius* in a new genus *Nilgiritragus*), but actually any of the three solutions is, under present circumstances, completely arbitrary.

It is surely abhorrent to maintain such an important taxonomic ranking, one whose name is the first word in a binomial, in this unfalsifiable limbo. Modifying previously expressed ideas going back to Willi Hennig half a century ago, Groves (2001a, 2004) argued that the genus category must be given some objectivity by linking it to time depth and suggested, by using the "principle of least violence," that this would be best put at about the Miocene–Pliocene boundary. Such time-depth information is not always—and even not usually—available, either from the fossil record or from the molecular clock; nonetheless, it is something to aim for, and future studies will doubtless modify some generic recognitions that we have used here.

As far as the tahr are concerned, Ropiquet & Hassanin (2005) gave the following mean separation times (under a local clock model): *Hemitragus* from *Capra*, 4.5 million years ago (Ma); *Arabitragus* from *Ammotragus*, 6 Ma; *Nilgiritragus* from *Ovis*, 4.5 Ma. Under the Groves regime, the genus *Arabitragus* would certainly be valid; the other two—including *Hemitragus* itself!—might be questionable. (Later, however, Ropiquet [2006:figure 18], in her preferred synthetic tree, put all these separations well into the Pliocene, in which case none of the genera could be maintained).

Groves (2001a, 2004) likewise argued that families, as well as genera, should have a time depth going back to the Oligocene–Miocene boundary.

In this book, no formal taxonomic conclusion should be drawn from the fact that, when a genus is large and relatively unwieldy in terms of the numbers of species it contains, we arrange the species according to *species-groups*. These are not intended to be monophyletic units, although they may be in many cases. Instead, they are simply grouping devices, the species within them being united on a phenetic basis.

The concept of geospecies, first used in ornithology, has been applied to primates by Grubb (2006), who defined the concept as follows: "Geospecies are lineages passing through an evolutionary history from the stage when they have lost allopatric sister taxa through extinction, to the stage when they have proliferated by cladogenesis, but not so far that daughter taxa have yet become sympatric." They are, thus, what have been called superspecies (monophyletic lineages of allopatric species), to which have been added species that stand on their own and do not belong to such diverse clusters. The problem with such concepts is always that they are rather fuzzy at the edges. Taxa may quite easily fail to be sympatric not because they have not diverged far enough, but because, however phylogenetically distinct they may be, they may be adapted to quite different ecosystems; or they may simply have retained their original ecological adaptations so as to exclude even distantly related taxa from sympatry. We do not here use the superspecies concept, preferring instead to use the phylogeny-neutral term species-groups, as explained above.

PHENOTYPIC PLASTICITY

In this book, one of the methods that we use a great deal is multivariate morphometrics, especially craniometrics. Geist (1989), however, has particularly urged that morphometrics is not a valid taxonomic tool, because of the ubiquity of phenotypic plasticity. In particular, he has cited the work of Franz Vogt who, in the 1930s, conducted experiments in red deer and roedeer breeding, rearing his unselected breeding groups on an especially high-quality diet (specifically, pressed sesame-seed cakes). Vogt found noticeable increases in body and antler size in both species, and these increases continued over four generations (presumably due to maternal effect); the differences were reversed in the red deer when they are were returned to less high-quality diets. The degree to which the conclusions from these experiments can be applied across the board, such as to the differences between "maintenance" and "dispersal" phenotypes, as Geist has propounded (both in his 1989 survey and elsewhere), is arguable, however. Deer, and also pigs, seem especially susceptible to the effects of different planes of nutrition and other factors (such as temperature) on growth, but these are much less evident in bovids, beyond some early effects, such as milk supply in single versus twin lambs (see the summaries in J. Hammond et al., 1983). In general, we take the "default" situation to be that wild animals, living in environments in which they can flourish and maintain viable populations, grow along a trajectory to achieve an adult size and shape where both are predictable from their genotype (within reasonably narrow limits). The argument maintaining that differences between two populations are entirely a result of environmental effects, or that environmental factors overwhelm genetic differences, needs to be substantiated.

MORPHOMETRICS

Both of us have taken substantial amounts of skull measurements (and sometimes collated flesh measurements) on most of the taxa treated in this book. We have worked out means, standard deviations, and absolute ranges for the measurements for geographic samples, or calculated one measurement as a percentage of another, or plotted two measurements together on a graph, or—whenever we could—included them in a multivariate analysis.

The ideal multivariate method is *discriminant analysis*. This is a method that requires specimens to be assigned to groups, and it then minimizes intragroup variation while maximizing intergroup variation. The major constraint is that there should not be too many variables (measurements, in this case) for the number of specimens per group: as a rule of thumb, most of the groups should have more specimens than there are variables, or we run the risk of a *type I error* (i.e., a false positive).

Sometimes discriminant analysis is not possible, and we use instead the rather unsatisfactory alternative of a *principal components analysis*, in which specimens are entered without grouping them. Whereas in discriminant analysis the different groups, whatever their size, are typically treated equally, so that small groups are given an equal chance to contribute to the final dispersion, in principal components analysis the less well-represented samples are "swallowed up" by the larger samples and fail to contribute equitably to the dispersion. Principal components analysis takes the raw measurements, which are all correlated with each other, and turns them into a number of independent variates; separating the input specimens into groupings is only incidental to the process, and groupings may not emerge at all. That is why, if at all possible, samples need to be grouped.

As long as sample sizes are large enough, the ideal is to compare different samples of restricted geographic origin with each other, aggregating those which turn out not to be discrete. Eventually, a picture is arrived at in which two or more of these aggregations may remain, and, if they are discrete and there is morphometric space between them, they then answer to the requirements of the PSC. If no discrete groups remain, then there is no morphometric evidence for different species.

In multivariate analysis, it is important that all the included measurements should be of the same kind, that is to say, either growth dependent or growth independent, and these two types of measurements should not be included simultaneously in the same analysis. *Growth-dependent measurements* are, in almost all of our cases, those of bone and/or horns. They increase with age, up to maturity at least (and beyond, in cases of indeterminate growth), and they bear an allometric relationship to each other. *Growth-independent measurements* are, in our cases, those of teeth. Teeth do not grow; they erupt and are then of a standard size (saving changes with wear, which usually means size reduction due to interproximal attrition), and they bear no allometric relationship with growth-dependent measurements. It may be of interest to make a bivariate plot between a measurement of skull size and a measurement of tooth size, but skull measurements and tooth measurements should on no account be combined in a multivariate analysis.

A BRIEF HISTORY OF THE TAXONOMY OF UNGULATES

Ungulate taxonomy has advanced, due to—or, in some cases, one might better say suffered under—a variety of sometimes colorful characters. Many of these also, of course, contributed to other areas in mammalogy, including primatology, and their accomplishments have been briefly recounted by Groves (2001a, 2008b), but a few of them will be further highlighted here.

Although Linnaeus (1758, 1766) described quite a few ungulates, the detailed description of species of ungulates did not really begin until the work of Pallas (1766, 1767–1777, 1811), who not only described new species on the basis of his travels throughout the Russian Empire, but gave scientific names to many others, based particularly on descriptions by Buffon, who had conscientiously refused to use Linnaean nomenclature. Pallas was followed, in particular, by Alfonse Desmarest (1822), who described new species from the growing collections of the Paris Museum (they had grown not only as a result of overseas collections by French explorers, but also by the activities of Étienne Geoffroy Saint-Hilaire and others who, as Napoleon conquered one country after another, had visited the museums and cabinets of the conquered countries and confiscated interesting specimens); and by Charles Hamilton Smith (1827), who likewise described many new species largely on the basis of the growing British Museum collections. Between them, these three authors described a surprisingly large proportion of the ungulate species that we recognize even today.

The British Museum collections continued to grow and were described by John Edward Gray—its assiduous, opinionated, and cranky mammal curator—who wrote catalog after catalog from the 1840s to 1870s (he died in 1873). In the 1830s and 1840s, the Museum's collections were enormously enriched by specimens sent by Brian Hodgson, first from Nepal and afterwards from a new base in Darjeeling. Unfortunately, his transfer from Nepal remained unremarked, so that many of these specimens, which were actually from Sikkim, Tibet, or elsewhere, have been labeled "Nepal," and it is only by looking at the date of acquisition and contextual information that these errors can be corrected. Hodgson himself described quite a large number of new species from Nepal, Sikkim, and Tibet; he also contributed his notes and color plates to the Museum's archives, and these are sometimes invaluable in filling in deficiencies of the collection (see, for example, Grubb, 1982b).

The latter half of the 19th century was an era of empire building by Western European nations; especially, this was the period of the notorious "scramble for Africa," which meant a further influx of large mammals—slain by explorers, empire builders, and the big game hunters who followed in their wake—into the British Museum (Natural History), which, by that time, had become a separate institution. These were described by Oldfield Thomas, Richard Lydekker, and others, with a growing contribution by Reginald Pocock who, after his retirement from the London Zoo (he took care to examine the animals that died there, particularly the ruminants, describing their skin glands, noses, and genitalia), worked on a voluntary basis at the Museum.

In France itself, mammalian taxonomy had almost dried up, but in Shanghai a French missionary, Pierre-Marie Heude, became passionately interested in natural history, especially large mammals, and was, whenever possible, diverting his attention from spreading the Word in favor of hunting, collecting live animals, and getting his brother missionaries and other colleagues to send him specimens—not only from China but from neighboring territories, which he loosely and inaccurately referred to as "the Chinese Empire": the Philippines, Indochina, Korea, the Russian Far East, and even Japan. He had not the least idea of species, and described hundreds of spurious ones—often half a dozen or more from exactly the same locality—in the midst of his few genuine ones. For a long time after his death (in 1905), his enormous collection remained underappreciated, and finally was lost to view, until it was rediscovered by CPG in 1996, and finally cataloged by us and our colleagues in 2000.

In the introduction to our catalog, we noted Heude's unchecked excesses of splitting and wrote: "Before we join in the near universal condemnation of this prolixity, it should be remembered that Heude stood at the threshold of a new era in mammalian taxonomy, when for the very first time large samples were becoming available. Used to samples of one or two specimens per region, from which one or two species would be described, the sudden embarrassment of riches proved overwhelming for many taxonomists. Heude was not the only one to suffer bemusement" (Braun et al., 2001:610), and we went on to instance C. Hart Merriam, the founder of American mammalogy, parts of whose record were at least as immoderate.

Despite these contributions, the period from about 1890 to 1924 was really the era of Paul Matschie. This extraordinary individual, perhaps the last notable holdout against the theory of evolution among professional zoologists, registered hundreds, even thousands of specimens, including (and especially) ungulates, that were sent to the Berlin Museum by explorers and hunters in the newly acquired German territories (which, in Africa, were Togo, Cameroon, Tanganyika ["German East Africa"], and Southwest Africa). He also paid regular visits to the Berlin Zoo and examined animals there. The result was an enormous profusion of new species and, when he came to adopt the concept, subspecies—and an overarching theory, outlined in his *Bemerkungen über die Verbreitung der Tiere in den Deutschen Schutzgebieten* (Matschie, 1910a). For him, animal distributions were centered on river valleys; a rather sensible idea, one might say, but Matschie took it to extremes. Every major river valley had to have its own species in every species-group, and so did many rather minor ones; Matschie strove mightily to identify these species and the differences between them. The boundaries of their ranges were, of course, the watersheds between the river systems.

Every so often there would be an exhibition of game trophies in Berlin, and on most occasions Matschie would attend it and publish a report on it. His reports would be peppered with descriptions of the specimens in the exhibition, illustrated by photographs of mounted heads and horns in rows. Often enough, he would declare many of the trophies to represent new taxa (species and subspecies) and duly name them, even though the specimens concerned were mostly destined to disappear into the private collections of those who had obtained them; in a very few cases, the collectors would generously donate their specimens to the Berlin Museum, where at least they would be available in perpetuity, but mostly the only remaining evidence of these type specimens would be the small photos in Matschie's reports in the *Deutsche Jäger-Zeitung*.

Matschie also had a young admirer, Ludwig Zukowsky, who was, at that time, an assistant in the Berlin Zoo, but often came to study specimens in the Museum. In what may have been his first published paper (Zukowsky, 1910), he took Matschie's watershed idea to extremes. For example, apparently on the basis of conversations with Matschie (we cannot trace that Matschie had previously published the idea), hybrids between different species were to be identified by being asymmetrical. Again, in the light of modern work on hybrid developmental instability (Ackermann et al., 2006), this is not so unreasonable, but in Zukowsky's view it meant that the specific characters on one side were those of one of the parent species, and those on the other side, of the other parent species. Where should one expect hybrids? In watersheds, of course; and, in his notorious 1910 paper, Zukowsky identified some hybrids that he claimed came from watersheds. Famously, he figured the horns of a buffalo which had been shot in the district of Bihe, which he said was in the watershed of the N-flowing Cuanza and the SW-flowing Cunene and SE-flowing Cubango (= Okavango) rivers. Bihe, now called Kuito (although the province of which it is the capital is still called Bie [*sic*]), is at 12° 22' S, 16° 55' E, in a highland area that is indeed a watershed, although somewhat to the E of the sources of all the above-mentioned rivers. The horns in question are strikingly asymmetrical, the right horn being directed mainly forward and upward, left horn much more outward and backward. Zukowsky identified it as a hybrid between the Cunene and Cubango buffaloes, and mysteriously seemed to know which horn represented which buffalo, because he made the right horn the type of *Bubalus caffer cunenensis* (subsp. nov.) and the left horn the type of *Bubalus caffer cubangensis* (subsp. nov.) In the same paper, he did it again, describing *Bubalus caffer sankurrensis* and *Bubalus caffer lomamiensis* on another asymmetrical specimen, and *Bubalus rufuensis* on the left horn of a specimen of which he identified the right horn as belonging to the already-described *Bubalus schillingsi* Matschie. Matschie and Zukowsky later coauthored a series of three papers on Lichtenstein's hartebeest, which they divided into a large number of different species, again naming some of the new species on the evidence of one horn or the other of asymmetrical specimens (Matschie & Zukowsky, 1916, 1917, 1922).

Matschie got more and more mystical with age, substituting his watershed hypothesis with one called the "elementary areas of distribution," in which he divided the landscape into rectangles whose sides were the diagonals between every 2° of latitude and longitude (Matschie, 1920). In one of his last papers (Matschie, 1922), he described the distributions of what he regarded as three different species of kiang, thus: *E. kiang* 35/80, *E. holdereri* 37/100, and *E. tafeli* 35/99.

Matschie's colleagues continued to respect him for his vast knowledge, but had long since ceased to take him seriously as a taxonomist. And so it was that, in part as a reaction, mammalian taxonomy entered a world in which lumping held sway. The Phylogenetic Species Concept, proposed in 1983 (but foreshadowed in the 1970s), took a long while to enter mammalogy, although in retrospect its logic should have been more obvious; indeed, to a degree it is a return to the tacit species concepts of Oldfield Thomas and Gerritt Miller. Although one of us (CPG) was feeling his way toward this concept in the late 1990s, and finally published a book employing it (Groves, 2001a), it was only later that Peter Grubb (PG), after discussions with CPG and after reading the work of Cotterill (2003, 2005), also accepted the value of it. Nonetheless, PG decided to be conservative in his approach to species listing; in his contributions to Wilson & Reeder (Grubb, 2005a, b), he took the position that he ought to follow a policy of adopting the classification from the latest revision in each case.

DOMESTICATION

Many species of ungulates have been domesticated, and the evidence tends to suggest that in many cases what we view as modern domestic "species" are a mixture of several different domestication events, or at least a mixture to which several different wild populations have contributed. Some of what has been proposed on this score may, however, be spurious, if the conclusions of Ho et al. (2005) are justified, since these authors argued that rates of molecular evolution are initially extremely high, and then slow down after 1 million years or so. Molecular-clock calculations of the separation time between two domestic breeds often go back well into the Pleistocene, implying that the ancestors of the breeds in question separated well before actual domestication, so that they must be descended from different wild ancestors. If the Ho et al. model holds, however, then such early dates will need to be drastically reduced, and multiple domestication models may not always be necessary.

The nomenclature of domestic "species" has been problematic. After long literature discussions, going back at least to Bohlken (1958), the International Commission on Zoological Nomenclature (ICZN) finally made a ruling (ICZN, Opinion 2027, 2003). A. Gentry et al. (2004) explained what this means for the nomenclature of domestic animals: the name given to wild populations of species take precedence over those given to domesticates. Nonetheless, it is often convenient to continue to refer to domestic animals by their own scientific names, regardless of whether they might, in fact, be of mixed (hybrid) origin.

The first paragraph of the ruling states that under the plenary power, "it is hereby ruled that the name for each of the wild species listed in (2) and (3) below is not invalidated by virtue of being pre-dated by a name based on a domestic form" (ICZN, Opinion 2027, 2003:81). The names in question are listed in a further paragraph. Reference to this ICZN Opinion is made throughout this book in the appropriate places. The Opinion ends with the following sentence: "The names listed in the ruling above, which are the first available names in use based on wild populations, apply to wild species and include those for their domestic derivatives if these are not distinguishable."

PG always regretted that, under pressure of time, he at first did not appreciate the full meaning of the Opinion, and continued to use the names given to domestic animals for their wild ancestors in his contributions to Wilson & Reeder (Grubb 2005a, b). It is most unfortunate that this two-volume book was commissioned by the Smithsonian Institution Press, who then suddenly ceased publishing, and major changes were not permitted in the interim as new publishers for the photo-ready manuscript were sought. Hence the work, as finally published (by the Johns Hopkins University Press), was perforce not quite up-to-date—including in its treatment of domesticated species and their wild forebears.

The process of domestication is complex, but in each case it has apparently followed along somewhat parallel lines. For the biology of domestication, and brain-size changes in particular, see especially Hemmer (1983) and Kruska (2005, 2007).

In this book, we have decided not to deal with domestic animals in any depth. For completeness' sake, however, here is a list of the scientific names applicable to domestic species, together with their probable wild ancestors:

Equus caballus Linnaeus, 1758, horse—from *Equus ferus* Boddaert, 1785

Equus asinus Linnaeus, 1758, donkey (ass, burro)—from *Equus africanus* von Heuglin & Fitzinger, 1866

Camelus bactrianus Linnaeus, 1758, Bactrian camel—from *Camelus ferus* Przewalski, 1878

Camelus dromedarius Linnaeus, 1758, Arabian camel (dromedary)—wild ancestor unknown

Lama glama (Linnaeus, 1758), llama—from *Lama guanicoe* (Müller, 1776)

Lama pacos (Linnaeus, 1758), alpaca—from *Lama guanicoe* (Müller, 1776) × *L. vicugna* (Molina, 1782)
Sus domesticus Erxleben, 1777, pig—from *Sus scrofa* Linnaeus, 1758, and other species
Bos taurus Linnaeus, 1758, European (humpless) cattle—from *Bos primigenius* Bojanus, 1827
Bos indicus Linnaeus, 1758, humped cattle (zebu)—from *Bos namadicus* Falconer, 1859
Bos frontalis Lambert, 1804, mithan—from *Bos gaurus* Hamilton Smith, 1827
Bos javanicus domesticus Gans, 1916, Bali cattle—from *Bos javanicus* d'Alton, 1823
Bos grunniens Linnaeus, 1766, yak—from *Bos mutus* (Przewalski, 1883)
Bubalus bubalis (Linnaeus, 1758), water buffalo—from *Bubalus arnee* (Kerr, 1792)
Capra hircus Linnaeus, 1758, goat—from *Capra aegagrus* Erxleben, 1777
Ovis aries Linnaeus, 1758, sheep—from *Ovis gmelini* Blyth, 1841

CONSERVATION

We unfortunately live in a time when the preservation of the world's biodiversity is in crisis. While we may only infer that substantial numbers of invertebrates are going extinct, unremarked, every year as their habitats disappear, ungulates are large and visible, and the continuing drastic declines in their numbers and ranges are all too obvious. We have said little about this under the different species' headings, but, as one of the major aims of this book is to document taxonomic diversity, we must draw attention to it here. Species treated in this book that are in imminent danger of extinction are listed below. We do not list subspecies, though some, like *Hippotragus niger variani*, are also highly endangered in their own right. We have selected only the most critically endangered:

Equus przewalskii—persists only in captivity, although most of the captive stock consists of what is called the B-line, known to be partly descended from at least two domestic horses; the A-line, almost certainly descended from pure-blooded Przewalski horses, numbers somewhat over a hundred individuals.
Equus africanus—the nominotypical subspecies is possibly extinct, though it may persist as a captive stock (possibly from the hybrid zone?) in private hands; subsp. *somaliensis*, perhaps a distinct species, is critically endangered in Somalia, but a few hundred are protected in Eritrea.
Rhinoceros unicornis—slowly increasing in number under protection.
Rhinoceros sondaicus—the second-rarest of all large mammals, only ±60 remain, but protected.
Dicerorhinus sumatrensis—only 200–300 remain, not well protected.
Diceros bicornis—in the 1970s and 1980s, suffered the most astonishing crash in numbers known for any large mammal, from nearly 100,000 to only 2500 individuals; now very slowly increasing under protection.
Ceratotherium cottoni—on the verge of extinction, only 10 known to survive, all in captivity or semicaptivity.
Choeropsis heslopi—not reported since 1943.
Dama mesopotamica—very low numbers in the wild; bred in captivity.
Rucervus schomburgki—presumed extinct since the 1930s.
Panolia (all species)—persist only in very small populations, except for those on Hainan Island; *P. eldii* from Manipur, whose numbers were once as low as 14, has, however, increased to more than 200 under protection.
Cervus pseudaxis—number unknown in the wild, but the amount is certainly very small; bred in captivity.
Cervus alfredi—number unknown in the wild, but the amount must be extremely small; bred in captivity.
Bos sauveli—probably extinct, but searches for it continue.
Bubalus arnee—survives only in small isolates.
Bubalus mindorensis—perhaps 200–300 survive.
Pseudoryx nghetinhensis—still illegally hunted, despite its small numbers; restricted range and nominal protection.
Gazella acaciae—only about 20 remain; now protected.
Gazella arabica—not seen since its original description, 180 years ago.
Gazella bilkis—may be extinct, or nearly so, because of overhunting.
Eudorcas rufina—not recorded since the 1880s, and presumed extinct.
Procapra przewalskii—only about 200 remain, in small isolated populations around Qinghai Lake and on the Buha River.

Alcelaphus tora—uncertain whether any still exist.
Beatragus hunteri—500–1000 remaining; declining in its original habitat; introduced into Tsavo East National Park, where about 100 persist.
Damaliscus selousi—unknown if any still exist.
Oryx dammah—extinct in the wild, though breeding well in captivity.
Oryx leucoryx—well represented in captivity, but populations reintroduced into the wild have been under constant threat from poaching.
Addax nasomaculatus—reduced to a few small, fragmentary populations.
Capra walie—found in small fragments, mainly in Semyen National Park, Ethiopia, where it has recently increased to about 500 under protection.
Cephalophus jentinki—threatened by hunting and habitat loss in its very small range.
Cephalophus adersi—survives in only very small numbers in both known populations (Zanzibar, and Sokoke Forest in Kenya).

BIOGEOGRAPHY OF UNGULATES

The pictures of the biogeography of ungulates that emerge from this book will take some time to analyze and digest, but we would like to draw attention to a few findings, some of which were unanticipated:

—Vrba's stenotypic versus eurytopic division—This, which in hindsight now seems so obvious, is abundantly confirmed. We return to this theme in our introduction to the Bovidae.
—The Sudanic grasslands ecosystem—Many species are spread throughout this ecosystem, from Senegal in the W to the borders of Ethiopia in the E, with little or no geographic variation, except sometimes for minor differentiation W and E of the Nile. This contrasts with many of the same species-groups in E and S Africa, where their ranges are restricted to isolates. An example would be the contrast in the *Damaliscus korrigum* group between the Sudanic *D. korrigum* (with the dubiously distinct *D. tiang* E of the Nile) and the five species dotted around the E African landscape.
—The Serengeti-Mara ecosystem—It contains a number of apparently unique species. It is therefore important not only as an example of a nearly intact E African ecosystem in its own right, but also as the sole repository of a range of endemic species.
—The lower Oubangui River—Wide as it is, this river appears to be insignificant as a faunal barrier; many forest species (in ungulates as in primates) go right across it as if it did not exist. There is some suggestion that its upper course, mostly a continuation of its major constituent, the Uele River, was at one time (during the Early Holocene?) part of the Shari-Logone system, and has since been captured by the Congo to form the lower Oubangui.
—The Cape region—It has its own range of species, separate from their congenerics farther N.
—Lower Yangtze River—There is a special ecosystem S of the lower Yangtze, apparently centered on Lake Po-yang, with its own species (as far as ungulates are concerned) of deer and pigs.
—Russian Far East—Finally, we would like to draw attention yet again, as has been done previously (especially in Russian works), to the separate faunal status of the Russian Far East, including at least parts of the Korean peninsula and the Manchurian provinces of China.

ORGANIZATION OF THE TAXONOMIC SECTIONS

In each case, after introducing the family, we discuss its constituent genera in turn, listing species (and subspecies, where applicable) either before or after we have detailed our own research and/or that of others. As noted in the introduction, we have not invariably listed complete synonymies.

PART I: PERISSODACTYLA

THERE ARE NO particular problems in the order Perissodactyla, at least at the higher levels: there are three families in the living fauna, and, as far as we know, no one has demurred from this assessment.

All perissodactyls are hindgut fermenters; none has a complex stomach, and all have a large caecum and colon.

From being the most diverse ungulates in the Oligocene, they began to decline in the Miocene, presumably in the face of competition from the rising artiodactyls, and today perissodactyls are greatly reduced in both diversity and abundance.

I

Equidae

EQUIDAE GRAY, 1821

The taxonomy of the Equidae is summarized by Groves & Ryder (2000). There is only a single living genus in the family, but this divides well into three subgenera.

Horses (subgenus *Equus*)

Two species of wild horses survived into modern times, but one (the tarpan) has been extinct since 1920, the other (the takhi or Przewalski horse) is extinct in the wild but survives in a flourishing captive stock (very little of which, however, is purebred). The tarpan is supposed to be the ancestor of domestic horses, though this needs to be tested genetically.

Domestic horses, even within a single breed, show high diversity in mtDNA, but much less in Y-chromosome DNA (Kavar & Dovč, 2008). The unrooted neighbor-joining tree in Kavar & Dovč(2008:figure 7) shows Przewalski sequences closely associated with one domestic cluster (C1), and a sequence from the Urals associated with a different cluster (C4); interestingly, wild horses from Germany and from Siberia were not (in most cases) closely associated with any domestic horse cluster.

Equus ferus Boddaert, 1785

tarpan

1785 *Equus ferus* Boddaert. Bobrovsk region, near Voronesh.

1826 *Equus sylvestris* Brincken. Bialowieza Forest.

1912 *Equus gmelini* Antonius. Bobrovsk region, near Voronesh.

1936 *Equus przewalskii sylvaticus* Vetulani. Bialowieza Forest.

A full synonymy is given, because there is still some misunderstanding about the applicability of the names. The names *ferus* and *gmelini* are objective synonyms, both being based on a description of steppe tarpans by Gmelin (see Groves, 1994); the names *sylvestris* and *sylvaticus* have different bases, but both refer to wild horses from the Bialowieza Forest.

The name *Equus ferus* Boddaert, 1785, was placed on the Official List of Specific Names in Zoology by ICZN, Opinion 2027 (2003).

Suggestions have been made that forest and steppe tarpan were distinct, but there is no objective evidence for this.

The history of the extinction of this species has been documented by Heptner et al. (1961), Groves (1994), and elsewhere.

Equus przewalskii Poliakov, 1881

Przewalski horse, takhi

1881 *Equus przewalskii* Poliakov. "In the steppes of eastern Dzungaria" (Harper, 1940).

1903 *Equus hagenbecki* Matschie. Ebi Spring, Njursu, and the Urungu River.

1909 *Equus equiferus* Hilzheimer. On this name, frequently but mistakenly ascribed to Pallas, see Harper (1940).

1946 *Equus caballus gutsenensis* Skorkowski. Gashun Oasis or Gutchan Mountains, Mongolia.

This species, and its former distribution and the history of knowledge of it, are described in Groves (1994).

The evidence of intermediate horses E of the Urals, on the basis of which Groves (1986) originally proposed to put this species and the last together as *Equus ferus*, is equivocal, and the differences between the (now extinct) European wild horse and the (still extant) Mongolian wild horse are clear cut (see summary in Groves, 1994).

Asses (subgenus *Asinus*)

The taxonomy of the Asian forms follows Groves & Mazák (1967), but is updated.

Equus africanus

The name *Equus africanus* Heuglin & Fitzinger, 1866, was placed on the Official List of Specific Names in Zoology by ICZN, Opinion 2027 (2003).

On this species, see Groves (1986, 2002) and Groves & Ryder (2000), and for the nomenclature, see Groves & Smeenk (2007). Its former distribution in the Middle East was discussed by Uerpmann (1987).

Equus ?africanus africanus von Heuglin & Fitzinger, 1866

Nubian wild ass

Long ears, 182–245 mm; shoulder height 115–121 mm; hoofs high, narrow. Dorsal stripe always present, nearly always complete from the mane to the tail-tuft. Leg-stripes, where present at all, restricted to a few bands at the fetlocks. Skull length usually less than in other subspecies; diastema short; postorbital constriction well marked; orbit placed high, generally interrupting the dorsal profile.

Formerly known from the Atbara region of Sudan, E to the Red Sea Hills, and probably into northernmost Eritrea; not known to survive in the wild.

Groves (1986, 2002) showed that there are slight differences between populations. In the Atbara population, the color is more buffy; the shoulder-stripe is thick, well marked, short; the diastema is shorter; the occipital crest is narrower; there are never traces of leg-stripes. In the Red Sea population, the color is grayer; the shoulder cross is nearly always thin and poorly expressed, and sometimes absent altogether; the diastema is longer; the occipital crest is broader.

Wild asses—probably truly wild, not feral—are found in some parts of the Sahara, and apparently resemble this subspecies more. They have not been described as a separate taxon.

Equus ?africanus somaliensis Noack, 1884

Somali wild ass

Ears shorter, 187–200 mm; shoulder height 120–125 mm; hoofs wider, lower. Dorsal stripe often absent, and, when present, generally incomplete and broken at some point along the dorsum; shoulder cross, when present, poorly expressed. Leg-stripes present from above the carpus and tarsus to the hoofs. Skull longer; diastema long; postorbital constriction less marked; orbit low-placed. Apparently longer-legged, shorter-bodied than *E. a. africanus.*

It is not clear whether the two "subspecies" are discrete and ought to be regarded as species.

Central (C) and S Eritrea, the Danakil region of Ethiopia, Djibouti, and N Somalia. There are slight differences between the Somali population and those farther W.

Beja-Pereira et al. (2004) sequenced mtDNA (control region) from what they deemed Nubian wild asses, from a zoo sample of Somali wild asses (Berlin Tierpark, from stock bred in the Hai Bar Reserve, Israel, originally captured in the Danakil region), and from domestic donkeys from all over the Old World. The DNA of the Somali and presumed Nubian wild asses formed two entirely different clusters, with very strong bootstrap support and Bayesian posterior probability. The DNA of domestic donkeys likewise formed two distinct clusters, one being intermixed with the reputed Nubian wild ass sequences, the other being close to, but separate from, the Somali wild ass sequences (both of these domestic groups were represented in all geographic regions except for the far W of Africa, where only the "Nubian" type was found).

On the face of it, this would be interpreted as meaning that Nubian and Somali wild asses are different species, and that some domestic donkeys are descended (at least in the female line) from the Nubian wild ass, while others are from a source similar but not identical to the Danakil population of the Somali wild ass, the two lineages being mostly intermixed in local stocks. But the true identity of what was reported as Nubian wild asses must be questioned. Beja-Pereira et al.'s (2004) samples (fresh feces) were obtained from "two different wild herds" in NE Sudan, 21.08° N, 36.20° E and 20.46° N, 36.49° E. These localities are somewhat to the N of the recorded localities of the Nubian wild ass, which is interesting and not necessarily a problem in itself, but, in fact, no evidence of the survival of Nubian wild asses at all has been forthcoming since the 1930s, the date of the collection of museum specimens by Powell-Cotton and by Mason, and of the capture of the Hellabrun-Catskill breeding stock, both apparently in the Red Sea Hills just on either side of the Sudan–Eritrea border. Therefore, it cannot be guaranteed that Beja-Pereira et al.'s samples really were from Nubian wild asses, as opposed to feral donkeys.

***Equus kiang* Moorcroft, 1841**

kiang

On this species, and its separation from *Equus hemionus*, with which it had previously been considered conspecific, see Groves & Mazák (1967). It needs to be reiterated that the kiang is strongly and consistently distinct from other Asian equids.

Equus kiang kiang Moorcroft, 1841

western kiang

Large size; short nasals; short toothrow; short diastema; color very dark, especially in winter, with dark areas predominating on the flanks.

Ladakh and neighboring parts of SW Tibet.

Equus kiang holdereri Matschie, 1911

eastern kiang

Somewhat larger; long nasals; long toothrow; long diastema; color lighter, less red, with the light area of the underside reaching nearly halfway up the flanks.

E Tibet and W plateau of Sichuan.

Equus kiang polyodon Hodgson, 1847

southern kiang

Very small in size; short occiput.

Plateau area of Sikkim and the region just to the N.

On this taxon, see Neumann-Denzau & Denzau (2003).

***Equus hemionus* Pallas, 1775**

onager, Asian wild ass

Dorsal outline of the skull very straight; skull low-crowned; dark color of the upperside extending well down the flanks, restricting the white of the underside; dark hoof-rings present; dorsal stripe extends down the tail to the tuft.

Equus hemionus hemionus Pallas, 1775

Mongolian wild ass

1775 *Equus hemionus* Pallas. Tarei Nor, Transbaikalia, 50° N, 115° E.

1911 *Equus (Asinus) hemionus bedfordi* Matschie. Supposedly from Kobdo, Mongolia.

1911 *Equus (Asinus) hemionus finschi* Matschie. NE of Zaisan Nor.

1911 *Equus (Asinus) hemionus luteus* Matschie. Surin Gol, Gansu.

Large size; sandy coloration, grading into the light fawn of the underside, especially in summer; light border to the dorsal stripe vague or absent, especially in the adults.

Mongolia, extending N into Transbaikalia and W into Xinjiang and SW Siberia in the Semipalatinsk area.

Groves & Mazák (1967) divided this taxon into two, from N and S Mongolia, respectively (the N form extending N into Transbaikalia and W into the Semipalatinsk region), differing in the degree to which they exhibit grading (desertlike) coloration. Neumann-Denzau & Denzau (1999) criticized this arrangement, arguing that all Mongolian hemiones are consubspecific. They also discussed the remarkably disruptively patterned taxon described as *Equus onager castaneus*, known only from a painting, leaving it "in limbo" for future investigation. For this reason, the full synonymy is given here, indicating, in agreement with Neumann-Denzau & Denzau (1999), that all wild asses from Mongolia and surrounding areas are essentially the same.

Based on 31 microsatellite loci (hereafter referred to as microsatellites), this taxon appears, curiously, to be somewhat closer to *E. kiang* than to *E. h. onager* (Krüger et al., 2005).

Equus hemionus kulan (Groves & Mazák, 1967)

kulan, Turkmenistani onager

Smaller in size; pale sandy yellow in summer, darker and more disruptively colored in winter; below the eye, the border between the colored area of the face and the white of the interramal region cutting diagonally across the jaw angle; occiput strongly elongated.

Turkmenistan; nowadays restricted to Badkhyz and a few populations derived from that reserve.

Whether *E. h. kulan* and *E. h. onager* are really subspecies of *Equus hemionus* needs to be reexamined.

Equus hemionus onager Boddaert, 1785

Iranian onager

Medium size; rather pale yellow-brown, less disruptively colored than *E. h. kulan* in summer; below the eye, the border between the colored area of the face and the white of the interramal region more or less following the line of the jaw angle; occipital region short.

Iranian plateau.

Based on 31 microsatellites, *E. h. onager* is closely related to *E. h. kulan*, but is distinct (Krüger et al., 2005).

***Equus khur* Lesson, 1827**

Indian wild ass, khur

Somewhat smaller in size; skull very high-crowned; nasals convex anteriorly, making the dorsal profile of the skull sinuous. Color of the upperside much restricted, with the white of the underparts reaching at least halfway up the flanks; no hoof-rings; dorsal stripe ending at or slightly beyond the base of the tail, not reaching the end.

Presently restricted to the Little Rann of Kutch, but formerly W and NW through the Thar Desert into Baluchistan, as far as Kandahar.

***Equus hemippus* I. Geoffroy, 1855**

Syrian wild ass, achdari

Very small in size; intergrading desert coloration; white of the underparts restricted to the belly; dorsal stripe extending to the tail-tuft; hoof-rings present. Skull high-crowned; dorsal outline concave.

Now extinct. Lived in the desert country of Iraq, Jordan, and Palestine until the 1930s.

Zebras (subgenus *Hippotigris*)

On zebras in general, see Groves & Bell (2004); on *Equus quagga*, see Grubb (1981).

An influential paper by Bard (1977) proposed that the striping differences between the species (or species-groups) of zebra can be explained if the pattern is laid down in the third week of development in *E. quagga*, in the fourth week in the *E. zebra* group, and in the fifth week in *E. grevyi*. We wonder if this idea might be profitably considered anew at a more detailed taxonomic level.

Leonard et al. (2005) extracted two short mtDNA sequences from four preserved specimens of "true quagga." All quagga specimens were found to be nested within the plains zebra clade; indeed, even within the S African subclade. Six of specimens of E African plains zebras (*E. q. boehmi*) formed a separate subclade to the S African specimens, while the seventh individual was part of the S subclade, suggesting that the N was the original stock, and that the S populations were progressively derived from those farther N, losing stripes as they moved S.

Here, we follow Groves & Bell (2004), wherein full synonymy will be found.

Equus zebra Linnaeus, 1758

Cape mountain zebra

Black stripes broader, wider than the white interspaces. Relatively small in size; males smaller than the females. Muzzle broader; diastema longer than in *E. hartmannae*.

Mountainous regions of the S Cape.

Equus hartmannae Matschie, 1898

Hartmann's zebra

Black stripes much narrower than the white interspaces. Size larger, with sexual dimorphism. Muzzle relatively narrow; diastema shorter. In morphological characters, this species appears, on the available evidence, to be diagnosably distinct from *E. zebra*.

Coastal hills of S Angola and N Namibia.

Using both microsatellites and the mitochondrial control region, Moodley & Harley (2005) could not confirm that the lineages of *E. hartmannae* and *E. zebra* were reciprocally monophyletic; nonetheless, all control region haplotypes were exclusive to one or the other (3 to *E. zebra*, 25 to *E. hartmannae*).

Equus grevyi Oustalet, 1882

Grévy's zebra

Kenya, NE of the Mt. Kenya region, into S Ethiopia and S Somalia, though whether it still occurs anywhere but in Kenya is unclear.

Groves & Bell (2004) could find no significant geographic variation in this species.

Equus quagga

plains zebra

On the general biology of this species (under the name *Equus burchellii*), see Grubb (1981).

Neuhaus & Ruckstuhl (2002) found year-round reproduction in this species in Etosha National Park, Namibia; they concluded that, in association with the harem mating system, this forced the synchronization of the time budget between the males and the females, and might explain the relative lack of sexual dimorphism in body size.

Reynolds & Bishop (2002) found that S and E populations differ notably in size; E African populations are far smaller than their fossil predecessors, both Late Pliocene and Early Pleistocene. There has been a size reduction in S Africa as well, but a much less dramatic one. The conclusion was that during Late Pliocene and Early Pleistocene times, the E African climate was drier and more seasonal; hence the strong size reduction of the E African zebras represents adaptation to a warmer, wetter environment, whereas the South African environment remained much the same.

There is very marked geographic variation in external features, as well as in skull size and shape and in the presence or absence of infundibula in the lower incisors (Groves & Bell, 2004). In contrast to this, Lorenzen et al. (2008b) were able to find no convincing geographic structuring in either the mitochondrial control region or in any of seven microsatellites, although there was some separation of northernmost (*E. q. borensis*) and southernmost (*E. q. quagga*) populations. Evidently a good new model sweeps the board.

Equus quagga quagga Boddaert, 1785

quagga

Striped with dark brown on a buffy white background on the head and neck; flanks darker, yellower, the stripes fading out along the flanks, sometimes reaching the haunches; legs white. Skull relatively broad; occiput narrow.

Formerly from S Cape and Free State N to the Vaal-Orange system; now extinct.

Equus quagga burchellii (Gray, 1824)

bontequagga, Burchell's zebra, Damara zebra, Zululand zebra

Striped with brownish black or black on an off-white ground on the head, the neck, and the flanks;

stripes reaching the haunches, and appearing variably down the upper segments of the limbs, sometimes with traces below the carpus and the tarsus. One or two shadow stripes between the main bold, broad stripes on the haunch. Skull small, narrow, especially the occiput; incisors relatively broad; diastema longer.

The size increases from the SE to the NW of the range, but the differences are slight.

Formerly found extensively N of the Vaal-Orange system, from Etosha and the Kaokoveld to Swaziland and KwaZulu-Natal; surviving at the two ends of the distribution, but extinct in the midportion. Traditionally, this extinct midportion has been deemed to be the "extinct true Burchell's zebra," and the surviving (indeed, flourishing) Damara zebra has been known as *Equus quagga antiquorum*—but there seems to be no rationale for such a division. As there has been such a lot of misunderstanding about this, it is necessary to reiterate yet again that the type locality of *Hippotigris antiquorum* Hamilton Smith, 1841, is the Mafeking district, very close to the Kuruman district from which *Asinus burchellii* was itself described; it is therefore very unlikely that the two names designate different taxa, and quite categorically the name *antiquorum* is not available for any putative Namibian subspecies ("Damara zebra") if such should prove separable. References to an "extinct true Burchell's Zebra" are inappropriate, as the subspecies survives not only in Namibia but also in KwaZulu-Natal.

Equus quagga chapmani Layard, 1865

Chapman's zebra

Striped with black on an off-white ground on the head, the neck, the flanks, the haunches, and the upper segments of the limbs, broken below the carpus and the tarsus, but generally extending to the hoofs. Shadow stripes prominent on the haunches and, usually, on the neck. Skull large; snout shorter, narrower.

"Transvaal" N to Zimbabwe, W to the Okavango, the Caprivi Strip, and S Angola.

Equus quagga crawshayi de Winton, 1896

Crawshay's zebra

Striped with black on a striking white ground on the head, the neck, the flanks, the haunches, and the whole of the limbs, extending unbroken to the hoofs. No shadow stripes. Stripes and interspaces strikingly narrow, with five or more stripes joining the belly-stripes (in other subspecies, the stripes nearly always fewer than five). Infundibulum on the lower incisor usually reduced to a cup, or absent altogether.

Zambia E of the Luangwa River, Malawi, SE Tanzania, and Mozambique, S at least to Gorongoza.

Equus quagga boehmi Matschie, 1892

Grant's zebra

Striped with black on a white ground on the head, the neck, the flanks, the haunches, and the whole of the limbs down to the hoofs. Shadow stripes poor or absent. Stripes and interspaces broad. Infundibula on the lower incisors usually reduced to a cup, or absent altogether. The smallest subspecies.

Zambia W of the Luangwa, W at least to Kariba, Shaba, Tanzania (except for the SE), SW Uganda, SW Kenya, E Kenya (E of the Rift Valley), and into southernmost Ethiopia, perhaps as far as the Juba River, Somalia.

The Juba population appears to be very small and maneless, with white ears; better evidence would probably establish it as a distinct taxon, for which the name *isabella* Ziccardi, 1959, is applicable.

Equus quagga borensis Lönnberg, 1921

half-maned zebra

Resembling *E. q. boehmi* externally, except that the mane is tufty or absent in the adult males, but usually present in the females and the subadults; backs of the ears white. Larger than *E. q. boehmi*, and more sexually dimorphic; rather short diastema. Infundibula on the lower incisors apparently absent.

NW Kenya, from Lake Baringo to Karamoja, and extreme SE Sudan.

2

Tapiridae

TAPIRIDAE GRAY, 1821

On the general biology of this family, see Eisenberg et al. (1987). A detailed anatomical description and analysis of the proboscis of the tapir has been published by Witmer et al. (1999).

All tapirs are externally dark, often with white rims to the ears, sometimes white around the mouth and around the hoofs; only in one genus, *Acrocodia*, is there what may be described as a color pattern on the body itself. Infants have a pattern of longitudinal whitish stripes, generally partly or almost entirely broken up into rows of elongated spots.

Although they are rather similar externally, the skulls of the four living species are very different from each other; there are a few differences in the dentition; and *Acrocodia* differs from other tapirs in the structure of the foot (Radinsky, 1963).

A phylogeny based on the mitochondrial COII gene (Ashley et al. 1996) placed *Tapirus terrestris* and *T. pinchaque* very close to each other and quite distant from the other two species that, in one of the two most parsimonious trees, actually formed a clade together (though with only 51% bootstrap support); in the other tree, they formed successive branches. Applying a molecular clock, and calibrating it to the date of separation between the Tapiridae and the Equidae, they suggested that (what we here call) *Acrocodia* separated from the neotropical tapirs 21 to 25 Ma, and *Tapirella* separated from the South American species 19 to 20 Ma, while the separation between *T. terrestris* and *T. pinchaque* was only about 3 Ma, which, as they point out, was not too long after the formation of the land bridge between North and South America and, presumably, the arrival of tapirs in South America.

Using the cytochrome *b* gene, Ruiz-García et al. (in press) found that the three American tapirs do, in fact, form a clade with respect to the Malayan tapir. They experimented with different calibration points, finding that the best was a split between *T. terrestris* and *T. pinchaque* at 3 Ma, and between the Asian and American clades at 18 Ma, with *T. bairdii* splitting from the other American species at some period in between.

Tapirus Brisson, 1762

This was one of the generic names which A. Gentry (1994) requested to be conserved, even though other names from Brisson (1762) were to be rejected.

The facial skeleton is low, so that the nasal cavity is fairly low; the dorsal profile of the cranium slopes upward from the nasals to the crown. The occiput is narrow, sloping backward. The dorsal spiral grooves on the nasals are narrow and sharply bordered posteriorly. The cartilaginous nasal septum is relatively small, and sits in something of a slot between the maxillae on either side. The anterior end of the premaxilla is not greatly curved down, so that the incisors occlude in the plane of the occlusal line of the cheekteeth. The ectoloph crest is notched. The MtI rudiment does not articulate with MtIV.

These tapirs are relatively slenderly built, with a slender and not excessively elongated proboscis. The body is uniformly gray, brown, or black, though there may well be (and usually are) white or whitish marks on the head, the throat, and the legs.

Tapirus terrestris (Linnaeus, 1758)

South American tapir

The skull has a very high, narrow sagittal crest, which rises posteriorly and gives the dorsal outline of the skull a strongly convex contour. Unusually, the sagittal crest does not develop from the fusion of the superior temporal lines, but emerges from the middle of the braincase very early in ontogeny, the temporal muscles being already well developed; with growth, the temporal muscles continue to enlarge, and the sagittal crest heightens (Holbrook, 2002). The nasal bones are narrow and reduced. On the upper molars, both the anterior and the posterior cingula are narrow.

The body color is brown, varying from fawn to grayish or nearly black, with somewhat lighter cheeks, chin, and throat, and there are often white rims to the ears. There is a short, but conspicuous, narrow mane from the forehead to the withers. The proboscis is relatively short.

Shoulder height 77–110 cm; weight 180–250 kg; greatest skull length 375–407 mm, varying geographically.

Geographic variation remains to be analyzed in detail, but the following patterns can be seen (collected data of CPG):

1. From E Brazil S to Minas Gerais and Paraguay, and W to Mato Grosso and Bolivia—Body color tends to be pale gray-brown to tawny; greatest skull length averages about 389 mm ($N=5$) in Paraguay, rising to 398 ($N=7$) to the NW, in Bolivia. If subspecific status for these four geographic forms should turn out to be warranted, this would be the nominotypical *Tapirus terrestris terrestris* (Linnaeus, 1758), the type locality being Pernambuco.
2. From Venezuela to the Guyanas, Brazil (Amazonas), E Ecuador, and Peru—Body color tends to be black-brown, with contrastingly light to white cheeks and throat. Relatively small in size; greatest skull length averages 382 mm ($N=32$). The name *Tapirus terrestris tapir* (Erxleben, 1777) may be available for this form if it is a distinct subspecies, as the distribution given by Erxleben is "from the Isthmus of Darién to the Amazon."
3. N Argentina and Rio Grande do Sul—These tapirs tend to be very dark, and large in size; greatest skull length averages 407 mm ($N=8$). The available name for this taxon would be *Tapirus terrestris spegazzinii* Ameghino, 1909 (type locality: Río Pescado, Dept. Orán, Salta, Argentina).
4. W Colombia—Dark to medium brown in color, and small in size; greatest skull length averages 381 mm ($N=5$). This putative taxon is *Tapirus terrestris columbianus* Hershkovitz, 1954.

Tapirus pinchaque (Roulin, 1829)

mountain tapir, woolly tapir

The braincase is low and flat; the sagittal crest forms with age, by the fusion of the superior temporal lines, remains low, and is not arched. The nasal bones are very long, and in the plane of the frontals, so that there is only a shallow rise posteriorly. The anterior cingulum on the upper molars is wider than in *T. terrestris*, as is the posterior cingulum, especially on the second molar, causing the tooth margins to be convex.

Body color is nearly black, with striking white lips and (usually) white ear rims. Pelage is long, thick, and coarse. The proboscis is fairly short.

T. pinchaque is shorter-legged than *T. terrestris*, but more compactly built, so that the weight is much the same. Height 75–80 cm, weight about 225–250 kg; greatest skull length 370–380 mm.

Confined to high altitudes (2000–4000 m) in Colombia and Ecuador.

Tapirella Palmer, 1903

The nasal bones are short, but continued anteriorly by the nasal septum, which is much more strongly ossified than in *Tapirus*; the edges of the maxillae are dorsally prolonged to support this massive ossification. The plane of the nasals and their continuation is abruptly stepped down from that of the frontals. The temporal lines do not fuse, but form a double sagittal crest. The maxillae and the premaxillae are slightly down-bent, so that the incisors occlude slightly below the plane of occlusion of the cheekteeth.

The anterior premolars are more molarized and wider than in other tapirs, and the cingula are narrow, so that the anterior margins of all three molars are straight. The ectoloph crest is not notched.

Tapirella bairdii (Gill, 1865)

Baird's tapir

The body color is dark brown; there is a short, barely expressed mane. The proboscis is greatly enlarged (supported by the septal ossification). The legs are long; the tail is also long, up to 13 cm (not above 10 cm in other species).

Height up to 120 cm; weight up to 300 kg; greatest skull length 409–433 mm.

From southernmost Mexico, Belize, and Guatemala SE to W Colombia, where it is apparently sympatric with *Tapirus terrestris*.

Acrocodia Goldman, 1913

The skull is very high-crowned; the nasal bones emerge high above the floor of the nasal fossa; the dorsal profile, from the nasals to the occiput, is straight; the superior temporal lines do not meet, and the dorsal table between them is low and wide; the occiput is wide and vertical. As in *Tapirus*, the cartilaginous nasal septum is small and sits in a slot between the two maxillae. The dorsal depressions of the nasals form deep, wide channels that are prolonged back into the frontals as a pair of laterally spiraled scrolls. There is a noticeable frontolacrimal tubercle. The anterior premolars are not well molarized; the upper molars have wide anterior and posterior cingula. The anterior ends of the premaxillae curve down, so that the incisors occlude at a level well below the occlusal line of the cheekteeth.

In the foot skeleton, the presumed rudiment of the first metatarsal is swung well laterally to articulate with MtIV as well as MtIII and the entocuneiform (Radinsky, 1963).

Heavily built, with a long, capacious proboscis.

Acrocodia indica (Desmarest, 1819)

Malay tapir

The general color is black, with the body, behind the shoulders and above the hips, white (except in a rare all-black mutant, known from the Palembang region). There is no mane.

Height 90–105 cm, weight 250–320 kg; greatest skull length 410–444 mm.

Sumatra (not the extreme N) and the Malay Peninsula. Recorded in Laos (Deuve, 1972), and may have survived in Borneo until well into the 20th century (Cranbrook & Piper, 2009).

3

Rhinocerotidae

RHINOCEROTIDAE GRAY, 1821

The most up-to-date survey of the status of living rhinos is by Amin et al. (2006). On rhinoceros taxonomy and evolution, see Groves (1997c).

It was Flower (1876) who laid the basis of modern rhinoceros taxonomy by reducing the number of recognized species to a provisional six (one of them regarded as extremely doubtful) and arranging them into three genera. Flower also listed and described the most cogent characters by which the species can be recognized, illustrating some of them. Groves (1971) elucidated the differences between the horns of the existing taxa.

There have been many attempts to decipher the exact branching order of the three major lineages: *Rhinoceros*, *Dicerorhinus*, and African rhinos (*Diceros* and *Ceratotherium* being universally accepted to be branches of the same major lineage). Groves (1983a), for example, on the basis of a detailed survey of skulls and teeth, put *Dicerorhinus* closer to *Rhinoceros*. The latest attempt (Willerslev et al., 2009) was unable to resolve the pattern: the separation times, regardless of whether *Rhinoceros*, *Dicerorhinus*, or the African lineage separated first, were all between 32 and 30.4 Ma (note that a mean of 60 Ma was specified for the age of the root [i.e., separation from other perissodactyls], and used as the calibration point—which some paleontologists might think outrageously early!).

Rhinoceros Linnaeus, 1758

A single pair of large, compressed, blocklike upper incisors, with (at least in the young adults) a much smaller lateral pair; mandibular central incisors very small, the laterals large, procumbent, and pointed, forming formidable tusks. Postglenoid and posttympanic processes united below the external auditory meatus. Occipital plane very wide, forming a flat-topped triangle in posterior view, and sloping forward in lateral view. Dorsal outline of the cranium deeply concave. Foramen magnum more or less triangular. Nasal bones anteriorly pointed. (Following Flower, 1876; Pocock, 1945; Chakraborty, 1972.) Externally, there are deep folds in the skin, as follows: along the angle of the jaw (much deeper in the males than in the females); two to three vertically around the neck, with one emerging posteriorly and running horizontally for some distance, about halfway up the thickness of the neck; vertically anterior to the foreleg; horizontally around the base of each foreleg; vertically posterior to the foreleg, travelling across behind the shoulder to meet its opposite number on the other side; vertically anterior to the hindleg, crossing in front of the croup to meet its opposite number, and continuous, with one going horizontally across the base of each hindleg; posterior to each hindleg, going vertically across behind the croup and above the tail; and, finally, one running horizontally from the posterior hindleg fold, level with lower surface of the tail base, and usually not reaching the anterior hindleg fold. Body hairs sparse, but locally detectable in certain lights and to the touch. One single nasal horn.

A survey of the Asian rhinos, incorporating much of what was known up to that date, is in Groves (1982a).

Rhinoceros unicornis Linnaeus, 1758

Indian one-horned rhinoceros, greater one-horned rhinoceros

1758 *Rhinoceros unicornis* Linnaeus. Rookmaaker (1998) showed that the sources for this species actually included *R. sondaicus*, as well. The lectotype is the rhino in Dürer's famous woodcut of 1515.

1779 *Rhinoceros rugosus* Blumenbach. For this name, see Rookmaaker (2004).

1875 *Rhinoceros jamrachi* Jamrach. Manipur.

1970 *Rhinoceros unicornis bengalensis* Kourist. Bengal.

Only synonyms not listed in Ellerman & Morrison-Scott (1951) are given here.

In the upper molars and premolars, the crochet and the crista become united with wear, cutting off an accessory valley from the medisinus. The posterior end of the vomer is thickened, fused to pterygoid plates on either side; the mesopterygoid fossa is narrow, the bases of the pterygoid plates almost meet posteriorly.

The skinfolds are deep, and hang loosely; the horizontal fold emerging posteriorly from the neck folds is short, petering out before reaching posterior foreleg fold. The body skin is studied with low knobs. Sexual dimorphism in the development of the mandibular and neck folds is considerable: adult males develop a noticeable "bib."

The upper lip has a median point, which is used for browsing, but more usually the point is "tucked in," so that the lip becomes a wide cropping organ.

Rookmaaker (1980) has traced the recent distribution of this species, at least as far as India, Bangladesh, and the Indochinese region are concerned. It was known in modern times from Bihar E through N Bengal and Assam, perhaps just entering Bangladesh, as far E as the Tirap Frontier tract (uncertain), and possibly even into N Laos.

Rhinoceros sondaicus Desmarest, 1822

Javan one-horned rhinoceros, lesser one-horned rhinoceros

The upper molar and premolar crochet and crista almost never fuse.

The posterior end of vomer is thin, free from pterygoids; the mesopterygoid fossa is wide. (Mainly following Flower, 1876; Pocock, 1945.)

The skinfolds are much "tighter" than in *R. unicornis*; the horizontal fold emerging posteriorly from the neck folds slants upward, crossing the withers to join its opposite number on the other side, forming a kind of "saddle." The skin has no raised knobs, but it is covered by a network of very fine cracks; the head is small in proportion to the bulk of the body; the females are slightly larger than the males, on average, and the males develop only a small "bib."

The upper lip is long and pointed, suitable for browsing; the premaxilla does not fuse to the maxilla until old age, extending the gape and increasing the mobility of the upper lip.

The horn in the male is smaller, and more slender, than in *R. unicornis*, and the supporting nasal bones are narrow; in the females, the horn is very reduced.

The characterizations of the subspecies follow Groves (1967a), and especially Groves & Guérin (1980); see also Groves & Chakraborty (1983). Some of the differences appear discrete, and nonoverlapping, and the Indochinese and Bengal taxa are theoretically candidates for recognition as full species, but caution is required, since only very small sample sizes are available.

Rhinoceros sondaicus sondaicus Desmarest, 1822

1822 *Rhinoceros sondaicus* Desmarest. Java; exact locality indeterminable (Rookmaaker, 1982).

1824 *Rhinoceros javanicus* E. Geoffroy & F. Cuvier. Java.

1827 *Rhinoceros camperis* Griffith. Java.

1829 *Rhinoceros javanus* G. Cuvier. Java.

1836 *Rhinoceros camperii* Jardine. Java.

1868 *Rhinoceros floweri* Gray. Sumatra.

1868 *Rhinoceros nasalis* Gray. "Borneo," probably Java.

1876 *Rhinoceros frontalis* von Martens. Error for *Rhinoceros nasalis*.

The full synonymy is given here, following Rookmaaker (1983).

Dentition relatively small; palatine bones 44%–53% of the total palate length (posterior to the incisive foramina); palatine bones >80% as wide as long, and usually longer; mandibular corpus not greatly deepening posteriorly, its depth at the third molar 111%–144% of that at the second premolar; on P^2, protoloph fusing with the ectoloph with wear; anterior premolars (probably actually milk molars) remaining in place for much of the animal's adult life; crochet on the premolars usually doubled; crista on the premolars usually much reduced on Sumatran and mainland individuals, but often quite absent on those from Java, and absent on the molars in all animals.

Java, Sumatra, and the Malay peninsula as far N as Tenasserim. Survives, as far as is known, only in Ujung Kulon National Park, westernmost Java.

Rhinoceros sondaicus annamiticus Heude, 1892

1892 *Rhinoceros annamiticus* Heude. Tay-ninh, 700 km from Saigon (see Braun et al., 2001).

Rookmaaker (1983) gave the corrected reference to this name.

Nuchal surface more inclined anteriorly; facial skeleton relatively low, with a comparatively deep dorsal concavity; palatine bones 39%–63% of the total palate length; palatine bones 36%–66% as wide as long; mandibular body as in nominotypical *R. s. sondaicus*; in P^2, protoloph fusing with the ectoloph with wear, as in nominotypical *R. s. sondaicus*; anterior premolars (probably actually milk molars) remaining in place for much of the animal's adult life; crochet on the premolars usually not doubled; crista usually absent on the premolars, but usually present, though small, on the molars. Long bones apparently longer and more slender than in nominotypical *R. s. sondaicus*; medium metapodials shorter and wider.

Rookmaaker (1980) noted that this subspecies is depicted on a relief in Angkor Wat in Cambodia and he gives evidence that it may still have existed in S Laos at the time of his writing. It possibly still occurs in Cat Tien National Park, NE of Saigon.

Rhinoceros sondaicus inermis Lesson, 1838

1838 *Rhinoceros inermis* Lesson. Sunderbans.

Dentition relatively large; palatine bones occupying a much larger proportion of the hard palate, >80% of the total palate length posterior to the incisive foramina; palatine bones 50%–80% as wide as long; mandibular corpus noticeably deepening posteriorly, its height at the third molar 167% of that at the second premolar; on P^2, protoloph remaining separate from the ectoloph, apparently throughout life; anterior premolars (probably actually milk molars) shedding before maturity; crochet on the premolars usually not doubled; crista present on the premolars, absent from the molars.

In females, the horn appears to have been absent altogether.

Now extinct. Rookmaaker (1980) has recorded its former distribution in India. *R. s. inermis* was known from Moraghat in the Bhutan Duars, probably the Sikkim Terai, the Sundarbans (where its more detailed distribution was given by Rookmaaker, 1997), and Chittagong. In Bangladesh, there is a record from Sylhet.

Dicerorhinus Gloger, 1841

The maxillary incisors are as in *Rhinoceros*; the mandibular central incisors are lacking, but the laterals are tusklike. There is no subaural closure. The occipital plane is narrow, somewhat rectangular in posterior view, and more or less vertical in lateral view. The dorsal outline of the cranium is not strongly concave, with a slight convexity on the frontals. The foramen magnum is pear-shaped, its dorsal rim reduced upward into a narrow prolongation. The nasal bones are anteriorly pointed. (Following Flower, 1876; Pocock, 1945; Chakraborty, 1972.)

A survey of the Asian rhinos, incorporating much of what was known up to that date, is in Groves (1982a).

Dicerorhinus sumatrensis Fischer, 1814

Sumatran rhinoceros

The name *Rhinoceros crossii* Gray, 1854, belongs to this species, but to which subspecies is unknown.

On this species in general, see Groves (1982a). A survey of the biology of *D. sumatrensis*, intensively studied at the Melaka Zoo, has been given by Zainuddin et al. (1990).

The skinfolds are tighter than in *Rhinoceros*, and only the fold behind the foreleg travels over the shoulder to join the one on the opposite side. There are two horns, with the frontal horn placed well behind the nasal horn; the bases are not continuous. The skin of the nose is heavily cornified, lacking the wrinkles anterior to the nostril characteristic of other rhinos, with only a single deep crease running between the anterior margins of the nostrils from side to side, allowing mobility for the upper lip.

Body hair is comparatively long and profuse, compared with other rhinos.

A population aggregation analysis of mtDNA by Amato et al. (1995), using six individuals from Sumatra, four from Borneo, and seven from West Malaysia, found extremely low diversity: only a single haplotype on Borneo, one on the mainland, and two in Sumatra. Two variable sites differentiated the Borneo samples from the samples from Sumatra plus the mainland.

Subspecies follow Groves (1967a); see also Groves & Chakraborty (1983).

Dicerorhinus sumatrensis sumatrensis (Fischer, 1814)

1814 *Rhinoceros sumatrensis* Fischer. Fort Marlborough, Bintuhan district, Sumatra.
1822 *Rhinoceros sumatranus* Raffles. Sumatra.
1873 *Ceratorhinus blythii* Gray. Pegu.
1873 *Ceratorhinus niger* Gray. Sunghi-Njong district, Malacca.
1874 *Rhinoceros malayanus* Newman, nom. nud.
The full synonymy is given here, following Rookmaaker (1983).

Size large; teeth medium to small; occiput low, narrow.

Sumatra and the Malay peninsula, N to Pegu. The only certain records referable to it in the Indochinese district seem to be Nong Het (Laos), and Lao-Dao and Nhatrang (Vietnam), but whether they were referable to this subspecies is unknown.

Dicerorhinus sumatrensis harrissoni (Groves, 1965)

1912 *Rhinoceros borniensis* Hose & McDougall, nom. nud.
1965 *Didermocerus sumatrensis harrissoni* Groves. Suan-Lambah, Sabah.

Size small; teeth small; occiput narrow, but proportionally high, and forwardly inclined.

Borneo.

Dicerorhinus sumatrensis lasiotis (Sclater, 1872)

1872 *Rhinoceros lasiotis* Sclater. About 16 hours (as the elephant marches) S of Chittagong (Harper, 1940).

Size very large; teeth proportionately very large; occiput broad and high.

The northernmost subspecies, now perhaps extinct. The westernmost locality was the Sankosh River, and Rookmaaker (1980) mapped other records in India and Bangladesh.

Diceros Gray, 1821

The incisors are rudimentary (possibly only deciduous) or absent, and the front of both jaws is abbreviated. There is no subaural closure. The occipital plane is narrow, extending backward in lateral view, and overhanging the occipital condyles posteriorly. The nasal bones are much thickened, and truncated anteriorly. There are two horns, with the frontal horn placed immediately behind the nasal horn; the bases of the two horns are sometimes continuous. The upper lip is long, pointed, and mobile. Body hairs are lacking, except on a few specific places (the tail tip, the ear rims, and the base of the horns).

Diceros bicornis (Linnaeus, 1758)

black rhinoceros

The basic outlines of our knowledge of geographic variation were summarized by Groves (1967b) and Rookmaaker & Groves (1978). Rookmaaker (1995) proposed that this taxonomy, with later published amendments, should be used pending future work; he registered a strong objection to the practice, at that time becoming increasingly common, of using subspecific names for what authors specifically stated were "ecotypes." Later (Rookmaaker, 2005), he also deplored the tendency to "redefine the subspecies of the black rhinoceros on an ad hoc or geographically limited basis," insisting that they must be allocated with reference to real data.

Rookmaaker (1983) showed that the names *Atelodus bicornis* var. *plesioceros* and *A. b. platyceros* Brandt, 1878, belong to this species, but to which subspecies is not known; he showed however, that the name *A. b. porrhoceros*, occurring in the same publication, referred to an animal from Upper Nubia; hence, it is a synonym of *D. b. brucii*.

Harley et al. (2005) showed that, in the main, genetic differentiation in microsatellites between the subspecies was reasonably good, and that all individuals examined could be ascribed to their own subspecies with some confidence, with the exception of *D. b. chobiensis*.

Diceros bicornis bicornis (Linnaeus, 1758)

1758 *Rhinoceros bicornis* Linnaeus. As shown by Rookmaaker (1998), Linnaeus based this name on specimens both of this species and of *Dicerorhinus sumatrensis*. A neotype, with the locality being the Cape of Good Hope, was validly designated by Zukowsky (1965).

1797 *Rhinoceros africanus* Blumenbach.

1836 *Rhinoceros keitloa* A. Smith. Mafeking.

1842 *Rhinoceros gordoni* Lesson. Near the sources of the Gamka River.

1845 *Rhinoceros camperi* Schinz; not Jardine, 1836. Cape of Good Hope.

1845 *Rhinoceros niger* Schinz. Tsondap, Nuuibeb Mountains, S Namibia.

As before, the full synonymy is given here, following Rookmaaker (1983), who investigated the basis of the names given.

Very large in size, with large teeth. Mandibular first premolar apparently always absent in the adults. Maxillary [second, third, and fourth] premolars commonly possessing a crista. Radius under 80% of the humerus length (in all other subspecies, it is over 80%). Foreleg slightly shorter than the hindleg; limbs apparently rather slender. Skin apparently smooth, not deeply folded.

Formerly from the Cape N to Kuruman, and apparently to S Namibia, but not the coastal strip E of the Drakensberg. This subspecies has long been extinct; this vital information was inadvertently absent from Groves (1967b), but was emphasized by Rookmaaker & Groves (1978) in their general review of this subspecies. Unfortunately, the failure by Groves (1967b) to note that it was extinct has resulted in the name becoming (spuriously) widespread among wildlife workers for the still-extant Namibian population.

Diceros bicornis chobiensis Zukowsky, 1965

1965 *Diceros bicornis angolensis* Zukowsky. Virui waterhole, Mossamedes, Angola.

1965 *Diceros bicornis chobiensis* Zukowsky. Konsumbia, parent streams of the Loma River, tributary of the Cuando River, SE Angola.

Somewhat smaller than nominotypical *D. b. bicornis*. No crista on the upper premolars, or very minute. Skin with deep body folds.

Okavango region.

Apparently only a single individual is known to survive. This individual was studied by Harley et al.

(2005), using microsatellites; it was not well differentiated from *D. b. minor.*

Diceros bicornis minor (Drummond, 1876)

1876 *Rhinoceros bicornis major* Drummond. Country SE of Zambezi.
1876 *Rhinoceros bicornis minor* Drummond. Zululand.
1893 *Rhinoceros bicornis holmwoodi* Sclater. Udulia, 50 mi. S of Speke Gulf.
1947 *Diceros bicornis punyana* Potter. Hluhluwe.
1965 *Diceros bicornis nyasae* Zukowsky; a conditional name, hence unavailable. N end of Lake Malawi.
1965 *Diceros bicornis rowumae* Zukowsky; a conditional name, hence unavailable. Inland from Mikindani, Tanzania.
1972 *Rhinoceros kulumane* Player.

Still smaller in size. Externally, characterized by a short, compact body, with well- marked skinfolds and a large head. Mandibular first premolar apparently always absent in the adults in E Africa, but retained in 60% of those from Hluhluwe. No crista on the upper premolars, or very minute. Foreleg slightly longer than the hindleg.

From KwaZulu-Natal N into NW Tanzania and the SW borders of Kenya.

Diceros bicornis occidentalis (Zukowsky, 1922)

1922 *Opsiceros occidentalis* Zukowsky. Kaokoveld-Cunene region.

Size similar to *D. b. minor*; much broader across the zygomata. External phenotype resembling *D. b. chobiensis* and *D. b. minor.* No crista on the upper premolars, or very minute.

N Namibia and S Angola.

This subspecies was not recognized by Groves (1967b), but the study of a few more specimens has shown that it can be largely distinguished from *D. b. minor.*

Harley et al. (2005) studied microsatellites from 53 individuals of this subspecies (but unfortunately used the designation "*D. b. bicornis*," for the reasons given above) from Namibia, and from populations reintroduced into South Africa. They were clearly assignable to their own subspecies, rather than to any other.

Diceros bicornis michaeli Zukowsky, 1965

1965 *Diceros bicornis michaeli* Zukowsky. Between Engaruka and Serengeti.
1965 *Diceros bicornis rendilis* Zukowsky. N Guaso Nyiro.

Still smaller (one of the smallest subspecies); relatively broad-skulled. External habitus the same as in *D. b. chobiensis*, *D. b. minor*, and *D. b. occidentalis*. Mandibular first premolar apparently always absent in the adults. No crista on the upper premolars, or very minute. Foreleg slightly longer than the hindleg.

NW Tanzania into E Kenya, including Tsavo and the N Guaso Nyiro.

Individuals of this subspecies studied by Harley et al. (2005) were differentiated, in microsatellites, from those of other subspecies he studied.

Diceros bicornis brucii (Lesson, 1842)

1842 *Rhinoceros brucii* Lesson. Tscherkin, between the Bahr Salaam and Atbara rivers.
1878 *Atelodus bicornis* var. *porrhoceros* Brandt.
1897 *Rhinoceros bicornis somaliensis* Potocki. "Somaliland."
? 1947 *Diceros bicornis palustris* Benzon. Near Aweng, N of the Lol River, Bahr-el-Ghazal.
1965 *Diceros bicornis atbarensis* Zukowsky. Anseba Valley, Eritrea.

Size as in *D. b. minor*; very narrow across the zygomata. Mandibular first premolar retained in the adults. Crochet on the maxillary premolars simple, not bifid (unlike most subspecies); these premolars commonly possessing a crista. External appearance, as documented in photographs, clearly different from those of *D. b. minor*, *D. b. occidentalis*, and *D. b. michaeli*, with the skinfolds much less marked.

Somalia, the Ogaden and N Ethiopia, Eritrea, and N Sudan E of the Nile; an isolated population, possibly attributable to this taxon, occurred in a small district of Bahr-el-Ghazal. Probably extinct.

Diceros bicornis ladoensis Groves, 1967

1965 *Diceros bicornis ladoensis* Zukowsky; a conditional name, hence unavailable under Zukowsky's authorship. Shambe, near Lado, southernmost Sudan.
1967 *Diceros bicornis ladoensis* Groves.

Larger than *D. b. minor*; broad-skulled, especially across the occipital crest. Mandibular first premolar apparently always absent in the adults. No crista on the upper premolars, or very minute. Foreleg slightly shorter than hindleg.

Kenya Rift Valley NW into S Sudan, E of the Nile.

Diceros bicornis longipes Zukowsky, 1949

1949 *Diceros bicornis longipes* Zukowsky. Mogrum, Chad.

A very small subspecies, equivalent in size to *D. b. michaeli*; shorter occipital crest; long distal limb segments;

wide, square base to the horns. Mandibular first premolar retained in the adults. Crochet on the maxillary premolars simple (as in *D. b. brucii*), not bifid (unlike most subspecies); these premolars commonly possessing a crista.

Formerly from SW Chad, Central African Republic (CAR), N Cameroon, and NE Nigeria. Probably extinct.

Microsatellites of a single individual of this subspecies were studied by Harley et al. (2005), who found it more strongly differentiated from the other subspecies he studied than these latter were from each other.

Ceratotherium Gray, 1868

The basic skull characters are as in *Diceros*. The incisors are rudimentary (possibly only deciduous) or absent; the front of both jaws is abbreviated. There is no subaural closure. The occipital plane is narrow, extending backward in lateral view, and overhanging the occipital condyles posteriorly. The nasal bones are much thickened, and truncated anteriorly.

Compared with *Diceros*, the cranium is extremely elongated; the occipital crest is enormously prolonged posteriorly; the dorsal outline of the cranium is still less concave; the protoloph and the metaloph on the upper cheekteeth are strongly curved backward, fusing with wear; the cheekteeth are higher-crowned, with much crown cement; the mandibular symphysis is very broad; the ascending ramus is more backwardly inclined; there is a pre-sacral eminence, formed by the anticlinal status of the 17th or 18th thoracic vertebra; there are two horns, the frontal horn placed somewhat behind the nasal horn; the bases of the two horns rarely touch; the mouth is broad and blunt, with scarcely any median prolongation of the upper lip; there is a muscular nuchal hump; copious subcutaneous fat causes atrophy of the body folds and the costal grooves; the prepuce is translucent, and the penis has eccrine as well as apocrine glands. (Following Groves, 1975b.) Body hair is much reduced, but still detectable in *C. simum*.

Tests on 30 samples from Umfolosi showed that, as far as microsatellites are concerned, there seemed to be a rather low degree of genetic variability, presumably due to the severe population decrease that had occurred by the beginning of the 20th century.

A hybrid between a black rhinoceros male and a southern white rhinoceros female was born in a large enclosure in South Africa, and verified by cytogenetics and microsatellite analysis (Robinson et al., 2005). It was intermediate in head shape between the two; its mouth was more like that of a white rhino, but its ear more like that of a black rhino. To judge by the photos, the shape of the dorsal outline of the body was intermediate between the two parents, and the deep costal grooving was more like that of a black rhino.

Groves et al. (2010) reviewed the differences between northern and southern white rhinos and came to the conclusion that they are strongly distinct species, with about 1 million years of separation. For conservation purposes, this conclusion is especially significant, and it is possible that, had the nature and consistency of the differences between the two been realized earlier (rather than being hidden as "mere subspecies"), more stringent efforts would have managed to save the critically endangered northern species.

Ceratotherium simum (Burchell, 1817)

southern white rhinoceros

1817 *Rhinoceros simus* Burchell.
1827 *Rhinoceros burchellii* Lesson.z
1827 *Rhinoceros canus* Griffith.
1847 *Rhinoceros oswelli* Elliot.
1866 *Rhinoceros kiaboaba* Murray.
1878 *Atelodus simus* var. *camptoceros* Brandt.
1878 *Atelodus simus* var. *prostheceros* Brandt.
The synonymy includes some names elucidated by Rookmaaker (1983).

This is the only species of rhinoceros that today can be considered reasonably "abundant"; it has been widely reintroduced over much of its former range S of the Zambezi, and beyond it (Kenya).

Ceratotherium cottoni (Lydekker, 1908)

northern white rhinoceros

1908 *Rhinoceros simus cottoni* Lydekker.

A very detailed description of this species, a comparison with the southern white rhino, and an argument as to why it has to be considered a distinct species have all recently been given by Groves et al. (2010) and thus will not be repeated here.

It is unfortunate that a study such as this was made only as this species arrived at the brink of extinction; one hopes that the remnants can still be persuaded to breed, and that it will not be "saved" by hybridizing it with *C. simum*.

PART II: ARTIODACTYLA

HERE, WE CONTINUE to refer to the even-toed ungulates as the order Artiodactyla, while acknowledging that the cetaceans belong in them (as a sister group to the Hippopotamidae; see especially Gatesy et al., 1999). In this, we follow Helgen (2003; see also Asher & Helgen, 2010), who argued that if the Cetacea were the sister group to the Artiodactyla as such, the name Cetartiodactyla would be appropriate; yet, as they are deeply nested within the Artiodactyla, there should be no change of ordinal name (on the precedent of, for example, the Carnivora, which retained that name even with the inclusion of Pinnipedia).

A succinct summary of the phylogeny, including major fossil representatives, has been given by A. W. Gentry & Hooker (1988). As far as ruminants are concerned, there seems to have been a very rapid radiation in the Late Oligocene or Early Miocene, making it very difficult to decipher the exact interrelationships of the living families (Kraus & Miyamoto, 1991).

Waddell et al. (1999) classified what they called the Cetartiodactyla (and we the Artiodactyla) into unranked nested categories. We, more old fashioned, retain suborders and infraorders.

Similarly, Skinner & Chimimba (2005) used the new divisions, but at the ordinal level. In the cohort Ferungulata, they recognized three superorders (Ferae for the orders Pholidota and Carnivora, Paraxonia [*recte* Mesaxonia] for the Perissodactyla, and Cetartiodactyla), dividing the latter into three orders: Suiformes, Whippomorpha, and Ruminantia (no Tylopoda, simply because there are no camels in S Africa). This incorporates the view, which we endorse, that as long as all groups are monophyletic, there is no need to make a classification that is strictly dichotomous.

We prefer to retain the Artiodactyla (as we call it) as an order, because the molecular evidence suggests that its subdivision into the modern crown-groups dates from the Cenozoic, and we classify the Artiodactyla into four suborders. Like Skinner & Chimimba (2005), we use an old name, Ancodonta, for just the hippos (among the living fauna, at any rate), but at the infraordinal level. We, like them, use one of Waddell et al.'s (1999) names, Whippomorpha, for Cetacea plus Ancodonta, but as a suborder. Our final difference from Skinner & Chimimba (2005) is that we use Suina for the pigs and peccaries, rather than Suiformes, because the latter name was used in older classifications for this group plus the hippos.

Our arrangement, therefore, is as follows:

Suborder Tylopoda
 Family Camelidae
Suborder Suina
 Family Tayassuidae
 Family Suidae
Suborder Whippomorpha
 Infraorder Cetacea
 Infraorder Ancodonta
 Family Hippopotamidae
Suborder Ruminantia
 Infraorder Tragulina
 Family Tragulidae
 Infraorder Pecora
 Family Moschidae
 Family Giraffidae
 Family Antilocapridae
 Family Cervidae
 Family Bovidae

4

Tylopoda

CAMELIDAE GRAY, 1821

Lama G. Cuvier, 1800
Guanacos

B. González et al. (2006) reviewed geographic variation in guanacos.

Lama guanicoe guanicoe Müller, 1776
guanaco

1776 *Camelus guanicoe* Müller. Placed on the Official List of Specific Names in Zoology by ICZN, Opinion 2027 (2003). Krumbiegel (1944) proposed to fix the type locality as Patagonia, but see below.
1782 *Lama huanacus* Molina. Type locality: Quillota, 32.54° S, 71.16° W (O. Thomas, 1917). Homonym of *guanicoe*, according to Cabrera (1961), for the following reason. The *International Code of Zoological Nomenclature*, Article 33.1, states: "A subsequent spelling of a name, if different from the original spelling, is either an emendation, . . . or an incorrect subsequent spelling, . . . or a mandatory change." Art. 33.2.3 states: "Any other emendation [i.e., not 'demonstrably intentional'] is an 'unjustified emendation' . . . and is a junior objective synonym of the name in its original spelling."
1880 *Palaeolama mesolithica* Gervais & Ameghino. "Prehistoric" remains from Cañada de Rocha, near Luján, Buenos Aires Province. According to Cabrera (1961), this takes precedence over *Lama guanicoe voglii* (below) if the guanaco of the pampas is distinct.
1944 *Lama guanicoe voglii* Krumbiegel. The pampas of Argentina.

Krumbiegel (1944) recognized four subspecies of guanaco: *L. g. guanicoe*, *L. g. huanacus*, *L. g. voglii*, and *L. g. cacsilensis*. He described nominotypical *L. g. guanicoe*, from Patagonia, as brownish red, with the head, the cheeks, and the nape light gray. B. González et al. (2006) surveyed the literature and reported that guanacos in Buenos Aires Province have "a cinnamon rufous half-line on the dorsum," with lighter flanks, and dirty white on the abdomen; continental populations of Patagonia are slightly reddish brown, those N of the Magellan Straits are yellowish, and those of Tierra del Fuego are dark reddish and long-coated. Accordingly, Krumbiegel's (1944) characterization is only partly correct. Those from Chile (*L. g. huanacus* of Krumbiegel, but this name cannot be used; see above) are described by Krumbiegel as having a black-gray head, with dark gray cheeks, and a gray to black nape. Krumbiegel described his new subspecies *L. g. voglii* as softer, lighter in color, sandy (compared with the strong red-brown in the Patagonian guanacos), with a light gray tone on the cheeks and the nape. He mentioned no specimens in museums; therefore, it is quite unclear what the evidence was supposed to be for the subspecies!

Lama guanicoe cacsilensis Lönnberg, 1913

Type locality: Cacsile, Nuñoa, 14.29° S, 70.39° W, Dept. of Puno, Peru.

This was described from a single skull of very small size. Wheeler (1995) described Peruvian guanacos as being ochery yellow in color.

This seems, for the moment, to be a valid subspecies (see below).

According to our discriminant analysis of all regions, with the males and the females combined, valid taxa would be (1) Tierra del Fuego, and (2) Patagonia = Chubut = Bolivia = Chile. These two groupings occupy separate regions in multivariate space. The small sample from NW Argentina is unclassified, as, in the discriminant analysis, it falls well outside any other range. It is not clear whether the type of *L. g. cacsilensis* represents a distinct population of small size, or whether it is simply an extremely small individual of the second taxon.

High nasal length and breadth, condylobasal lengths, and diastema length are what distinguish NW Argentina and Chile specimens from the other samples; high biorbital breadth and diastema length largely distinguish Patagonia and Chubut specimens from the other samples. The males have long canine teeth, which the females do not, but there is no indication of any metrical differences between the sexes.

Table 1 Skull measurements for the *Lama guanicoe* group

	Gt l	Cb l	Nas l	Nas br	Biorb
Tierra del Fuego					
Mean	307.71	288.50	81.57	61.57	149.00
N	7	6	7	7	7
Std dev	8.139	6.626	5.682	5.682	5.132
Min	298	281	75	54	139
Max	318	299	89	71	155
Patagonia, Chubut, Bolivia					
Mean	295.80	274.52	71.48	56.52	141.25
N	55	23	23	23	56
Std dev	13.469	11.253	5.367	3.616	5.734
Min	271	253	61	50	120
Max	331	304	81	62	152
C Chile					
Mean	281.00	273.60	68.40	59.00	141.11
N	9	5	5	5	9
Std dev	14.900	17.358	5.550	4.416	5.968
Min	260	244	60	53	128
Max	302	285	74	64	147
cacsilensis					
Mean	261.00	244.00	—	—	128.00
N	1	1	—	—	1
NW Argentina					
Mean	318.00	297.50	79.00	61.50	145.67
N	3	2	2	2	3
Std dev	3.606	—	—	—	1.155
Min	315	290	77	60	145
Max	322	305	81	63	147

Table 1 gives the measurements for the different geographic samples. Those from NW Argentina are very large (mean skull length 318.0 mm, *N* = 3), followed by Tierra del Fuego (307.7 mm, *N* = 7), the S Argentina and Bolivia group (295.8 mm, *N* = 55), then C Chile (281 mm, *N* = 9), with the only known specimen of *L. g. cacsilensis* being the smallest (at 261 mm), though this is equaled in size by the very smallest specimen from C Chile, which measures 260 mm. Condylobasal length is not entirely consistent with this: the mean for the Chile series is equal to that of S Argentina / Bolivia, whereas the greatest length is very much smaller, indicating a lesser development of the occipital crest. Similarly, the Chile sample has shorter but broader nasals, and the biorbital breadth is equal. The NW Argentina sample is comparatively narrow. None of the ranges of variation are actually exclusive.

Two mitochondrial sequences (complete cytochrome *b* and partial control region) were studied by Marín et al. (2008). Sequences from the range of *L. g. cacsilensis* were almost (but not entirely) exclusive, but those of the other three putative ("Krumbiegel") subspecies were intermixed. There was mixture along the borders between *L. g. cacsilensis* and the others. The authors suggested that the S (non-*cacsilensis*) guanacos underwent a postglacial range expansion.

Wheeler (1995, 1998) said that, even though there are no feral llama or alpaca living in the C Andes, there is (or was) a group of guanaco-llama hybrids living wild in Cordoba Province, Argentina; these resemble guanacos, though some have white areas of varying distribution, especially on the head and the neck.

Vicuñas

We do not accept that vicuñas are generically distinct from *L. guanicoe*. The incisors grow from persistent pulps, unlike in guanacos (and llamas, but like some—yet not all—alpacas), but this need not indicate a deep separation and, according to the molecular data (see below), does not.

Lama vicugna (Molina, 1782)

1782 *Camellus* [*sic*] *vicugna* Molina. Andes of Coquimbo and Copiapó, Chile. This name was placed on the Official List of Specific Names in Zoology by ICZN, Opinion 2027 (2003).

1944 *Lama vicugna elfridae* Krumbiegel. Type locality: unknown.

Wheeler (1995) described the Chilean vicuña as lacking long chest hairs, and having a light beige pelage, with much white on the body, coming halfway up the flanks and all the way to the crest of the ilium, as well as to the anterior portion of the hindlegs.

Krumbiegel's *L. v. elfridae* was described on the basis of living animals and was chiefly distinguished by its much larger size—guanaco-sized, the shoulder height 92–110 cm (*N* = 5), whereas he stated that small vicuña ("*L. v. vicugna*"; Krumbiegel evidently was unaware of *L. mensalis*) are 73–83 cm (*N* = 8). Three animals of known locality were also referred by Krumbiegel to *L. v. elfridae*—one from Jujuy, and two thought to be from Salta; a specimen cited by Lydekker from Catamarca was also provisionally referred to *L. v. elfridae*. When we use only animals of known locality, these shoulder height figures become 100–110 cm (*N* = 2), compared with 73–78 cm (*N* = 4). Wheeler (1995) described the S vicuña as having a shoulder height of 90 cm on average, which puts it on the edge of the range given by Krumbiegel (1944) for his total *L. v. elfridae* sample, though admittedly below that for individuals of known locality.

Krumbiegel also described *L. v. elfridae* as being darker red-brown than the more sandy-colored "nominate" form of vicuña, with the legs red-brown, and the white strongly set off; whereas in the small race, their color is grading. He described the mane in winter as being unnoticeable, while in summer it was 20 cm long, compared with 35 cm in the small "nominate" race.

Later, without explanation, Krumbiegel (1952) regarded *elfridae* as a full species, even seeming to imply that it might be closer to the guanaco.

Krumbiegel, in fact, correctly described the differences between the two species of vicuña: large size, short brisket mane, dark color, white legs. But because he thought that the name *vicugna* denoted the Peruvian species, he described what is, in effect, the larger S species as new. Therefore, it is evident that *L. v. elfridae* is a synonym of *L. vicugna*.

Lama mensalis Thomas, 1917

1917 *Lama vicugna mensalis* Thomas. Incapirca, Junin, Peru.

Described by Wheeler (1995) as being slightly smaller (70 cm high at the withers, compared with 90 cm in nominotypical *L. v. vicugna*); teeth much smaller; color more strongly fulvous. Wheeler (1995) may have been the first person to focus on the prime distinction of Peruvian vicuña: they have a long mane on the chest and the lower throat. The color, as she described it, is dark cinnamon, with white on the underside, the inner aspects of the legs, the underside of the tail, and the lower portion of the face; the eye-rings and ear rims are white.

Elsewhere in Peru, on the pampas, larger vicuña apparently exist; Wheeler (1995) cited a study by Paucar et al. (which we have not seen) giving average heights of 90.4 cm for the males and 86.5 cm for the females from Pampa Galeras, Peru.

The differences between the Chilean and Peruvian vicuñas are great, and, on the evidence, consistent (see our skulls analyses and the DNA data below). We therefore separate them at the specific level.

There were too few specimens of most regional groups for a discriminant analysis, so we used a principal components analysis: Peru specimens (including the type of *L. v. mensalis*) fall outside the general Peru–Bolivia scatter, and a young adult from Junin falls at the edge of it. The Catamarca skull (the one which Krumbiegel provisionally referred to *L. v. elfridae*) is close to the one skull from Chile (which has no exact locality).

Table 2 Skull measurements for the *Lama vicugna* group

	Gt l	Cb l	Teeth	Biorb
Catamarca				
Mean	244.00	224.00	64.00	121.00
N	1	1	1	1
Chile				
Mean	236.00	221.00	66.50	116.50
N	2	1	2	2
Min	234	—	66	115
Max	238	—	67	118
N Peru				
Mean	235.75	216.50	54.33	108.33
N	4	2	6	6
Std dev	9.878	—	6.593	24.792
Min	221	211	47	58
Max	242	222	64	123
Peru–Bolivia border area				
Mean	228.14	211.14	56.14	118.00
N	7	7	7	8
Std dev	5.900	5.146	3.338	5.127
Min	217	202	52	110
Max	236	219	61	126

The Chile and Catamarca vicuñas are of great absolute size, contrasting with their low nasal length and diastema length.

Table 2 gives skull measurements of the geographic samples. The skull from Catamarca is 244 mm in greatest length, larger than any other. The greatest skull length means of the Chile and N Peru specimens are very close (approximately 236 mm), but the difference in values for condylobasal length is greater (n.b., very small sample sizes!), indicating that the Chile sample has less development of the occipital crest. The toothrow is much longer in vicuñas from Catamarca and Chile than in the two Peru samples. The N Peru sample has a narrow biorbital breadth, on average (note, however, the very wide range of variation).

Using the mitochondrial control region, Marín et al. (2007) found N and S vicuñas to be enormously different, with just a small amount of marginal mixing between them. The N population extended from Catac in C Peru, SE to Ingenio on the Bolivian border (not far from La Paz), and S along the Chile–Bolivia border to Salar Surire. The S population extended from Inta and Cineguillas in far NW Argentina, S along the Argentina–Chile border to San Juan. The authors hypothesized that the barrier would be what is called the "dry diagonal" between the N summer rainfall and S winter rainfall areas; a few individuals in the S population are, however, found within the dry diagonal. The N population, but not the S, shows the signatures of fairly recent expansion from a refugium.

Camelus Linnaeus, 1758

Two quite distinct domestic camel species are known: *Camelus bactrianus* (the two-humped, or Bactrian, camel) and *Camelus dromedarius* (the one-humped, or Arabian, camel). The wild ancestry of the former is presumably from an extinct species similar, in some respects, to *C. ferus*; the wild ancestry of the latter is unknown.

Analysis of the mitochondrial cytochrome *b* gene (Stanley et al., 1994) showed a 10.3% difference between the two domestic species of camel, much greater than between any two taxa of *Lama*. If the two genera separated 11 Ma, as they interpret it, then speciation in *Camelus* would have begun in the early Pliocene.

Camelus ferus Przewalski, 1883

1883 *Camelus bactrianus ferus* Przewalski. Border of the Kum-Tagh, E of Lob Nor, N of the Altyn-Tagh, Xinjiang.

The name *C. ferus* Przewalski, 1878, was placed on the Official List of Specific Names in Zoology by ICZN, Opinion 2027 (2003).

On this species, the only living truly wild camel, see Schaller (1998), who concluded that it is not as closely related to domestic Bactrian camels as has been thought.

5

Suidae

The Suiformes (as presently restricted). The two living families. Genera of the Suidae, including the following:

—The realization of several species of *Babyrousa*.
—*Porcula*, the pygmy hog—Much more distinct than Colin Groves had thought; why he could have got it wrong.
—Warthogs—The rediscovery of the desert warthog; how Griff Ewer confirmed the reality of a neglected myth in the fossil record, and how Peter Grubb's perspicacity brought it into the living fauna.
—*Potamochoerus*—Peter Grubb discovered that the bushpig and the red river hog are thoroughly distinct species; the biogeographic implications of this; what is known of their geographic variation.

SUIDAE GRAY, 1821

The differences in the heads of pigs (Suidae) and peccaries (Tayassuidae), especially relating to their function and their dentition, were reviewed by Herring (1972). In pigs, the upper canine of the males turns outward and/or upward; it is useful in visual display and in protection of the face during male–male combat; in peccaries, the upper canine is not sexually dimorphic, points downward in the usual mammalian fashion, and is a weapon.

On the Suidae, see Groves (1981d, 1997a, 2007); Groves & Grubb (1993a, b); and Grubb (1993a, b)

The cytogenetics of the family have been reviewed by Bosma et al. (1991).

A phylogenetic tree, using the mitochondrial cytochrome *b* gene, placed *Babyrousa* as sister to *Phacochoerus* plus *Sus* (Randi et al., 1996). Within *Sus*, *S. barbatus* was sister to *S. scrofa*; within the latter, the Maremma wild pig, other European wild pigs, and Asian wild and domestic pigs occupied different positions on the tree, as calculated by different methods.

Babyrousa Perry, 1811

Stomach larger than that of the domestic pig, with a large diverticulum ventriculi; mucus-producing cardiac glands occupying over 70% of the stomach area, versus 33% in the pig (Leus et al., 1999).

Three animals from Sulawesi had $2n=31$; most chromosomes look like those of the domestic pig, but there appear to be some translocations, and five pairs are very unlike those of *Sus* (Bosma et al. 1991).

What had hitherto been classed as three subspecies of a single species by Groves (1980) were later raised to species rank by Meijaard & Groves (2002).

Babyrousa babyrussa (Linnaeus, 1758)

1758 *Sus babyrussa* Linnaeus. "Borneo"; corrected to Buru by Groves (1980).
1811 *Babyrousa quadricornis* Perry.
1827 *Babirussa alfurus* Lesson. Buru.
1920 *Babirussa babyrussa frosti* Thomas. Taliabu, Sula Islands.

Long, thick body hair; tail-tuft well developed. Skull short, cheekteeth small; frontal furrows deep, sharp-edged. Upper canine of the males short, slender; alveolus forwardly rotated, with the lower canine crossing it in lateral view; upper canines generally divergent or parallel or weakly convergent.

Buru; Taliabu and Sulabesi in the Sula Islands.

Babyrousa togeanensis Sody, 1949

1949 *Babirussa babyrussa togeanensis* Sody. Malenge, Togian Islands.

Body hair present, but less long and dense than in *babyrussa*, and paler on the underparts; tail-tuft well developed. Skull very large, but the cheekteeth small, especially the third molars; frontal furrows shallow, with beveled edges. Upper canine of the males short, slender, somewhat rotated forward, always converging.

Known only from Malenge.

Babyrousa celebensis Deninger, 1909

1909 *Babirussa celebensis* Deninger. Pulau Lembeh, N peninsula of Sulawesi.

1964 *Babyrousa babyrussa merkusi* de Beaufort, nom. nud.

Body hair short, sparse (appearing naked); tail-tuft small, sparse. Skull large, with large teeth; frontal furrows shallow, with beveled edges. Upper canine of the males long, thick, vertically implanted, with the lower canine not crossing it in lateral view; almost always converging.

N peninsula of Sulawesi.

Babyrousa from C and SE Sulawesi are unclassified.

Hylochoerus Thomas, 1904

giant forest hog

Pelage with very stout bristles, always black in color; piglets weakly striped, if at all. Nasal disk extremely broad. Zygomata thickened and pneumatized, especially in the males, supporting huge, funguslike infraorbital warts. No canine apophyses. Cheekteeth hypsodont, with much crown cement; anterior premolars lost with age. *Hylochoerus* has the lowest diploid chromosome number among the Suidae, $2n = 32$, although it is based on a single specimen (Bosma et al., 1991).

Grubb (1993b) commented on previous taxonomic arrangements of the genus, and, while placing the taxa all in one species, initiated the basic arrangement into three taxa which we are able to reiterate below, on the basis of new multivariate analyses.

We find that the three described western taxa are all largely distinct in discriminant analysis, although their dispersions approach each other closely, especially *H. ituriensis* and *H. rimator*. The described taxa (*H. ituriensis*, *H. rimator*, and *H. ivoriensis*) are distinguished by size, from high to low scores, especially emphasizing facial length (preorbital and palatal lengths) and mastoid breadth. The sample of *H. rimator* has a low value for greatest skull length and preorbital length.

When we compare the three E taxa (*H. ituriensis* and *H. rimator* with E Africa), we find no differences between Uganda and Kenya, or, in Kenya, between those from E and W of the Rift Valley. But the two easternmost samples as a unit (*H. meinertzhageni*) are 100% distinguishable from those from C Africa (*H. ituriensis* and *H. rimator*), *H. meinertzhageni* being of great size, as represented by greatest skull length, preorbital length, and palatal length, and by breadth across the canine apophyses.

A specimen from Mt. Kahuzi identifies with the E African sample, not with *H. ituriensis*.

Table 3 confirms that the males of *H. meinertzhageni* are very much larger than any other species, as are the females (to some extent). *H. ivoriensis* is not distinguished by any single measurement, only by a combination (as we found in our discriminant analyses).

Therefore, our arrangement is that we recognize three species, of which the first—the E species—is dramatically different from the other two.

Hylochoerus meinertzhageni Thomas, 1904

1904 *Hylochoerus meinertzhageni* Thomas. Kakamega Forest, Kenya.

By far the largest of the three species, especially in the males. Tusks in the males flaring very widely; enamel pillars on the cheekteeth more widely separated, with more cement on the crown (Grubb, 1993b).

Mountains W of the Albertine [Western] Rift, Rwanda, northernmost Tanzania, Uganda, the Imatong Mountains in S Sudan, the Kenya highlands, and, apparently, the Ethiopian highlands (as summarized by Grubb, 1993b).

The discoverer of this species was the notorious Richard Meinertzhagen, recently shown to have lived a life marked throughout by deceit and humbug, ranging from continual self-promotion to scientific fraud, and even to sometimes murderous treachery (Garfield, 2006). His discovery of the giant forest hog, early in his career, was, however, genuine enough; presumably, when you are in the presence of something startlingly new, there is no need for pretense.

Hylochoerus rimator Thomas, 1906

1906 *Hylochoerus ituriensis* Matschie. Ituri Forest.

1906 *Hylochoerus rimator* Thomas. Dja River, Cameroon.

1909 *Hylochoerus gigliolii* Balducci. Upper Congo.

Smaller than *H. meinertzhageni*; tusks less flared in the males; less crown cement on the cheekteeth.

Cameroon–Nigeria border through W-C Africa to the Ituri Forest, N into the CAR and S Sudan, W of the Nile. Apparently absent from S of the Congo River.

Hylochoerus ivoriensis Bouet & Neuville, 1930

1930 *Hylochoerus ivoriensis* Bouet & Neuville. Bolobo, Liberia.

Somewhat smaller than *H. rimator*; rostrum and palate short; narrow across the mastoids.

W African forests, from Liberia to Ghana.

Table 3 Skull measurements for *Hylochoerus*

	Gt l	Bas l	Pal l	Mast br	Pal br	Zyg	Preorb l	Canine br
Males								
ivoriensis								
Mean	377.69	322.64	232.00	142.73	76.39	202.87	247.76	113.47
N	16	14	17	13	18	15	17	19
Std dev	12.695	10.058	7.778	8.308	2.873	17.435	8.182	9.076
Min	355	305	214	134	72	179	227	99
Max	397	335	243	163	82	235	258	127
rimator								
Mean	363.22	325.71	235.90	152.33	79.00	203.56	245.10	112.00
N	9	7	10	9	10	9	10	8
Std dev	14.438	11.295	9.905	7.348	3.559	18.304	11.865	10.198
Min	341	317	221	145	74	171	223	95
Max	388	349	252	166	84	231	261	125
ituriensis								
Mean	388.32	343.17	248.59	159.61	79.18	213.94	263.96	122.30
N	28	24	28	27	28	27	28	27
Std dev	13.644	11.251	9.871	12.069	4.701	15.577	9.738	8.826
Min	362	317	230	115	71	187	246	102
Max	415	363	269	177	88	243	282	139
meinertzhageni								
Mean	427.65	372.60	276.83	174.06	86.16	245.75	291.29	147.24
N	17	15	18	17	16	18	17	17
Std dev	14.904	18.007	15.085	12.671	4.162	20.611	14.347	12.286
Min	407	328	248	155	78	199	262	127
Max	461	407	310	195	93	281	318	170
Mt. Kahuzi								
Mean	425.00	382.00	—	—	94.00	277.00	—	—
N	1	1	—	—	1	1	—	—
Females								
ivoriensis								
Mean	352.29	314.25	221.00	133.33	69.71	174.33	235.14	95.71
N	7	4	6	6	7	6	7	7
Std dev	15.119	12.312	9.423	5.241	1.496	4.502	11.466	5.648
Min	333	299	210	127	67	169	220	85
Max	372	329	234	141	72	180	252	102
rimator								
Mean	357.00	320.00	230.50	142.20	75.67	174.33	241.67	93.67
N	5	5	6	5	6	6	6	6
Std dev	17.306	12.288	8.216	3.899	2.582	6.743	9.933	6.250
Min	334	303	222	137	72	164	226	85
Max	377	331	245	147	79	183	252	100
ituriensis								
Mean	365.25	325.42	235.33	149.79	77.03	181.43	249.38	103.13
N	20	19	20	19	20	20	20	20
Std dev	13.932	10.516	9.069	5.075	4.544	8.580	11.011	5.094
Min	331	305	214	141	72	165	227	91
Max	391	349	252	161	89	193	272	109
meinertzhageni								
Mean	393.50	348.93	259.65	155.79	79.22	197.24	270.57	112.09
N	14	15	13	17	18	17	15	17
Std dev	14.378	16.184	14.185	11.754	6.141	13.722	14.294	10.007
Min	375	315	225	137	64	177	235	86
Max	427	377	278	185	88	235	299	134

Porcula Hodgson, 1847

The resurrection of this genus was recommended on morphological grounds by Ghosh (1988), and has been supported molecularly (Funk et al., 2007).

Only three pairs of mammae, rather than the six of *Sus*; tail rudimentary, about 30 cm, with only 10 or fewer vertebrae (compared with over 20 in *Sus*). Nasal shorter and broader than in *Sus*; premaxilla rectangular, rather than long and narrow, with a tapering nasal branch; orbit situated above the second molar, rather than above the third molar; zygomatic arch not inflated; parietal longer, narrower; dorsal margin of the skull arched, not straight; maxilla laterally notched.

Porcula salvania Hodgson, 1847

pygmy hog

$2n = 38$; karyotype said to resemble that of the domestic pig, except for certain banding differences (Bosma et al., 1991).

No mane; ear small, oval, virtually hairless; withers lower than the rump.

Phacochoerus F. Cuvier, 1817

Distinguished by the disproportionately large head, slender limbs, and extreme unguligrady. Orbits placed well back on the cranium, and high up; braincase extremely shortened. A pair of deep sphenoidal pits behind the internal nares. Cheekteeth strongly hypsodont, with numerous, close-packed enamel pillars; premolars tending to be shed during adult life. Maxilla deep; ascending ramus of the mandible elongated; canine sheaths very wide. Upper canines very long, retaining their points throughout life; females also have long canines. Pelage very sparse, retaining only a prominent, long dorsal crest; often a fringe of white hairs on the cheeks, especially in the young. Long, conical warts on the cheeks and on the snout.

It was Ewer (1957) who first proposed, on the grounds of fossil material, that there were two distinct species of warthog in the Pleistocene, and that these probably corresponded to different Holocene species as well. This was investigated by Grubb (1993b), who found not only that the differences (in the presence/absence of incisors and the timing of the eruption of the third molars) are consistent and valid, and that there are other important differences in the skull, but also that warthogs still existing in NE Africa belong to the otherwise extinct Cape species.

Randi et al. (2002) obtained mtDNA and five different nuclear loci from both species from E Africa. The differences in both cases were very great, suggesting separation during the Late Pliocene.

Phacochoerus aethiopicus (Pallas, 1767)

desert warthog

Skull shorter and broader than in *P. africanus*. Zygomatic arch greatly pneumatized, swollen into a hollow knob just in front of the suture with the squamous temporal. Sphenoidal pits enormously expanded. Upper incisors absent, even in the young; lower incisors rudimentary or absent. Posterior molars not developing roots until the entire crown has come into wear.

Externally, the warts on the cheeks hook-shaped; tips of the ears bent backward; suborbital areas are swollen, pouchlike; head appearing more egg-shaped (d'Huart & Grubb, 2005).

Phacochoerus aethiopicus aethiopicus (Pallas, 1767)

In historic times, this subspecies may have been approximately restricted to the Karoo; Grubb (1993b) mentioned early specimens from the E Cape, between the Sondags and Boesmans rivers, and the upper Orange River.

Phacochoerus aethiopicus delamerei Lönnberg, 1909

In NE Africa, the distribution of this species has been mapped by d'Huart & Grubb (2001). It is known from Somalia, NW almost as far as the borders of Djibouti, the Ogaden, the Webi Shebeyli and Juba rivers, and NE Kenya W to about 37° E along the N Guaso Nyiro and 39° E along the Ethiopian border, SE to the Lamu district. Only in the Berbera district of Somalia is it approximately sympatric, or parapatric, with *P. africanus*.

Phacochoerus africanus (Gmelin, 1788)

common warthog

$2n = 34$ (reported as *P. aethiopicus*); chromosome 1 appears to be a fusion between *Sus* numbers 13 and 16, and chromosome 3 appears to be a fusion between *Sus* numbers 15 and 17 (Bosma et al., 1991).

Skull longer, less broad. Zygomatic arch robust, but less pneumatized. Sphenoidal pits distinct, but not greatly expanded. Two upper incisors present, but sometimes lost in aged individuals; six lower incisors. Posterior molars developing roots at the time of eruption.

Externally, the facial warts conical; ears leaf-shaped; suborbital region less enormously swollen; head more diabolo-shaped (d'Huart & Grubb, 2005).

In NE Africa, *P. africanus* seems to occur in less arid country than *P. aethiopicus*—hence not, in general, E of about 38° E in Kenya or 42° E in Ethiopia—but it also occurs in the Berbera district of Somalia, and here it may be sympatric or parapatric with *P. aethiopicus* (d'Huart & Grubb, 2001).

Potamochoerus Gray, 1852

Most similar to *Sus*, but longer-bodied and shorter-limbed. Canine apophyses greatly enlarged, with an additional exostosis above them; together, these support a large rostral wart. No infraorbital warts. Postorbital part of the skull elongated, low-crowned; rostrum relatively short. Cheekteeth brachyodont, with thick enamel. Size relatively small, with reduced sexual dimorphism.

Potamochoerus larvatus (F. Cuvier, 1822)

bushpig

Body hairs bristly, very long, relatively sparse, forming a nuchodorsal crest. Head contrasting in color with the body, but not forming a facial mask. Body color often strikingly polymorphic within a single population.

Five individuals (reported in error as *P. porcus*) from Kenya and Zimbabwe had $2n = 34$, with apparent banding differences from other genera.

Grubb (1993b) proposed the following three subspecies.

Potamochoerus larvatus hassama (Heuglin, 1863)

Face white, often with blackish markings in the females and the young males. Adult males black or off-white. Size small; skull length 327–353 mm (females), 341–377 mm (males).

Ethiopia, S Sudan, E DRC, Rwanda, Burundi, Kenya, Uganda, N Tanzania; inhabits densely vegetated highland areas (Grubb, 1993b).

Potamochoerus larvatus somaliensis de Beaux, 1924

Provisionally recognized as distinct by Grubb (1993b). Similar to *P. l. hassama*, but larger.

Forests along the Tana, Juba, and Shebeyli rivers.

Potamochoerus larvatus larvatus (F. Cuvier, 1822)

1822 *Sus larvatus* F. Cuvier. Madagascar.

1831 *Sus koiropotamus* Desmoulins. South Africa.

Face mostly gray, with a broad blackish band over the muzzle. Usually, body brown or reddish brown, grading into a darker color on the belly and the limbs; occasionally entirely blackish or off-white in the males. Skull larger, 345–395 mm (females), 367–415 mm (males).

Left bank of the lower Congo River, Katanga, Tanzania (N at least to Kilosa), S into Namibia and N South Africa; a population in the S Cape may be isolated by a hiatus in KwaZulu in Natal and Transkei. *P. l. larvatus* is widespread (presumably introduced) in Madagascar and on Mayotte, and in Madagascar it appears to have differentiated into slightly different E and W forms. It should be noted that the type specimen of *P l. larvatus* was from Madagascar.

Potamochoerus porcus (Linnaeus, 1758)

red river hog

Bright reddish orange; white dorsal line—not a bristly crest—beginning from behind the head; face with a black mask, although with a white muzzle, white eye-rings, and long white cheek whiskers. Ear extremely elongated, with a long terminal tuft. Pelage bristly on the snout and much of the face, but short, soft, and dense on the body and the forehead. Relatively small; skull length 269–378 mm (females), 327–405 mm (males).

Rainforest and gallery forest from Senegal to W-C Africa and the DRC. There is little or no detectable geographic variation (Grubb, 1993b).

Sus Linnaeus, 1758

Mainly plesiomorphic, compared with other genera of the family; large canines in the males, the upper ones curling upward, and the lower ones directed laterally; tips of the upper canines wear as they erupt against the lower canines, so the tips of the upper canines are blunted, while those of the lower canines are sharpened. Canines much smaller in the females, pointing downward in the upper jaw and upward and forward in the lower jaw. Large canine apophyses in the maxilla in the males. Dentition not reduced; only the third molars elongated, with multiple conical cusps and cuspules. Externally, the body bristly, with underwool in temperate-zone forms. Facial warts sometimes present.

There is a great deal of polymorphism in the karyotype within this genus, even (and especially) among taxa hitherto referred to *Sus scrofa*. The domestic pig, as represented by the Swedish landrace, has $2n = 38$; the Y chromosome is the smallest of the set, and is metacentric (Hansen-Melander & Melander, 1974).

Based on cytochrome *b* sequences, Mona et al. (2007) found two major clades in this genus: one containing *Sus barbatus* and *S. verrucosus*, the other containing *S. philippensis* and the *scrofa* group. Unexpectedly, samples of *S. celebensis* occurred in both clades.

Domestic pigs undoubtedly come from several different wild sources. This was argued long ago by Groves (1981d), and is amply supported by molecular studies (Larson, Dobney, et al., 2007, and other sources).

Sus scrofa Group

"wild boar"

Distinguished by the relatively short facial skeleton; comparatively shallow preorbital fossa; and the shape of the male's lower canine, with the inferior surface narrower than the posterior surface. No facial warts.

All members of this group were placed in a single species by Groves (1981d), but the differences between some of the taxa in the group are sharper and more consistent than previously recognized.

Larson, Dobney, et al. (2007), using the control region of mtDNA, found that this species-group has a "basal comb" (which also included samples of *S. celebensis*), of which one branch contained all of the Eurasian samples. The basic division within Eurasia was India versus the rest; then a basal cluster of general Eurasian series; with a final, distinctive clade of European and Middle Eastern samples. This certainly implies a Southeast Asian diversification for the genus, especially as the "basal comb" included samples of *S. verrucosus* and *S. barbatus*. Some of the same authors (Larson, Cucchi, et al., 2007) concentrated on the Southeast Asian and Pacific region, finding links to both island and mainland Southeast Asia; they also studied the shape of the third lower molar, distinguishing a Sulawesi/Philippine form, a New Guinea / Flores form, and a China / Southeast Asia form. These molar differences were later revisited and refined by Cucchi et al. (2009).

Europe

Wild pigs from Spain ($N=7$) and Italy ($N=1$) were included in an analysis based on mitochondrial cytochrome *b* and control region sequences by Alves et al. (2003), centered mainly around a comparison between native European and Chinese / Chinese-derived domestic breeds. The Spanish wild pigs nested within the European domestic breeds, while the Italian sample was sister to these; the Chinese and Chinese-derived breeds formed an entirely separate lineage.

$2n=$usually 36. Totaling the figures for Austria, Germany, the Netherlands, Switzerland, France, and Spain given by Bosma et al. (1991, table 1) gives 112 with $2n=36$, 20 with $2n=37$, and 3 with $2n=38$. Chromosomes involved in Robertsonian translocation are always numbers 15 and 17. The same chromosome polymorphism in the same translocations occur in Lithuania, Belarus, and Central Russia, but no individual numbers were given, except that the total number of individuals was 15; and nine individuals from the former Yugoslavia had $2n=38$ (summarized in Bosma et al., 1991). Note that these figures are different from those available to Groves (1981d), who thought that chromosomal differences, together with an average size difference, supported the separation of an E European *S. s. attila* from a W European *S. s. scrofa*.

Western, Central, and Northern Asia

Diploid chromosome numbers may be $2n=36$, 37, or 38 in Azerbaijan ($N=8$), Kyrgyzstan ($N=37$) and the Russian Far East ($N=20$). The distribution of these figures is not given, but the Robertsonian translocation involved is between chromosomes 16 and 17, that is, different from Europe (Bosma et al., 1991).

Japan

Bosma et al. (1991) recorded only two karyotyped individuals from Japan (apparently from the main Japanese islands), both having $2n=38$.

Endo et al. (1995) produced the first detailed report on the Iriomote wild pigs, which are very small in size. On the basis of an admittedly small amount of material, there appears to be no sexual size difference; three males with third molars in wear averaged 263 mm in "profile length"; and three females, 261 mm (calculated from their table 4).

In a later study, Endo et al. (2001) compared mandibles from different parts of the main Japanese islands; there was a decline in size from N to S.

Indochina

Groves et al. (1997) announced the apparent rediscovery of *S. bucculentus*, based on the skull of a young individual that resembled the type specimen. The authors noted, however, that the form of lower canine in the type specimen was not, in fact, precisely as had been indicated by Heude in his original description. Groves & Schaller (2000) gave further details, reiterating that the type specimen was not as similar to *S. verrucosus* as had been formerly supposed.

Robins et al. (2006) found that this same subadult specimen ascribed to *S. bucculentus* had mtDNA almost identical to that of pigs from the Solomon Islands. Likewise, Mona et al. (2007), using cytochrome *b*, found that this specimen was nested deep within the *S. scrofa* group.

When all the male Indochinese skulls in the Heude collection were studied by CPG in October 2009, it was evident that they formed a graded series, with the type of *S. bucculentus* at one extreme of, in particular, depth of the preorbital fossa and enlargement of the malar in the facial region. Their sizes were consistent within a comparatively narrow range. The shape of the third lower molar was that of Larson, Dobney, et al.'s (2007) "China" type.

A discriminant analysis, using eight craniometric variables, was done comparing samples from Burma, Indochina, and the Kienté and Ningkuo region of S-C China. Skulls from S-C China are well outside the range of those from Burma and Indochina; these Chinese skulls are low in greatest skull length and palate length, with relatively wide bizygomatic breadth and postorbital constriction breadth.

The conclusion must be that the skulls from Burma and Indochina are the same taxon, and in subsequent analyses they were combined; but S-C China (Kienté, Ningkuo) is different.

When we added in the S China specimens, we found that a single specimen from Zhejiang belongs in the S-C China sample, whereas two from Fujian went along with Indochina/Burma. Shaanxi and Anhui are both at the edge of the Burma/Indochina range, but not beyond it, whereas the S-C China sample is still quite distinct.

We added in our two skulls from Sichuan: both of these skulls fall into the same Burma/Indochina group, with their long palate and fairly high greatest length, contrasted with low bizygomatic breadth and postorbital constriction (as before).

We next compared all the Chinese samples with each other, adding in those from Korea and the Russian Far East. The Anhui and Shaanxi samples fall together with Korea, but the Heilongjiang and Russian Far East sample is entirely separate. A skull from Fu Song (42.18° N, 127.17° E) falls with Heilongjiang/Russia.

Shanxi identifies with the Shaanxi group.

Finally, we performed an analysis combining the groups which fell together (as summarized above).

We find a single taxon from all of Burma, Indochina, Korea, and China, except in the swamps SW of Shanghai (the S-C China sample), where there is a second taxon, and Heilongjiang (plus Fu Song in Jilin, plus the Russian Far East), where there is a third. Overwhelmingly, this differentiation is based on size, with the face longer and the bizygomatic narrower in the larger taxa (table 4). We propose provisionally that these represent distinct species, as follows:

1. Burma/China—*Sus moupinensis* Milne-Edwards, 1871. Synonyms as in Ellerman & Morrison-Scott (1951), plus *Sus canescens* Heude, 1897, *Sus collinus* Heude, 1892, *Sus coreanus* Heude, 1897, *Sus meles* Sowerby, 1917—plus *Sus bucculentus* Heude, 1892!
2. S-C China—*Sus chirodontus* Heude, 1888. Synonymy as in Ellerman & Morrison-Scott (1951).
3. Heilongjiang / Far East—*Sus ussuricus* Heude, 1888. Synonymy as in Ellerman & Morrison-Scott (1951).

Southeast Asia

Four individuals from Java all had $2n = 38$ (Bosma et al., 1991). Because of the uncertainty now surrounding Southeast Asian wild pigs, particularly since the studies of Larson, Cucchi, et al. (2007) and Cucchi et al. (2009), we will not venture a new arrangement of wild *scrofa*-like pigs. For the moment, we propose that they be classified as *S. vittatus*, recognizing that they do, nonetheless, all share a morphology which is diagnosably distinct from other members of the group.

While we cannot venture a new classification of the species of the *S. scrofa* group, we do have fresh evidence, beyond that reported by Groves (1981d), for E and NE Asia, and mainland Southeast Asia. Nonetheless, we will still make some comments on the 1981 arrangement, since Groves's (1981d) subspecies were of varying quality. Some of them were actually diagnosably distinct; hence recognizable as species. Some of them were very poor; hence fit only to be pushed into synonymy. Finally, some of them remain indeterminable, and are listed below with a query:

Sus scrofa Linnaeus, 1758 (synonyms *Sus attila*, *Sus lybicus*, *Sus algira*)—Wild pigs from (most of) Europe, the Middle East as far as the Zagros Range, and North Africa.

Table 4 Skull measurements for *Sus* from mainland Southeast Asia, China, Korea, and the Russian Far East

	Gt l	Cb l	Bizyg	Occ br	Occ ht	Pal l	Molars	M3 br	Lm3 l
Males									
Burma/Indochina *moupinensis*									
Mean	379.11	330.52	156.27	76.93	102.36	229.38	73.910	39.471	—
N	27	25	30	27	25	29	29	21	—
Std. dev	14.766	13.326	9.009	8.471	8.371	9.770	4.6876	3.4104	—
Min	354	312	136	61	84	216	63.0	32.8	—
Max	418	364	176	97	116	256	83.0	47.3	—
chirodontus									
Mean	417.50	365.75	157.80	84.00	109.75	258.90	79.700	44.725	—
N	10	8	10	10	8	10	10	4	—
Std. dev	7.276	7.126	5.007	5.558	7.833	6.757	2.7101	2.0451	—
Min	408	352	152	73	99	249	74.0	41.8	—
Max	432	378	167	90	122	270	82.0	46.4	—
N China / Korea *moupinensis*									
Mean	397.55	350.45	158.73	78.10	105.09	245.55	75.709	40.467	—
N	11	11	11	10	11	11	11	6	—
Std. dev	18.272	19.351	11.791	11.100	6.848	16.979	3.9612	2.6440	—
Min	362	317	142	66	94	222	70.0	38.0	—
Max	430	380	186	96	115	276	84.0	44.6	—
ussuricus									
Mean	452.63	399.50	172.75	93.00	120.67	280.00	87.875	48.350	—
N	8	8	8	3	3	8	8	6	—
Std. dev	20.646	19.034	7.517	5.292	3.215	12.649	4.0861	1.9316	—
Min	423	378	163	89	117	263	79.0	46.7	—
Max	482	431	188	99	123	301	91.0	50.8	—
Females									
Burma/Indochina *moupinensis*									
Mean	358.00	317.00	142.33	77.33	90.00	216.00	74.500	—	35.000
N	2	2	3	3	3	2	2	—	1
Std. dev	—	—	4.509	11.150	8.660	—	—	—	—
Min	354	308	138	69	80	212	69.0	—	—
Max	362	326	147	90	95	220	80.0	—	—
chirodontus									
Mean	357.00	314.67	138.00	69.67	91.00	218.00	74.667	—	—
N	3	3	3	3	3	3	3	—	—
Std. dev	10.536	13.614	3.606	10.599	5.568	6.245	3.5119	—	—
Min	347	304	134	60	86	211	71.0	—	—
Max	368	330	141	81	97	223	78.0	—	—
N China / Korea *moupinensis*									
Mean	384.00	336.67	148.33	72.33	92.00	232.67	75.667	—	34.833
N	3	3	3	3	3	3	3	—	3
Std. dev	24.637	18.903	6.028	4.041	8.888	16.563	2.8868	—	3.8423
Min	361	322	142	70	85	217	74.0	—	30.4
Max	410	358	154	77	102	250	79.0	—	37.2
ussuricus									
Mean	405.80	366.67	152.60	—	—	251.00	86.750	—	44.120
N	5	3	5	—	—	5	4	—	5
Std. dev	5.933	11.015	8.620	—	—	6.124	2.8723	—	2.7842
Min	402	356	139	—	—	244	85.0	—	40.5
Max	416	378	160	—	—	257	91.0	—	47.3

? *Sus meridionalis* Forsyth Major, 1882—The Sardinian/Corsican wild pigs may or may not be identical to those from S Spain, as proposed by Groves (1981d), but certainly their affinities need further study.

? *Sus nigripes* Blanford, 1875—We have seen no further information on this pig (from Kyrgyzstan and Xinjiang) since that recounted (on the basis of the literature, and of a single specimen) 30 years ago by Groves (1981d).

? *Sus sibiricus* Staffe, 1922—Likewise, the affinities of this pig, from Mongolia and Transbaikalia, remain totally obscure, and further study is needed. We have seen no specimens; the type description suggests close affinity, perhaps identity, with *S. nigripes* (if that is itself distinct).

Sus leucomystax Temminck, 1842—The Japanese wild pig is distinct from any from the mainland of Asia.

Sus taevanus Swinhoe, 1863—The Taiwan wild pig, a very small-sized taxon, is not identifiable—on the admittedly slim evidence so far available—with any other pig.

? *Sus riukiuanus* Kuroda, 1924—The Ryukyu pig, as noted above, is distinguished by its very small size from those on the main Japanese islands, but it needs close comparison with the Taiwan pig.

Sus davidi Groves, 1981—This species, from Rajasthan, Pakistan, and Iran E of the Zagros Mountains, is distinct, on both cranial and external characters, from those to the E and W of its range.

Sus cristatus Wagner, 1839—The distinctive Indian wild pig is diagnosably separate from other *Sus* taxa. On the basis of evidence accumulated since, Groves (1981d) was wrong to include the Burmese pig in the same taxon: it has a much shorter third molar, a less distinct mane, and a much deeper preorbital fossa. Instead, its affinities lie with those from China and Indochina, as noted above. On the other hand, it was probably not justified to distinguish *S. affinis* from S India, which differs only in average size.

Taking into consideration the three species distinguished in the new craniometric analyses (*S. moupinensis*, *S. chirodontus*, and *S. ussuricus*), plus *S. vittatus*, we are left with a minimum of eight species (where previous arrangements have recognized only one), but with a large number of subspecies; whether any of the four species listed above with a query against them might be recognizable, in some sense at least, cannot at the moment be determined.

Warty Pigs

Sus celebensis Müller, 1840

Sulawesi warty pig

$2n=38$; the Y chromosome is submetacentric and larger than in the *S. scrofa* group, in which it is metacentric (Bosma et al., 1991).

Small in size, with short legs, short ears, and a long, simply tufted tail; three facial wart pairs in the males, of which the preorbital is the most developed, at least until old age, when the gonial wart hypertrophies. Color usually black, often with white or yellow hairs intermixed, but some animals predominantly red-brown or yellowish; underside becoming light yellow with age, contrasting sharply with the upperside. Often a yellow snout band. Prime males with conspicuous "toupée" on the crown, abrading with age.

In the molecular plus morphological study by Larson, Cucchi, et al. (2007), the third lower molar morphology of *S. celebensis* was largely similar to that from the Philippines, but overlapping with the New Guinea sample; samples from Flores fall in both the *S. celebensis* and New Guinea ranges. Further detailed study by Cucchi et al. (2009) showed that the samples from Sulawesi, the Philippines, and New Guinea are fully separable from each other.

Sulawesi and offshore islands; widely introduced elsewhere in Indonesia, particularly the W Sumatran islands, Maluku, and Nusatenggara.

An unexpected finding is that molecular samples of this species assort both with the *barbatus/verrucosus* clade and with the *scrofa/philippensis* clade (Mona et al., 2007), although no haplotypes are actually identical with any from elsewhere. There is indeed some suggestion from the morphology that specimens from N and S Sulawesi may be somewhat different, although related to one another, rather than nonmonophyletic, as the molecular evidence suggests. An explanation for this could be that the species is at least partly of hybrid origin; nuclear DNA sequences are needed to determine the full story.

Sus verrucosus Müller, 1840

Javan warty pig

$2n=38$; the Y chromosome is large and submetacentric; the banding pattern of chromosome 10 is specific.

Skull distinctive, with a greatly elongated facial skeleton. Sexual dimorphism extreme, the males being more than twice the weight of the females; small size of the canines in the females striking. Long-limbed; ears large; tail long, simply tufted; head large and heavy in the living animal. Three pairs of warts on the face in the males: infraorbital (the largest), gonial (marked by a tuft of long hair in the young, before the wart itself has developed), and rostral. Body color agouti-reddish. Skull length in the adult males 408–429 mm. Shape of the third lower molar similar to that of *S. barbatus*, as one would expect, but both, unexpectedly, also close to Southeast Asian *S. scrofa* (Cucchi et al., 2009).

Confined to Java and, formerly, Madura.

Sus blouchi Groves, 1981

Bawean warty pig

S. blouchi is closely related to *S. verrucosus*, but absolutely different: much smaller (skull length in an adult male only 354 mm), and the pelage yellowish, not reddish.

Confined to Bawean, where its survival prospects seem, ironically, rather better than those of its endangered sister species on Java.

Sus barbatus Müller, 1838

bearded pig

$2n = 38$; the karyotype is said to be similar to the domestic pig.

Facial skeleton even more elongated; warts much smaller, with only two pairs (no gonial wart). Long dorsal mane. Females are larger, relative to the males, than in *S. verrucosus*. Exceptionally long-legged; ears relatively small; tail long, with a large terminal tuft divided into distinct anterior and posterior parts.

The most recognizable characteristic of this species is a bushy cheek-beard. Groves (1981d) distinguished two subspecies, one from Borneo and one from Sumatra plus the Malay peninsula, largely on the supposedly different forms of the beards, but specimens seen since then by CPG do not sustain this difference.

Borneo, Sumatra, Bangka, and the Malay peninsula. Those from Bangka and S Sumatra are considerably smaller, at least on average, than those from elsewhere, a difference which needs further investigation, as does the possibility of differences between migratory and nonmigratory populations.

Philippine Group

Groves (1997a) revised the Philippine wild pigs, dividing them into three species: *S. barbatus*, the Philippine subspecies being *S. b. ahoenobarbus*; *S. philippensis*, with subspecies *S. p. philippensis* (Luzon group), *S. p. mindanensis* (Mindanao group), and *S. p. oliveri* (Mindoro group); and *S. cebifrons*, with subspecies *S. c. cebifrons* and *S. c. negrinus*. Attention was also drawn to the possibility of the existence of further taxa, especially one from Jolo, in the Sulu Archipelago.

We performed a discriminant analysis on skulls of Philippine pigs of the *philippensis* group. On average, we could distinguish the three groups. Mindoro (now *S. oliveri*; see below) has a long palate and wide bizygomatic breadth, contrasting with a short greatest length and narrow postorbital constriction. Mindanao and Mindoro are both distinguished from Luzon by a long palate and wide zygomatic breadth, contrasting with their low greatest length and especially low occipital breadth.

In tooth measurements, Mindoro is absolutely distinct from the other two, which overlap slightly. Mindoro has a wide M^2 and a fairly long P^3, contrasting especially with a short M^1. Mindoro and Mindanao differ from Luzon in a wide M^3 and a fairly long P^4 and M^2, contrasting with a narrow P^4. Leyte clusters with Mindanao, and Catanduanes with Luzon.

Comparing members of the *S. philippensis* group with a sample from Negros (*S. cebifrons*), we found that, as expected, Negros clusters apart from the other two.

We next compared samples of all three Philippine groups (*S. barbatus*, *S. philippensis*, and *S. cebifrons*), incorporating *S. b. ahoenobarbus* (now *S. ahoenobarbus*; see below), and were able to include two skulls from the S islands, Tawtawi and Jolo, both of which clustered with those from Palawan.

When we dropped *S. cebifrons*, as being irrelevant to finding the position of the pigs from the S islands, and brought in *S. celebensis*, Tawitawi appeared closest to Mindanao, while Jolo was far from anything. Palawan differs greatly from the others in wide bizygomatic breadth and a (fairly) long greatest length, contrasting with an extremely narrow and fairly low occiput. Jolo differs from all the others in its very short nasals and narrow postorbital constriction, contrasting with a long palate and a high occiput.

Reducing the variable list enabled us to bring in two incomplete specimens from Jolo and make a group out of it, instead of merely including a single specimen as an unknown. One Jolo skull groups with Mindanao; the others are separate.

Ungrouped specimens fall as follows: Tawitawi is close to Jolo (and not too far from Mindanao), while Basilan is close to Mindanao.

Table 5 Skull measurements for Philippine pigs

	Gt l	Cb l	Bizyg	Occ br	Occ ht	Nas l	Pal l
Mindoro							
Mean	340.00	292.33	151.00	85.00	103.00	168.33	208.33
N	3	3	3	3	3	3	3
Std dev	9.539	12.055	2.646	6.557	5.568	6.110	6.658
Min	334	281	149	78	98	163	204
Max	351	305	154	91	109	175	216
Luzon							
Mean	325.20	277.77	139.00	81.07	97.23	164.33	195.93
N	15	13	15	15	13	15	15
Std dev	18.339	9.194	6.990	6.923	10.787	13.916	10.660
Min	287	265	125	72	82	139	172
Max	352	294	151	91	118	188	210
Mindanao							
Mean	347.78	289.38	146.22	80.67	100.67	176.50	207.22
N	9	8	9	9	6	8	9
Std dev	22.554	11.413	12.538	6.305	15.552	17.163	9.680
Min	313	276	126	71	81	155	195
Max	385	307	162	90	117	207	220
Negros							
Mean	315.33	271.50	142.25	81.00	101.00	149.67	185.25
N	6	4	6	4	4	6	6
Std dev	20.887	21.579	7.973	8.602	8.206	15.240	14.511
Min	295	254	132	70	90	134	174
Max	356	303	155	90	109	179	214
Palawan							
Mean	369.00	322.25	147.00	68.00	96.50	181.75	239.25
N	4	4	4	4	4	4	4
Std dev	10.100	9.179	7.958	5.477	8.185	7.544	8.958
Min	362	314	139	62	89	171	231
Max	384	334	158	74	108	187	252
Jolo							
Mean	332.00	292.67	142.67	75.00	99.00	166.67	204.00
N	1	3	3	1	1	3	3
Std dev	—	7.506	3.215	—	—	15.308	1.000
Min	—	285	139	—	—	149	203
Max	—	300	145	—	—	176	205
Tawitawi							
Mean	343.00	301.00	139.00	77.00	101.00	171.00	216.00

Palawan has a very long palate, contrasting with short nasals and a fairly short basal length; Jolo differs in short nasals and is narrow across the postorbital processes, with a long condylobasal length.

In a discriminant analysis of the lower dentition of several Philippine pigs, all are distinct. Palawan has a long P^2 compared with others. Jolo has a long P^2, a short M^1.

Table 5 gives the univariate statistics for skull measurements, and table 6 for mandibular tooth measurements. The skull for Jolo is small, like Luzon, with a narrower, lower-crowned occiput and a wider palate; compared with Palawan, it is smaller and relatively broader, with a wider, higher-crowned occiput. Tawitawi, compared with Jolo, is slightly bigger and narrower; compared with Palawan, it is smaller, with a wider, higher-crowned occiput; compared with Mindanao, it has a somewhat shorter greatest length compared with condylobasal length, a narrower bizygomatic breadth and occiput, and shorter nasals. As far as the teeth are concerned, there is a general trend for *S. celebensis* and *S. cebifrons* to be small in size, and *S. ahoenobarbus* to be large, but the molars and fourth premolar of the Jolo pig are more or less as large as those of *S. ahoenobarbus*, while the anterior premolar of

Table 6 Length measurements of mandibular cheekteeth for Philippine pigs and *Sus celebensis*

	Lp1 l	Lp2 l	Lp3 l	Lp4 l	Lm1 l	Lm2 l	Lm3 l
Mindoro							
Mean	7.900	10.733	11.800	12.900	13.400	18.067	29.333
N	2	3	3	3	3	3	3
Std dev	1.5556	.1155	.2000	.5292	1.9313	1.1590	2.6026
Min	6.8	10.6	11.6	12.5	11.7	17.0	26.8
Max	9.0	10.8	12.0	13.5	15.5	19.3	32.0
Luzon							
Mean	8.860	11.000	11.643	12.686	13.343	16.563	26.738
N	5	7	7	7	7	8	8
Std dev	1.2402	.4041	.5884	1.0746	.7613	.6844	1.3469
Min	8.0	10.5	11.0	11.3	12.0	15.8	25.5
Max	11.0	11.8	12.4	14.0	14.1	17.6	29.8
Mindanao							
Mean	8.083	10.650	11.633	12.100	13.700	17.783	26.600
N	6	6	6	6	6	6	5
Std dev	.3545	.7287	.7033	.7155	.7266	.5913	.9301
Min	7.6	9.8	11.0	11.0	12.7	16.6	25.7
Max	8.5	11.5	12.7	12.9	14.9	18.2	28.0
Negros							
Mean	6.967	9.400	10.867	10.800	13.133	16.167	25.067
N	3	2	3	3	3	3	3
Std dev	2.1733	.4243	.5686	.1732	.6110	.2082	2.2301
Min	5.0	9.1	10.4	10.7	12.6	16.0	23.4
Max	9.3	9.7	11.5	11.0	13.8	16.4	27.6
Palawan							
Mean	7.400	13.050	14.050	14.250	17.100	19.975	32.125
N	2	2	2	2	2	4	4
Std dev	.5657	.0707	.3536	.2121	.1414	1.1266	1.9788
Min	7.0	13.0	13.8	14.1	17.0	19.0	29.9
Max	7.8	13.1	14.3	14.4	17.2	21.0	34.4
Sulawesi							
Mean	6.175	9.371	10.314	11.286	12.650	16.571	26.429
N	4	7	7	7	6	7	7
Std dev	.9251	.4821	.9173	.9512	1.0075	.6075	2.3705
Min	5.0	9.0	9.0	10.0	11.0	16.0	24.0
Max	7.2	10.0	11.0	13.0	13.9	17.5	30.0
Jolo							
Mean	8.333	11.667	12.000	13.500	14.333	17.667	32.333
N	3	3	3	3	3	3	3
Std dev	.5774	.5774	.0000	.5000	1.1547	1.5275	.5774
Min	8.0	11.0	12.0	13.0	13.0	16.0	32.0
Max	9.0	12.0	12.0	14.0	15.0	19.0	33.0

S. philippensis, *S. oliveri*, and the Jolo pig are the largest in the series.

Sus ahoenobarbus Huet, 1888

Palawan wild pig

Very small size; darker in color than *S. barbatus*.

S. ahoenobarbus was originally classed as a subspecies of *S. barbatus*, as, for example by Groves (1981).

Palawan, Balabac, and the Calamianes group.

Sus philippensis Nehring, 1886

Philippine wild pig

Black, sometimes with a pale snout band and red patches in the mane. Large crown crest (or "toupée") in prime males. Inferior surface of the lower canine exceptionally broad.

Luzon, Mindanao, and offshore islands.

Probably there are at least two subspecies to be included here (as in Groves, 1981d, 1997a); they differ on average, but we can find no absolute differentiation between them.

Sus oliveri Groves, 1997

Mindoro wild pig

Long palate; wide bizygomatic breadth; wide M^2; fairly long P^3, compared with a short M^1.

S. oliveri was originally described as a subspecies of *S. philippensis*, but it is, in fact, absolutely distinct.

Mindoro.

Sus cebifrons Heude, 1888

Visayan wild pig

Mane long, extending along the back to the rump; large facial warts. Skull very short, exceptionally high-crowned.

Breeding groups from both Negros (*S. c. negrinus* Sanborn, 1952) and Panay (no taxonomic designation) have been established in the Philippines, and from photographs kindly supplied to CPG by William Oliver, they are undoubtedly different in external appearance. There is, however, a problem in representing this difference taxonomically: the prior available name for any Visayan pig is *Sus cebifrons*, described from Cebu, a population now apparently extinct and known by a single adult male skull, but not by its external appearance. It seems, then, that we have to await the demise of one or more of the adult males of the Panay breeding group before we can place this population, determining whether it is identical to the Cebu form, or distinct from both this and the Negros form.

TAYASSUIDAE PALMER, 1897

The most striking difference between this family (peccaries) and the Suidae (pigs) concerns, first of all, the canines, which in the peccaries point downward in both sexes in the normal mammalian fashion; and, second, the lack of lateral false hoofs on the hindfeet. The dentition in the Tayassuidae is very heavy, and the molars are shorter than in the Suidae; the whole facial skeleton forms a kind of compact, load-bearing cylinder.

Herring (1974) found that most sutures of the palate and the facial skeleton fuse very early in peccaries, and suggested that this relates to rooting and mastication, with concomitant needs to reinforce the snout.

On this family, see Grubb & Groves (1993). On the differences between *Tayassu* and *Pecari*, see Woodburne (1968); on *Catagonus*, see Wetzel (1977).

The problems and topics concerning this family include:

—The two long-recognized genera, and their tortuous nomenclatural history.
—The shock of discovering, 30 years ago, that the genus *Catagonus*, described as a fossil, is part of the living fauna.
—Geographic variation—how many species in each genus?
—The problems raised by the description of the new species *Pecari maximus* in 2007.

Chromosomes of peccaries were studied by Benirschke & Kumamoto (1989) and Vassart, Pinton, et al. (1994).

Gongora & Moran (2005), using three mitochondrial and five nuclear sequences, found that *Tayassu* and *Catagonus* form a clade (with 100% bootstrap, in the case of the mitochondrial sequences) with respect to *Pecari*.

Tayassu Fischer von Waldheim, 1814

white-lipped peccaries

Chromosomes of white lipped peccaries from Brazil had $2n = 26$, but the X chromosome was found to be polymorphic (Vassart, Pinton, et al., 1994): that from São Paulo was telocentric, whereas that from S Amazonia was subtelocentric.

We recognize here only a single species, with four subspecies, although, at present, we do not know their full distributions, nor can we place our small samples from E Brazil.

Tayassu pecari pecari (Link, 1795)

1795 *Sus pecari* Link. "Paraguay"; but, according to Hershkovitz (1963), Cayenne.
1921 *Tayassu pecari beebei* Anthony. Kartabo, Guyana.

Skins dark, but much more extensively light straw on the underparts than the other geographic samples; dark color invading the underside only on the chest. White hairs on the body; longer, light bases on the dark hairs, with the skins more brindled than other taxa; intermixed hairs may be whitish, straw-colored, or more red-brown, giving the generally darker color. White on the face fairly developed, but not unduly extensive.

We have seen specimens from Guyana and Suriname.

Tayassu pecari aequatoris (Lönnberg, 1921)

1921 *Dicotyles pecari aequatoris* Lönnberg. Gualea, Pichincha Province, Ecuador.

Median dorsal mane very long; skins vary, especially in the length of the light bases of the hairs; some hairs with light tips, especially on the light nuchal mane. Less white on the face than in nominotypical *T. p. pecari*. Size relatively small.

E Colombia (Bauco and Bogotá), Ecuador, Peru, Bolivia; details of the distribution remain to be worked out.

Tayassu pecari albirostris (Illiger, 1815)

1815 *Sus albirostris* Illiger. Paraguay.
1817 *Dicotyles labiatus* G. Cuvier.

Noticeable straw color in the axil and the groin; face markings extensively white. Hairs with straw-colored bases, the rest being black or brown, making it dark pepper-and-salt to nearly black overall color (the light hair bases very short in skins with the latter tone). Skull with relatively narrow bicanine apophyses.

Paraguay, Mato Grosso, and Amazonas.

Tayassu pecari ringens Merriam, 1901

1901 *Tayassu albirostris ringens* Merriam. Apazote, Campeche (Yucatán).
1912 *Tayassu albirostris spiradens* Goldman. Talamanca, Costa Rica.

Distinguished by the greater extension of the whitish face markings (the light area including the whole muzzle, from the tip to midway between the nose and the eyes, and extending backward along the sides of the lower jaw below the ears). Ill-defined white band above each pair of hind hoofs. Underparts grizzled black and fulvous. Skull relatively small, distinguished by the wide canine apophyses and a short, broad occipital crest.

For our craniometric study, we analyzed N and S groups separately. Among the N groups, Central America is separate, but the Guyanas (Guyana and Suriname) and Colombia are intermixed, differing only on average.

Central American skulls have wide canine apophyses and a fairly broad occiput, but a narrow bi-temporomandibular joint width, and a relatively low greatest length compared with condylobasal length.

Among groups in N South America, there is little discrimination: the SE Peru sample overlaps with the Guyanas, while NE Peru is slightly more separate. The Guyanas have a very high greatest length and high bicanine apophyses and palate breadth, contrasting with low condylobasal length and bi-temporomandibular width; NE Peru is the opposite; SE Peru is intermediate.

The two Peruvian samples and the one from Bolivia again overlap, but they differ on average. When comparing samples from Brazil with those from the Guyanas, it is, curiously, the sample from Pará which is somewhat separate.

Table 7 shows that Amazonia and Mato Grosso / Paraguay have relatively narrow bicanine apophyses compared with others; all the South American samples have a narrow occiput, compared with Central American ones; the South Americans (except Colombia and Venezuela) tend to be larger than the Central Americans; SE Brazil has a short greatest length compared with condylobasal length (i.e., a less projecting occipital crest); and Pará has large teeth.

We conclude that, thus far, craniometry contributes rather little toward understanding geographic variation in white-lipped peccaries, and a great deal more material is needed.

Pecari Reichenbach, 1835

collared peccaries

Cope (1889) distinguished two species of collared peccary, as follows:

1. *D. tajassu* (from S Brazil)—malar crest terminating above the infraorbital foramen; nasal bones rounded in cross section; first superior premolar tritubercular or rounded in outline, premolariform; molars not wrinkled.
2. *D. angulatus*, sp. nov. (from Texas)—malar crest continuing forward to the base of the canine alveolus; nasal bones pinched or angulate on the middle line; first superior premolar quadritubercular, with intermediate tubercles, and quadrate in outline, molariform; molars wrinkled.

He also noted the following:

> The characters cited are constant, although the amount of angulation of the nasal bones in *D. angulatus* is subject to some variation. Another character, and generally constant, is the form of the fossa above the diastema. In *D. tajassu* it is a narrow groove; in *D. angulatus* it is a wide fossa (Cope, 1889:147).

These comparisons were based on 16 skulls from S Brazil and 6 from Texas. We find that, despite Cope's small sample sizes, the main differences he elucidated are, indeed, absolutely consistent.

Table 7 Measurements for *Tayassu* samples

	Gt l	Cb l	Bi-THJ	Bican apoph	Teeth	Pal br ext	Occ br
Guatemala, Belize							
Mean	279.00	247.00	113.50	63.33	78.67	53.50	43.17
N	6	6	6	6	6	6	6
Std dev	5.967	4.336	3.886	1.633	2.160	1.049	2.927
Min	270	239	106	61	76	52	40
Max	287	252	116	66	82	55	48
Costa Rica							
Mean	274.00	244.00	117.00	60.00	73.00	52.00	39.00
N	1	1	1	1	1	1	1
Colombia							
Mean	279.25	243.63	116.75	61.37	77.13	53.13	37.75
N	8	8	8	8	8	8	8
Std dev	8.155	6.232	4.367	7.110	2.850	2.031	4.713
Min	270	237	112	53	73	51	32
Max	291	257	124	70	83	57	45
Venezuela							
Mean	270.50	241.00	111.50	59.50	73.50	51.00	36.00
N	2	2	2	2	2	2	2
Min	269	239	111	59	72	50	35
Max	272	243	112	60	75	52	37
Guyana, Suriname							
Mean	292.33	247.63	119.78	63.56	79.67	54.56	42.44
N	9	8	9	9	9	9	9
Std dev	10.100	6.140	5.094	4.927	2.739	1.333	2.404
Min	278	241	111	56	77	53	39
Max	307	259	130	73	84	56	46
NE Peru							
Mean	282.33	249.00	120.00	59.00	79.00	53.67	38.33
N	6	6	6	6	6	6	6
Std dev	5.715	7.099	2.966	2.191	4.243	1.366	4.633
Min	277	238	117	57	72	52	35
Max	291	258	125	63	85	56	47
SE Peru							
Mean	281.71	244.67	117.63	60.37	77.13	53.75	39.86
N	7	6	8	8	8	8	7
Std dev	7.064	8.430	2.326	3.662	3.682	1.753	3.237
Min	273	234	114	55	69	52	36
Max	291	256	120	66	80	56	44
Bolivia							
Mean	281.50	245.20	116.63	64.38	78.50	54.88	41.13
N	8	5	8	8	8	8	8
Std dev	12.917	11.077	5.553	3.739	4.140	1.553	3.907
Min	255	230	110	61	69	53	35
Max	295	257	124	70	82	58	46
Amazonia							
Mean	288.50	251.00	116.50	58.00	79.00	52.50	41.50
N	2	2	2	2	2	2	2
Min	285	248	116	56	78	52	38
Max	292	254	117	60	80	53	45
Mato Grosso, Paraguay							
Mean	288.00	249.00	118.50	59.50	78.50	53.00	40.50
N	2	2	2	2	2	2	2
Min	284	246	116	59	77	52	40
Max	292	252	121	60	80	54	41

(continued)

Table 7 (continued)

	Gt l	Cb l	Bi-THJ	Bican apoph	Teeth	Pal br ext	Occ br
Pará							
Mean	286.00	254.33	119.67	64.33	81.00	54.67	45.00
N	3	3	3	3	3	3	3
Std dev	7.211	2.309	8.145	8.737	1.000	3.786	5.292
Min	280	253	114	57	80	52	41
Max	294	257	129	74	82	59	51
SE Brazil							
Mean	285.80	245.20	119.80	64.40	77.20	53.80	42.60
N	5	5	5	5	5	5	5
Std dev	11.167	4.658	2.049	2.881	2.588	.837	5.505
Min	278	238	118	61	73	53	34
Max	305	249	123	68	80	55	49

Lydekker (1915a) accepted Cope's two species, but, curiously, Lydekker did not mention the most striking distinguishing character, that of the malar crest, mentioning instead only the characters of the basal angulation of the nasals, the palatal ridge, and the preorbital depression.

We refer here to what Cope called *D. angulatus* and *D. tajassu* as the "N" and "S" types, respectively. Despite the distinctiveness of the two morphological types, most later authors (probably following Cabrera, 1961) have united them, recognizing only a single species of collared peccary.

The recent description of a new species of collared peccary, *P. maximus* (see van Roosmalen et al., 2006), is difficult to test, given the small number of available specimens from Amazonia; it is not easy to use the measurements published in that paper, because very few of them correspond to those used in the present study. In their phylogenetic tree, employing several mtDNA sequences and two short interspersed nuclear elements (SINEs), *P. maximus* was sister to three other collared peccary specimens: one from Arizona, one from Colombia, and one of unknown origin. As Colombia is the country where "N" and "S" types both occur, this may be telling us only that *P. maximus* belongs to the "S" group, which we can indeed detect from the illustrations accompanying the type description. Much more work is needed on this question.

We find that "N" and "S" skull types are characteristic of Central American and South American collared peccaries, respectively, and that they meet in Colombia. Therefore, it is worthwhile examining their distribution within that country. In Colombia, skulls of "N" type are from the following:

Socorré (7.51° N, 76.17° W)
Unguia (8.02° N, 77.05° W)
Coloso (9.29° N, 75.21° W)
Los Micos (4.58° N, 75.49° W)
Rio Sandó, 100–160 m (5.13° N, 77.05° W)
Old Magdalena Valley

Skulls of "S" type are from the following:

Rio Mecaya (0.28° N, 75.20° W)
Rio Guapaya (ca. 5.00° N, 75.00° W)
Rio Saija, Cauca, 100 m
César
Narino, Candelilla, 200 m (1.27° N, 78.40° W)

Collared peccaries from French Guiana (Vassart, Pinton, et al., 1994) had $2n = 35$, like those previously studied from North America, but there were differences in details: the X chromosome of the North American specimens was metacentric, while that of the French Guiana specimens was acrocentric; and the North American animals had six medium-sized metacentric autosomes, while those from French Guiana had only five (one extra autosome therefore being acrocentric).

In striking agreement with the cranial data, both the mitochondrial and four nuclear sequences divided North American (including Central American) and South American lineages cleanly, with specimens from Colombia being allocated to both (Gongora et al., 2006). In the mitochondrial tree, two specimens known to be from N Colombia, two from C Colombia, and one from

S Colombia assorted with the N clade; while one from SE Colombia, one from N Columbia, one from S Columbia, one known merely as being from "Colombia," and one known to be from Caqueta assorted with the S clade. There was no geographic sorting in the N clade, but within the S clade there were significantly different clusters from Colombia, (N) Bolivia, and Argentina. Interestingly, the nuclear network showed that those Colombian specimens were that were part of the N clade were well within it, not at its root, implying a reinvasion into Colombia after the two lineages had been separated for some while.

Therefore, it is clear that there are at least two species involved in this genus; we find evidence, in fact, for at least three (two "N," one "S"), but there are many gaps in our dataset.

Pecari tajacu (Linnaeus, 1758)

The "S skull-type" species: malar crest directed upward, ending above the infraorbital foramen; nasal bones more rounded in cross section; first upper premolar premolariform; molars not wrinkled.

At present we cannot find more than one species in South America (Colombia excepted).

Pecari tajacu tajacu (Linnaeus, 1758)

1758 *Sus tajacu* Linnaeus. Pernambuco, according to Cabrera (1961).

Stated to have the general color being grizzled gray or tawny. Six skins from Chapada and Miranda, Mato Grosso, somewhat similar to the Paraguay and Jujuy series, but the dorsal stripe sometimes less marked; collar often quite faint, but the collar in one skin even better expressed than in the Jujuy skin.

Skull relatively large, long, narrow, vaulted.

We have seen specimens of this subspecies from Mato Grosso, N Argentina, and Paraguay.

Pecari tajacu patira Kerr, 1792

1792 *Sus patira* and *Sus minor* Kerr. French Guyana.

1816 *Dicotyles torquatus* G. Cuvier.

1917 *Pecari tajacu modestus* Cabrera. Rio Napo.

1921 *Pecari tajacu macrocephalus* Anthony. Guyana.

Skins from the Magdalena River and Rio César, Colombia, and six skins from Guyana, are dark, due to a predominance of dark bands on the hairs. Expression of the collar very good. Dark dorsal stripe not expressed, except down the nape. Skull relatively small, short, wide, depressed.

Unknowns

We presume that the following specimens belong to previously unnamed taxa:

—Two skins from Paraguay are very distinctive, with the hairs multibanded straw/black, giving an overall fairly light tone, but with a solid, thick, black dorsal stripe. Collar barely marked amid all the long, light hair bases. One from Sierra Santa Barbara, Jujuy, with a less striking differentiation, but the dorsal stripe still well marked; collar much better marked.

—One skin from San Francisco de Caup, Minas Gerais, is different from the others. No marked dorsal stripe; collar virtually absent; predominance of the whitish bands on the hairs.

Pecari angulatus (Cope, 1889)

1889 *Dicotyles angulatus* Cope. Guadelupe Valley, Texas.

1897 *Dicotyles angulatus sonoriensis* Mearns. San Bernardino Valley, Sonora, Mexico.

1901 *Tayassu angulatus humeralis* Merriam. Armería, Colima, Mexico.

One of two "N skull-type" species. Malar crest slopes forward, ending above the canine alveolus; nasal bones pinched or angulated; first upper premolar quadritubercular, with intermediate tubercles, molariform; molars wrinkled.

Hair banding blackish and gray or white; head lighter in tone; collar white, but often not marked; sometimes with a black dorsal stripe. Skull larger and narrower than in *P. tajacu*.

Evidently a species characteristic of the dry zone of N Mexico and the southernmost United States.

Pecari crassus (Merriam, 1901)

1901 *Tayassu angulatus crassus* Merriam. Metlaltoyuca, Puebla, Mexico.

1901 *Tayassu angulatus yucatanensis* Merriam. Tunkás, Yucatán.

1901 *Tayassu nanus* Merriam. Cozumel Island, off Yucatán.

1902 *Tayassu crusnigrum* Bangs. Boquete, Chiriquí, Panama.

1917 *Pecari angulatus bangsi* Goldman. Boca de Cupe, Darión, Panama.

1926 *Pecari angulatus nelsoni* Goldman. Huehuetán, Chiapas.

1926 *Pecari angulatus nigrescens* Goldman. Chamelecón, Cortés, Honduras.

"N skull-type," like *P. angulatus*. Coat coarser; light hair bands wide, making the overall color gray or white; black dorsal stripe; head lighter; collar various, often indistinct.

A skin from Vera Cruz, however, is darker, only slightly less dark than those from Guyana (see above); there is no dorsal stripe, even on the nuchal area; no noticeable collar.

Presumably, this is the species of the humid rain forest zone of S Mexico and Central America, extending into NW Colombia and Ecuador W of the Andes.

The following two names may be synonymous with *P. crassus*:

1. ?*Pecari tajacu torvus* (Bangs, 1898)
   —Type locality: Santa Marta, Colombia.
   —This name would have priority, if it is truly identical with the North American humid forest species.
2. ?*Pecari tajacu niger* (J. A. Allen, 1913)
   —Type locality: Esmeraldas, Ecuador.
   —General color is stated to be a nearly uniform black.

In an analysis of collared peccary skulls from North and Central America, Oaxaca plus Veracruz (humid zone) versus Texas and N and C Mexico (arid zone) overlap only slightly. The Texas and C Mexico skulls are larger, with a somewhat broader occiput and narrower palates.

Whether taking the skulls of the "N" and "S" crest types as a whole, or restricting the comparison to those in Colombia (table 8), the "N type" are larger and narrower. Compared with Guatemala/Belize, the Columbian "N type" skulls are not distinguished morphometrically, whereas the Columbian "S types" ones are clearly distinguished by being smaller and broader. Comparing the two Colombian types with a sample from E Peru, however, the difference remains, but the Peruvian sample is again different from either of the Columbian types. E Peru skulls have a larger condylobasal length, especially compared with the greatest length, and are broader across the temporomandibular joints but narrower across the canine apophyses; they have a somewhat broader palate, but a narrower occiput. The "S type" Colombian skulls, compared with the "N type," are far smaller, wider across the temporomandibular joints, and have a larger condylobasal length. Evidently there is more differentiation in South America than we are able to document.

Table 8 Discriminant analysis of skulls for *Pecari* with N versus S facial crest types in Colombia

		Predicted group membership		
		N	S	Total
Original count	N	15	0	15
	S	0	4	4
Original %	N	100.0	—	100.0
	S	—	100.0	100.0

	Function
	1
Gt l	−1.866
Cb l	.085
Bi-THJ	.774
Bican apoph	1.056
Pal br ext	−.729
Occ br	.812

E Peru and the Peru highlands groups separate well: the lowland skulls, compared with the highland ones, have a longer condylobasal length, but a shorter greatest length; and are wider across the palate and the temporomandibular joints, but narrower across the canine apophyses.

A skull from Trinidad is part of the Colombia "S" scatter.

The Guyanas overlap with the C Peru highlands, and they only just overlap with E Peru lowlands; compared with the latter, they have a larger greatest length, but a smaller condylobasal length, and they are narrower. All of these skulls, as well as the Peru highlands ones, differ from the Colombia "S" type in their somewhat larger size, but they are narrower across the canine apophyses.

Guyana and a single Amazonas skull have a larger condylobasal length (especially relative to greatest length), and they are narrower across the canine apophyses and the occiput. Mato Grosso differs from Pará in being wider across the temporomandibular joints; Guyana is more like Mato Grosso in this.

Pará, compared with Mato Grosso, Bolivia, and Amazonas, is smaller, and wider across the canine apophyses and the occiput, compared with the temporomandibular joints and the palate. Bolivian skulls have a larger greatest length, especially compared with condylobasal length; they also have a very broad palate and are very narrow across the temporomandibular joints. The single Amazonas specimen is an extreme example of the Mato Grosso morphology.

Mato Grosso overlaps with Paraguay; Bolivia and São Paulo seem distinct. Bolivian skulls, as before, have a long greatest length and a short condylobasal length, a broad palate, and narrow bi-temporomandibular and canine apophyses breadths. São Paulo seems to differ from Mato Grosso ($N=2$) in having a short greatest length compared with a fairly long condylobasal length, and a narrow width across the temporomandibular joints and the palate, compared with a greater width across the canine apophyses and the occiput.

E Peru and Amazonas (including a skull from "north of Xavantina, Serra do Roncador," which is in the state of Mato Grosso but well N of the other Mato Grosso specimens) are separate from C Peru and from the main Mato Grosso / Bolivia sample, with a long condylobasal length and a short greatest length, and a broad width across the temporomandibular joints, but a narrow one across the canine apophyses.

Our conclusion is that, in our limited sample, Amazonas and E Peru form a single group; Mato Grosso and Bolivia, together with C Peru, form another group.

Table 9 gives univariate statistics for the skulls of collared peccaries. There are no great differences in size (except for the small size of the single available skull from Trinidad). The São Paulo (including Santa Catarina) sample stands out in its narrow width across the temporomandibular joints. Samples from the United States and the more arid N parts of Mexico

Table 9 Measurements for *Pecari* samples

	Gt l	Cb l	Bi-THJ	Bican apoph	Teeth	Pal br ext	Occ br
USA							
Mean	236.81	201.31	101.63	59.38	64.38	46.31	35.00
N	16	16	16	16	16	16	16
Std dev	7.378	6.610	5.760	4.080	2.918	1.537	3.706
Min	221	187	93	54	59	43	26
Max	247	210	116	71	70	50	41
Durango, Sinaloa							
Mean	245.00	204.17	103.33	58.50	62.50	44.00	37.50
N	6	6	6	6	6	6	6
Std dev	8.198	5.419	3.266	2.665	1.761	1.897	3.564
Min	231	198	99	54	60	42	32
Max	254	214	108	61	65	46	43
Veracruz							
Mean	239.00	204.50	104.50	60.00	66.00	46.00	32.00
N	2	2	2	2	2	2	2
Min	238	203	104	57	64	46	32
Max	240	206	105	63	68	46	32
Oaxaca, Tabasco, Guatemala, Belize							
Mean	240.78	203.75	101.11	55.56	60.56	43.22	32.44
N	9	8	9	9	9	9	9
Std dev	7.807	6.692	5.600	3.167	3.127	1.481	4.304
Min	229	193	93	49	56	41	27
Max	254	216	109	59	66	46	38
Colombia, "N type"							
Mean	236.00	200.35	97.33	55.06	63.11	45.11	34.37
N	16	17	18	18	18	18	16
Std dev	6.552	6.614	3.218	8.299	3.660	2.220	2.306
Min	227	183	92	48	55	42	29
Max	248	211	104	85	69	49	37

(continued)

Table 9 (continued)

	Gt l	Cb l	Bi-THJ	Bican apoph	Teeth	Pal br ext	Occ br
Colombia, "S type"							
Mean	229.50	199.50	98.50	55.00	60.50	40.00	34.50
N	2	2	2	2	2	2	2
Min	226	199	96	52	58	34	34
Max	233	200	101	58	63	46	35
E Peru, E Ecuador							
Mean	240.69	209.80	103.14	53.50	66.21	46.86	35.57
N	13	10	14	14	14	14	14
Std dev	7.620	7.193	5.318	4.468	3.068	2.282	2.901
Min	227	202	95	48	61	44	31
Max	254	225	113	63	71	50	41
C Peru							
Mean	242.20	205.80	100.00	51.00	62.60	45.80	32.40
N	5	5	5	5	5	5	5
Std dev	7.596	7.050	4.301	5.916	2.302	1.924	2.074
Min	231	197	94	44	59	44	31
Max	251	216	106	58	65	49	36
Trinidad							
Mean	217.00	189.00	100.00	49.00	62.00	42.00	28.00
N	1	1	1	1	1	1	1
Guyana, Suriname, Roraima							
Mean	244.07	207.93	102.79	53.64	64.86	44.64	30.86
N	14	14	14	14	14	14	14
Std dev	5.427	5.015	6.278	4.272	2.770	1.906	4.092
Min	233	202	87	48	61	41	22
Max	252	219	115	61	70	48	35
Pará							
Mean	233.00	197.80	95.00	50.40	61.20	42.60	31.80
N	5	5	5	5	5	5	5
Std dev	7.778	7.430	2.345	3.975	4.438	.548	3.701
Min	224	188	93	47	56	42	26
Max	245	208	99	56	67	43	36
Amazonas							
Mean	252.17	216.00	100.00	49.00	66.00	45.00	36.00
N	6	1	1	1	1	1	1
Std dev	11.907	—	—	—	—	—	—
Min	230	—	—	—	—	—	—
Max	262	—	—	—	—	—	—
Mato Grosso, E Bolivia							
Mean	231.38	194.25	99.75	53.00	61.50	44.00	29.63
N	8	8	8	8	8	8	8
Std dev	10.127	9.528	4.400	4.957	3.817	2.000	4.809
Min	209	174	94	45	56	40	21
Max	240	206	106	60	67	46	36
N Argentina, Paraguay							
Mean	231.00	196.86	102.57	54.88	64.71	45.25	33.20
N	5	7	7	8	7	8	5
Std dev	10.320	7.381	3.823	3.091	2.812	2.188	5.263
Min	222	187	98	50	60	42	28
Max	247	208	110	59	67	50	42
São Paulo, Santa Catarina							
Mean	232.00	201.50	95.00	52.50	65.50	45.00	35.50
N	2	2	2	2	2	2	2
Min	229	199	93	52	62	45	35
Max	235	204	97	53	69	45	36

are noticeable for the wide distance across the canine apophyses.

Catagonus Ameghino, 1904

$2n = 20$; all chromosomes are metacentric, with unusually large heterochromatin blocks (Benirschke et al., 1985).

Catagonus wagneri (Rusconi, 1930)

On this species, see Wetzel (1977). The discovery of such a large species of living mammal—one, moreover, that was previously known only as a fossil—was one of the great biological surprises of the 20th century.

6

Hippopotamidae

The Whippomorpha concept. The aquatic adaptations of the Hippopotamidae: are they really those of whales writ small, or is it just the "Aha" effect? Is there significant geographic variation in the common hippo? Is the pygmy hippo really a survivor of the fossil genus *Hexaprotodon*? The pygmy hippo of the Niger Delta: a distinct species.

Two decades of molecular research have now clearly established that cetaceans (whales and dolphins) are not only part of the Artiodactyla, but are a sister group to hippos. This sister-group relationship was formalized by Waddell et al. (1999) under the name Whippomorpha. Skinner & Chimimba (2005) split the Artiodactyla into several distinct orders, of which Whippomorpha was one. We gave our reasons (see Part 2) for preferring to retain the order Artiodactyla and recognize Whippomorpha at the subordinal level, dividing the suborder into two infraorders, Cetacea (not treated here) and Ancodonta, the latter including (in the living fauna) only the family Hippopotamidae.

HIPPOPOTAMIDAE GRAY, 1821

Some of the characteristics of members of this family, such as copious body fat, relative hairlessness, compact build, and underwater communication, could be seen either as ad hoc aquatic adaptations in the family, or, nowadays, as inherited from their common ancestor with Cetacea. We simply know too little of the fossil history of hippos and their relatives, but in the light of the relatively new Whippomorpha concept, it is irresistible to try to find such potential evidence of affinities, and of a very ancient aquatic heritage!

Hippopotamus Linnaeus, 1758

Hippopotamus amphibius is the only survivor of a once-diverse radiation.

Hippopotamus amphibius Linnaeus, 1758

common hippopotamus

Grubb (1993a) recognized subspecies of this species, but was inclined to doubt whether there was really much geographic differentiation; he recommended further morphological study, and this is still needed.

Mitochondrial control region sequences (Okello et al., 2005) showed low differentiation among hippopotamus populations from Uganda, Kenya, Tanzania, and Zambia, with only weak geographic structuring and evidence of comparatively recent population expansion; the most ancestral haplotypes were clustered around Murchison Falls National Park, and the authors related this expansion to the changing river patterns consequent on the E African tectonic activity of the Middle Pleistocene. We may note that, during the Early Pleistocene, there was at least one other species, *Hippopotamus gorgops*, that was probably a competitor and was much commoner in the fossil record; presumably a rapid expansion of *Hippopotamus amphibius* would have followed on the extinction of its competitor.

Choeropsis Leidy, 1853

Very small in size; anterior ends of the nasals comparatively long and curved downward; orbit placed well below the dorsal outline; palatine with a strong posterior nasal spine; tympanic bulla inflated; lateral notch posterior to the basilar tubercles; cranial roof curved downward posteriorly; only two mandibular incisors.

As a result of the work of Coryndon (1978 and elsewhere), it became usual to assign the living pygmy hippo to the genus *Hexaprotodon*. Subsequent to the description of further hippos, however, Boisserie (2005) reinstituted the genus *Choeropsis*, which he considered the sister to all other hippos (except probably the fossil genus *Saotherium*). The diagnosis of the genus given above depends partly on his analysis.

G. Corbet (1969) described the Niger Delta taxon as a subspecies of *C. liberiensis*, but noted that 9 of the 11 character states by which it differed were discrete.

While the sample size was small ($N=2$–4, compared with $N=10$–14 [$N=3$ in the case of the nasal measurements] of nominotypical *C. l. liberiensis*), the sheer number of the differences, and the distinct gaps between the two taxa in many of them, give confidence that the two really are absolutely distinguishable, with, as Corbet put it, "no indication of clinal variation within the range of *liberiensis* affecting the characters that distinguish" *C. heslopi*. We therefore have little hesitation in recognizing two species in this genus.

Choeropsis liberiensis (Morton, 1844)

western pygmy hippopotamus

Dorsal margin of the zygomatic process of the squamosal strongly sinuous, forming a rounded 90° concavity; nasals broad, the smallest width >15% of the greatest length, the greatest width 32%–37% of the greatest length; bullae angular, the anterior muscular processes rudimentary; frontal bones flat, or slightly convex in the midline; very long (6–10 mm) median process on the posterior margin of the palate; P^2 wide, its width 67%–80 % of its length, with a prominent posteromedian ridge; posteromedian ridge on P^3; these two premolars long, their length 46%–60 % of the total toothrow; median nasal tips usually short; ventral grooves on the malars usually shallow. (Following G. Corbet, 1969.)

Sierra Leone to Ivory Coast.

Choeropsis heslopi Corbet, 1969

Niger Delta pygmy hippopotamus

1969 *Choeropsis liberiensis heslopi* Corbet. Omoku, Owerri Province, Nigeria (5° 19' N, 6° 40' E).

Dorsal margin of the zygomatic process of the squamosal not so strongly sinuous, forming only a slight concavity; nasals narrow, the smallest width <15% of the greatest length, the greatest width 22%–26% of the greatest length; bullae rounded, the anterior muscular processes prominent; wide median concavity in the frontal bones; no (or only a very rudimentary) median process on the posterior margin of the palate; P^2 narrow, its width 54%–58% of its length, with no posteromedian ridge; no posteromedian ridge on P^3; these two premolars relatively short, their length only 42%–45% of the total toothrow; median nasals tips long, extending about 10 mm beyond the adjacent concavities; deep ventral grooves on the malars. (Following Corbet, 1969.)

Niger Delta. Unfortunately, pygmy hippos appear not to have been recorded in the Niger Delta since the 1940s.

It appears that one of the dwarf hippos that are now known from Madagascar may have persisted until the late 20th century (Burney & Ramilisonina, 1999).

7

Tragulidae and Moschidae

The Tragulidae: astonishing diversity in these small-sized, little-known animals. *Tragulus*: why not to try to shove 50 "subspecies" into just two species. *Moschiola* as an example of the astonishing biodiversity of Sri Lanka. *Hyemoschus* and its relationship to other mouse deer and to the fossil genus *Dorcatherium*.

TRAGULIDAE MILNE EDWARDS, 1864

The lacrimal orifice is single, elongated, and situated just inside the rim of the orbit (Leinders & Heintz, 1980).

Mouse deer (chevrotains) are sister to all other remnants, and on this basis we separate them into an infraorder Tragulina; other ruminant families belong to an infraorder Pecora.

Tragulus Brisson, 1762

This was one of the generic names which A. Gentry (1994) requested to be conserved, even though other names from Brisson (1762) were to be rejected.

Chromosomes were studied by Kim et al. (2004), who compared the results with earlier studies. All *Tragulus* appear to have $2n = 32$. *T. kanchil* (reported as *T. javanicus*) from Pahang and Selangor have a metacentric Y chromosome; in the same species from Sabah, it is subtelocentric. In *T. williamsoni* from Yunnan, the Y was submetacentric, and larger than in the Malaysian males. In *T. napu* from Pulau Tioman [Tioman Island], the Y was subtelocentric (like *T. kanchil* from Sabah) but larger, and the short arm of the X chromosome was much shorter than in the *javanicus* group.

Our arrangement follows Meijaard & Groves (2004a), and we here give only brief summaries of the findings of that revision.

Tragulus napu Group

The large Malay chevrotains (see Meijaard & Groves, 2004a).

Large size; relatively small auditory bullae (negative allometry?). Usually a distinct, characteristic pelage.

Tragulus napu (F. Cuvier, 1822)

Tragulus napu napu (F. Cuvier, 1822)

1822 *Moscus napu* F. Cuvier. S Sumatra.
1906 *Tragulus bancanus* Lyon. Bangka.
1935 *Tragulus javanicus abjectus* Chasen. Sirhassen Island.

Sumatra, except for the N (Medan and presumably farther N); Malay Peninsula; Pangkor Island; Langkawi; Serasan; Bangka; Borneo; and Pulau Laut [Laut Island].

Tragulus napu neubronneri Sody, 1931
Maintained provisionally by Meijaard & Groves (2004a) as a separate subspecies, as the occipital height and the mandibular toothrow length appear to be considerably greater than in *T. n. napu*.

N Sumatra.

Tragulus napu niasis Lyon, 1916
Maintained provisionally by Meijaard & Groves (2004a) as a separate subspecies, distinguished from *T. n. napu* by the shorter hindfoot.

Nias.

Tragulus napu terutus Thomas & Wroughton, 1909
Distinguished by the long, narrow auditory bullae; the wide skull; and the high occiput.

Terutan.

Tragulus napu bunguranensis Miller, 1901
Distinguished from other members of the species by its large size. Meijaard & Groves (2004a) consider that this is possibly a distinct species.

Bunguran, in the N Natuna Islands.

Tragulus napu rufulus Miller, 1900

1900 *Tragulus rufulus* Miller. Tioman Island.
1903 *Tragulus formosus* Miller. Bintang.
1903 *Tragulus pretiosus* Miller. Lingga.

Distinguished by the very bright reddish color; also by the nape stripe being absent or, occasionally, present but vague.

Many islands of the Riau Archipelago (Batam, Galang, Sekikir, Bulan, Bintang, Lingga, Bakong, Sebangka), and Pulau Tioman [Tioman Island].

Tragulus napu banguei Chasen & Kloss, 1931
Very small in size; throat pattern distinctive.
Banggi and Balembangan.

Incertae sedis

- *Moschus stanleyanus* Gray, 1836 (locality unknown)
- *Tragulus brevipes* Miller, 1903
- *Tragulus napu amoenus* Miller, 1903 (= *Tragulus napu jugularis* Miller, 1903; these two taxa are evidently color morphs of the same species)
- *Tragulus napu batuanus* Miller, 1903
- *Tragulus pinius* Lyon, 1916
- *Tragulus russulus* Miller, 1903

Tragulus nigricans Thomas, 1892

Upper parts washed with black; neck mixed black and fulvous.

Balabac.

The following taxa are provisionally recognized as subspecies of this species by Meijaard & Groves (2004a):

- *Tragulus billitonus* Lyon, 1906-the allocation to this species-group is uncertain
- *Tragulus napu hendersoni* Chasen, 1940—the allocation to this species-group is uncertain
- *Tragulus nigrocinctus* Miller, 1906—could be a distinct species (Meijaard & Groves, 2004a)

Tragulus versicolor Group

Skull very small, but relatively wide; relatively wide, rounded bullae (possibly allometric); very long nasals. Hair very coarse; marked contrast between the ochraceous-buff foreparts and the clear gray posterior half (starting behind the shoulders); no dark median line on the white underside; no dark lines from the eye to the nose; absence of the dark collar that is almost universal in other *Tragulus*. On the throat, the lateral reddish lines converge posteriorly but do not touch, with the central white throat line being continuous with the white on the underparts.

Meijaard & Groves (2004a) consider this Indochinese species as quite distinct from either of the other two species-groups, and it is not, as is commonly supposed, a member—even, a subspecies!—of the *T. napu* group.

Tragulus versicolor Thomas, 1910

See above (for the species-group).

Tragulus javanicus Group

The small Malay chevrotains (see Meijaard & Groves, 2004a).

Tragulus javanicus (Osbeck, 1765)

Meijaard & Groves (2004a) suggested that, as had been proposed by earlier authors, there may indeed be two distinct mouse deer on Java, but this needs substantial further work.

Java.

Tragulus kanchil (Raffles, 1821)

Tragulus kanchil kanchil (Raffles, 1821)
Sumatra.

Tragulus kanchil hosei Bonhote, 1903

1903 *Tragulus kanchil hosei* Bonhote. Baram, Sarawak.
1935 ?*Tragulus kanchil abruptus* Chasen. Pulau Subi [Subi Island].

NW, W, E, and S of Borneo; Subi Island(?).

Tragulus kanchil klossi Chasen, 1934
Larger, at least on average, than other *T. kanchil*, and apparently absolutely larger than *T. k. hosei*; pelage mottled, more like *T. napu*.

At least the C and E parts of Sabah, and the N area of East Kalimantan. The closest locality for *T. k. hosei* is the Baram River, E Sarawak; there are no specimens available for any mouse deer from S Kalimantan.

Possibly a distinct species, according to Meijaard & Groves (2004), but this needs further study.

Tragulus kanchil ravulus Miller, 1903
Paler than *T. k. ravus*, with a less defined nape stripe; auditory bullae short, narrow; nasals short; short mandible; narrower braincase.

Adang and Rawi islands.

Tragulus kanchil affinis Gray, 1861

1861 *Tragulus affinis* Gray. Cambodia.
1903 *Tragulus kanchil pierrei* Bonhote. Bienhoa, S Vietnam.

Craniometrically not distinct from other mainland taxa, but externally pale, without blackening on the neck.

Vietnam, Laos, Cambodia, and SE and E Thailand.

Tragulus kanchil ravus Miller, 1902

1902 *Tragulus ravus* Miller. Trang, southernmost Thailand.
1903 *Tragulus lancavensis* Miller. Langkawi Island.

Paler than *T. k. fulviventer*, with a much less distinct nape stripe.

SW Thailand, C and N Malay Peninsula, Langkawi and Butang islands.

Tragulus kanchil augustiae Kloss, 1918
Nape stripe well defined.

Tenasserim, from 10° N to about 16° 30' N.

Tragulus kanchil fulviventer Gray, 1836

1940 *Tragulus javanicus pumilus* Chasen

Distinguished by its pale fulvous underparts.

Malay Peninsula, approximately S of the Thai–Malay border (including Singapore, the type locality).

Tragulus kanchil everetti Bonhote, 1903

1903 *Tragulus kanchil everetti* Bonhote. Pulau Bunguran [Bunguran Island] (N Natuna Islands).
1903 *Tragulus natunae* Miller. Bunguran.

Resembling *T. k. hosei*, but a much deeper red on the flanks; teeth larger; bullae smaller.

The following taxa are provisionally recognized as subspecies of this species by Meijaard & Groves (2004a):

Tragulus kanchil anambensis Chasen & Kloss, 1928
Tragulus kanchil carimatae Miller, 1906
Tragulus kanchil fulvicollis Lyon, 1908
Tragulus kanchil insularis Chasen, 1940
Tragulus kanchil lampensis Miller, 1903
Tragulus kanchil luteicollisi Lyon, 1906
Tragulus kanchil pallidus Miller, 1901
Tragulus kanchil pidonis Chasen, 1940
Tragulus kanchil rubeus Miller, 1903
Tragulus kanchil siantanicus Chasen & Kloss, 1928
Tragulus kanchil subrufus Miller, 1903

Tragulus williamsoni Kloss, 1916

Very much larger than any other member of this species-group; color fairly similar to *T. k. affinis*.

Known securely only by the type specimen, from Meh Lem, Muang Pre, Song Forest, far N Thailand, 18° 25' N, 100° 23' E.

Moschiola Gray, 1853

Type species: *Meminna indica* (Gray, 1843), by monotypy.

This genus, formerly considered monotypic, was recently revised by Groves & Meijaard (2005). There are three (possibly four) distinct species.

Moschiola indica (Gray, 1852)

1852 *Meminna indica* Gray. E Ghats, India.

Dull brown; stripes and spots very clear, white; upper row of spots forming a continuous stripe on the shoulder, but disintegrating into spots halfway back on the body; other longitudinal stripes broken completely into rows of elongated spots; three spot-rows across the croup and the rump partly fused into stripes; crown and forehead dark brown; underside pale beige, becoming creamy medially. Size very large; long hindlegs; hindfoot length >140 mm. Skull broad, especially across the rostrum; zygomatic breadth >50 mm; occipital height (measured from the basion) >27.8 mm; width across the canine alveoli >15.2 mm; least maxillary breadth >14.5 mm; braincase breadth >34.2 mm.

Much of India, and formerly—but perhaps not surviving in recent times?—also in Nepal (Baral et al., 2009).

Moschiola meminna (Erxleben, 1777)

1777 *Moschus meminna* Erxleben. Wariayapola, 7° 37' N, 80° 13' E, 460 m, North Central Province, Sri Lanka.

Color and pattern very like *M. indica*, but the upper row of spots not extensively fused into a continuous stripe on the shoulder. Size small; hindlegs relatively long; hindfoot length 120–144.5 mm. Zygomatic breadth <50 mm; occipital height <27.5 mm; width across the canine alveoli <15 mm; braincase breadth <33 mm.

Dry zone of Sri Lanka.

Moschiola kathygre Groves & Meijaard, 2005

2005 *Moschiola kathygre* Groves & Meijaard. Kumbalgamuwa, 7° 06' N, 80° 51' E, 747 m, Kandy district, Central Province, Sri Lanka.

Color a much warmer, more ochery brown; spots on the stripes yellowed, not white; at least two tolerably complete longitudinal stripes along the flanks, with an elongated spots row between them and two spot rows above them; upper stripe curving around on the shoulder, becoming continuous with the anterior transverse stripes; two bold stripes over the haunch, and a third one farther back under the tail; haunch more densely spotted; crown and nose less darkened; on the underside, a sharp differentiation between a white median strip and the pale ochery of the rest of the underside; lower halves of the hindlimbs darker. Size small; hindlegs relatively short; hindfoot length <115 mm. Rostrum breadth narrower, relative to the skull length. Zygomatic breadth <51 mm; occipital height 25.1–29.2 mm; width across the canine alveoli <15; braincase breadth <33 mm.

Wet zone of Sri Lanka, into the highlands, at least to the region of Kandy.

An undescribed species from Nuwara Elieya, in the cloud forest at over 2000 m, is known from a single skull, but other specimens are in the process of study. At any rate, the diversity of mouse deer in Sri Lanka, associated with the climatic zones, duplicates that of many other vertebrates.

Hyemoschus Gray, 1845

African mouse deer, water-chevrotains.

Table 10 Measurements for *Hyemoschus* samples

	Gt l	Nas l	Nas br post	Nas br ant
W Africa				
Mean	143.618	45.367	19.350	10.850
N	6	6	6	6
Std dev	3.5217	3.9073	2.8219	1.3751
Min	139.1	39.0	16.4	9.5
Max	148.0	49.3	23.9	13.0
Ghana				
Mean	144.000	—	25.500	—
N	2	—	2	—
Min	137.7	—	22.2	—
Max	150.3	—	28.8	—
Okuni				
Mean	136.800	40.900	23.300	15.900
N	1	1	1	1
Batouri				
Mean	143.833	45.267	22.700	11.933
N	3	3	3	3
Std dev	.8622	3.4962	.7211	.5859
Min	142.9	43.1	22.1	11.5
Max	144.6	49.3	23.5	12.6
Ituri/Uele				
Mean	143.567	46.500	19.933	11.400
N	3	1	3	2
Std dev	2.8885	—	4.8087	—
Min	141.8	—	15.3	11.3
Max	146.9	—	24.9	11.5

Hyemoschus aquaticus (Ogilby, 1841)

Although subspecies have been described (*Hyemoschus aquaticus aquaticus* from Sierra Leone; *Hyemoschus aquaticus batesi* Lydekker, 1906, from Efulen, Cameroon; *Hyemoschus aquaticus cottoni* Lydekker, 1906, from the Ituri Forest), the small number of available samples suggest little in the way of craniometric differentiation. A single skull from Okuni, in SE Nigeria, has unusually short nasals that widen anteriorly. The nasals of the westernmost (Sierra Leone, Gambia, Senegal) and easternmost (Uele, Ituri Forest) samples are narrow, compared with those of intervening localities.

MOSCHIDAE GRAY 1821

As in the Tragulidae, the lacrimal orifice is single, elongated, and situated just inside the rim of the orbit, but on its medial border there is a "small protuberance" resembling either an incipient or a former state of division of the orifice (Leinders & Heintz, 1980). The anterior sagittal gully of the metatarsus is closed by a bridge distally, as in the Cervidae. The polarity of the character states distinguishing the Moschidae from other (non-tragulid) ruminants is questionable (Janis & Scott, 1987).

A gall bladder is present.

A detailed, morphology-based cladogram placed *Moschus* as sister to the clade containing the Giraffidae and the Bovidae and their fossil relatives, not with the Cervidae, as had been previously "traditional" (A. W. Gentry & Hooker, 1988). Hassanin & Douzery (2003), using both mitochondrial and molecular sequences, placed the Moschidae firmly as sister to the Bovidae, with very strong support.

Moschus Linnaeus, 1758

musk deer

Our arrangement follows Groves et al. (1995), but first we must acknowledge a competing classification.

Sokolov & Prikhod'ko (1997) lumped all musk deer into a single species, *M. moschiferus*, dividing it into two groups, one with four subspecies and one with three, as follows:

1. *sibirica* group
   Siberian *Moschus moschiferus moschiferus*—the largest skull, decreasing in size from W to E
   Verkhoyansk (E Siberia)
   M. m. arcticus—smaller skull, especially shorter face, narrow interorbital distance, and short diastema

Far Eastern *M. m. turowi*—skull still smaller, facial part shorter, toothrows shorter, short mandible
Sakhalin *M. m. sachalinensis*—intermediate between Siberia and the Far East, with a long upper toothrow, relatively narrow braincase and interorbital distance, and short mandible

2. *himalaica* group
Korean *M. m. parvipes*—skull smallest, short face, narrow interorbital distance
Chinese *M. m. chrysogaster*—braincase shorter
Himalayan *M. m. leucogaster*—supraorbital ridges well developed, the edges protruding beyond the line of the frontals

While, in many respects, the authors' work is very careful—studying the inheritance of color features, and reviewing the growth of the skull and testing different regions for significant differences—this revision is very clearly overlumped as to species, but apparently oversplit as to subspecies. For example, inspection of their table 2 shows that the skulls of the Verkhoyansk sample are statistically different from those from the three subregions of the main Siberian sample, but the differences are, in very few cases, even as much as one standard deviation apart. The same is true for comparisons between the Far Eastern and Sakhalin samples. On the other hand, the differences between Siberia, on the one hand, and the Far East plus Sakhalin, on the other, do seem substantial, the latter being very much smaller: the mean greatest skull length for the four Siberian populations is 152.4–154.9 mm (the standard deviation in the largest sample [65 individuals from E Sayan] being 3.52), whereas that for the Far East is 147.3 mm, and that for Sakhalin is 149.8 mm. The sample sizes for the *himalaica* group are admittedly much smaller, with only those taken from Sheng (1992) being numerous; the measurements for the Korean sample look very much like those from the Far East, and it is not clear why they are placed in the S group rather than in the N group (it does not appear to fit at least one aspect of the definition of the S group; namely, that the facial part of the skull is greater than 50% of the greatest skull length).

The same authors published a second paper in the following year, dealing with chromosomes and pelage (Sokolov & Prikhod'ko, 1998). Karyotypes seem to be the same in all musk deer. The authors carefully described the pelage of the Siberian musk deer, and those few skins from China and the Himalayas available to them, but they again produced a lumped taxonomy, stating, for example, that *M. berezovskii* is a color morph of what they referred to as *Moschus moschiferus chrysogaster*, the distinction between them being often due to "mistakes in ageing." They did, however, now accept that Korean musk deer are closer to those from the Far East and Sakhalin.

Moschus moschiferus Linnaeus, 1758

Pelage softer, less quilly, than in other species. Individual hairs dark, with a white subterminal band. Underside lighter, grayish brown. Neck brownish; head more grayish, sometimes either lighter or darker than the body. Ears dark brown or black, lighter at the base. A pair of narrow creamy or white stripes (made up of white-tipped hairs) from the chin down the neck to the chest. Light spots variably visible, prominent in the young. Lacrimal length shorter than its height; braincase elongated; orbits tubular.

Moschus moschiferus moschiferus Linnaeus, 1758

1758 *Moschus moschiferus* Linnaeus. "Tataria versus Chinam"; Harper (1940) restricted this to the Russian Altai.
1779 *Moschus sibiricus* Pallas. Above Abakan, by one of the upper tributaries of the Yenisey River.
1929 *Moschus moschiferus arcticus* Flerov. Mt. Toulaja, 460 km N of Verkhoyansk.

Overall color predominantly dull brown. Skull length >145 mm, usually >150 mm.

E Altai and Yenisey, Mongolia; NE Nei Mongol; NW Heilongjiang.

Moschus moschiferus parvipes Hollister, 1911

1911 *Moschus moschiferus parvipes* Hollister. Mok-po, Korea.
1945 *Moschus moschiferus turowi* Tsalkin. Sikhote-Alin, Russian Far East.

Darker than nominotypical *M. m. moschiferus*. Skull length 141–151 mm; bizygomatic breadth 60–68.6 mm; lacrimal height 20–21.9 mm; lacrimal length 12.8–14 mm. Russian Far East, North Korea, S Heilongjing, E Jilin, and E Liaoning

Moschus moschiferus sachalinensis Flerov, 1929

1929 *Moschus moschiferus sachalinensis* Flerov. Sakhalin.

Color dark, as in *M. m. parvipes*. Skull length 147–151 mm; bizygomatic breadth 67.5–70.3 mm. Lacrimal height 19.5–20.3 mm; lacrimal length 14–15 mm.

Moschus chrysogaster **Hodgson, 1839**

The largest species; skull length >150 mm, averaging 155.1 mm. Overall color striated yellowish gray; paler on the flanks; underside reddish creamy gray. Individual hairs brown, with a red-yellow subterminal band. Ears light brown, tipped with yellow, gray inside. Throat with a single broad, ill-defined, creamy longitudinal band. Legs lighter than the body on the lower segments, grayish yellow, somewhat darkened down the front surfaces. Rump paler than the back, yellowish. Young animals spotted. Faint dorsal stripe. Hair bases long, milky gray or brown. Winter hairs 34–42 mm on the withers, 55–64 mm on the rump.

Lacrimal length much longer than its height. Limbs elongated; metacarpal length 109–118 mm; metatarsal length 128–138 mm.

On this species, including Hodgson's own documentation, see Grubb (1982b).

Moschus chrysogaster chrysogaster Hodgson, 1839

1839 *Moschus chrysogaster* Hodgson. Probably from the Tibetan plateau, N of the Himalayas.

Skull length 144–160 mm. More reddish tones; red-gold underside, inner surfaces of the limbs, and midline of the throat; orange eye-ring; Throat with white stripes; interramal region white. Limbs dark brown, becoming light brown on the lower segments; ears dark brown, orange-gold internally, at the base, and along the rims. Black patch on the buttocks. Rostrum less than half the length of the skull.

Alpine zone of Nepal, Sikkim, Bhutan, and S Tibet; 2800–4000 m.

Moschus chrysogaster sifanicus Buechner, 1891

1891 *Moschus sifanicus* Buechner.

Size averaging slightly larger; skull length 150–165 mm. Throat with a broad creamy band. Limbs lighter; ears with yellow tips. Rostrum length more than half the skull length.

S Gansu, W Sichuan, Qinghai, S Ningxia, SE Tibet, N Yunnan; 3500–4800 m.

Moschus leucogaster **Hodgson, 1839**

1839 *Moschus leucogaster* Hodgson. Probably from the Himalayan slopes of Nepal.

1839 *Moschus saturatus* Hodgson. Probably from the Himalayan slopes of Nepal.

Skull length 153–160 mm. Brownish yellow, weakly striated. Head gray-brown. Ears brown, rims and inside grayish white; poorly expressed grayish eye-ring. Legs dark. Rump dark. Bases of the dorsal hairs pure white. No neck-stripe; throat all dark. Interramal region grayish white; underside, from the chest to the groin, grayish white. Juveniles spotted. Lacrimal length longer than its height.

Nepal, Bhutan, and Sikkim; 2500–4000 m.

On this species, including Hodgson's documentation of it, see Grubb (1982b), who, at that time, treated it as a subspecies of *M. chrysogaster.*

There is possibly a separate subspecies (or species?) from Zhangmu (28°N, 87° E), in the forest zone of S Tibet: it is dark, without hair-banding, but the bases of the hairs are white; the ears, the limbs, and the neck are all dark; the posterior aspect of the rump is orange-white; and the nasals are anteriorly expanded.

From the Khumjung Glacier district in Nepal come skins related to the species, but differing slightly: the neck is somewhat paler; the throat is pale, with a poorly marked stripe on either side; the chin and the interramal region are creamy white; the ears are gray basally, and black terminally; the buttocks are yellow-brown; the legs are mostly black; the chest is black; and the belly is grayer. In Groves et al.'s (1995) multivariate analysis, the combined Zhangmu/Khumjung sample separates from the single available specimen of *M. leucogaster.*

There is a further unnamed form, related to this species, known from Kulu in NW Uttar Pradesh, at 3000 m or more, and from part of Nepal. This, which we call the "pepper-and-salt" form, is brown or red-agouti, sometimes forming a saddle, often with a light spot; throat light, this color often in a diffuse median band; lower limb segments lighter than the body, because of the white speckling; buttocks and tail paler; ears dark gray or brown, often with a whitish border, and white inside. Hair bases yellow.

Moschus cupreus **Grubb, 1982**

1982 *Moschus leucogaster cupreus* Grubb. Kashmir, at over 3000 m.

Skull length 150–155 mm. Gray-brown, often vaguely spotted, with a conspicuous, coppery-reddish, unspeckled dorsal saddle; rump very dark gray; underside light gray; throat white; lower segments of the limbs whitish. Ears dark brown, white at the base, with frosted rims. Hairs with long white bases; hairs 33–38 mm long on the withers, 37–58 mm on the rump. Lacrimal length longer than its height.

Kashmir.

On this species, see Grubb (1982b), who, at that time, treated it as a subspecies of *M. chrysogaster*. In Groves et al.'s (1995) discriminant analysis, this species is quite separate from all the others, but it is somewhat closer to *M. chrysogaster* than to the other taxa.

Moschus fuscus Li Z-x., 1981

1981 *Moschus fuscus* Li. Babo, Gongshan, 3500 m.

Blackish brown all over; much darker than any other species, with no neck or throat markings, but very occasionally traces of yellowish spotting on the body. Neck sometimes lighter than the body; throat often with two incomplete yellow "collars." Underside dark. Rump with ochery tones, but the buttocks black. Juvenile hairs agouti-banded; adult hairs solid-colored, except for very short white bases. Hairs long, 32–46 mm on the withers, 51–63 mm on the rump. Skull length 135–150 mm; nasals <46 mm long. Muzzle short, less than half the length of the skull; lacrimal length longer than its height. Limbs more elongated than in *M. berezovskii*, despite the smaller skull; metacarpal length 88–101 mm, metatarsal length 126–135 mm, with the hindlimbs even more elongated than in *M. berezovskii*. Hoof elongated, 27–29 mm in one specimen (maximum in other species, 24 mm).

NW Yunnan (Gongshan, 26° 30' N, 98° 50' E), SE Tibet (Zayu, 28° 25' N, 97° 06' E), and northernmost Burma (Dchpu L'kha; Adung-Seingku); 2800–4200 m.

There is a possibility that this species is the one that has been found at over 4000 m, in the Khumbu region of Mt. Everest and Tserping in Nepal, Sikkim, and Bhutan. Both the adults and the juveniles resemble *M. fuscus* from China and Burma, except that the Khumbu specimens have a slightly lighter underside and interramal region.

Moschus anhuiensis Wang et al., 1982

1982 *Moschus moschiferus anhuiensis* Wang et al.

A small species; skull length 141–149 mm (mean 143.8 mm). Gray-brown in color, with dense (but not well-marked) light spots on the body. Neck-stripe present. Lacrimal length shorter than its height. Ears darker than the body. Hairs long, 38 mm on the withers, 54 mm on the rump.

First proposed as a subspecies of *M. moschiferus*, this taxon was initially transferred to *M. berezovskii* (Groves & Feng, 1986), then given species rank. Using principal components analysis, Li M. et al. (1999) found that skulls of this species fall far from those of any other, and, using cytochrome *b*, they found consistent differences from other species.

SW Anhui; below 500 m.

Moschus berezovskii Flerov, 1929

Hair bases short, gray-white; throat with three wide longitudinal stripes, white to orange in color, the lateral pair running from the jaw angles to the brisket, the central one shorter, and all three sometimes broken; haunch yellower; rump nearly black; limbs dark down the front surface. Underside yellow. Ear orange at the base, black at the tip, white inside. Length of the hairs 73–87 mm on the withers, 45–68 mm on the rump (varying according to season). Limbs relatively short; metacarpal length 73–87 mm; metatarsal length 103–118 mm (hence the hindlimbs relatively longer than in most other taxa). Lacrimal length shorter than its height.

Moschus berezovskii berezovskii Flerov, 1929

1929 *Moschus berezovskii* Flerov.

Skull length 142–153 mm. Color very dark, olive-brown, with orange tones.

Sichuan, S parts of Shaanxi, Gansu, and Ningxia, SE Qinghai, and SE Tibet; between 500 and 2500 m.

Moschus berezovskii caobangis Dao, 1969

1969 *Moschus moschiferus caobangis* Dao. Caobang Province, Vietnam; S Yunnan (Mile and Langcang) and Guangxi; 50–400 m.

The smallest musk deer; skull length 120–135 mm. Color very light; limbs dark only from the knee and the hocks downward; buttocks and ear tips still black. Hair shorter, 21–30 mm on the withers, 38–47 mm on the rump.

We have not seen specimens of the following two taxa:

Moschus berezovskii bijiangensis Wang & Li, 1993. NW Yunnan to Zayu, SE Tibet. Described as being paler than nominotypical *M. b. berezovskii*, fulvus-brown, with a grayish white ventral surface, the toothrow averaging slightly shorter.

Moschus berezovskii yunguiensis Wang & Li, 1993. Yunnan-Guizhou plateau margins, E to Ichang. Described as being somewhat smaller than nominotypical *M. b. berezovskii*.

8

Antilocapridae

The Antilocapridae: last survivors of a once diverse family.

ANTILOCAPRIDAE OWEN 1841

The lacrimal duct opens into two orifices; the inferior one, as in the Cervidae, is situated on the orbital rim, but the superior one is inside the rim (Leinders & Heintz, 1980). As in the Cervidae, the anterior sagittal gully of the metatarsus is closed by a distal bridge. The hindfoot articular structure has converged on the bovid type, resulting in a marked "pogostick effect" (Leinders, 1979).

A gall bladder is present.

On pronghorns, and particularly on the nature of their appendages, see O'Gara (1990).

The position of the family Antilocapridae, whether sister to either the Cervidae or the Bovidae, or to a clade containing both, is still not quite clear. The combined mitochondrial and nuclear results of Hassanin & Douzery (2003) suggest that it branched off at about the same time as the Giraffidae, or slightly earlier; in other words, it could even be the earliest branch from the basic ruminant stem.

Antilocapra Ord, 1818

Antilocapra americana (Ord, 1815)

O'Gara (1990) has given a detailed account of this species, including whether it contains significant geographic variation (which he tended to doubt).

9

Giraffidae

Giraffidae. PG's use of giraffe cranial variation to illustrate progressive clines. Working out the complex geographic variation of modern giraffes: skin pattern, and its relation to African biogeography. The okapi: early taxonomic misunderstandings, although it is still enigmatic after all this time.

GIRAFFIDAE GRAY 1821

Usually there is one single, very small lacrimal orifice inside the rim of the orbit, although there may be no orifice at all (Leinders & Heintz, 1980).

Molecular data (see, for example, Hassanin & Douzery, 2003) tend to place the Giraffidae as sister to the Cervidae plus the Moschidae plus the Bovidae, the position of the Antilocapridae being less secure.

Giraffa Brisson, 1762

This was one of the generic names which A. Gentry (1994) requested to be conserved, even though other names from Brisson (1762) were to be rejected.

It was shown by Solounias (1999) that to say that the giraffe has seven cervical vertebrae is to oversimplify. Essentially, there is an extra cervical vertebra, but what was C8 became thoracicized (the morphology of the vertebra in question shows this). This has to do with the position of the neck on the thorax: it enters the thorax from a superior direction above the scapulae, instead of from an anterior direction in front of them, and the glenohumeral joint protrudes anterior to the thorax, making it possible to breathe while the neck is angled downward between the forelimbs for drinking.

A remarkable morphocline runs from the S species (such as *G. giraffa*), via *G. tippelskirchi*, to the N species (such as *G. camelopardalis*), relating to the development of the medial ossicone, which gets larger as one goes N, and is associated with the greater development of the hyperossification of the dorsal skull roof.

Our revisionary work is outlined below, followed by our taxonomic arrangement.

Skull Comparisons of Males and Females

We compared the three named taxa of N Kenya, Uganda, and NE DRC (*G. reticulata*, *G. rothschildi*, *G. congoensis*) by discriminant analysis, using just four variables. We found that there are overlaps, but, on the whole *G. reticulata* separates from the other two, which do not separate well from each other. The differences are based mainly on size, especially the orbital rim to frontal distance (representing the height of the frontal horn), followed by the greatest skull length, then (to a much lesser degree) the mastoid breadth and the condylobasal length.

The three C and W taxa—*G. congoensis*, *G. rothschildi*, and the C African sample ascribed to *G. peralta*—separate completely. Again, the main discriminator is the orbital rim to frontal distance.

We then ran a discriminant analysis on the S groups: *G. tippelskirchi*, *G. thornicrofti*, *G. giraffa*, and *G. angolensis*. Three specimens of *G. giraffa* were intermixed with the *G. angolensis* sample, and one was within the *G. tippelskirchi* dispersion. We found that the single available specimen of *G. thornicrofti* is outside the range of any of the other three. The difference in the first case relies on small contrasts between a large orbital rim to frontal distance, on the one hand, and, on the other, small preorbital length, palate length, and biorbital breadth. The specimen of *G. thornicrofti* has a low orbital rim to frontal distance, and a low biorbital breadth.

Univariate statistics for males and females are given in tables 11 and 12. In size, *G. reticulata* is noticeably smaller than the other species, while *G. camelopardalis* and *G. giraffa* are noticeably larger. The differences are not as substantial in the condylobasal length as they are for the greatest length, denoting a greater development of the occipital crest in larger species. *G. giraffa* has a noticeably wide palate; *G. reticulata* is very narrow across the zygomatic arches, and also across the orbits; *G. thornicrofti* also has a very small biorbital breadth (n.b., only one specimen). The N giraffes all have a larger orbital margin

Table 11 Measurements for giraffe species: males

	Gt l	Cb l	Pal l	Teeth	Pal br	Zyg br	Biorb	Orb rim fr
reticulata								
Mean	634.78	590.33	340.00	142.73	156.88	215.62	269.33	193.50
N	9	12	9	11	8	8	15	10
Std dev	20.687	20.366	15.819	3.379	12.800	3.662	15.089	18.858
Min	600	566	311	136	148	210	242	161
Max	662	640	357	147	188	222	298	216
camelopardalis								
Mean	703.29	642.57	380.14	151.21	165.00	256.10	313.00	263.29
N	7	7	7	7	7	5	6	7
Std dev	24.669	25.376	21.098	3.740	7.095	8.721	15.518	17.802
Min	678	605	350	146	153	248	294	250
Max	746	675	410	155	173	269	339	289
antiquorum								
Mean	676.73	634.78	365.00	148.33	152.50	235.92	289.00	228.36
N	11	9	10	12	2	12	13	11
Std dev	26.811	22.225	15.026	5.245	.707	13.153	22.884	20.101
Min	636	605	344	142	152	215	250	198
Max	727	673	391	158	153	256	322	255
peralta								
Mean	668.50	612.50	349.25	152.75	153.00	246.33	293.50	193.25
N	4	4	4	4	1	3	4	4
Std dev	15.927	10.878	15.435	5.679	—	35.726	13.077	30.544
Min	655	601	334	145	153	220	283	166
Max	690	627	364	157	153	287	311	224
tippelskirchi								
Mean	686.22	637.00	372.73	150.09	157.55	229.77	290.21	185.11
N	18	14	15	17	11	15	19	19
Std dev	38.374	23.085	20.600	6.949	9.606	10.479	21.168	25.486
Min	625	598	336	139	144	212	254	132
Max	752	683	407	162	172	241	331	237
thornicrofti								
Mean	667.00	634.00	380.00	155.00	150.00	—	275.00	150.00
N	1	1	1	1	1	—	1	1
giraffa								
Mean	698.00	642.75	384.00	152.80	169.00	234.50	293.80	171.33
N	4	4	3	5	3	2	5	3
Std dev	20.543	16.317	6.557	2.775	2.646	13.435	15.563	19.502
Min	670	620	378	150	167	225	276	149
Max	718	656	391	157	172	244	312	185
angolensis								
Mean	687.89	639.57	383.88	150.94	155.69	236.71	291.56	164.00
N	9	7	8	9	8	7	9	9
Std dev	27.479	26.676	14.116	6.635	12.056	7.653	17.249	14.950
Min	647	615	358	142	132	225	269	147
Max	737	690	401	162	170	246	320	194

to frontal distance, indicating the development of a large frontal horn; this is far larger in *G. camelopardalis* than in any of the other species, even other N species.

The small size and lack of sexual size difference in *G. reticulata* is noticeable; the other species lacking a sexual size difference is *G. peralta*, in which the females are much the largest (relative to their males) of any giraffe species.

The dramatically large size of the frontal horn (indicated by the orbital rim to frontal distance) of the *G. camelopardalis* males is followed successively by somewhat smaller sizes in *G. antiquorum*, *G. peralta* plus *G.*

Table 12 Measurements for giraffe species: females

	Gt l	Cb l	Pal l	Teeth	Pal br	Zyg br	Biorb	Orb rim fr
reticulata								
Mean	623.40	582.67	344.40	141.80	148.40	211.00	247.17	144.80
N	5	6	5	5	5	3	6	5
Std dev	26.773	26.319	11.283	9.338	3.847	6.928	18.137	25.791
Min	590	538	334	130	145	207	232	126
Max	659	605	361	152	155	219	283	190
camelopardalis								
Mean	616.67	589.50	354.00	142.67	159.00	200.00	263.00	142.00
N	3	2	3	3	2	1	3	3
Std dev	19.035	—	9.165	2.082	—	—	20.809	7.810
Min	597	584	344	141	155	—	239	133
Max	635	595	362	145	163	—	276	147
antiquorum								
Mean	619.50	576.00	340.00	148.00	148.00	229.00	260.50	142.50
N	2	2	2	2	1	1	2	2
Min	606	566	328	145	—	—	255	141
Max	633	586	352	151	—	—	266	144
peralta								
Mean	607.00	574.50	338.50	143.50	—	206.80	256.25	137.25
N	4	2	2	6	—	5	4	4
Std dev	12.910	6.364	7.778	6.535	—	14.584	25.303	10.782
Min	592	570	333	133	—	186	229	123
Max	622	579	344	153	—	226	290	147
tippelskirchi								
Mean	666.50	640.67	369.67	145.75	151.67	—	270.00	182.67
N	4	3	3	4	3	—	3	3
Std dev	53.681	38.031	21.385	5.058	8.083	—	13.892	11.590
Min	602	598	345	140	147	—	254	172
Max	716	671	383	152	161	—	279	195
thornicrofti								
Mean	631.00	588.20	356.60	139.17	155.88	219.10	259.33	134.00
N	6	5	5	6	4	5	6	6
Std dev	12.458	19.601	9.965	4.167	1.436	7.603	11.325	7.321
Min	615	565	342	135	154	212	239	121
Max	650	610	368	146	157	228	268	142
giraffa								
Mean	620.00	610.00	—	149.00	—	216.50	261.50	—
N	1	1	—	2	—	2	2	—
Min	—	—	—	148	—	216	255	—
Max	—	—	—	150	—	217	268	—
angolensis								
Mean	625.00	588.00	352.00	148.20	163.00	223.60	266.00	134.00
N	5	5	4	5	4	5	5	5
Std dev	8.515	14.612	12.193	4.324	3.367	5.550	4.899	3.391
Min	619	572	342	143	159	217	260	130
Max	640	609	368	155	167	231	272	138

reticulata, *G. tippelskirchi*, and, finally, the S species (n.b., very small female samples in these latter). The females of *G. peralta* have rather large frontal horns, like the males, mirroring the general lack of sexual dimorphism in this species.

External Characters

We obtained a large number of photos of wild giraffes, as well as some photos of captive giraffes of known origin: from the literature, from our own photographs, and from the massive collection of Mr. Velizar Simeonovski, for whose generosity we are extremely grateful. We compared giraffes from different areas for a number of features:

ground color
color of blotches on the neck and the body
extent of spotting on the legs
color of the lower limb segments
pattern above the eye, and along the dorsum nasi
pattern below the eye, and along the lateral margin of the lower jaw

In an attempt to formalize aspects of color-pattern variation, we took measurements from the photographs, as follows:

1. We drew a line between the elbow and the stifle, extending it forward to the anterior outline of the body and backward to the posterior outline of the rump, and counted the number of spots anterior to the elbow (that is, along the base of the foreleg), between the elbow and the stifle (i.e., along the side of the abdomen), and posterior to the stifle (along the base of the hindleg).
2. We drew a line across the base of the neck, starting at the bump corresponding to the tip of the neural spine of the second rib-bearing vertebra (usually identified as T2, but the "true" T1 according to Solounias, 1999) and extending forward and downward at the widest section. We counted the number of spots/blotches that were crossed by this line, and measured the line itself and the length of each blotch. Adding up the scores of the blotches, we expressed this as a percentage of the line measurement, thus calculating the proportion of the pelage in this region that was covered by the blotches, rather than the ground color.
3. We inspected the outline of each of the blotches that were cut by this line (see item #2 above), and counted the number of "bites" (emarginations) out of the edges of each, in an attempt to document how stellate each blotch was.

In the second and third cases, we chose the base of the neck as being the region where we could take these measurements with the greatest degree of objectivity, and we then inspected the pattern over the whole neck and body, in order to make sure that the neck-base pattern was fully representative of the pattern as a whole.

Finally, we measured the length of the head and its height, the latter measurement extending from the forehead (including the median horn, if present) to the jaw angle, thus expressing height as a percentage of length.

These measurements, of course, could be taken only on photographs depicting giraffes standing directly laterally to the camera. In addition, we were careful to measure each individual only once; if different photographs appeared to depict the same individual, all photos but one were excluded.

DNA Revolution

The study of mtDNA and microsatellites in giraffes by Brown et al. (2007) caused quite a stir in the field, because the authors proposed that there are a number of species of giraffes, not merely one, as had been assumed for so long. Six taxa were studied by them, with each represented by numerous samples obtained (noninvasively) from the wild: *G. reticulata*, *G. rothschildi*, *G. peralta*, *G. tippelskirchi*, *G. giraffa*, and *G. angolensis*. Each formed a monophyletic clade, characterized by numerous substitutions, and there were subclades within the N and S groupings that were also strongly distinct. Within *G. tippelskirchi*, there were, in fact, somewhat distinctive clades E and W of the Rift Valley.

Their phylogeny was as follows:

(*angolensis* ((*tippelskirchi*, *giraffa*) (*reticulata* (*peralta*, *rothschildi*)))),

although the position of *angolensis* as sister to all the other taxa may have been an artifact of the tree-building method; more likely, it was sister just to the other two S taxa.

The authors emphasized that there was no evidence that even neighboring populations of adjacent taxa were exchanging genes, and they concluded that the six taxa were behaving more like species than like subspecies. The entire study found only two *rothschildi*/*reticulata* hybrids and one *tippelskirchi*/*reticulata* hybrid (meaning pelage types with the "wrong" mtDNA).

Separation times, according to different models of the substitution rate, were around a quarter to nearly half a million years within the N and S groups, and half to over one and one-half million years between the two groups.

An almost simultaneous study by Hassanin et al. (2007)—this time including not only wild giraffes but also zoo specimens of known origin—added further populations, as well as a further taxon, *G. antiquorum*.

The cladogram again divided into N and S groups. The two S subclades were (1) *G. angolensis* from the Lisbon Zoo, and (2) *G. tippelskirchi* from the Basel Zoo plus *G. giraffa* from the Thoiry Zoo and from the wild. The two N clades were (1) reputed *G. peralta* from the Vincennes Zoo (the female founder was ultimately from Negry, N CAR) and from Waza and Boubandjida national parks, Cameroon, plus *G. antiquorum* from the Antwerp Zoo (ultimately from near Khartoum) and from Zakouma, Chad, and (2) *G. peralta* from Kouré, Niger, plus zoo specimens of *G. reticulata* and *G. rothschildi*.

Our morphological data strikingly corroborate the DNA data and enable us to produce new synonymies.

Northern Giraffes

The median horn in the adult male is well expressed, so that the head height is usually greater than 60% of the head length. The borders of the blotches are entire, or with (at most) 1–2 extremely small emarginations. There are always 3–4 blotches on the base of the neck. The limbs below the knees and the hocks are white to ochery fawn, with very little spotting.

Giraffa camelopardalis (Linnaeus, 1758)

Nubian giraffe, Rothschild's giraffe

1758 *Cervus camelopardalis* Linnaeus. Sennar; restricted by Harper (1940).

1903 *Giraffa camelopardalis rothschildi* Lydekker. Guasin-Gishu Plateau.

1904 *Giraffa camelopardalis cottoni* Lydekker. Koten Plain, S Doborsa country, NE Uganda.

Spots above the eye and along the dorsum nasi tending to be fused; spots below the eye fairly large. Spots 3 on the lower flanks (between the elbow and the stifle), 3–4 on the base of the foreleg anterior to the elbow, and 4–5 on the base of the hindleg behind the stifle. Blotches covering 66%–77% of the base of neck, the rest being ground color. Ground color dirty creamy; blotches liver-brown, with darkened centers in the adult males, the centers tending to become blackish with age. Spots on the legs barely reaching the knees and the hocks, or not at all.

Formerly from the latitude of Khartoum S through E Sudan and W Ethiopia (W of the highlands); persisting in NW Kenya and E Uganda.

As far as the evidence goes, there seems to be no essential difference between the (presumably extinct?) N populations and those that survive today in Kenya and Uganda under the name *G. rothschildi*; a possible difference is the extreme darkening with age in the extant E African population, although we know little about the Nubian aging process.

Giraffa reticulata de Winton, 1899

reticulated giraffe

Very dark above the eye and along the dorsum nasi, with individual spots hardly (or not at all) visible; very large blotches below the eye. Spots 3 on the lower flanks, 2–3 on the base of the foreleg, 4 on the base of the hindleg. Pattern essentially dominated by rectangular blotches, with the ground color reduced to a network of fine lines; 86%–90% of the base of the neck covered by blotches, which never have emarginations; neck blotches tending to be in longitudinal rows. Ground color white, blotches liver-red, darkening to nearly black in old males. Spots on the legs usually reaching only to the knees, occasionally halfway down the shanks; below the spotted area, the legs white to ochery fawn.

Intermediacy between *G. reticulata* and neighboring species was recorded by Stott (1959) and Stott & Selsor (1981): on the S bank of the Tana River, 9 mi. SW of Garissa, out of a group of four giraffes, one resembled *G. reticulata*, and one *G. tippelskirchi*, while the other two appeared intermediate; on the other hand, to the N of the river, all giraffes were *G. reticulata*. Intermediates between *G. reticulata* and what the authors referred to as *G. rothschildi* were seen on the Laikipia Plateau, "along the Maralal–Rumuruti Road," and between Rumuruti to the S and the Loroghi Plateau to the N, as far as the desert 24 km SE of the S tip of Lake Turkana, and "east to the dry bed of the Barsaloi River."

Giraffa antiquorum (Swainson, 1835)

Kordofan giraffe

1835 *Camelopardalis antiquorum* Swainson. Bagger el Homer, Kordofan, about 10° N, 28° E (as fixed by Harper, 1940).

1903 *Giraffa camelopardalis congoensis* Lydekker. Dungu, Uele Valley, NE DRC.

Smudgily spotted, not very dark, above the eye; spotting below the eye, reaching to the mouth corner. Blotches 3–6 on the lower flanks, 3–4 on the base of the foreleg, 6–7 on the base of the hindleg (corroborating the description "broken up into a number of very small and irregular" spots [Lydekker & Blaine, 1914b]). Blotches covering 56%–72% of the base of the neck; blotches entire, or with (at most) one very small emargination. Ground color off-white to pale yellowish; spots medium to dark brown, paler on the body than on the neck. Spots on the legs reaching down to the knee and the hock, sometimes slightly below; legs white below the spotted area.

The Paris Zoo breeding group belongs to this species; they were identified by Hassanin et al. (2007) as *G. peralta*, but they correspond in appearance to *G. antiquorum*. Thus the distribution extends from Kordofan S to the N CAR and to Zakouma in Chad. Giraffes in the Garamba National Park, NE DRC, also belong to this species.

Giraffa peralta Thomas, 1908

West African giraffe

1908 *Giraffa camelopardalis peralta* Thomas. Lokojya, Niger–Benue junction, Nigeria.

1971 *Giraffa camelopardalis renatae* Krumbiegel. Hinterland of Lagos, and between Parakau and Moschi (°9° N, 3° E).

Sparsely spotted or smudgy gray, not very dark, above the eye; spotting below the eye not reaching forward to the mouth region. Blotches 4–5 on the lower flank, 3–5 on the base of the foreleg, 5–9 on the base of the hindleg. Ground color predominant in the pattern; blotches covering 50%–71% of the base of the neck; blotches entire (rounded, rather than blocklike), or with very small emarginations. Ground color white, gray-white, or pale yellow; spots gray to dark liver-brown, becoming black-brown with age. Spots on the legs reaching (at most) down to the knee and the hock, the remainder of the leg being white.

Formerly from West Africa, as far E as N Cameroon; giraffes in Waza National Park belong to this species. Apart from the Cameroon populations, the only surviving population of *G. peralta* appears to be in Niger.

Southern Giraffes

The median horn in the adult male is poorly expressed or absent, so that the head height is usually less than 60% of the head length. The blotches are stellate, with up to 8 emarginations, which often reach deeply into the blotch. The number of blotches on base of the neck varies. The limbs below the knees and the hocks tend to be darker, often spotted.

Giraffa tippelskirchi Matschie 1898

Maasai giraffe, vine-leaf giraffe

1898 *Giraffa schillingsi* Matschie. Taveta, SE Kenya.

1898 *Giraffa tippelskirchi* Matschie. Lake Eyasi, NW Tanzania.

Entire upper face dark; spotting below the eye reaching as far forward as the mouth corner or beyond. Blotches 3–5 on the lower flanks, 3–5 on the base of the foreleg, 5–9 on the base of the hindleg (very occasionally more than this number). Blotches 3–4, occasionally as many as 6, on the base of the neck; blotches covering 60%–80% of the base of the neck; borders of the blotches deeply emarginate, with up to 8 "bites" out of the larger blotches. Ground color creamy to fawn; spots medium liver-brown, darkening with age. Spots on the legs indistinct, reaching halfway down the shanks or to the fetlocks; lower leg light fawn; fetlocks contrastingly white.

This is the only S giraffe with noticeable sexual dimorphism in the development of the median horn: head height is 52%–58% of its length in the males, but only 40%–55% in the females (the female figure is similar to that of the N giraffes).

Giraffes in Tsavo East National Park appeared to be much more variable than those in Tsavo West National Park, Mikumi, Tarangire, or Ruaha. Those in Serengeti are also variable; at these two ends of the range they differ on average, especially in the number of deep "bites" in the neck spots—up to eight in Tsavo East, but often only one, and that not very deep, in Serengeti. In other words, the Tsavo East giraffes are often the most "extreme" in the species. Recall that Brown et al. (2007) found a deep division between haplotypes of *G. tippelskirchi* from E and W of the Rift Valley, so it may be that further study of the external characters is required to test this.

Giraffa thornicrofti Lydekker, 1911

Luangwa giraffe

Upper half of the face tending not to be very dark, the spots often only partly fused; spotting below the eye nearly (or quite) as far forward as the mouth angle. Blotches 5–7 on the lower flanks, 4–5 on the base of the foreleg, 7–9 on the base of the hindleg. Blotches 5–8 on the base of the neck, covering 53%–77% of the neck; borders of the blotches often deeply emarginate, with 1–5 deep "bites." Ground color creamy gray; spots medium to light brown on the neck, darker on the body. Legs spotted only to the hock and just above the knee; lower legs pale fawn or ochery; fetlocks not contrasting.

Confined to the Luangwa Valley, E Zambia.

This species is the only one whose mtDNA has thus far not been studied.

Giraffa giraffa (Boddaert, 1785)

Cape giraffe

1785 *Camelopardalis giraffa* Boddaert. S of Warmbad, Orange River (Rookmaaker, 1989).

1842 *Camelopardalis capensis* Lesson. Gamma River, Great Namaqualand, about 27° S, 18° E.

1898 *Giraffa infumata* Noack. Barotseland, Zambia.

1904 *Giraffa camelopardalis wardi* Lydekker. N Transvaal.

Upper face and dorsum nasi dark to the end of the snout; spotting below the eye reaching as far forward as the mouth. Blotches 4–5 on the lower flanks, 3–6 on the base of the forelegs, 6–8 on the base of the hindlegs. Blotches 3–5 on the base of the neck, covering 59%–83% of the neck; borders of the blotches with 1–13 emarginations, one or two of them being very deep. Ground color off-white or fawny white; spots medium to dark brown. Legs spotted, reaching all the way down, but fading toward the fetlocks, on a pale fawn ground.

N South Africa, Zimbabwe, Botswana.

Giraffes in Kruger National Park are, on average, more closely spotted than those in the Okavango, Botswana.

Giraffa angolensis Lydekker, 1903

Angolan giraffe

Upper face dark to the end of the snout; partly fused spots below the eye extending to the mouth. Blotches 3–4 on the lower flanks, 3–4 on the base of the foreleg, 4–6 on the base of the hindleg. Blotches 3–4 on the base of the neck, covering 61%–80% of the neck; borders of the blotches with 1–6 emarginations, one or two of which are very deep. Pattern with a "washed-out" appearance; ground color gray-white; spots gray-brown on the neck, darker on the body. Legs spotted, reaching to about halfway between the knee and the hock; lower shanks pale to very pale fawn.

Okapia Lankester, 1901

okapi

The okapi does not share in the bizarre reshuffling of the cervical/thoracic vertebral junction seen in the giraffe (Solounias, 1999).

Okapia johnstoni (Sclater, 1901)

On the general biology of this species, see Gijzen (1959) and Lindsey et al. (1999).

There have been a few attempts to split up this species, all of them based on early, inadequate information. Nonetheless, Gijzen (1959) did draw attention to locality records S of the Grand Cuvette (the great bend of the Congo River), and specimens from these localities clearly need to be examined and compared with those from the region of the Ituri Forest and westward, from which the species is well known.

10

Cervidae

CERVIDAE GOLDFÜSS, 1820

The lacrimal duct almost invariably has two orifices on the rim of the orbit, the inferior one being slightly anterior to the superior one (Leinders & Heintz, 1980).

The distal end of the anterior sagittal gully of the metatarsus is closed by a bridge.

Leinders (1979) drew attention to differences in the hindfeet between this family and the Bovidae: in the Cervidae, the sagittal crests on the distal articular surfaces of the metatarsus are less prominent and less proximally extended; there is a nonarticulatory surface at the posterior end of the proximal articular facet of the medial phalanx ("plateau postarticulaire"); and the proximal articular facet of the distal phalanx is much less posteriorly extended and lacks the anterior extensor process ("éminence pyramidale"). This, according to Leinders (1979), results in less flexibility of these joints; that is, there is less of a "pogostick effect" than in the Bovidae.

The gall bladder is absent.

On antlers, their nature, and how they differ from the horns of the Bovidae and the pronghorns of the Antilocapridae, see Bubenik (1990). The Cervidae, unlike the Bovidae, possess a number of clearly derived character states, quite apart from the kind of cranial appendages (Janis & Scott, 1987).

The major outlines of the taxonomy of this family have become much clearer lately because of molecular phylogenetic studies, but, at the same time, these have muddied previous assumptions about the interrelationships of some groups, particularly the genus *Cervus* and its close relatives (the Cervini). The controversies over how many genera exist in the Cervini, and the plethora of genera presently recognized in South America, throw into stark relief the need for objective criteria for the recognition of higher categories in taxonomy. This fairly long chapter goes through the two subfamilies—each with two or more tribes, and each of them with two or more genera—and considers the number of species (and, where necessary, subspecies) in each.

Groves & Grubb's (1987) division of the family into three subfamilies—the Hydropotinae, the Odocoileinae and the Cervinae—was based on the lack of antlers in the first, and the way the lateral metacarpals have been retained in the other two: the Odocoileinae have kept the distal ends (telemetacarpal); and the Cervinae, the proximal (plesiometacarpal). In plesiometacarpal deer, the postglenoid foramen is entirely within the squamosal; in telemetacarpal deer, the petrous forms its medial border (Bouvrain et al., 1989); behavioral features separate the two, as well (Cap et al., 2002). Both of these latter studies also showed that *Hydropotes*, antlerless though it is, is part of the odocoileine clade.

Miyamoto et al. (1990), Li & Sheng (1998), Randi, Mucci, et al. (1998), Hassanin & Douzery (2003), Pitra et al. (2004), and Gilbert et al. (2006), using different DNA sequences, confirmed the basic division and showed that *Hydropotes* is part of the telemetacarpal clade.

A more recent study of male vocal behavior (Cap et al., 2008) resulted in a cladogram almost exactly duplicating the molecular phylogeny: muntjac are associated with Old World deer; *Capreolus* and *Hydropotes* are in a special clade (*contra* Cap et al.'s (2002) phylogeny, in which *Hydropotes* was the sister to other members of the Capreolinae); and so on.

We use the amended nomenclature of Grubb (2000b).

Subfamily Capreolinae Brookes, 1828

The name Odocoileinae, used by Groves & Grubb (1987), is antedated by Capreolinae, and the name Neocervinae Carette, 1922, still frequently used for this group, is not available, because it is not based on a genus-group name (Grubb, 2000b).

Karyotype evolution (Abril, Sarria-Perea, et al., 2010) seems to make the Rangiferini paraphyletic, in that an initial split separates the *Mazama americana* group from the rest of the subfamily. Evidently there has been some homoplasy; inspection of these authors' figure 2 shows that the source of the homoplasy lies in a sequence of two pericentric inversions that are known to have considerable potential for parallel occurrence.

Tribe Rangiferini Brookes, 1828

This name, not Odocoileinae, is the prior available name for the New World deer (Grubb, 2000b). *Rangifer* is likely to be the sister taxon to the other genera (Pitra et al., 2004). The tribe is distinguished from the Capreolini and the Alceini by the derived condition of the vomer (completely dividing the choanae), by contact between the stylohyoid and the paroccipital, and by the widely separated antler pedicels (Webb, 2000).

Detailed information on the neotropical representatives of this tribe will be found in Duarte & González (2010); their evolution, both in the fossil record and in their adaptive biology, is treated by Merino & Rossi (2010).

Webb (2000) proposed the following morphological cladogram:

((*Pudu* (*Rangifer* (*Hippocamelus*))) (*Mazama* (*Ozotoceros* (*Blastocerus*, *Odocoileus*)))).

Pitra et al. (2004) proposed the following molecular phylogeny:

(*Rangifer* (*Blastocerus*, *Pudu*) (*Odocoileus*, *Mazama*)).

(The remaining two genera were not available.) Nonetheless, it now appears that the genus *Mazama* is polyphyletic. Duarte et al. (2008) found that mtDNA splits *Mazama* into two clear groups that assort not with each other but, respectively, with *Odocoileus* (the *M. americana* group) and with the other South American deer (the *M. gouazoubira* group). The same conclusions, on larger samples, have been reached by S. González, Duarte, et al. (2010). In fact, the *Mazama* sample used by Pitra et al. (2004) was a red brocket (*M. americana* group), so, to that extent, their conclusions are entirely consistent with those of other authors.

We have found that cranial characters strongly support this division.

We retain the old divisions for the moment, but, ultimately, it will almost certainly be necessary to arrange the neotropical deer quite differently. For the red brockets, the name *Mazama* Rafinesque, 1817, is available, with the type species *Mazama pita*, which, although it cannot be definitively allocated (see below), certainly refers to a member of the *M. americana* group. The generic name *Subulo* Hamilton Smith, 1827, has no type species, but the first of the three species standing in that genus was *Cervus rufus*, likewise a member of the *M. americana* group and probably referring to the same animal as *M. pita*. The name *Odocoileus* Rafinesque, 1832, postdates *Mazama*, and, given the relatively short time span inferred for the group as a whole, this name will probably have to be regarded as a junior synonym of *Mazama* (to the dismay of North American mammalogists).

For the brown brockets, the earliest available names are *Passalites* Gloger, 1841, and *Nanelaphus* Fitzinger, 1873. Both are based on *Cervus nemorivagus* F. Cuvier, 1817, and both have been unused for well over a century, certainly rating as *nomina oblita*. As, however, the brown brockets did not split very long ago from the other (more purely) neotropical deer, the earliest name for the entire group should probably be *Hippocamelus* Leuckart, 1816. Lydekker found this name "highly objectionable" (either because of its mixed derivation, the first half Greek and the second half Latin, or else because of its general meaninglessness,

Table 13 Skull characters for the two groups of South American deer

	M. americana-Odocoileus group	*M. gouazoubira-Hippocamelus-Pudu-Blastoceros-Ozotoceros* group
Ethmoid vacuity	Very large	Restricted
Frontolacrimal suture length	Much shorter than the frontal margin of the ethmoid vacuity	About equal to the length of the frontal margin of the ethmoid vacuity
Maxillolacrimal suture length	Equal to or less than the maxillary margin of the ethmoid vacuity	Much longer than the maxillary margin of the ethmoid vacuity
Median tips of the nasals	Much shorter than the lateral prongs	Reach anterior to the lateral prongs
Lateral ridges of the diastema	Turn in anterior to P^2, then curve out very slowly and slightly laterally to the incisive foramina	Deeply bowed inward at P^2, then bowed outward, again laterally to the incisive foramina
I_1	More broadened at the tip; I_2 is also broadened, so it partly fills in the gap	Only slightly broadened at the tip, separated by a V-shaped gap from the others
Antler texture	Roughened	Ridged or fluted (exception—*Ozotocerus*)

i.e., "horse-camel"); so that he could use it for as few species as possible, he broke up what would otherwise have been a single genus into four!

Odocoileus Rafinesque, 1832

white-tailed deer, black-tailed deer, and mule deer

Grubb (2000b) reviewed the use of the name *americanus* Erxleben, 1777, in combination with generic names *Moschus* (now = *Mazama americana*) and *Cervus* (now regarded as an unavailable senior synonym of *Odocoileus virginianus*).

Odocoileus is a very difficult genus taxonomically. In North America there are two species-groups, each currently being classed as a single species, but this is almost certainly overlumped. One problem is that white-tailed deer (*O. virginianus*—if this really is a single species, rather than a species complex) and mule deer (*O.* cf. *hemionus*) are known to interbreed in West Texas. It was discovered 20 years ago that in this region, the two share a mtDNA restriction type characteristic of white-tailed deer, suggesting that interbreeding had occurred between male mule deer and female white-tailed deer, and that in the hybrid herds, the white-tailed phenotypic characteristics have been lost by generations of backcrossing with mule deer (Carr et al., 1986). A subsequent study (Cathey et al. 1998) confirmed that F_1 hybrids are rare and found that Y-chromosome DNA, in contrast with mtDNA, assorted more clearly along species lines. Some implications of this hybridization, which actually occurs in both directions in different populations, have been discussed by Bradley et al. (2003).

Within what is commonly classified as *Odocoileus hemionus*, a recent study (Latch et al., 2009) found that there are two distinct clades, corresponding to what are known as mule deer and black-tailed deer. These had originally been described as separate species, *O. hemionus* and *O. columbianus*, respectively; they are indeed quite distinct and easily recognizable phenotypically, but they had been united into one species early in the 20th century on the grounds that they have been known to interbreed. The DNA finding seems to be good evidence that the two should be re-split; there were, of course, some instances of the "wrong" DNA clade in a given population on the borders of the ranges of the two, but this should not be taken as evidence that they are one and the same. Within *O. hemionus* proper, there were seven distinct mtDNA haplotypes, but these cross-cut described subspecies (whose phenotypic distinctiveness needs to be retested anyway).

There is a decline in size of the deer identified as *O. virginianus* down the E coast of North America, culminating in the diminutive deer of the Florida Keys (currently known as *Odocoileus virginianus clavium*). A study of mtDNA of deer in Florida, including the Florida Keys deer, found three different haplotypes, but these corresponded only poorly with the described subspecies, including *O. v. clavium* (Ellsworth et al., 1994). This brought up the question, frequently posed in recent years, of whether putative subspecies would be better defined as possessors of unique mitochondrial haplotypes rather than, as traditionally defined, being based on morphological characters. It has been argued that the function of the subspecies category is to delimit geographically restricted lineages. Against this, we would maintain, first, that of course mtDNA is inherited only matrilineally, and that the depicted lineages might be entirely different if Y-chromosome DNA would be studied (as in the West Texas hybridization study cited above); and second, that, in any case, morphological characters are themselves (broadly speaking) heritable, and that there is value in continuing to recognize gene pools that are strongly divergent as a whole.

An analogous situation recurs in Venezuela, at the S end of the distribution of white-tailed deer. Molina & Molinari (1999) recognized three distinct species in Venezuela (and suggested that there was evidence for a fourth). Their three species are *O. margaritae* (Margarita Island), *O. lasiotis* (Mérida, Andean highlands), and *O. cariacou* (from the rest of the range). In contrast, a study of mtDNA by Moscarella et al. (2003) identified four clades, but these corresponded rather poorly with the proposed species. It is clearly time for a new look, utilizing both methods and using the same samples for both. For the moment, we leave the taxonomy of *Odocoileus* strictly alone, recognizing that we have neither the data nor the experience to deal with it; and we strongly urge North American taxonomists—as well as South American ones—to look closely at the genus. The neotropical forms are described by Gallina et al. (2010).

Blastocerus Wagner, 1844

Grubb (2000b) argued that this name is available (antedating the same name by Gray, 1850), and he designated *Cervus paludosus* Desmarest, 1822 (= *Cervus dichotomus* Illiger, 1811), as the type.

Skull characters are typical for those of the other neotropical endemics and the *Mazama gouazoubira* group (see below), with a few modifications. The supraorbital sulcus is extremely long and broad, and it contains four fairly large foramina. The unworn antlers are heavily rugose.

***Blastocerus dichotomus* (Illiger, 1815)**

marsh deer

For a full account of this species, see Piovezan et al. (2010).

A study using mtDNA of four populations in the La Plata Basin (the La Plata Delta, Iberá, S Bolivia, and the Pantanal of SW Brazil) found considerable differences among them (Márquez et al., 2006). Control region haplotypes were unique to each region: two cytochrome *b* haplotypes were shared between the Pantanal and Iberá samples, but S Bolivia and the La Plata Delta each had a unique haplotype, that of the Delta being the most divergent. The authors pointed to the somewhat divergent biogeographic nature of the Paraná Delta, and estimated the time of separation of this population from the others at about 200,000 years ago, which suggests the need for a taxonomic reappraisal.

Ozotoceros Ameghino, 1891

> The prior available name for this genus has frequently been held to be *Blastoceros* Fitzinger, 1860, a virtual homonym of *Blastocerus* Gray, 1850 (*recte* Wagner, 1844; see above). Fitzinger was aware of this name, and his version of it is, therefore, to be regarded as an unjustified emendation; hence, it is unavailable as a name for the present genus (Grubb, 2000b).

$2n = 68$, based on specimens from Uruguay and Paraguay. The X chromosome is metacentric.

The skull characters are typically those of the neotropical endemics (see above). The antlers are curious, with little excrescences all over.

***Ozotocerus bezoarticus* (Linnaeus, 1758)**

pampas deer

For a full account of this species, see S. González, Cosse, et al. (2010).

The question of possible taxonomic variation within this species, or even between related species, has been much discussed. Using craniometrics, S. González et al. (1991) found that there are differences, at least on average, between samples ascribed to *O. b. celer* (from the Pampas), a sample from Uruguay, and samples ascribed to *O. b. leucogaster* and *O. b. bezoarticus*. The authors ended on a cautionary note: "A wide sample of specimens from the entire distribution area would reveal the true subspecifical systematics." Later, some of the same authors (S. González et al., 2002) more definitively divided the species into five subspecies, discussed below.

Ozotocerus bezoarticus bezoarticus (Linnaeus, 1758)

S. González et al. (2002) had no comparative material for this subspecies, which appears to be known from a single skull.

Ozotocerus bezoarticus celer Cabrera, 1943

Mean condylobasal length 214.8 mm (males), 212.4 mm (females).

According to S. González, Cosse, et al. (2010), this is the subspecies found in Buenos Aires Province.

Ozotocerus bezoarticus leucogaster (Goldfüss, 1817)

Mean condylobasal length 215.0 mm (males), 208.1 mm (females); this is based on 9 males and 20 females, so the apparent size difference between the sexes may be real. In the discriminant analysis of S. González et al. (2002), *O. b. leucogaster* overlaps greatly with *O. b. celer*, even in the analysis of the female skulls, which are much better differentiated than those of the males.

According to S. González et al. (2002), this is the subspecies found in Mato Grosso and the Argentinian Chaco.

Ozotocerus bezoarticus uruguayensis González et al., 2002

> Type locality: Uruguay, Rocha Department, Sierra de Los Ajos, 33° 50' S, 54° 01' W.

Color light fawn-brown to tawny, bay, and dark cinnamon. Antorbital pit unusually long and deep. Breadth of the frontal 73 mm. Mean condylobasal length 213.8 mm (males), 218.6 (females), but the sample sizes are small, so the apparent larger size of the females may, in this case, be illusory.

E grasslands of Rocha Department.

In S. González et al.'s (2002) discriminant analysis of the female skulls, the range of variation of *O. b. uruguayensis* falls beyond those of others, but the admittedly small sample size dictates caution.

Ozotocerus bezoarticus arerunguaensis González et al., 2002.

> Type locality: Uruguay, Salto Department, Arerunguá, El Tapado, 31° 41' S, 56° 43' W.

Color from fawn-brown to tawny and cinnamon. Antorbital pit small. Breadth of the frontal 58.5 mm. This seems to be the smallest subspecies; mean condylobasal length 208.2 mm (males), 205.3 (females), although the ranges of variation are very wide. In the discriminant analysis of the female skulls (S. González et al., 2002), the range of variation of *O. b. arerunguaensis* has a small overlap with that of *O. b. celer*, but it does not overlap with any other; interestingly, *O. b. arerunguaensis* seems furthest from the other Uruguayan subspecies, *O. b. uruguayensis*.

NW grassland, near the type locality.

Hippocamelus Leuckart, 1816

The two species presently allocated to this genus are extremely different.

Hippocamelus antisensis (d'Orbigny, 1834)

taruka

For a full account of this species, see Barrio (2010).

Second lower incisor hardly widened at all. Sharp supraorbital sulci, containing three large foramina. Lacrimal fossa uniquely wide. Posterior ends of the nasals (somewhat widened in all genera) remarkably expanded here, and then rather suddenly constricted anteriorly. Unworn antlers deeply fluted, with a long anterior branch, somewhat over half the length of the posterior branch. Pelage coarse, grayish, and white underneath, from the chin down the front to the chest; rest of the underside dark brown, but the entire inner aspects of the legs, and the outer sides of the legs below the carpus and tarsus, strikingly white; rump extensively white; tail white below.

Along the Andes Mountains, extending from N Peru through W Bolivia into N Argentina and northernmost Chile, between 2000 and 3000 m in the S and as much as 3800–5000 m in the N part of the range (Barrio, 2010).

Hippocamelus bisulcus (Molina, 1782)

huemul

For a full account of this species, see Vila et al. (2010).

$2n = 70$.

Second lower incisor somewhat wider. Supraorbital sulci shallow or flat, each containing just one large foramen. Lacrimal fossa fairly deep, but narrow. Posterior ends of the nasals not strongly expanded. Unworn antlers fluted, but not to the extent seen in *H. antisensis*; antlers shorter than in *H. antisensis*, especially the posterior branch. Pelage coarse, dark brown, lighter and more grayish yellow in winter than in summer. Only the underside of the tail, the groin, and spots above the eyes and beside the mouth white.

Andes of S Chile and Argentina, extending to the S limit of the mainland but not to Tierra del Fuego; ranging from sea level to 2000 m (Vila et al., 2010).

Mazama Rafinesque, 1817

brockets

Mazama is an even more taxonomically complex genus than *Odocoileus*, with far more species than the four or so that had been previously assigned to it (see, for example, Medellín et al., 1998, who identified a further distinct species for Yucatán).

For the moment, we will retain both red and brown brockets in the genus *Mazama*, while anticipating that a totally different classification will have to be adopted in the future. Meanwhile, we would like to pay tribute to the first reviser, J. A. Allen, whose 1915 monograph was a model of thoroughness and has a remarkably modern aspect.

Group Forming a Clade with *Odocoileus* (Red Brockets)

According to Duarte et al., 2008, this group divides into four subclades: (1) ten individuals of *M. americana*, (2) *Odocoileus*, (3) one individual called "*Mazama* sp.", and (4) nine individuals of *M. americana*, one of *M.* sp., *M. bororo*, and *M. nana*.

Eisenberg (1989), Redford & Eisenberg (1992), Eisenberg & Redford (1999), and Varela et al. (2010) gave the following measurements for members of this group, which they mainly treated under the species name *Mazama americana*: weight 34.9 kg, head plus body 1154 mm, ear 103 mm, metatarsus 233 mm, tail 136 mm, height 642 mm (Brazil); total length 1200 mm, tail 126 mm, hindfoot 674 mm, ear 99 mm, weight 29 kg (presumably from the N distribution?); total length 1209 mm, head plus body 1082 mm, tail 127 mm, hindfoot 674 mm, ear 99 mm (Argentina); total length 1200 mm, tail 150 mm, weight 28.9 kg ($N = 11$, Paraguay). These measurements therefore vary geographically, but all red brockets can generally be distinguished from others by their large size and, of course, their bright red-brown color, with a white or creamy underside. The hindlegs are often very dark, almost black, below the heel to the hoof.

Only 14% (in a Suriname sample) have canines.

For a full account of this species-group, see Varela et al. (2010).

Duarte (1996, and elsewhere) has recognized three species in the *americana* group: *Mazama americana*, *M. nana*, and *M. bororo*. Differentiating these three phenotypically is problematic, so they will simply be treated according to locality in our analysis. It is clear that, even shorn of these more obviously extraneous components, the species *Mazama americana* is taxonomically heterogeneous. Karyotypes vary enormously; for example, Abril, Vogliotti et al. (2010) have illustrated two living females that appear almost identical externally yet differ strongly in karyotype: the one from Paraná has $2n=52$, the one from Rondônia has $2n=42$, and there is yet more chromosome polymorphism within Brazil (Varela et al., 2010). On the other hand, there is very little protein polymorphism (Garcia & de Oliveira, 1910).

We lack a sufficient number of specimens to make large sample sizes for multivariate analysis, except in the case of the three montane samples, which we find are 100% differentiated. Skull lengths (greatest and condylobasal) are high in the Peruvian samples, low in Colombia (with nasal breadth); nasal breadth and biorbital breadth are high, and toothrow length low, in the Peruvian Andes group; these characters are the opposite in the Ecuador–Peru border mountains, with Colombia in between.

For other geographic regions, although sample sizes are often small, it is nonetheless worth giving a table of univariate measurements (table 14), because there are striking differences among the samples; available names for particular regional samples, where they exist, are indicated. Notable variations are as follows: The Darien specimen and the *zetta* sample are the smallest in size, though the Darien skull is broad; *M. americana* and the La Macarena skull are unusually large. The Roca Nova skull (*M. jucunda*) is below the size of any other. The *M. whiteleyi* mean skull length is average for the group, but the toothrow is short. The Meta sample has long nasals for its skull size.

The antlers are long in *M. whiteleyi*, *M. trinitatis* and La Macarena, medium-sized in *M. fuscata*, *M. zetta* and the São Paulo skull (which may represent *M. nana*); and very small in *M. zamora*, *M. americana*, and Meta.

We offer the following provisional arrangement, using relevant species names simply for guidance, and listing what, from our data seem, to be their outstanding characteristics.

Mazama bororo Duarte, 1996

1919 *Mazama bororo* Miranda-Ribeiro; conditional name, made available by its definitive use by Duarte (1996).

For a full account of this species, see Vogliotti & Duarte (2010), whose description and comparisons are followed below.

$2n=32–34$.

Medium-sized. Resembles *M. americana* in the reddish color, position of the white body areas, and the body mass. Red body color, however, more homogeneous; no dark area on the hindlimbs, except on a line from the posterior region of the ankle to the final third of the metatarsal. White half-moon mark at the base of the ear more marked than in *M. americana*.

In many respects, *M. bororo* resembles captive-bred hybrids between *M. americana* and *M. nana*.

Atlantic forest of Brazil, the Serra do Mar coastal forests ecoregion (Vogliotti & Duarte, 2010).

Weight 25.0 kg, head plus body length 1061 mm, ear 94 mm, metatarsus 220 mm, (tail, no data), height 575 mm. This compares with *M. nana* × *M. americana* hybrids as follows: weight 22.0 kg, head plus body length 1054 mm, ear 90 mm, metatarsus 205 mm, tail 112 mm, height 562 mm.

Mazama nana (Lesson, 1842)

$2n=36–39$.

For a full account of this species, see Abril, Vogliotti, et al. (2010).

Medium-sized. Weight 15.2 kg, head plus body 900 mm, ear 91.2 mm, metatarsus 178 mm, tail 88 mm, height 482 mm (Duarte et al., 2008); but Eisenberg & Redford (1996) give total length 853 mm, head plus body 776 mm, tail 78 mm (all specimens, $N=10$), ear 83 mm ($N=9$), weight 8.2 kg ($N=1$).

Striking chestnut-brown; venter snow-white. Males with spike antlers. Relatively short-legged. Large preorbital glands.

SE Brazil, the far NE of Argentina, and adjacent easternmost Paraguay; in moist forest, often in hilly terrain.

Mazama whitelyi (Gray, 1873)

1898 *Mazama tschudii* Lydekker, in part; not Wagner, 1855.

Medium-sized (in our admittedly small sample), with a very short toothrow. Antlers long.

S Peru.

Mazama americana (Erxleben, 1777)

We propose that the following names are synonyms of *M. americana*:

1817 *Cervus rufus* F. Cuvier, in part; not Illiger, 1815.

? 1850 *Coassus auritus* Gray.

1872 *Homelaphus inornatus* Gray.

1915 *Mazama americana juruana* J. A. Allen.

1915 *Mazama americana tumatumari* J. A. Allen, in part; skin only.

Table 14 Univariate statistics for samples of the *Mazama americana* group

	Gt l	Cb l	Teeth	Biorb	Nas l	Nas post br	Antler l
Darien *temama*	198.00	188.00	61.00	94.00	62.00	26.00	—
N	1	1	1	1	1	1	
Manabi *fuscata* = *gualea*	212.00	199.00	65.00	94.00	—	28.00	69
N	1	1	1	1	—	1	1
Mts of Colombia *zetta*							
Mean	202.13	190.50	59.78	86.44	61.00	25.67	67.33
N	8	8	9	9	9	9	3
Std dev	8.659	9.150	1.986	3.812	7.762	3.873	22.546
Min	193	177	57	81	50	19	44
Max	215	203	63	90	72	29	89
Zamora *zamora*							
Mean	210.00	199.60	65.20	88.40	65.00	20.40	20
N	5	5	5	5	5	5	1
Std dev	5.292	4.722	2.775	5.177	3.162	1.817	—
Min	203	193	62	81	60	18	—
Max	217	204	69	95	68	22	—
Peruvian Andes *whitelyi*							
Mean	214.75	203.50	60.38	93.38	62.75	24.62	90.33
N	8	8	8	8	8	8	3
Std dev	4.621	5.555	3.543	3.503	4.334	1.923	16.503
Min	208	197	55	89	56	22	74
Max	222	212	65	99	69	28	107
S Bolivia *sarae*	—	—	58.00	78.00	—	26.00	—
N	—	—	1	1	—	1	—
La Macarena	223.00	210.00	64.00	100.00	72.00	31.00	110
N	1	1	1	1	1	1	1
Meta							
Mean	214.67	204.00	65.00	91.33	70.50	22.50	9
N	3	3	3	3	2	2	—
Std dev	5.033	6.928	1.732	4.163	—	—	—
Min	210	200	63	88	69	22	—
Max	220	212	66	96	72	23	—
Guyana Shield *americana* = *tumatumari* = *juruana*							
Mean	227.00	216.00	66.50	94.00	67.00	24.00	20
N	2	2	2	2	2	2	—
Min	227	214	61	90	66	24	—
Max	227	218	72	98	68	24	—
Trinidad *trinitatis*							
Mean	210.67	201.67	64.00	91.67	61.67	24.00	92.5
N	3	3	3	3	3	3	2
Std dev	1.528	2.517	2.646	2.517	2.082	.000	—
Min	209	199	62	89	60	24	65
Max	212	204	67	94	64	24	120
Pará *bororo*?							
Mean	216.67	204.00	65.33	91.67	70.33	23.00	—
N	3	3	3	3	3	3	—
Std dev	4.163	5.000	3.786	5.859	3.215	1.732	—
Min	212	199	61	85	68	22	—
Max	220	209	68	96	74	25	—
Roca Nova *jucunda*	—	178.00	58.00	80.00	59.00	19.00	—
N	—	1	1	1	1	1	—
São Paulo *nana*?	207.00	199.00	64.00	91.00	61.00	25.00	56
N	1	1	1	1	1	1	1

Karyotype: $2n=68$–70; but in Brazil, $2n=48$–53.

Antlers small. Preorbital gland small or absent. Gestation 218 and 225 days (two records).

Guyanas, S and E Venezuela, NE Brazil.

Mazama jucunda Thomas, 1913

1884 *Mazama rufus* Nehring; not Illiger, 1815.

Very small in size (in our material).

SE Brazil, from São Paulo to Rio Grande do Sul.

Mazama zamora J. A. Allen, 1915

Apparently medium-sized; antlers very small.

SE Colombia, Ecuador E of the Andes, and NE Peru.

Mazama zetta Thomas, 1913

Small (in our dataset); antlers medium-sized.

Inter-Andean valleys of Colombia.

Mazama gualea J. A. Allen, 1915

1915 *Mazama fuscata* J. A. Allen.

A poorly defined, medium-sized species or population.

Ecuador, W of the Andes, and possibly SW Colombia.

Mazama temama (Kerr, 1792)

We propose that the following names are synonyms of *M. temama*:

1860 *Cervus sartorii* Saussure. S Mexico.

1913 *Mazama tema reperticia* Goldman. Panama, probably the extreme N of Colombia.

1914 *Mazama tema cerasina* Hollister. E Chiapas.

? 1959 *Mazama americana carrikeri* Hershkovitz. 3000–3900 m, Sierra Nevada de Santa Marta, N Colombia.

This is a well-authenticated species; for a full account, see Bello-Gutiérrez et al. (2010).

Karyotype: $2n=50$ (males), 49 (females).

Very small. Skull broad. Antler length 50–91 mm (Mexico, $N=10$), 81–96 mm (Central America and Colombia, $N=3$).

Extending from SE Mexico through Central America to NW Colombia (Bello-Gutiérrez et al., 2010).

Mazama trinitatis J. A. Allen, 1915

Light cinnamon-rufous, paler sides and belly; tail white below. Size medium to large; antlers long.

Trinidad.

Incertae sedis

The following names cannot be allocated at present:

Cervus rufus Illiger, 1815. From Mato Grosso, across Paraguay and extreme N of Argentina, to the Rio Bermejo. Synonyms—*Mazama pita* Rafinesque, 1817; *Cervus dolichurus* Wagner, 1844; *Mazama rufa toba* Lönnberg, 1919.

Mazama rufa rosii Lönnberg, 1919. N Argentina, "from Rio Bermejo to the provinces of Tucumán, Santiago del Estero, and Santa Fe."

Mazama sarae Thomas, 1925. Mountains of extreme S Bolivia and the neighboring part of Argentina.

The following two samples evidently represent distinct taxa, but no names appear to be available:

1. medium-sized; long nasals—Meta (antlers very small)
2. large; very broad, long nasals—La Macarena (antlers long)

Other Species

Two remaining species, *M. bricenii* and *M. rufina* (described below), which are almost certainly related to *M. americana*, were not examined by Duarte et al. (2008).

We have only very small samples of these two ($N=3$ and 4, respectively; see table 15). Even with a very high ratio of variable number to sample size (which risks a type I error), they did not separate in our discriminant analysis. Given the reported differences between them, however, we do not recommend uniting them, and table 15 shows that *M. bricenii* has smaller teeth than *M. rufina*. A specimen from Cundinamarca may represent a range extension for *M. rufina*.

Mazama bricenii Thomas, 1898

For a full account of this species, see Lizcano, Yerena, et al. (2010).

Eisenberg & Redford (1999) gave a skull length of 159 mm; antlers 5–5.5 cm. Reddish chestnut, with dark brown cheeks; underparts lighter brown, but the ventral surface of the tail white.

Andes of W Venezuela and neighboring Colombia.

Mazama rufina (Pucheran, 1852)

For a full account of this species, see Lizcano, Álvarez, et al. (2010).

Rich chestnut-brown above. Face dark brown. Well-developed preorbital glands. Dark face; outer side of the legs blackish; ears with some white hair on the

Table 15 Univariate statistics for *Mazama rufina* and *M. bricenii*

	Gt l	Cb l	Teeth	Biorb	Nas l	Nas post l	Antler l
Colombia/Ecuador *rufina*							
Mean	163.50	156.25	53.25	71.00	44.75	19.00	69.00
N	4	4	4	4	4	4	1
Std dev	4.203	3.500	2.630	1.826	4.349	1.414	—
Min	158	151	51	69	41	18	—
Max	168	158	56	73	51	21	—
Mérida *bricenii*							
Mean	160.67	153.33	50.00	67.33	42.67	20.33	—
N	3	3	3	3	3	3	—
Std dev	6.028	7.638	1.000	4.041	2.517	1.155	—
Min	155	145	49	65	40	19	—
Max	167	160	51	72	45	21	—
Cundinamarca							
Mean	158.00	149.00	49.00	71.00	41.00	19.00	—
N	1	1	1	1	1	1	—

inside. Fur short and thick, with a woolly undercoat. Antlers rarely more than 8 cm.

Premontane forests of S Colombia and Ecuador; dense undergrowth of premontane to montane forests, at 1500–3500 m.

Group Forming a Clade with *Hippocamelus*, *Blastocerus*, and *Ozotoceros* (Brown Brockets)

According to Duarte et al. (2008), this divides into four subclades: (1) *Mazama gouazoubira* and *Hippocamelus bisulcus* (support values for these two in the same clade are not very high); (2) *Blastocerus*; (3) *Hippocamelus antisensis* and *Ozotocerus*; and (4) *Mazama nemorivaga*.

Mazama gouazoubira Group

For a full account of this group, see Black-Décima et al. (2010).

$2n=63$–64, differing from all other New World deer in having an acrocentric X chromosome.

Total length 1034 mm ($N=32$), head plus body 924 mm, tail 110 ($N=33$), hindfoot 268 mm ($N=29$), ear 107.5 mm ($N=21$), weight 16.3 kg ($N=20$, Argentina, Paraguay, Uruguay).

Eisenberg (1989), Redford & Eisenberg (1992), Medellín et al. (1998), and Eisenberg & Redford (1999) gave the following information for this species-group, which they all treated as a single species:

—smaller and slighter than *M. americana*
—gray to gray-brown coat; predominantly gray in the S parts of the range
—supraorbital foramina minute or absent (*M. permira*; see below) or variable, with associated sulci usually more than 20 mm long (Colombia and Venezuela); nasals straight (*M. permira*) or slightly and evenly convex in lateral profile (Colombia and Venezuela); narrow postorbital constriction; squamosal root of the zygomata broadly arched (in lateral profile) above the glenoid fossa; mesopterygoid fossa V-shaped
—pedicels slender; antlers short to long, straight, ridged at the base when unworn

Gestation 206 days.

Panama S to N Argentina, including the Chaco; S to Catamarca.

We analyzed brown brockets in the following two groupings:

1. Montane populations. Our samples fall into three groups—(1) Santa Marta, Cauca, Bolivia, Mérida (up to 2000 m); (2) Loreto (Peru–Ecuador border), 100–200 m; (3) Meta (about 180 m)—with a specimen from Pasque (La Macarena) that is different from all the others.
2. Lowland populations. The Atlantic coast groups (Pará, São Paulo, Uruguay) fall close together, while the other lowland populations are separate.

We offer the following tentative revision of the *M. gouazoubira* group, but it is extremely provisional, and under the same constraints as for the *M. americana* group.

Mazama cita Osgood, 1912

1833 *Cervus humboldtii* Wiegmann, nom. nud.
1913 *Mazama sheila* Thomas.

Table 16 Univariate statistics for *Mazama gouazoubira*

	Gt l	Cb l	Teeth	Biorb	Nas l	Nas post br	Antler l
Santa Marta *sanctaemartae*							
Mean	180.69	171.69	55.69	78.92	55.23	23.62	106.00
N	13	13	13	13	13	13	1
Std dev	3.772	4.803	1.843	4.051	3.632	2.815	—
Min	175	165	53	72	49	19	—
Max	187	180	60	85	61	28	—
Cauca 2000 m							
Mean	178.67	167.67	59.00	82.33	52.67	22.00	44.00
N	3	3	3	3	3	3	1
Std dev	.577	1.528	2.000	2.517	2.887	.000	—
Min	178	166	57	80	51	22	—
Max	179	169	61	85	56	22	—
Loreto 100 m							
Mean	180.50	169.50	55.00	74.50	49.50	16.00	67.00
N	2	2	2	2	2	2	1
Min	174	162	54	71	42	15	67
Max	187	177	56	78	57	17	67
Pasco 1000 m *tschudii*							
Mean	—	—	60.00	94.00	—	24.00	90.00
N	—	—	1	1	—	1	1
Bolivia 50 m							
Mean	180.00	172.00	54.50	78.00	55.50	21.00	—
N	2	2	2	2	2	2	—
Min	178	170	53	76	54	21	—
Max	182	174	56	80	57	21	—
Mérida, Zulia *cita* = *sheila*							
Mean	181.50	173.75	57.75	78.25	53.25	21.50	53.50
N	4	4	4	4	4	4	2
Std dev	7.326	7.274	2.754	4.787	3.862	2.380	—
Min	171	163	55	74	51	20	33
Max	188	179	61	85	59	25	74
Pasque							
Mean	165.00	157.00	55.00	75.00	45.00	19.00	21.00
N	1	1	1	1	1	1	1
Meta, Caqueta 180 m *murelia*							
Mean	172.75	162.00	52.75	72.00	49.00	17.50	55.00
N	4	4	4	4	4	4	1
Std dev	2.754	2.309	1.258	1.826	2.449	1.291	—
Min	170	160	51	70	46	16	—
Max	176	164	54	74	52	19	—
Suriname *nemorivagus*							
Mean	176.25	167.00	50.25	78.25	52.75	17.25	67.50
N	4	4	4	4	4	4	2
Std dev	5.679	5.944	.500	1.708	.957	1.258	—
Min	172	162	50	76	52	16	58
Max	184	175	51	80	54	19	77
Pará *superciliaris*							
Mean	175.88	166.38	50.88	74.38	53.88	20.13	49.50
N	8	8	8	8	8	8	2
Std dev	8.983	9.501	6.010	6.739	5.384	2.588	—
Min	167	159	44	65	48	16	36
Max	194	186	64	86	64	24	63

Table 16 (continued)

	Gt l	Cb l	Teeth	Biorb	Nas l	Nas post br	Antler l
Mato Grosso 142 m *simplicicornis*							
Mean	183.80	177.40	55.60	76.20	51.60	20.80	92.00
N	5	5	5	5	5	5	1
Std dev	6.380	5.941	2.074	3.114	6.348	1.789	—
Min	174	171	54	71	44	20	—
Max	190	186	59	79	59	24	—
São Paulo							
Mean	185.50	175.50	63.00	85.00	57.50	22.50	—
N	2	2	2	2	2	2	—
Min	185	175	63	85	57	22	—
Max	186	176	63	85	58	23	—
Uruguay							
Mean	182.50	173.00	52.00	82.50	51.50	22.50	134.50
N	2	2	2	2	2	2	2
Min	180	171	49	80	51	19	123
Max	185	175	55	85	52	26	146

Large size; medium to short antlers.

Thomas and J. A. Allen referred *M. sheila* to the *americana* group, as it is bright rufous; but the skull has characteristics of the *M. gouazoubira* group. The color of *M. cita* has been noted as being different from other brown brockets.

N of Venezuela.

Mazama gouazoubira (G. Fischer, 1814)

1815 *Cervus simplicicornis* Illiger.
1817 *Mazama bira* Rafinesque.
1817 *Cervus nemorivagus* F. Cuvier, in part.
1879 *Nanelaphus namby* Fitzinger, 1873, nom. nud.
1883 *Nanelaphus nambi* Pelzeln.
1919 *Mazama argentina* Lönnberg.
1919 *Mazama simplicicornis* var. *kozeritzi* Miranda-Ribeiro.
1923 *Azarina fusca* Larrañaga.
1951 *Mazama gouazoubira* Hershkovitz.

Large size; antlers fairly long.

S of Brazil, from the state of São Paulo and the Planalto de Mato Grosso; Paraguay; N of Argentina as far as Tucumán, Santiago del Estero, and Entre Rios; Uruguay.

Mazama murelia J. A. Allen, 1915

Small size; short toothrow; narrow biorbital breadth; antlers medium to short.

SE of Colombia and Ecuador, E of the Andes.

Mazama nemorivaga (F. Cuvier, 1817)

For a full account of this species, see Rossi et al. (2010), who separated it from *M. gouazoubira* as a distinct species.

Medium-sized; very short toothrow; antlers medium length.

Guyanas, SE of Venezuela, and N of Brazil.

Mazama permira Kellogg, 1946

Type locality: Isla San Jose, Panama.

We have no experience of this putative species.

Mazama sanctaemartae J. A. Allen, 1915

Large size; long antlers.

N of Colombia.

Mazama superciliaris (Gray, 1852)

Medium-sized; toothrow very short; narrow biorbital breadth; broad nasals; antlers medium to short.

C and E Brazil, from Amazonas to Serra dos Parecis (in Mato Grosso), and the state of Espiritu Santo.

Mazama tschudii (Wagner, 1855)

Size apparently medium; long toothrow; antlers fairly long.

Peruvian Cordillera.

The following samples may represent distinct taxa, which appear to be without names:

Uruguay—size large, with wide biorbital breadth, and antlers very long
Cauca—also large and broad, but antlers short
Bolivia—size large, less broad
São Paulo—size large, with large teeth and long nasals
Loreto—size large, with narrow biorbital breadth, short narrow nasals, and antlers medium
Pasque—size very small, antlers very short

Mazama rondoni Miranda-Ribeiro, 1914

$2n=68$–70.

We have no experience of this putative species.

Mazama pandora Merriam, 1901

For a full account of this species, see Weber & Medellín (2010).

Distinguished as a species in a meticulous study by Medellín et al. (1998), from whom the description below is largely taken.

Brown to gray-brown, venter paler to whitish. Large patch of long, dark, stiff hairs on the forehead. Premaxillae comparatively broad, spatulate; postorbital constriction broad; posterior half of the nasals conspicuously humped; supraorbital foramina large, usually opening into long, prominent grooves; posterior margin of the palate usually U-shaped; dorsal margin of the squamosal root of the zygoma narrowly arched above the glenoid fossa. Frontal region broad, especially in the males. Large auditory bulla. Condylobasal length 161–177 mm ($N=8$). Antler pedicels massive; antlers long, divergent, usually curved, sometimes converging at the tips. Antlers heavy, fluted along almost the whole length when unworn. Antler length 112–142 mm ($N=6$).

Yucatán peninsula; in S and coastal Campeche, it is sympatric with *M. americana*.

Mazama chunyi Hershkovitz, 1959

For a full account of this species, see Rumiz & Pardo (2010).

Head plus body 70 cm; height 38 cm; antlers 3.5 cm; weight averaging 11 kg.

Short-tailed; brown, with a dark head, legs, and the backs of the ears; underside of the tail white. Pale; white markings on the ear margins and the tip of the muzzle. Lacks the dark brown face of *M. rufina*.

E premontane forests of Andean Peru and Bolivia; 1500–3200 m.

Pudu Gray, 1852

pudu

The skull characters are typically those of the neotropical endemics, with modifications. The supraorbital sulci are short, or nearly flat, and contain two large foramina. For details of the cranial and postcranial morphology, see the monograph by Hershkovitz (1982); for a full account of their biology, see Escamilo et al. (2010) and Jiménez (2010). The two species are *Pudu puda* (Molina, 1782) and *Pudu mephistophiles* de Winton, 1896.

Rangifer Hamilton Smith, 1827

reindeer (Europe), caribou (North America)

Rangifer are currently regarded as a single species, *R. tarandus*, which is broadly divided into woodland, tundra, and High Arctic groups of subspecies. There is a convincing cline of shorter limbs related to lower temperature, covering all *Rangifer*, whether woodland, tundra, or Arctic (Klein et al., 1987)—giving the Arctic island forms, in particular, a distinctive appearance—but there are many other characters, varying within the genus, that are not especially related to climate. The woodland forms (*R. t. fennicus* in the Old World, *R. t. caribou* in the New World, in Banfield's [1961] classic monograph) are very different in appearance—antler form, color, and build—from the tundra forms (*R. t. tarandus* in the Old World, *R. t. granti* and *R. t. groenlandicus* in the New World). There is said to be some seasonal overlap in range when tundra reindeer / caribou enter the N fringes of the coniferous forest, but how much interbreeding there may be is not known. This situation suggests that speciation is well underway (or complete), but before we can contemplate a full taxonomic arrangement, we need to know whether Old and New World woodland forms fall into a clade separate from the tundra forms, or are separately derived from them (or, even, if the tundra forms are derived from the woodland forms).

On the other hand, the monophyly of the three High Arctic forms has already been investigated: the two New World forms, *R. t. pearyi* and *R. t. eogroenlandicus*, do indeed form a clade, one related to the nearest tundra subspecies, *R. t. groenlandicus* (known as barren-ground caribou in Canada); while the Svalbard reindeer, *R. t. platyrhynchus*, though morphologically not dissimilar to them, is a clear derivative of Old World tundra reindeer (Gravlund et al., 1998). Tundra and High Arctic reindeer are said to be seasonally sympatric on some of the Canadian High Arctic islands; they are strikingly different in appearance, but in this case one wonders how much the differences might be due to phenotypic plasticity.

The taxonomy and distribution of the caribou of the Canadian Arctic have been reviewed by D. Thomas & Everson (1982). The Boothia Peninsula has *R. t. groenlandicus*. *R. t. pearyi* occurs on the W Queen Elizabeth Islands and Parry Islands, and on Somerset Island and Prince of Wales Island; the caribou of these two island groups have some differences between them. The main differences found in the study were that *R. t. pearyi* is smaller, with a much shorter muzzle and much lighter coloration. Strictly speaking, however, the more S island populations were considered to be "intergrades"; these apparently spent the winter on the N Boothia Peninsula and the summer on Prince of Wales and Somerset islands.

There have been several attempts to test the woodland versus barren-ground division in North America, with different results. Cronin (1992) found shared genotypes, whereas Røed et al. (1991) showed that they differed in transferrins, and that woodland caribou were geographically variable as well (see also van Staaden et al. [1995], who likewise studied transferrin frequencies). Flagstad & Røed (2003), on their minimum spanning network, discovered that almost all samples of caribou formed a single cluster, whereas *R. t. groenlandicus* and *R. t. granti* clustered in amongst the undifferentiated *R. tarandus* sample, but tended to be within their own subclusters. Courtois et al. (2003) similarly found low gene flow between woodland, barren-ground, and mountain caribou in NE Quebec, using eight microsatellites.

Geist (1998) did not think that the division between the two was necessarily the most fundamental; he argued, rather, that tundra caribou have evolved more than once from woodland forms, although in Europe it may be the other way round (i.e., European forest reindeer derived from tundra forms). In addition, he noted that there are fundamental differences between Eurasian and American taxa in color, especially the much greater extension of the white neck color onto the shoulder, and the presence of an extensive white area on the flanks, separated by a clear dark line from a broadly white underside, both characteristic of Eurasian taxa.

Indeed, population subdivision has been important throughout *Rangifer* history, as was shown by Côté et al. (2002), who found that even within Svalbard, in that very distinctive taxon, there is some subdivision of populations.

In sum, there is a clear need for a thoroughgoing revision of this genus.

Tribe Capreolini Brookes, 1828

This tribe contains two genera: *Capreolus*, the roedeer, and *Hydropotes*, the Chinese water-deer.

Capreolus Gray, 1821

roedeer

Lister et al. (1998) described the morphology of roedeer as a whole and affirmed that there are two quite distinct species. Randi, Pierpaoli, et al. (1998) studied control region haplotypes of European and Siberian populations of both species, and found them to be 100% different; the split between them goes back 2 to 3 Ma. Moreover, there are two clusters within each of the species, opening the possibility that the taxonomic diversity is still underestimated (Randi, Pierpaoli, et al., 1998). The genus needs a thoroughgoing taxonomic revision, which we are not yet equipped to undertake.

The western roedeer *Capreolus capreolus* seems to have extended its range at the expense of the E *C. pygargus* in modern times.

Capreolus capreolus (Linnaeus, 1758)

western roedeer

Karyotype: $2n = 70$; there are no B chromosomes (Baskevich & Danilkin, 1991).

A general survey of roedeer in Europe was given by von Lehmann & Sägesser (1981), including a brief comparison with *C. pygargus*: the skulls of European roedeer are shorter but relatively broader and higher than those of *C. pygargus*; and the ascending branch of the premaxilla makes a point contact, or none at all, with the nasal, whereas in *C. pygargus* it makes a broad contact.

Von Lehmann & Sägesser (1981) distinguished two summer pelage types in *C. capreolus*: "continental" (paler; yellowish ochery-colored; longer, stronger hair, with a dark gray to black base) and "Atlantic" (deep red; thin hair with a white base). The Atlantic type is seen in Belgium, the Netherlands, and the lower Rhine region; the continental type is in, for example, the Rothaar Mountains (in N Germany, E of Cologne). Both types occur in France and Spain. Scottish animals mix the deep red color of the Atlantic type with the dark hair bases and strong hair structure of the continental type. Ear length is less in the N than in the S. Roedeer from Spain and Switzerland seem to be somewhat bigger than those from Germany; those from the Carpathians and Croatia also seem rather large. The phalanges divide into an E type, which are thinner and longer, and a shorter, thicker W type. Von Lehmann & Sägesser (1981) went on to list 12 subspecies, but without details such as sample size.

Lister et al. (1998) noted that E European roedeer are larger than W European roedeer: sample means for E Europe are 207–230 mm for the males, and for W Europe 194–207 mm (the females are correspondingly smaller). Markov et al. (1985) performed a multivariate analysis on the limited geographic samples available to them, finding a sharp difference between the French and Bulgarian samples, on the one hand, and Ukrainian samples and those from the Baltic countries, on the other. Geist (1998) likewise commented on geographic forms of roedeer, with particular reference in this case to a distinctive, apparently unnamed, taxon in Garganta, S Spain. Randi, Pierpaoli et al. (1998) identified genetically unique features in the Italian roedeer *C. c. italicus* versus other European samples.

In a study of roedeer in both France and Sweden, Vanpé et al. (2007) found that antler size was correlated with age and with body mass, but it was not affected by environmental conditions (except for population density, and that only in one of the three populations). The authors concluded that antler size is therefore an honest signal of quality.

Capreolus pygargus (Pallas, 1771)

Siberian roedeer

The basic karyotype is $2n=70$, as in *C. capreolus*, but there are varying numbers of B chromosomes. According to Baskevich & Danilkin (1991), $2n$ (B chromosomes) = 1–4 in W Russia as far as Cis-Baikal (Irkutsk), but $2n=2$–14 in the Altai, $2n=6$–8 in Kazakhstan and Kyrgyzstan, $2n=5$–8 in Mongolia, $2n=8$–12 in Transbaikalia, and $2n=5$–10 in the Far East. The authors treated the first (low number) populations as nominotypical *C. pygargus pygargus*, and the second (mostly high number) series of populations as *C. pygargus tianschanicus*. It is possible that these could indeed be treated as different subspecies, though the Altai would evidently be a transition zone between them, rather than being true *C. p. tianschanicus*, as they proposed.

In Randi, Pierpaoli, et al.'s (1998) study, there was a consistent difference between specimens from the Kurgan region of W Siberia and those from the Amur region of the Far East. Similarly, animals from Cheju Island, Korea, formed a distinct clade from those from the Kurgan and Amur regions (Koh & Randi, 2001).

Hydropotes Swinhoe, 1870

This genus, according to multiple datasets (molecular and others), is nested deep within the Cervidae, and is sister to *Capreolus*. Accordingly, the old problem of whether *Hydropotes* is primitively antlerless, or has lost them, is decided in favor of the second proposition.

Hydropotes inermis Swinhoe, 1870

Chinese water-deer

There seems to be no more detailed study of the biology of this species than the Wikipedia article (http://en.wikipedia.org/wiki/Water_deer).

A recent study (Hu et al., 2006) could find no phylogeographic structure in the mtDNA sequences which they studied.

Tribe Alceini Brookes, 1828

In North America, modern moose appeared only at the end of the Pleistocene, when they replaced a late-surviving population of the precursor species *Alces latifrons* (classed by Azzaroli [1985] as a distinct, endemic North American species, *A. scotti*).

Alces Gray, 1821

moose (North America and international), elk (Europe)

Most authors place all living *Alces* in a single species, but Boyeskorov (1997, 1999) has convincingly argued that there are, in fact, two species: *Alces alces* from Europe and W Siberia, and *A. americanus* from E Siberia and North America. The boundary seems to be the Enisey River. They differ in chromosome number, color and color pattern, body proportions, skull characters, and antler form. The small moose from the Manchuria/Primoriye region, which Boyeskorov referred to as *A. americanus cameloides*, has single-palm antlers like *A. alces*, rather than the double-palm antlers of the larger *A. americanus*, and it may rank as a third species. The status of the recently extinct Caucasus moose has yet to be settled.

By contrast, Hundertmark, Shields, Bowyer, et al. (2002) found no such division in cytochrome *b* sequencing; rather, there was a strong division between the three haplotypes in Europe and the four in Asia, with the single haplotype found in North America being slightly closer to that of Europe. The authors did not, however, give the localities of their Asian haplotypes.

In contrast, the localities of origin were given in a different study by many of the same authors, using the control region (Hundertmark, Shields, Udina, et al.,

2002); here there was some, although very little, diversity in North America. The minimum-spanning tree (their figure 3) showed three phylogroups, with several substitutions between each pair: the first contained all the North American samples, plus two from the Russian Far East and two from Yakutia; the second, the bulk of the E Asian samples; and the third, the European and three E Asian samples. The authors suggested that the strong differences—for example, in chromosomes—between the E and W moose must be of rather recent origin.

***Alces alces* (Linnaeus, 1758)**

This, the moose from Europe and W Asia, at least as far E as the Altai, has $2n=68$ (Boyeskorov et al., 1993; Boyeskorov, 1997).The localities where this chromosome number has been recorded were mapped by Boyeskorov (1999).

Boyeskorov (1999) summarized the overall differences between the two species. The W species has an evenly brownish overall color, with the legs whitish. The body size index (body length as a percentage of shoulder height) is 80.8%–84.5%. In the skull, the posterior end of the nasal branch of the premaxilla is "spoonlike," although a very small proportion (3.57%) have the pointed "American" type, and an intermediate type occurs in 11.71%. The rostrum is relatively short. The antlers have a single palm (80.3%), rarely a double palm, which is known as the butterfly type (13.6%), while 6.2% had virtually no palmation. The antler shaft is relatively short, averaging 91.3 mm.

Earlier, Geist (1987) considered the differences between the two types of moose in a different manner. The "European type," he found, has "noticeably smaller hindquarters" and their antlers are "best described as being of a three-pronged plan: those of the American-type moose of a four-pronged plan"; this principle was illustrated in his figure 2.

***Alces americana* (Clinton, 1822)**

Moose with $2n=70$ occur in North America and, as far as Eurasia is concerned, have been confirmed in C Sakha, on the Kolyma River (Boyeskorov et al., 1993; Boyeskorov, 1997).

As summarized by Boyeskorov (1999), the overall color is from light brown to black, with legs from light to dark brown or gray. The body size index is 87.8%–89.6%, but a NE Siberian moose is 94.3%. The posterior end of the nasal branch of the premaxilla is narrow and pointed; the "spoonlike" type occurs in only 0.57% of the E Siberian specimens, and the intermediate type in 4.57% of them, but apparently never in moose from America or the Far East. The rostrum is relatively long; the choanae are broad. Over 90% of the antlers have a double palm, the exception being those from the Far East and E Mongolia, which normally have almost unpalmated antlers. The antler shaft is longer, averaging 120 mm. There are said to be differences in vocalization, in some proteins, and in sex pheromones.

The division between the two species goes back to Flerov (1931), who distinguished them only at the subspecific level. For E Siberia, he recognized *Alces alces pfizenmayeri* Zukowsky, 1910, which, he perceptively suggested, "may turn out to be identical with those of Alaska *A. a. gigas* Miller." He also recognized what he called "*Alces alces* subsp.?" from the Amur, which has the premaxillary characters of E elks but is of small size; in fact, the available name for this kind of moose is *A. a. cameloides* (Milne-Edwards, 1867).

Geist (1987) also noted that *A. americana* has a much larger bell, with a longer rope; the face of the cow is reddish brown, whereas in the bull the rostrum is black. In principle, there is a light saddle on the back, but this is less conspicuous in E North America.

The control region shows low diversity in North America (Hundertmark et al., 2003) and forms a star phylogeny, indicating a recent expansion of populations (the fossil record depicts the modern moose entering North America only after the end of the Pleistocene).The validity of the described subspecies will not, however, depend on the inferred late date of their diversification, but rather on whether the differences between them (see Peterson, 1950) characterize a sufficient proportion of their respective populations.

Subfamily Cervinae Goldfuss, 1820

The plesiometacarpal deer; the brow-tine deer (*pace Elaphurus*!).

On the Old World deer in general, see Groves & Grubb (1987), Geist (1998), and Grubb (200b).

Tribe Muntiacini Knottnerus-Meyer, 1907

The name Cervulini Sclater, 1870 (originally Cervulinae), has technically priority over Muntiacini, but the former has now been placed on ICZN's Official Index of Rejected and Invalid Family-Group Names.

The reputed diversity of species in *Muntiacus*, with several new species described over the past 10 years or so (beginning with the giant muntjac, originally placed in a separate genus and named *Megamuntiacus vuquangensis*), is getting out of hand. It is surely time to gather the available specimens of little black muntjac in one place, compare them morphologically and morphometrically, and run several mtDNA and nDNA sequences. That said, we now agree that the revision by Groves & Grubb (1990) was somewhat overlumped.

Elaphodus Milne-Edwards, 1871

tufted deer

1871 *Elaphodus* Milne-Edwards. Type species: *E. cephalophus* Milne-Edwards, 1871.

1874 *Lophotragus* Swinhoe. Type species: *L. michianus* Swinhoe, 1874.

Chromosomes 48 (at least in the females).

Antler pedicels short, very slender, not extending as more than weak ridges on the face, and continuing along the upper margins of the orbits; extremely large, well-defined preorbital fossa; preorbital fissure poorly developed or absent; ears broad, rounded; lateral hoofs very long; frontal tuft short but very thick.

Elaphodus cephalophus Milne-Edwards, 1871

1871 *Elaphodus cephalophus* Milne-Edwards. Baoxing, Sichuan.

1874 *Lophotragus michianus* Swinhoe. Ningpo, Chejiang.

1904 *Elaphodus ichangensis* Lydekker. Ichang, Hubei.

1904 *Elaphodus michianus fociensis* Lydekker. "Kohwang near Ching Feng Ling, about 100 miles northwest of Foochow" (see Groves & Grubb, 1990).

There is a good deal of geographic variability in *Elaphodus*, but with so little available material, Groves & Grubb (1990) were unable to make much of it.

As noted by Groves & Grubb (1990), there is a considerable size difference between tufted deer from the W China mountains and N Burma (skull length 181–202 mm), and those from the coastal mountains of E China (skull length 171–181 mm); but the type of *Elaphodus michianus fociensis* is large, like the W ones. Clearly, a larger series of specimens must be studied before any conclusion can be reached as to taxonomic diversity.

Muntiacus Rafinesque, 1815

muntjac

1815 *Muntiacus* Rafinesque. Type species: *Cervus muntjak* Zimmermann, 1780. This name has been validated by ICZN, Opinion 460.

1816 *Cervulus* de Blainville. Type species: *Cervus muntjak* Zimmermann, 1780.

1827 *Stylocerus* Hamilton Smith. Type species: *Cervus muntjak* Zimmermann, 1780.

1836 *Prox* Ogilby. Type species: *Prox moschatus* Ogilby, 1836.

1843 *Muntjacus* Gray. Type species: *Muntjacus vaginalis* Gray, 1843 (= *Cervus vaginalis* Boddaert, 1785).

1923 *Procops* Pocock. Type species: *Cervulus feae* Thomas & Doria, 1889.

Antlers longer than in tufted deer, with long pedicels, extending down onto the face as thick straight "ribs" ending on the upper orbital margins, but not following them; preorbital fossa smaller, less well defined; preorbital fissure well developed; ears narrower, more pointed; lateral hoofs shorter; frontal tuft, if present, less thick or matlike.

Muntiacus is notorious for having the widest range of chromosome numbers in any mammalian genus, from $2n=46$ in *M. reevesi* to $2n=6/7$ in *M. vaginalis*, with several different numbers in between (*M. crinifrons* and *M. gongshanensis*, $2n=8/9$; *M. feae*, $2n=13/14$). The homologies between them were studied by Yang et al. (1995) through chromosome painting; these authors showed, for example, that the chromosomes of *M. crinifrons* and *M. gongshanensis*, despite having the same diploid number, actually have several rearrangements between them.

Lan et al. (1995) compared the four Chinese species of *Muntiacus* using restriction sites of mtDNA. Whether using parsimony or distance methods, "Indian muntjac" (here, *M. vaginalis*) was sister to the others, followed by a split between *M. reevesi* on one branch, and *M. gongshanensis* and *M. crinifrons* on the other.

Amato et al. (2000), however, on the basis of four mitochondrial sequences, placed *M. reevesi* as sister to the other (mostly mainland) species, followed by a split between *M. truongsonensis*, *M. putaoensis*, *M. rooseveltorum*, and *M. vuquangensis*, on the one hand, and *M. muntjak*, *M. gongshanensis*, *M. crinifrons*, and *M. feae*, on the other.

Pitra et al. (2004) calculated that *M. reevesi* separated from *M. crinifrons* plus *M. vaginalis* somewhat over 3 Ma, while these latter two would have separated some 2.5 Ma.

There is still much to be learned about this genus. The necessity of breaking up the former *M. muntjak* into several different species, in addition to the plethora of new species that have been described over the last 20 years or so, necessitate the genus being divided in some way; we have divided the species into four informal groups, which do not pretend to be homogeneous or monophyletic groups.

Muntiacus muntjak Group (Red or Indian Muntjac)

Large muntjac with low chromosome numbers.

It should be noted that karyotypes are known only for *M. muntjak* and *M. vaginalis*; while it may well be that the other mainland taxa will have identical karyotypes to those of *M. vaginalis*, it should not be assumed.

Muntiacus muntjak (Zimmermann, 1780)

1780 *Cervus muntjak* Zimmermann. Java.

Synonymy as in Groves & Grubb (1990), under *Muntiacus muntjak muntjak*.

2n = 8 in the females, 9 in the males.

Antlers long, >80 mm, usually >100. Black stripes on the pedicels exceptionally thick and deep black. Color deep rufous, the nape and the mid-dorsal region contrastingly dusky; forehead light cinnamon-rufous, contrasting with a dusky face; ear-backs dark; thighs and shoulders somewhat contrastingly dark chestnut-brown, becoming darker and grayer down the shanks; throat creamy white, becoming light cinnamon-rufous on the rest of the underparts; pubic region and the front of the thighs creamy white. Skull relatively narrow.

Java, Bali, Lombok, Borneo, Bangka, Lampung, and coastal areas of E Sumatra; Malay Peninsula N to about Trang.

Groves & Grubb (1990) anticipated the separation of the Malaysian/Indonesian and the general mainland forms of red muntjac at the species level.

Muntiacus vaginalis (Boddaert, 1785)

We have some doubts whether the two subspecies placed under this name are correctly assigned to a single species, although certain characters (shorter antlers, darker midback, gray nape, orange-brown forehead and occiput, etc.) do unite them. Their colors, on the other hand, are rather different.

Muntiacus vaginalis vaginalis (Boddaert, 1785)

1785 *Cervus vaginalis* Boddaert. Bengal.

1827 *Cervus moschatus* Hamilton Smith; not de Blainville, 1816. Nepal.

1833 *Cervus ratwa* Hodgson. Nepal.

1839 *Cervus melas* Ogilby. Himalayas.

1844 *Cervus stylocerus* Schinz.

1846 *Prox ratva* and *Prox albipes* Sundevall.

1846 *Stylocerus muntjac* Cantor.

2n = 6 (females), 7 (males).

Antlers generally shorter, 90–125 mm. Dark reddish in general, somewhat darker on the midback; nape slightly grayer; forehead and occiput light orange-brown, the rest of the face grayish; ear-backs reddish at the base, the remaining two-thirds dark gray; limbs brown to gray; underside paler; groin and the line on the front of the hindlegs white.

Nepal, most of E India, to W Burma.

Muntiacus vaginalis curvostylis (Gray, 1872)

1872 *Cervulus curvostylis* Gray. Pachebon, Thailand.

1904 *Cervulus muntjac grandicornis* Lydekker. Thouagyen Forest, Amherst district, Burma.

1928 *Muntiacus muntjak annamensis* Kloss. Langbian.

1988 *Muntiacus muntjak menglalis* Wang & Groves. Pujiao, Mengla County, Xishuanbana, Yunnan.

Much paler than the nominotypical *M. v. vaginalis*; legs hardly (or not at all) darker than the body.

S and E Burma, Thailand, Laos, Vietnam (except the extreme N), southernmost Yunnan.

Muntiacus malabaricus Lydekker, 1915

1915 *Muntiacus muntjak malabaricus* Lydekker. Nagarahole.

Smaller in size than *M. vaginalis*, with shorter antlers (<95 mm) and short pedicels; much paler reddish, with gray tones on the nape and the back; limbs colored as in the body; underside drab; white area on the lower limbs prominent, extending around to the front of the pasterns, with only a narrow band down the limbs.

Sri Lanka and W Ghats.

Muntiacus aureus (Hamilton Smith, 1826)

1826 *Cervus aureus* Hamilton Smith. "Some part of southern India," according to Lydekker (1915a).

1844 *Cervus albipes* Wagner. "Bombay and Poona."

1872 *Cervus tamulicus* Gray. Deccan.

Smaller than the other Indian taxa. Pale yellowish, with a gray nape; forehead and occiput pale orange-brown, the rest of the face light gray-orange; ear-backs orange at the base, becoming gray and dark gray on the tips and the rims; limbs colored like the body; underside slightly paler; white line down the front of the thighs to the hocks; antlers short, <100 mm.

NW and C India; and on the lower Chindwin in Burma.

Groves & Grubb (1990) discussed this species, noting its curiously disjunct distribution, but there seems to be no doubt about its homogeneity, or its striking differences from others of this group. Presumably a former continuous range has been interrupted by the spread of *M. vaginalis*.

Muntiacus nigripes G. M. Allen, 1930

1930 *Muntiacus muntjak nigripes* G. M. Allen. Nodoa, Hainan.
1988 *Muntiacus muntjak yunnanensis* Ma & Wang. Wokang Dashan, Menglai, Cangyuan County, W Yunnan, 2200 m.

Color bright orange to deep red; forehead and occiput deep brown or black in color; ear-backs nearly black, except at the bases; shanks, or the whole of the limbs, gray to dark chestnut to blackish brown; chin and throat, axillae, and groin pure white, sending a white line down the inside of the hindlimbs to beyond the hocks; white patch on the pasterns; rest of the underside a paler version of the upperside. Antlers short, but the pedicels comparatively long.

Hainan, northernmost Vietnam (including the Annamites; see Groves & Schaller, 2000), and Yunnan, N of about 23° 10' N.

The two described forms—one on the mainland, one on Hainan—differ in size and average color, and it may be that they should be separated.

Incertae sedis

Muntiacus muntjak guangdongensis Li & Xu, 1996. Dinghushan Reserve, Gaoyiao County, Guangdong Province, 850 m.

We have not seen specimens of this taxon. The description states that it is "of medium size"; greatest skull length 191–205 mm, with an exceptionally wide interorbital width. Pale brown-yellow in color, darkening on the midback; flanks and underparts paler to creamy white; black stripes running up the antler pedicels. In about 50% of the individuals, the anterior part of the forelimbs and the sides of the hindlimbs dark brown. (Following Li J-x. & Xu, 1996.)

Muntiacus crinifrons Group

Large muntjac with intermediate chromosome numbers.

Muntiacus crinifrons (Sclater, 1885)
hairy-fronted muntjac

1885 *Cervulus crinifrons* Sclater. Ningpo.

2*n* = 8 (females), 9 (males).

Antlers short, up to 52 mm, often unbranched; pedicels long (40–70 mm) and slender, sometimes depressed below the line of the facial profile; greatest skull length 205–224 mm; skull with a convex facial skeleton and a depression between the face and the interorbital-frontal region; nasals somewhat compressed, the posterior ends narrow; ethmoid fissure very small, narrow; forehead very narrow between the frontal ridges; preorbital fossa restricted to the lower half of the lacrimal bone. Lateral hoofs long; tail very long, with a long white fringe restricting the black dorsum to a narrow median area; color black, with some red tones; frontal tuft bright red, very long, thick, hiding the antler pedicels, the red color extending back to the ear-backs and the occiput; underside only slightly lighter than the upperside.

W Zhejiang, SE Anhui, extending to Pucheng County, Fujian.

Muntiacus gongshanensis Ma, 1990
Gongshan muntjac

1990 *Muntiacus gongshanensis* Ma. Mijiao (27° 35' N, 98° 47' E), Puladi, Gongshan County, E slope of the N sector of Gaoligong Mountain, NW Yunnan.
Although the 1990 paper describing this species was by Ma, Wang, and Shi, the new species was attributed to Ma alone.

A medium-sized species; skull length 194.8–205.4 mm. No forehead tuft. Chin glands rather small. Antler length 71.7 mm; pedicels 34.5 mm. Body color and upperside of the tail dark brown. Neck hair not reversed; forehead and crown reddish, blackened in the females. On the skull, the "condylolateral process" (as the authors called what seems to be the paroccipital process) larger than in the other species; rostrum long, narrow; orbital rims well defined, with a constriction between the orbit and the lower margin of the preorbital fossa; zygomatic arch angled.

Amato et al. (2000) could find no consistent differences between this species and *M. crinifrons* in the four mtDNA sequences they studied. Externally, however, *M. gongshanensis* does differ—it lacks a forehead tuft, and the tail is said to be not as dark—and the karyotype, though basically the same as in *M. crinifrons* (8 in the females, 9 in the males), differs in the nucleolar organizing regions, being on chromosomes 1 and 4, instead of on chromosomes 2 and 4 (Ma et al., 1990); there are also several other chromosome differences.

The distribution is given as the Gaoligong and Biluo mountains. It is doubtless this species, rather than *M. crinifrons* (as reported), that occurs in the

Hkakabo-Razi area in Burma, on the border of China. at about 30° N, 97° E (Rabinowitz et al., 1998), and it is evidently the "northern *M. feae*" described by Groves & Grubb (1990) from N Burma, W Yunnan, S Tibet, and possibly even Darjiling.

Muntiacus feae (Thomas & Doria, 1889)

Fea's muntjac

Grubb (1977) used the version *feai*, but this was unjustified (Grubb, 1990a, 2000b).

$2n = 12$, 13, 14 (females), with the same X-autosome fusion as seen in the *M. muntjak* group (Soma et al., 1983). $2n = 14$ (males).

Antlers short, fairly stout, usually branched, regularly cast; pedicels long, very slender; skull broad, with the maxillae bulging outward; rostrum short, deep; preorbital fossa large; ethmoid fissure very large; nasals very flat, expanded at the posterior ends; tail long, black, with a white underside and a long white lateral fringe; lateral hoofs long; dull agouti-olive-brown, with contrasting deep blackish limbs; white ring around the hoofs. Hair reversed on the midline of the neck. Females larger than the males, with a long yellow frontal tuft with black stripes; males similar, but without a tuft.

According to Amato et al. (2000), this species is sister to the *crinifrons/gongshanensis* group.

S Thailand, from about 9° N to more than 14° N.

"New Muntjac" Group

Muntjac of varying size, of unknown karyotype.

Amato et al. (2000) found that, at least on mtDNA sequencing, the species of this group do indeed form a monophyletic clade.

Muntiacus vuquangensis (Tuoc et al., 1994)

giant muntjac

1994 *Megamuntiacus vuquangensis* Tuoc et al.

Notes and descriptions of this species were given by Schaller & Vrba (1996) and Groves & Schaller (2000).

Largest species of muntjac; only species with the antlers considerably longer than the pedicels.

Its distribution was mapped by Groves & Schaller (2000).

Though placed in a separate genus, *Megamuntiacus*, when described, Amato et al. (2000), on the basis of mtDNA, found it to belong in a clade with the other Indochinese species, but it was their sister group. This result was somewhat surprising, given the strong differences between *M. vuquangensis* and the small muntjac species, and it is possible that this is another case of nuclear swamping—in this instance, of one of the small Indochinese muntjac by a large species (proto-*vuquangensis*).

Muntiacus rooseveltorum Osgood, 1932

1932 *Muntiacus rooseveltorum* Osgood. Muong Yo, N Laos (12° 30' N, 102° E).

A small species; agouti-brown pelage; ochraceous crown; white throat; very large mental glands (the glands on the inferior surface of the mandibular symphysis).

Its known distribution, confined to Laos, was mapped by Groves & Schaller (2000).

DNA from the skin of the type specimen and from new specimens from Laos placed this species as sister to *putaoensis/truongsonensis*.

Muntiacus putaoensis Amato et al., 1999

1999 *Muntiacus putaoensis* Amato et al. Atanga village, 30 km E of Putao (27° 21' N, 97° 24' E), extreme N Burma.

Very small in size; pelage chestnut, including the upper surface of the tail; pedicels short (under 40 mm) and thin; antlers very short (8–35 mm), unbranched. Skull length averaging 175 mm; braincase width 45–53 mm, mean 47 mm (Amato et al., 1999; Rabinowitz et al., 1999).

A phylogenetic tree based on three mitochondrial sequences (Amato et al., 1999) showed this species as closely related to *M. truongsonensis*, but somewhat different; these two belonged in a clade with *M. rooseveltorum*, and the next closest species was *M. vuquangensis*. These four species formed a clade opposed to one containing the *M. muntjak* group plus *M. feae*, *M. gongshanensis*, and *M. crinifrons*; sister to all these species was *M. reevesi*, although the bootstrap value of the non-*reevesi* clade was only 35%.

Muntiacus truongsonensis (Giao et al., 1998t)

1997 *Caninmuntiacus truongsonensis* Giao et al. Hien district, W Quang Nam Province, Vietnam.

A small, black muntjac; tail short, fairly flat, with long, white lateral hairs; pedicels shorter and narrower than the *M. muntjak* group; antlers very short and unbranched; female canines almost as long as those of the males.

Because all available cranial material was based on trophy skulls, it was not possible to measure skull length; braincase width averages about 47 mm.

Antler pedicels average 36.7 mm, shorter than any other species except *M. putaoensis*; antler beam averages 10.9 mm, comparable only to *M. crinifrons* or *M. putaoensis*.

Its known distribution was mapped by Groves & Schaller (2000); it appears to be restricted to the S part of the Annamite range.

According to Amato et al. (2000), *M. truongsonensis* is sister to *M. putaoensis*. In Rabinowitz et al.'s (1999) table 3, the only significant difference between this species and *M. putaoensis* would seem to be the greatest width across the nasals, 22 mm versus 16 ±0.15 mm in *M. putaoensis*.

Muntiacus puhoatensis Chau, 1997

1997 *Muntiacus puhoatensis* Chau. Puhoat area, Que Phong district, Nghe An Province, Vietnam.

The diagnosis stated that it is a very small species of *Muntiacus*. Weight estimated at 8–15 kg (12 kg on average). Antlers short (25–26 mm) and unbranched. Horn pedicels short (ca. 27–28 mm) and parallel; pedicel with a diameter of 7–8 mm, and covered with long, thick, yellow and dark brown hair; underparts and undertail mixed brown and white.

The habitat is said to be closed-canopy evergreen forest, on steep slopes above about 700 m.

Muntiacus reevesi Group

Small muntjac of high chromosome number (where known).

Muntiacus reevesi (Ogilby, 1839)

Chinese muntjac

1839 *Cervus reevesi* Ogilby. Canton.
1871 *Cervulus lachrymans* Milne-Edwards. Baoxing.
1872 *Cervulus sclateri* Swinhoe. Ningpo.
1875 *Cervulus micrurus* Sclater. Taiwan.
1905 *Cervulus sinensis* Hillzheimer. Hwai Mountains, Anhui.
1906 *Cervulus reevesi pingshianicus* Hillzheimer. Pingshiang, C China.
1910 *Cervulus bridgemani* Lydekker. Hwai Mountains.
1914 *Muntiacus lachrymans teesdalei* Wroughton. Tatung, Yangtze Valley.

$2n = 46$ in both sexes.

Size small; antlers short, 40–80 mm, often unbranched, but irregularly shed; pedicels long, broad; preorbital fossa very large, often reaching back to the orbital margin; preorbital fissure very long and narrow; frontal tuft absent; mostly reddish yellow, with some gray overtones; head more reddish than the body, relatively large for the body size. In the females, the frontal region is black, this zone narrowing between the ears and continuing as a nuchal stripe; occiput orange. In the males, the frontal region bright orange, edged with black stripes, continuing onto the pedicels; orange color extending onto the occiput and the ears; nuchal stripe not connected to the head markings.

China, from about the Yangtze River S to Guangdong, including Taiwan. Groves & Grubb (1990) remarked on how poorly the Taiwan population was differentiated from any on the mainland.

Muntiacus atherodes Groves & Grubb, 1982

Bornean yellow muntjac

1982 *Muntiacus atherodes* Groves & Grubb. Near Forest Camp 1, Cocoa Research Station, Tawau, Sabah.

Very small; light orange-yellow in color, with a dark, diffuse dorsal stripe; no frontal tuft; frontal region blackish, extending onto the inner halves of the pedicels in the males, and continuous with the nuchal stripe, not divided into separate dark stripes; occiput not contrastingly orange-colored; antlers tiny, spike-like, on very short, slender pedicels. Antlers apparently not shedding, as a rule; very few specimens with burrs (indicating that the antlers have been shed) known.

Borneo.

On this species, and on the history of knowledge of the sympatry of the two species in Borneo, see Groves & Grubb (1982).

Muntiacus montanus Robinson & Kloss, 1918

1918 *Muntiacus muntjak montanus* Robinson & Kloss. S Kering, G[unung or Mt.] Kerinci, 2200 m.

Very small in size; greatest skull length 179–198 mm; antlers usually unbranched, usually not shed, short, <100 mm; pedicels very long, 96–102 mm, always considerably longer than the antlers. Color very dark, a dark chestnut, with blackish speckles; midback darkened; forehead, pedicels, occiput, and bases of the ears orange-brown, the rest of the face brown-gray; backs of the ears dark; sides and shoulders brown, becoming dark brown on the shanks; throat olive-buff, the rest of the underside reddish white (extending up onto the lower flanks), changing to whitish on the groin; feet white, this tone going up the front of the lower half of the limbs; tail blackish brown above; stripe on the pedicels thick, black.

Highlands of Sumatra, from Kerinci N into Aceh.

Groves & Grubb (1990) considered, and then tentatively rejected, the idea that this could be a distinct species, analogous to *M. atherodes* of Borneo. Reconsideration, and inspection by both of us of photos of live animals (from photo traps), leads us to definitively place it as a distinct species, but we still do not know whether it is related to *M. atherodes*, or *M. muntjak*, or perhaps occupies an isolated position.

Tribe Cervini Goldfuss, 1820

The phylogeny of the Cervini has been most recently explored by Pitra et al. (2004). Certainly four genera (*Axis*, *Rucervus*, *Dama*, and *Cervus*) should be recognized in the living fauna, but the recent discovery of a 7- to 9-million-year-old muntjac (Dong et al., 2004) indicates that the calibration chosen by Pitra et al. (2004) for their molecular clock—just 7 Ma for the muntjac-cervin split—has evidently been set too late, and we here propose to recognize two further genera, *Panolia* and *Elaphurus*.

Dama Frisch, 1775

fallow-deer

This genus is immediately distinguishable by the tendency to palmation in the antlers; the strongly marked pygal (rump) bands; the presence of white spots during all seasons; and the pendulous penis, whose prepuce is not closely adherent to the body wall. The nasal bones resemble those of *Rucervus*. There are no upper canines (except as an anomaly). The lower central incisors are widened, as in *Axis*.

Dama is closer to *Cervus* than are *Rucervus* and *Axis*, according to Pitra et al. (2004), who calculated the date of its separation from the *Cervus* group at over 5 Ma. It contains two living species, the European fallow-deer (*D. dama*) and the Persian one (*D. mesopotamica*); these are quite distinct, despite the urge of many authors to make them subspecies of a single species. The fossil record of the genus does not begin until the Middle Pleistocene, with *D. clactoniana*, of which the modern *D. mesopotamica* is essentially a size-reduced version, although Pfeffer (1999) (plausibly, in our view) also ascribed some Late Pliocene fossils to the *Dama* stem.

Mitochondrial D-loop sequence phylogeny showed that *D. mesopotamica* and *D. dama* differ drastically (Masseti et al., 2008); the authors estimated a divergence time of over 400,000 years.

Dama dama (Linnaeus, 1758)

common fallow-deer

Populations of this species from Rhodes and from Turkey (the latter being the last remaining population of the species in its original homeland) form distinct clades, with surprisingly high heterozygosity (Masseti et al., 2008). *Dama schaeferi* Hillzheimer, 1926, described from N Africa, was actually from Italy, and it was an ordinary fallow deer (Kock, 2000).

Dama mesopotamica (Brooke, 1875)

Mesopotamian fallow-deer

On this species, see the Wikipedia article (http://en.wikipedia.org/wiki/Persian_fallow_deer).

Axis Hamilton Smith, 1827

The slender antlers are built on a plesiomorphic three-point plan; the anterior ends of the nasal bones are bifid, the median prong being equal to or longer than the lateral one, so that together they form a concave anterior margin; the posterior ends form a shallow, blunt wedge into the frontals. There are no upper canines; and the lower central incisors are very wide, their width exceeding the combined width of the two lateral incisors plus the canine.

Pitra et al. (2004) found *Axis* to be the sister genus to *Rucervus*, from which it separated some 5 Ma (but, in our estimation, these dates should be pushed back considerably; see above); their common clade separated from the *Cervus*/*Dama* clade more than 6 Ma.

Axis axis Group

Axis axis (Erxleben, 1777)

chital, axis, spotted deer

For the full synonymy of this species, see Groves (2003).

No differences seem to be apparent between Sri Lankan and mainland populations.

Axis porcinus Group

In Pitra et al.'s (2004) analysis, *Axis porcinus* was placed in a clade with *Cervus timorensis*. This position is unlikely, and doubtless results from a misidentification, as argued by Gilbert et al. (2006), who found, as would have been expected, that this group sits alongside *A. axis*.

Cranial measurements of the described taxa are compared in table 17. *A. annamiticus* is the largest species, and *A. kuhlii* is the smallest (but there is a

Table 17 Univariate statistics for members of the *Axis porcinus* group

	Gt l	Preorb	Biorb	Braincase
porcinus				
Mean	230.50	121.60	104.00	69.00
N	6	5	6	5
Std dev	7.817	3.647	2.366	4.000
Min	219	118	101	64
Max	241	126	108	75
annamiticus				
Mean	236.50	122.00	107.50	70.25
N	2	2	4	4
Std dev	9.192	9.899	3.317	4.193
Min	230	115	104	64
Max	243	129	112	73
kuhlii				
Mean	223.50	112.50	102.00	70.00
N	2	2	3	3
Std dev	14.849	4.950	3.606	6.000
Min	213	109	98	64
Max	234	116	105	76
calamianensis				
Mean	229.50	120.50	97.00	63.00
N	2	2	2	4
Std dev	2.121	.707	.000	2.449
Min	228	120	97	60
Max	231	121	97	65

great deal of individual variation in size). *A. kuhlii* has a very broad skull, and *A. calamianensis* a very narrow one.

Axis porcinus (Zimmermann, 1780)

Indian hog-deer

From Sind E into Nepal and NE India; Sri Lanka, where it may be introduced.

Data on growth, adult body measurements, and antlers from the introduced population on Sunday Island, off Wilsons Promontory (Victoria, Australia), were provided by Presidente & Draisma (1980). Adult males weighed approximately 50 kg, females 35 kg. Antlers achieve full size, at a mean of about 300 mm, around 4 years of age, about two years before the body-weight plateau is reached.

Axis annamiticus (Heude, 1888)

Indochina hog-deer

Axis porcinus and *A. annamiticus* are cranially very similar, the latter being very slightly larger, but externally they are very different. *A. annamiticus* is much brighter; more ochery, less gray and buffy; and not speckled (this is due to the fainter banding on the hairs). In these features, it resembles *A. calamianensis*.

Axis kuhlii (Müller & Schlegel, 1845)

Bawean deer

Notes on *A. kuhlii* were provided by van Bemmel (1948), who was the first to seriously support its close affinity with *A. porcinus*.

Unlike *A. porcinus*, the fawns are, at most, only faintly spotted.

Axis calamianensis Heude, 1888

Calamian deer

Characterized by a shorter tail than other members of the genus; a narrow skull; a relatively large size; and long, slender antlers.

Rucervus Hodgson, 1838

swamp deer, barasingha

Rucervus contains the swamp deer (traditionally a single species, *R. duvaucelii*) of Nepal and India, and the presumed-extinct Schomburgk's deer (*R. schomburgki*) of Thailand. Pitra et al. (2004) extracted DNA from a Schomburgk's deer specimen and showed that it is indeed the sister species of the swamp deer, not the representative of a separate genus (or subgenus) *Thaocervus*, as had sometimes been thought. They found that *Rucervus* is most closely related (if distantly) to *Axis*.

Cranially, *Rucervus* resembles *Dama*—and is readily distinguished from the other Cervini—by the form of the nasofrontal suture: the nasal bones together make a deep, acute-angled wedge back into the frontals. At their free ends, each has two prongs, a very long lateral one and a rudimentary median one, as in *Axis*. There are no upper canines.

Rucervus duvaucelii (G. Cuvier, 1823)

western swamp deer, barasingha

1823 *Cervus duvaucelii* G. Cuvier. Kumaun.

For the full synonymy of this species, see Groves (1983c).

Nasals short, relative to the snout length; rostrum not deep; antlers long, slender, and not compressed or palmated; little or no sexual dimorphism in size. Tail relatively long and slim, with prominent white hairs on the underside; ears very large and rounded, with thick, white hair on the inside; in moult, the white spots in the dorsal region very prominent. Feet splayed, with bare "heels."

Rucervus ranjitsinhi (Groves, 1983)

eastern swamp deer

1983 *Cervus duvaucelii ranjitsinhi* Groves. Guwahati, Assam.

Elongated nasals; snout short, deep; maxilla narrow; antlers short, thick, branching low down, with an especially shortened anterior branch; antlers somewhat compressed, tending to be palmated. Heavily built; females notably smaller than the males; ears small, pointed, with very little white hair on the inside; tail short; in moult, the white spots in the dorsal region less prominent. Feet splayed, with bare "heels."

Rucervus branderi Pocock, 1943

hard-ground barasingha

1943 *Rucervus duvauceli branderi* Pocock. Mandla, C India.

Distinguished from *R. duvaucelii* and *R. ranjitsinhi* by the "well-knit" feet, with hairy pasterns. Size smaller; nasals long; snout short, but the nose not deep; maxilla somewhat broadened; antlers more as in *R. duvaucelii*, but extremely long, many branched, with a long brow tine; branching high up the beam; anterior branch especially long.

Rucervus schomburgki (Blyth, 1863)

Schomburgk's deer

Dark brown; large, multitined antlers.

A presumably extinct deer, known only from the swampy country of C Thailand.

Panolia Gray, 1843

Eld's deer, brow-antlered deer

The Eld's deer complex constitutes a separate genus, *Panolia*, whose affinities, based on mtDNA, are not with *Rucervus* (to which it has usually been thought to be related) but with *Elaphurus*, from which it separated some 3.5 Ma, according to Pitra et al. (2004). More likely, this separation would have been considerably earlier (see above). It is cranially distinct from other Cervini, especially as the posterior ends of the nasal bones form only a blunt, shallow wedge into the frontals; whereas anteriorly, each nasal is single-pointed, the points diverging from one another, leaving a midline, V-shaped gap—the form usually seen in *Cervus*. The antlers of *Panolia* are highly distinctive, the brow tine and the beam forming a continuous, almost unbroken arc. Traditionally only a single species, *Panolia* (formerly *Cervus* or *Rucervus*) *eldii*, has now been recognized. Balakrishnan et al. (2003) showed, however, that on mtDNA there is a deep split between a W clade (from Burma and Manipur) and an E clade (from the Indochinese region and Hainan). There are also considerable morphological differences between the two, as well as between the dryland-living thamin of Burma and the critically endangered sangai, confined to the floating reed beds (phumdi) of Logtak Lake in Manipur. Groves (2006a) suggested that Eld's deer be reexamined, with a view to ascertaining whether they ought to be reclassified into two or even three distinct species.

While there is a great deal of variability, the skulls of *P. thamin* have a slight tendency to be narrower than those of the other two taxa.

Table 18 shows that *P. siamensis*, on average, is considerably smaller than the other two species; the table also confirms the relative narrowness of *P. siamensis* and *P. thamin*.

Panolia eldii (McClelland, 1842)

sangai, Manipur Eld's deer

Large; short antlers; bare pasterns (swamp-living).

Although still exceedingly rare, *P. eldii* has undergone a dramatic improvement in its fortunes since a low point in 1975, when only 14 individuals were counted; in the 2003 survey, the number was 180 (Khaute, 2010).

Panolia thamin Thomas, 1918

thamin

Narrow-skulled, compared with other taxa; large in size; dark brown in color.

Burma and W Thailand.

Table 18 Univariate statistics for taxa of *Panolia* (adults)

	Gt l	Biorb
eldii		
Mean	329.00	131.33
N	3	3
Std dev	12.124	7.371
Min	316	123
Max	340	137
siamensis		
Mean	305.33	127.67
N	3	3
Std dev	17.786	4.041
Min	286	123
Max	321	130
thamin		
Mean	333.07	127.82
N	15	17
Std dev	12.372	7.376
Min	310	117
Max	353	140

Panolia siamensis **(Lydekker, 1915)**

eastern Eld's deer

Smaller than other taxa; lighter in color; spots visible along the median dorsal region.

E Thailand, Cambodia, S Laos, perhaps Vietnam, Hainan.

Elaphurus Milne-Edwards, 1866

The apparent affinities of the E Asian genus *Elaphurus* (Père David's deer) vary according to which system is being studied. As listed by Meijaard & Groves (2004b:table 7), its morphological features (except for its unique antler conformation) generally recall those of *Cervus*; protein electrophoresis and an nDNA sequence (k-casein) similarly align it with *Cervus*; yet mtDNA puts it in a clade with *Panolia*. At present, the best explanation of this is that *Elaphurus* originates from an ancient hybridization between stem representatives of *Cervus* (males) and *Panolia* (females) (Meijaard & Groves, 2004b; Pitra et al., 2004). *Elaphurus* is known as far back as the late Pliocene (*E. bifurcatus*) and, according to the molecular clock of Pitra et al. (2004), its separation from *Panolia* would date from about 3.5 Ma, although (as noted above) we would recalculate the separation as being earlier. Deer hybridize readily, at least within the same genus, and it may be that other species will also turn out to be of hybrid origin, but no case at present seems as plausible as that of Père David's deer. Y-chromosome DNA analysis in the Cervidae would be of enormous interest.

Elaphurus davidianus **Milne-Edwards, 1866**

Père David's deer

Cao (1978) analyzed Holocene sites where remains of this species have been found. His map shows a former distribution through what is effectively the coastal plain of N and C China, from the latitude of Beijing to Ningbo, somewhat S of the (present-day) mouth of the Yangtze. He concluded that Père David's deer were domesticated during the Zhou dynasty (1045–256 BC), but still existed in the wild in reduced abundance during the Han dynasty (206 BC–220 AD), after which they became extinct in the wild, surviving only in the Imperial Hunting Park.

Cervus Linnaeus, 1758

The genus *Cervus* is cranially most similar to *Panolia*, except that the posterior nasal bones form a distinct, but relatively short, point. The males possess small canines. Even shorn of some of its erstwhile components, the genus is large and unwieldy. As argued by Pitra et al. (2004), specializations of the display organs (antlers, mane, rump-patch, voice) are indicators of habitat and reproductive seasonality, rather than of phylogenetic affinity. The three well-separated clades—whose separation, according to Pitra et al. (2004), took place about 3.5 Ma (but see our remarks above)—are as follows:

1. Tropical clade—*Cervus timorensis* (rusa), and *C. unicolor* (sambar) and its relatives.
2. W temperate clade—*C. elaphus* and its relatives. Pitra et al. (2004) suggested that the geographically isolated C Asian *C. yarkandensis* (including *C. bactrianus*?) constitutes a distinct species; the spotted form *C. maral*, from Turkey and Iran, and the small, secondarily simplified N African deer (introduced to Corsica and Sardinia), are further candidates for species status.
3. E temperate clade—recently surveyed by Groves (2006a). This includes *C. albirostris* (white-lipped deer), and a subclade containing the *C. nippon* group (sika) and the large Sino-Rosso-American deer (wapiti and shou), which form a progressive cline running SW–NE from *C. wallichii* (Tibetan shou) via *C. macneilli* (Sichuan shou) and *C. xanthopygus* (izubra, from Primoriye, Manchuria, E Mongolia) to *C. canadensis* of the C Asian mountains and North America. One of the most unexpected, but most consistently corroborated, findings of molecular studies on deer has been that the wapiti and shou are not E subspecies of *C. elaphus*, as had always been assumed, but form an entirely separate clade, to which sika and white-lipped deer also belong.

The first wide-scale analysis of the genus using mitochondrial control region was by Randi et al. (2001). The maximum likelihood tree was as follows:

(*alfredi* ((*elaphus*—Europe) ((*timorensis*, *unicolor*) (*nippon*, *canadensis*))).

This appears to be the first recognition that taxa hitherto included within *Cervus elaphus* might not be monophyletic. For convenience, and because we suggest it may be more complex than simple nonmonophyly, we continue to refer to a catchall "*Cervus elaphus* group," which includes red deer, shou, and wapiti.

What follows is largely based on our independently written summaries (Grubb, 2004b; Groves, 2006a), plus some new research.

***Cervus elaphus* Group**

Perhaps the most easily identified feature that varies among members of this group is the rump-patch. Broadly speaking, there are three types:

1. European / W Asian type—true red deer. This includes Yarkand and Iranian deer. Red-brown / white rump-patch. Tail length in most populations 190–220 mm; hair length 5–8 mm; tail length without hair 14–16 mm; tail circumference with hair 13.5–16.5 mm. In Iran these figures are less; one specimen has measurements, respectively, of 150 mm, 4 mm, 11 mm, and 13 mm.
2. Tibetan group—commonly called shou. White rump-patch, with a 2–5 cm broad dark rim. Tail much shorter, rather thinner than in group 1. Tail measurements (for *C. macneilli* and *C. alashanicus*) are length 100–120 mm, hair 3—4 mm, length without hair 7–8 mm, circumference 12.5–13.5 mm.
3. E Asian / American group—wapiti. Mostly with a light brown / cream, two-colored rump-patch (Altai, Tianshan to America); but the izubra, *C. xanthopygus*, has a uniform golden-yellow rump-patch. Tail intermediate in length, but in hair length and circumference more like group 1. Tail measurements are length 130–160 mm, hair 4–11 mm, length without hair 3.5–11 mm, circumference 14–16.5 mm (including *C. xanthopygus*). Cap et al. (2008) continued to place this species in a clade with *C. elaphus*, which excluded *C. nippon*, although the bootstrap support was only 67%, much lower than for the other cervid clades in the authors' male-vocalization-based tree. Schonewald (1994) had previously argued strongly for the status of wapiti as a species distinct from European red deer, citing voice; skull size and shape (premaxilla length, interorbital width, palate depth, nasal breadth); the pattern of sexual dimorphism; and the apparent reproductive barrier between them in New Zealand, where they were introduced separately and have now met.

Antlers tend to be characteristic for each of these three groups, but they are more variable than the rump-patches. In almost all taxa, one can find individuals with one antler elaphoid and the other wapitoid. Geist (1998) also gave behavioral characters distinguishing the major species-groups.

Cranial material was studied by one or the other of us in Philadelphia, London, and the Oswald collection near Munich. Our studies were heavily weighted toward the more enigmatic taxa, especially those in E Asia (the shou group, *C. alashanicus*, *C. xanthopygus*); the largest sample of group 1 specimens was that in the Oswald collection from Spain. Our multivariate analysis found a broad division between those from E Europe and Iran, on the one hand, and those from W Europe plus, unexpectedly, C Asia, on the other. This, in particular, separates *Cervus elaphus* from *Cervus pannoniensis* (see below).

When the two largest samples, from Spain and Hungary, were simply contrasted in a two-way analysis, there was a slight overlap (1 of 21 specimens from Spain is misclassified with the 8 specimens from Hungary). This again separated *Cervus pannoniensis* from W European red deer, although it says nothing about whether the putative taxon *C. hispanicus* is distinct or not, as no other W samples were available.

Among group 2 specimens (shou), we find almost complete separation: the *C. hanglu*, *C. wallichii*, and so-called *kansuensis* (= *C. macneilli*; see below) samples are all discrete, mainly on the basis of overall size and nasal length.

Next, we examined the position of *C. alashanicus*: is it more like a shou or a wapiti? (Wapiti are represented in our sample by *C. xanthopygus*, our only reasonable wapiti sample). As different variables were available for different samples, first we used those that would maximize the representation of *C. kansuensis*. The three are largely discrete, mainly differing in skull breadth and, to a lesser extent, in nasal length.

Then we selected variables to maximize the representation of *C. alashanicus*. Again, all three groups are separate, on the basis of size, palate length, and palate breadth. We take this as evidence that *C. alashanicus* is neither a shou nor a wapiti, but different from either.

Finally, we compared different samples of wapiti (mainly Asian; we had only a single skull of an American wapiti) with each other. Our samples of *C. xanthopygus* and *C. asiaticus* overlap, whereas *C. songaricus* and *C. canadensis* are distinct, based on a contrast between the skull breadth and the nasal breadth.

Three phylogenetic studies based on mtDNA exist. In the first, Li M. et al. (1998) studied five Chinese taxa (*C. songaricus*, *C. macneilli*, *C. xanthopygus*, *C. kansuensis*, and *C. alashanicus*), using cytochrome *b*, and found considerable levels of divergence. Although bootstrap values were relatively low, they were not, apparently, significantly different from a five-way split, although they favored the view that *C. songaricus* diverged first from the other four; they put the times of divergence around one-half to three-quarters Ma.

Liu et al. (2003), studying a different mix of Chinese taxa, again using the cytochrome *b* gene, found the following:

> ((*sibiricus, songaricus*) (*xanthopygus* (*wallichii* (*macneilli, kansuensis*)))).

This was on a 5' partial sequence; but the whole sequence, available for fewer samples, reversed the order of the final clade, placing *C. kansuensis* as sister to *C. macneilli* plus *C. wallichii* (but the bootstrap values were not high in either case).

Ludt et al. (2004) sequenced cytochrome *b* in *Cervus elaphus* (as traditionally defined), finding that they divide into two extremely divergent clades, a W and an E. They are not each other's closest relatives, with the E clade being part of a major branch with the sika group and *Cervus albirostris*, exactly as Randi et al. (2001) had found earlier. Divisions within the two "*elaphus*" clades were as follows:

1. The W clade divides into a number of distinct subclades. The basic division is between the C Asian group (*Cervus bactrianus, C. yarkandensis*) and the European group. The European group in turn divides into
   —the "African group" (*C. barbarus, C. corsicanus*),
   —Middle East (*C. maral*, from Turkey and Iran),
   —Balkans group (Austria, Bulgaria, Hungary, Romania, and former Yugoslavia, i.e., *C. pannoniensis*), and
   —W Europe (Sweden, Norway, Spain, France, Britain, Germany, Poland, and Crimea, i.e., *C. elaphus*).
2. The E clade has three major divisions, namely
   —E Asia (*C. xanthopygus, C. alashanicus*),
   —S Asia (*C. wallichii, C. macneilli, C. kansuensis*), and
   —N Asia / America (*C. songaricus, C. sibiricus, C. canadensis*).

Bootstrap values are 95%–100% for all of these clades. The Balkan and Middle Eastern groups cluster together with 90% bootstrap confidence, and the non–N African group within the European subclade clusters together with 80% bootstrap support. Both Randi et al. (2001) and Ludt et al. (2004) concluded that (at least?) two species should be recognized, but many fewer subspecies.

We now present our taxonomic arrangement, in which craniometric and molecular results are in agreement.

Cervus elaphus Linnaeus, 1758

West European red deer

We have insufficient evidence to assess the validity of Swedish, Norwegian, British, Spanish, and European continental taxa within this species. Geist (1998) argued for the distinctness of *C. e. hispanicus* Hilzheimer, 1909, from S Spain, both in pelage and in a 90° inward direction of the fifth tine, associated with the curvature of the whole antler (resembling *C. e. angulatus*, a 250,000 BP fossil from Germany). For Pitra et al. (2004), however, the Spanish red deer is an integral member of the W European group.

The last remaining indigenous red deer of Italy, from Mesola Wood in the Po delta, were described by Mattioli et al. (2003). Their affinities to other living red deer are unclear, but their value for understanding red deer taxonomy and phylogeny is considerable.

Cervus pannoniensis Banwell, 1997

East European red deer

Botezat (1903) named two kinds of red deer in the Carpathians of Bukowena (on the Ukraine/Romania border): *C. campestris* (a large gray deer from the upland plains) and *C. montanus* (a smaller dark deer of the slopes and valley bottoms). Both names are unavailable; as pointed out by Grubb (2000b), they are both *nomina nuda*, as well as being unavailable because of the previously published *Cervus macrotis* var. *montanus* (Caton, 1881), a synonym of *Odocoileus hemionus*, and *Cervus campestris* F. Cuvier, 1817, a synonym of *Odocoileus cariacou*, respectively. Dobroruka (1960) was able to distinguish a darker, maned deer in the W Carpathians (which he called *C. hippelaphus*) and a grayer, maneless deer farther E (which he called *C. montanus*), but, unlike Botezat, he did not specify habitat differences. Could Botezat's and Dobroruka's two types be W and E European forms, meeting in the Carpathians?

Hartl et al. (1995) and Ludt et al. (2004) distinguished the deer of W Europe (including Poland) from those of Austria, Bulgaria, Hungary, Yugoslavia, Romania, and Turkish Thrace. For the latter, Banwell (1997) provided the new name *C. pannoniensis*, giving a detailed description, and Grubb (2004b) designated a skull with antlers in Château Chambord (from the Zala district, S Hungary) as the lectotype.

Cervus maral Gray, 1850
Turkish/Persian red deer

Type locality: N Persia.

Geist (1998), O'Gara (2002), and Oswald (2002) thought that the Balkan and Middle Eastern deer are the same, but, in fact, they differ genetically, and *C. maral* and *C. pannoniensis*, though related, are quite distinct in color, spotting, mane, and antlers (see Banwell, 1997, 1998, 2009). In the scheme of Pitra et al. (2004), *C. maral* separated from the European and N African clades around 1.5 Ma.

C. e. brauneri, described from Crimea, is an enigmatic taxon. It is a very small form, according to Heptner et al. (1961), and Ludt et al. (2004) found Crimean deer to have W European affinities. In the early to middle 20th century, N Caucasus deer were introduced into Crimea, and the present status of indigenous deer there is unclear.

Cervus corsicanus Erxleben, 1777
Corsico-Sardinian / North African red deer, Barbary stag

The Corsico-Sardinian deer is an island dwarf with reduced antlers, almost certainly derived from an Early Holocene human introduction. The N African *C. barbarus* Bennett, 1833, is very close. While not a dwarf, "*C. barbarus*" has short tines and no bez tine in 95% of the cases; molecularly, it is closely related to the Corsico-Sardinian deer (Pitra et al., 2004), and N Africa may thus be the source of the island populations, if the Barbary stag is not itself introduced from somewhere else (the Mesola deer [see above] should be carefully looked at in this regard). Pitra et al.'s (2004) molecular clock puts their joint separation from the European group at a little over 1 Ma. In one of the only major departures from the molecular tree, Cap et al. (2008) found that this species, according to the male vocalizations, is sister to a clade containing *C. elaphus*, *C. canadensis*, and *C. nippon*.

Cervus yarkandensis Blanford, 1892
Yarkand stag

A distinctive deer, with a straight antler beam and a simple, three-tined crown (see Dolan, 1988; Banwell, 2009).

Tarim Basin in Xinjiang.

Cervus bactrianus Lydekker, 1900
Bactrian stag

C. bactrianus is described as very similar to *C. yarkandensis*, with similar antlers and the same rump-patch form; they live in a similar semidesert habitat in saxaul and tamarisk forests, and in riparian poplar woods and reed beds.

The Syr Darya and Amu Darya basins.

Pitra et al. (2004) confirmed that *C. bactrianus* and *C. yarkandensis* are, in fact, distinct—becoming separated about 1 Ma—and placed the separation of their joint lineage from that of other "true red deer" at nearly 3 Ma.

In a female in the Cologne Zoo, a karyotype of $2n = 68$ was recorded by Schreiber (1994), who contrasted it with karyotypes of $2n = 66$ and 67 in what were reputed to be the same taxon in the Beijing Zoo; (given that they were in a Chinese zoo, these latter animals may have been *C. yarkandensis*).

Cervus hanglu Wagner, 1844
Kashmir stag

Dark liver-brown in winter, with the legs and the chest rather darker, and the face, the neck, and the back lighter; lighter in summer, sharply contrasting with the dark chest and the dark limbs. Like *C. wallichii*, the summer coat is simply a faded version of the winter coat, and their antlers are fairly similar; however, *C. hanglu* is smaller than *C. wallichii*. Only a short mane. Rump-patch narrow, hardly extending beyond the tail root, invaded dorsally by a wedge of body color sometimes extending onto the tail, and bordered laterally by a black band confined to the lower part. Chin, lips, and inside of the ears white; belly, groin, and inner surfaces of the hindlegs whitish. Metatarsal gland creamy to light red. Dark curly hair between the antler pedicels. Only a few pairs of antlers (9 out of 71) have 13 or more tines, 34/71 have 11 or 12 tines, and 28/71 have 10 or fewer. The rutting call, in the words of Geist (1998), begins like red deer, with a closed-lips roar, and then becomes a wapiti-like bugle as the lips are opened out.

Vale of Kashmir and neighboring regions.

While this species is dark, with a conservative rump-patch like *C. macneilli*, its mtDNA lineage is W (specifically, close to *C. yarkandensis*).

Cervus wallichii G. Cuvier, 1823
Tibetan shou

Yellowish brown in winter, dark gray-brown in summer, when the winter coat has faded (there is apparently no special summer coat). Rump-patch white, usually small, but extending well up onto the croup, often divided by a dark line, its border poorly marked (larger rump-patches appear to predominate at the E and W ends of the distribution, and smaller ones toward the middle). Mane poor. Tail very short. Antlers simple, usually 5-pointed, angulated at the middle of the beam; of 14 pairs of antlers, only 2 had more than 10 tines. Nasal region broad, presumably reflecting an enlarged nasal cavity.

S Tibet, from near Lake Mansarowar, (formerly) S into Sikkim, and E to the Tsari district.

Cervus macneilli **Lydekker, 1909**
Sichuan shou

$2n = 68$.

Stags in winter pale creamy gray, with reddish tones and a broad, reddish dorsal stripe; hinds redder. In summer, somewhat darker and redder. Rump-patch white, less extensive, divided by a black line sometimes extending onto the upper tail surface; characteristic black border (4–5 cm wide) to the rump-patch, expanded dorsally into a broad black patch extending up onto the croup. Tail somewhat longer than in *C. wallichii*. Antlers usually simple in structure, and somewhat angulated in the middle of the beam, but often less markedly so than in *C. wallichii*. Antlers with disproportionately small third tines, like the more wide-spreading antlers of *C. wallichii*, but of a six-point plan, unlike *wallichii*. Oswald (pers. comm. to PG) has found that 45% of the antlers have elaphoid crowns. Massive hoofs.

Sichuan and Gansu, at 3000–4500 m.

In Gansu, a separate taxon has been described, *Cervus kansuensis* Pocock, 1912. We have considered in some detail whether this is really a valid taxon, and concluded that it is probably not, for the reasons given below.

The type of *C. macneilli* is from Litang (30° 01' N, 100° 16' E). We have seen specimens ($N = 19$) from Yekundo (33° 01' N, 96° 44' E), Gur La Pass (Yangtze-Mekong watershed, 31° 50' N, 96° 40' E), Dzogchen Gomba (32° 07' N, 98° 54' E), and Xiao Sumang (33° 30' N, 97° E).

The type of *C. kansuensis* is from Taochow (34° 41' N, 103° 21' 52' E) in Gansu. We have seen specimens like it from the Peling Mountains; Longshou Shan, Gansu; Ikhe-Gol (36° 10' N, 97° 30' E); Qinghai Nan Shan (37° N, 100° E); and the N Tatung Range (37° 15' N, 100° E). There are differences in color, at least on average, from "true" *C. macneilli*, but not in the rump-patch or antlers, which are the same. Oswald (pers. comm. to PG) has found deer of this type from the E slopes of the Tibetan plateau (Qilian Shan) up to the mountain chain bordering directly onto the Alashan Desert; Oswald stated that 40% have an elaphoid crown, which is similar to "true" *C. macneilli*. *C. kansuensis* lives at somewhat lower altitudes (2500–3000 m).

Dolan (1988) kept *C. kansuensis* separate from *C. macneilli*; Geist (1998) argued that they are actually synonymous, and that what Dolan had illustrated as *C. macneilli* (from the Chengdu Zoo) were actually a group of *C. wallichii* with a dark median line on the tail, as they differ from *C. macneilli* in the much greater dorsal extension of the rump-patch and the absence of a dark dorsal rim. The Chengdu Zoo group were from E Nangjing Shan, on the Sichuan side, S of Batang and N of Chubalung (Dolan, pers. comm. to PG), and so considerably to the S of any *C. macneilli*; if Geist's argument is correct, which seems plausible, then this represents a considerable E extension of the range of *C. wallichii*.

According to Geist (1998), Dolan's "*kansuensis*" is the true *C. macneilli* (while G. M. Allen's description of "*kansuensis*" actually refers to *C. alashanicus*).

The molecular analysis of Pitra et al. (2004) placed *macneilli*/*kansuensis* and *wallichii* close together, suggesting that their lineage separated from that of their closest relatives, *C. xanthopygus* and *C. canadensis*, somewhat over 1 Ma.

Cervus alashanicus **Bobrinskii & Flerov, 1935**
Alashan stag

Oswald (pers. comm. to PG) identified the type locality of this taxon as Helanshan, Ninxia.

Our description of the species is largely from specimens in the Oswald collection (agreeing with the type description). Tail and large rump-patch brownish white or russet-orange, with a fairly wide, indistinct, brownish stripe approaching the root of the tail; almost no darkening along the edge in the type specimen, but mostly a distinct, dark brown margin (almost interrupted anteriorly on each side), continuous with a well-marked median stripe reaching to the tail base. Dark spinal stripe; mane whitish. Antlers with a five-tine plan.

There may be intermediate populations between this species and *C. macneilli*. Six specimens from Daqing Shan, Inner Mongolia, show the female and immature male pelage; they are identical with female *C. macneilli*, except that the rump-patch is orange and the amount of dark brown margin varies from none to broad and complete. Adult males from this locality have the dark margin incomplete medially, with a separate median dark stripe to the base of the tail. A female from River Long Chang, Shanxi, has the black edge almost complete, but thinner in the midline, with a prominent median stripe to the base of the tail.

Cervus xanthopygus Milne-Edwards, 1867

Manchurian wapiti

Stags creamy fawn in winter, with a dorsal stripe; in summer, light yellow-red, with a dark dorsal stripe; legs uniform dark brown; neck light brown, the hairs with reddish tips. Rump-patch reddish yellow, its shape and size resembling *C. canadensis*, but with a complete blackish brown, dark border and divided by a thin, reddish brown stripe. Hinds a more uniform gray-brown / light sandy; no mane; large, whitish yellow rump-patch, edged with dark brown only on the haunches, with no median line; legs only slightly darker than the trunk, with a narrow dark stripe down the front surface; neck brown, with yellow hair tips. Tail very short. Antlers complex, the beam somewhat curved, with seven points.

Yakutia, Amuria, and Manchuria.

Cervus canadensis Erxleben, 1777

Asian/American wapiti, elk (North America)

Gray-brown or buff in winter; dark brown in summer. Rump-patch creamy, very extensive, flaring laterally on the upper part of the rump, then contracting to be confined to the inner margins of the upper thighs; dark border confined to the lower part of the margin (on the thighs). Antlers complex in structure, the beam curved, P3 posterior to A3, not mesial to it (as in red deer); distal part of the antler flattened.

Whether the Asian and American populations can be distinguished has been much discussed. The names *Cervus sibiricus* Schreber, 1784, and *Cervus songaricus* Severtzov, 1873, are available for the Altai and Tianshan populations, respectively, should they be regarded as valid taxa. We describe the more poorly known Asian wapiti as follows:

1. Tianshan
   - —Stags with head and neck brownish gray, with longer hair than the rest of the body, each hair with black and pale gray-brown rings; shoulders, back, side, and thighs brownish gray, shaded with yellow, considerably lighter than the neck; yellow rump-patch separated from the body color by a dark gray stripe, sharply marked toward the tail, and gradually shading off into the gray of the back; tail light yellow generally, with a grayish brown line down the middle; breast, belly, and legs clear dark brown.
   - —Hinds dark brown; rump-patch smaller, light brown, no stripe; neck hair hardly longer than elsewhere.
   - —Specimens from Trans-Iliisk Alatau, near the E end of Issik-Kur (syntypes of *C. songaricus*); Alexandrovsk Mountains; lower Kok-an Valley, E Tianshan; Tekes Valley; Lake Balikol (46° N, 91° E), E of Urumshi.
2. Altai
   - —Two types of male coloration occur, one like the American wapiti, the other grayer, more reddish on the head, the neck, the belly, and the feet, the head and the neck an intense brown tone, with some reddish color; dark brown spine stripe only in the shoulder region; rump-patch large and, like the tail, bright reddish cream, with a dark brown rim above, an almost black stripe on the sides, and a black median stripe reaching partway to the tail (this seems variable; some have none).
   - —Females in winter more monochromatic, with a distinctly rimmed, white or reddish rump-patch and a slight darkening along the spine.
   - —Specimens from Altai; W Mongolia; deer farms; 95 km NE of Ulan Bator; Barguzin Ridge, Transbaikalia.

Our conclusion must be that—though they differ on average from each other and (especially those in the Tianshan) from North American wapiti, so possibly a separate subspecific status for each of the three could be supported—they all belong to *C. canadensis*.

Schonewald (1994) examined differences among North American populations, finding that they were definable, but that this needs more study.

Cervus albirostris Przewalski, 1883

white-lipped deer

For the general biology and distribution of this species, see Leslie (2009), who assigned it to a separate genus, *Przewalskium* Flerov, 1930.

This deer is not related to *Rusa*, as Flerov (1952) thought, but is part of the *canadensis-nippon* clade, from which it separated somewhat over 2.5 Ma, according to Pitra et al. (2004).

Cervus nippon Group

sika

On this group, see Banwell's (1999) detailed treatment.

There has been considerable interest in the nature and frequency of hybridization of *C. nippon* with native

C. elaphus in places to which the former has been introduced in Europe (including the British Isles). Perhaps the most detailed study is that of Goodman et al. (1999), who concluded that they are not, in fact, panmictic, the degree of introgression being apparently less than would be expected from current hybridization rates; there is significant linkage disequilibrium within both populations.

In their study of hybridization between red deer and sika in England, Lowe & Gardiner (1975) studied a number of different groups of sika. Specimens from the mainland (Manchuria and farther S in China) always separated very clearly from those from Japan (including Kerama). In their figure 4, a specimen from Taiwan assorted more closely with those from Manchuria.

Imaizumi (1970), in the course of describing the new species *Cervus pulchellus*, proposed a novel method of differentiating species in this group: he graphed skull measurements against "temperature index" (an index derived from the mean temperatures of the localities of taxa, modified by altitude). The rationale was that a given species would respond to temperature consistently, whereas different species would respond differently. Hence the deer from most of the Japanese islands qualified as one species; those from the mainland and Hokkaido as a second; those from Taiwan as a third; and the new species from Tsushima, *C. pulchellus*, as a fourth.

Mitochondrial D-loop sequences from sika throughout the Japanese islands were compared by Nagata et al. (1999). Both the neighbor-joining and the maximum parsimony methods divided them very strongly (in neighbor-joining, at >90% bootstrap) into two groups: N and S. The boundary was around the "neck" of land in S Honshu constricted by Biwa Ko; samples from Wadayama, Hyogo Prefecture, belong to the N group, and those from Yamaguchi to the S group. Several things are interesting about the composition of these two clades. First of all, samples from Hokkaido were nested well within the N clade, not sister to N Honshu. Within the S clade, samples from Tsushima assorted with those from Yamaguchi; those from the Goto Islands, with Kyushu; while those from Yakushima and Kerama formed distinct subclades, not, however, very strongly distinguished from the rest.

Randi et al. (2001) found that, in fact, Japanese and mainland plus Taiwanese sika did not form a clade when using partial control region sequences; the Japanese clade was sister to *C. canadensis*, whereas the non-Japanese clade was sister to the combination of Japan plus *C. canadensis*.

Wu H. et al. (2004) compared the mitochondrial control region in sika from four regions in China: Zhejiang, Jiangxi, Sichuan, and "the northeast of China," this last sample being taken from "domestic" (i.e., deer farm) populations. They fell into two very distinct clusters: one consisting solely of the Zhejiang sample ($N=12$), the other containing all the rest. The second cluster was again divided into two subclusters, one consisting of 7 out of the 11 specimens from Jiangxi plus 3 of the "NE" specimens, the other consisting of the remaining 4 from Jiangxi, all of the 7 from Sichuan, and the remaining 13 from the "NE." This could be taken to mean that the "domestic" populations were derived from several sources, including Sichuan, Jiangxi, and the indigenous "N" population, or that all of these populations had spread rather recently and had not achieved reciprocal monophyly.

Pitra et al. (2004) dated the separation of sika from the *C. canadensis* group at about 2.5 Ma; within the sika, they found mainland sika (as represented by *C. sichuanicus*) separating from the island forms around 2 Ma; and the *yesoensis*/"*centralis*" clade (equivalent to the N group of Nagata et al., 1999) was calculated to separate from the S clade—containing *C. nippon* from the S part of the main Japanese islands and the small island forms (*C. mageshimae, C. keramae, C. pulchellus*)—about 1.5 Ma.

Our craniometric studies rely heavily on the Heude collection in the Academia Sinica, Beijing, of which the largest sample is from the Goto Archipelago. They give the following results: Honshu, Goto, Yakushima, Mageshima, and Tsushima samples differ largely by size, especially in nasal length and mastoid breadth, and, to a lesser degree, in biorbital breadth and palate length (braincase length hardly differs among the samples). The order, from high to low, is N Honshu; Goto / S Honshu; Tsushima; and Yakushima/Mageshima. Another sample ordering, using nasal length and palate length, separates Tsushima, Yakushima, and Mageshima from Honshu and (especially) Goto. In the main, the small S Honshu sample is close to Goto.

When we turn to a comparison of the mainland (mainly E China) versus Japan, the long facial skeleton of the E China sample is distinct from all Japanese samples, but this is not strongly correlated with the overall size difference that separates Hokkaido / N Honshu from Goto. The skull of a hybrid between a Goto and a Hokkaido deer (bred by Heude) is intermediate, though closer to Goto, confirming that the differences between them do have a genetic basis. Yakushima and Mageshima skulls are closer to those from Honshu, whereas Tsushima is at the edge of the E China range of variation, and seems to be basically a mainland taxon. Sichuan and Vietnam skulls are both closest to E China.

When we look at the skulls of the females, we find the following results: Of the two skulls from Kerama, one is closest to Goto, while the measurements of the other actually fall within the Goto range. Two from Taiwan are intermediate between Goto and E China. Of the two Sichuan skulls, one falls within the E China range (just), while the other is intermediate between E China and the Far East sample. In absolute size, the sequence is Far East, E China, Goto. The Taiwan samples have a particularly high biorbital breadth compared with the skull length.

Summarizing all this by sample (table 19 gives the univariate statistics for male skulls, and table 20 for female skulls):

—Hokkaido does not differ from N Honshu, but S Honshu is smaller (absolutely so in greatest length and nasal length). We have here the basis of a division between the N and S parts of the main Japanese islands, with the division cutting across Honshu, exactly where the DNA data divide.

—Goto hardly differs from S Honshu (in variables where the sample size for S Honshu is reasonable), and Kerama, Yakushima, and Mageshima generally fall within the range of Goto. On this showing, they can all be placed in the S Japan taxon.

—Tsushima is larger in size than Goto, more like a small N Honshu, but it has absolutely short nasals and a short palate, and narrow mastoids. It is clearly a distinct taxon, though more of mainland than of Japan affinity.

—E China skulls are also much bigger than ones from Goto, their size being the same as N Honshu / Hokkaido, but long and relatively narrow. Skulls from the Far East, compared with E China, are very large, wide across the condyles, with a broad palate, and, in the females, with narrow orbits and small teeth; Beijing, Shansi, and Sichuan are also very big. If these and the Far East specimens are taken as one sample, they are absolutely wider in interorbital breadth, though (in the females, at least) Sichuan is somewhat smaller, with narrower mastoids.

—Vietnam skulls are slightly smaller than E China ones, and have narrow nasals, narrow orbits, and a short palate (none, however, being absolutely different)

—Solo (the enigmatic *Sikelaphus soloensis* Heude, from Jolo in the Philippines!) has the longest (but narrow) nasals, extremely large teeth, and a very wide (yet not absolutely the widest) palate.

The suggested species that emerge from these analyses are as follows:

1. Hokkaido + N Honshu.
2. S Honshu = Goto = Yakushima = Mageshima = Kerama—this taxon is absolutely smaller than taxon 1 in greatest length and nasal length.
3. Tsushima—slightly larger than S Honshu, with absolutely short nasals and a short palate, and narrow across the mastoids.
4. E China—as large as N Honshu, but with narrow orbits, a wide braincase, and smaller teeth. Vietnam is at the lower end of the range of E China.
5. Far East = Beijing = Shanxi—very large compared with E China, with a broad palate, wide across the condyles and with a wide interorbital breadth.
6. Sichuan—smaller than the Far East, with short nasals; larger on average than E China, but with narrow mastoids.
7. Solo—in the size range of N Honshu, with long, narrow nasals, the largest teeth of all, and a relatively wide palate.

Molecular data, as has been discussed above, make exactly the same divisions in cases where samples have been available.

Cervus nippon Temminck, 1836

For the full synonymy of this species, see Groves & Smeenk (1978).

In all Japanese sika, the antler velvet is black. Body color tends to be dark, and there is always a pale facial chevron. The nighttime vocalization of the stag in rut is described as being like a donkey's bray (Banwell, 1999).

Size small, distinguishing it from more N Japanese deer; very large neck mane; yellowish white hair around the metatarsal glands; lighter-colored face; darker winter coat; constant presence of a row of spots on either side of the median dorsal line.

S part of Japan: Honshu S from Biwa Ko, and smaller islands to the S.

We provisionally recognize a few subspecies, listed below.

Table 19 Univariate statistics for male skulls of the *Cervus nippon* group

	Gt l	Nas l	Nas br	Biorb	Interorb	Brain-case	Pal l	Mastoid	Teeth	Pal br	Span condyles
Hokkaido											
Mean	293.75	102.50	37.00	130.25	93.00	79.75	167.50	104.25	90.25	86.75	58.25
N	4	4	4	4	4	4	4	4	4	4	4
Std dev	5.560	5.686	5.598	4.113	3.266	4.113	6.403	3.775	4.349	3.096	1.258
Min	290	94	32	125	89	75	163	99	84	84	57
Max	302	106	45	135	97	85	177	108	94	91	60
N Honshu											
Mean	292.67	102.17	36.17	133.00	90.00	78.80	166.50	105.40	89.00	87.40	58.00
N	6	6	6	6	5	5	4	5	5	5	4
Std dev	12.011	5.913	3.430	6.325	7.106	5.848	5.916	6.986	1.732	2.608	2.944
Min	277	94	32	122	79	75	160	94	86	84	55
Max	305	109	42	140	98	89	172	113	90	90	61
S Honshu											
Mean	246.80	78.80	30.50	111.50	84.00	79.00	157.00	96.00	79.00	84.00	59.00
N	5	5	2	2	1	1	1	1	1	1	1
Std dev	14.550	5.891	6.364	16.263	—	—	—	—	—	—	—
Min	236	72	26	100	—	—	—	—	—	—	—
Max	271	86	35	123	—	—	—	—	—	—	—
Goto Islands											
Mean	245.00	75.10	28.70	109.60	69.80	70.80	136.89	85.10	72.60	73.00	48.70
N	9	10	10	10	10	10	9	10	10	10	10
Std dev	9.124	3.695	2.003	2.836	3.425	2.150	5.947	3.071	4.222	2.357	1.829
Min	233	71	26	105	65	68	130	80	67	70	46
Max	261	81	31	113	77	75	145	90	80	78	51
Yakushima											
Mean	231.00	72.00	23.00	105.00	72.00	71.00	134.00	70.00	—	—	46.00
N	2	—	—	1	—	—	—	—	—	—	—
Min	219	—	—	—	—	—	—	—	—	—	—
Max	242	—	—	—	—	—	—	—	—	—	—
Mageshima											
Mean	252	77	28	—	80	74	137	77	—	—	50
N	—	—	—	—	—	—	—	—	—	—	—
Tsushima											
Mean	271.00	83.00	29.00	—	76.00	73.25	148.50	80.25	—	—	56.00
N	1	4	1	—	4	4	4	4	—	—	1
Std dev	—	2.944	—	—	2.708	1.258	3.873	1.708	—	—	—
Min	—	79	—	—	74	72	143	78	—	—	—
Max	—	86	—	—	80	75	152	82	—	—	—
Far East											
Mean	—	—	—	140.00	92.00	84.00	—	111.00	88.00	91.00	60.00
N	—	—	—	1	1	1	—	1	1	1	1
Beijing											
Mean	304.00	105.00	—	—	120.00	—	—	99.00	—	—	—
N	1	1	—	—	1	—	—	1	—	—	—
Shansi											
Mean	—	125.00	—	—	100.00	—	—	84.00	99.00	—	—
N	—	1	—	—	1	—	—	1	1	—	—
Sichuan											
Mean	325.00	114.00	—	—	99.00	—	—	85.00	—	—	—
N	1	1	—	—	1	—	—	1	—	—	—
E China											
Mean	290.93	97.50	33.69	124.00	81.69	79.88	166.71	101.38	85.31	84.94	56.47
N	15	16	16	16	16	16	14	16	16	16	15
Std dev	18.553	10.838	3.554	10.152	8.867	4.365	10.306	10.639	3.842	5.144	3.182
Min	261	76	27	103	63	74	149	81	79	76	51
Max	323	115	39	138	91	90	186	115	92	95	62

Table 19 (continued)

	Gt l	Nas l	Nas br	Biorb	Interorb	Brain-case	Pal l	Mastoid	Teeth	Pal br	Span condyles
Vietnam											
Mean	272.00	77.50	25.00	109.50	68.00	74.50	142.50	85.67	85.00	78.00	49.50
N	2	2	2	2	2	2	2	3	2	2	2
Min	255	65	21	99	59	69	117	74	79	72	43
Max	289	90	29	120	77	80	168	101	91	84	56
Solo											
Mean	—	111.00	33.00	123.00	87.00	79.00	—	—	97.00	89.00	—
N	—	1	1	1	1	1	—	—	1	1	—

Table 20 Univariate statistics for female skulls of the *Cervus nippon* group

	Gt l	Nas l	Nas br	Biorb	Interorb	Brain-case	Pal l	Mastoid	Teeth	Pal br	Span condyles
Goto Islands											
Mean	226.75	69.13	24.25	94.38	55.88	66.87	127.57	70.75	68.75	67.88	44.63
N	8	8	8	8	8	8	7	8	8	8	8
Std dev	5.392	3.271	2.121	2.387	1.727	1.553	5.855	2.188	3.615	2.416	1.188
Min	220	64	22	91	53	64	121	68	64	64	43
Max	237	74	27	97	58	69	136	74	76	71	46
Kerama											
Mean	221.50	67.00	23.50	92.00	—	—	—	—	—	—	—
N	2	2	2	2	—	—	—	—	—	—	—
Min	210	63	22	88	—	—	—	—	—	—	—
Max	233	71	25	96	—	—	—	—	—	—	—
Tsushima											
Mean	243.50	76.50	31.00	105.00	66.50	72.50	138.00	80.00	73.00	80.50	43.50
N	2	2	2	2	2	2	2	2	2	2	2
Min	243	71	27	104	65	72	136	80	70	79	42
Max	244	82	35	106	68	73	140	80	76	82	45
Far East											
Mean	303.67	109.00	35.33	122.00	80.67	79.67	176.67	98.00	86.33	87.67	57.00
N	3	3	3	3	3	3	3	3	3	3	3
Std dev	9.074	6.083	2.517	2.646	4.041	3.055	7.024	1.000	3.512	.577	1.000
Min	294	105	33	120	77	77	170	97	83	87	56
Max	312	116	38	125	85	83	184	99	90	88	58
Sichuan											
Mean	285.00	93.50	—	—	74.00	—	—	88.00	—	—	—
N	2	2	—	—	2	—	—	2	—	—	—
Min	279	89	—	—	72	—	—	87	—	—	—
Max	291	98	—	—	76	—	—	89	—	—	—
E China											
Mean	273.83	91.00	29.50	107.50	70.67	74.83	161.50	87.00	83.83	78.33	53.50
N	6	6	6	6	6	6	6	6	6	6	6
Std dev	7.278	5.215	2.074	2.588	3.559	2.639	6.745	4.050	5.419	1.633	2.429
Min	267	84	27	103	67	72	153	82	77	76	50
Max	283	100	32	111	76	79	170	93	90	80	57

Cervus nippon nippon Temminck, 1838

1836 *Cervus nippon* Temminck. "Isles of Japan"; probably the surrounds of Nagasaki (Groves & Smeenk, 1978).
1884 *Sika orthopus* Heude. Kobe.
1884 *Sika schlegeli* (and other names) Heude. Goto Islands.
1888 *Sika paschalis* (and other names) Heude. Goto Islands.
1898 *Sikaillus daimius* (and other names) Heude. Goto Islands.

S Honshu, Shikoku, and Kyushu; Goto Islands.

Cervus nippon mageshimae Kuroda & Okada, 1951

1951 *Cervus nippon mageshimae* Kuroda & Okada. Mageshima.

Larger than the deer on Yaku.

Mageshima & Tanegashima.

Cervus nippon yakushimae Kuroda & Okada, 1951

1951 *Cervus nippon yakushimae* Kuroda & Okada. Yakushima.

On the differences between these small-island sika, see Banwell (1999).

Cervus nippon keramae Kuroda, 1924

1924 *Cervus nippon keramae* Kuroda. Kerama Island.

Smaller in size; dark, with no spots. As its describer himself admitted, it may be that the Kerama population derives from a medieval introduction.

Cervus aplodontus (Heude, 1884)

Males with a small neck mane, or none; underside dark, like the back; often an indistinct row of light spots along either side of the dorsal stripe. Light eye-rings; facial chevron present in summer, absent in winter. Rump-patch large, rimmed with black; tail very short, dark. Metatarsal glands dark, elongated, surrounded by light gray fur. Winter pelage dark gray; summer pelage reddish; males spotted in summer, the females without spots.

N Japan, N from Biwa Ko.

Provisionally, we recognize the N Honshu and Hokkaido deer as separate subspecies, but we are not especially convinced of this distinction.

Cervus aplodontus aplodontus (Heude, 1884)

1884 *Sika aplodontus* Heude. N of Tokyo.
1884 *Sika mitratus* Heude. Tokyo.
1884 *Sika xendaiensis* Heude. Sendai.
1897 *Sika ellipticus* Heude. Sendai.
1897 *Sika minoensis* Heude. Mino, W of Tokyo.
For the full synonymy of this subspecies, see Groves & Smeenk (1978).

Cervus aplodontus yesoensis (Heude, 1884)

1884 *Sika yesoensis* (and other names) Heude. Hokkaido.
1897 *Sika dolichorhinus* Heude. Hokkaido.
1897 *Sika rutilus* Heude. Hakodate.
For the full synonymy of this subspecies, see Groves & Smeenk (1978).

Cervus pulchellus Imaizumi, 1970

1970 *Cervus pulchellus* Imaizumi. Tsushima Island.

Distinguished by skull characters; by the long basal segment to the antlers; and by the whitish belly.

Cervus hortulorum Swinhoe, 1864

1864 *Cervus hortulorum* Swinhoe. Imperial Summer Palace, Beijing.
1864 *Cervus mantchuricus* Swinhoe. Newchwang (= Jingkou), Manchuria.
1874 *Cervus euopis* Sclater. Newchwang (= Jingkou), Manchuria.
1876 *Cervus dybowskii* Taczanowski. S Ussuri district, Manchuria.
1882 *Cervus cyclorhinus* and *Cervus hyemalis* Heude. Shandong Province.
1884 *Sika grassianus* Heude. Tsinglo-hsien (= Echeng), N Shanxi
1884 *Sika microspilus* Heude. Mukden (= Shenyang), Manchuria.
1894 *Sika imperialis* Heude. Songari Valley, Manchuria.

Specimens of what were identified as *Cervus nippon hortulorum*, from Altai deer farms, had $2n = 66$ (Graphodatsky & Radjabli, 1985).

Tail short, 8% of the head plus body length. Neck mane reduced or absent. Winter coat light red-brown to gray; rump-patch large.

N China and the Russian Far East.

As Geist (1998) has emphasized, the problem with the nomenclature of this species is that almost none of the type specimens were obtained in the wild. The type of *C. hortulorum*, for example, was from the grounds of the Imperial Summer Palace; those of *C. mantchuricus* and *C. euopis* were obtained in the harbor city of Jingkou; and so on. The only present-day wild populations apparently supported by names are *C. dybowskii* and *C. imperialis*, both from the Ussuri region; the similarity of the type of *C. hortulorum* to the Ussuri specimens does, however, suggest that it represents the same population. Possibly *C. cyclorhinus* and *C. hyemalis*, whose types could not be found by Braun et al. (2001), were also obtained in the wild.

Cervus sichuanicus Guo et al., 1978

1978 *Cervus nippon sichuanicus* Guo et al. Ruoergai district, Sichuan, 2900 m.

Described as a large species, with a short tail; no mane; median dorsal stripe distinct, dark, bordered on each side by a row of distinct white spots.

Sichuan.

Cervus taiouanus Blyth, 1860

Synonyms include several names by Heude, 1882 and 1884.

Clearly spotted in winter in both sexes; in summer, bright chestnut, with a prominent dorsal stripe; tail light fawn below, not white as in other species; inside of the ears fawn; males strongly maned. Antlers rather weak, but with 10 points; color of the velvet pinkish, not black as in other sika. Apparently long-bodied, compared with other sika.

Taiwan.

Cervus pseudaxis Gervais, 1841

1841 *Cervus pseudaxis* Gervais in Eydoux & Souleyet. No locality; thought to be Vietnam.

1871 *Cervus mandarinus* Milne-Edwards. Summer Palace, Beijing (see below).

1873 *Cervus kopschi* Swinhoe. Kien-chang (= Aicheng), Jiangxi.

1882 *Cervus frinianus* (and other names) Heude. Lake Poyang.

1884 *Sika brachyrhinus* (and other names) Heude. Lake Poyang.

1888 *Sika rivierianus* Heude. Lake Poyang.

1894 *Sika dugenneanus* Heude. Mountains of Phu-Lang-Thuong, Tonkin, Vietnam.

Externally, somewhat resembling *C. taiouanus*. Summer coat bright, winter coat gray to black, spotted in both seasons; dorsal stripe dark; underside light brown. Tail relatively long, 13% of the head plus body length, with a broad, dark dorsal portion; rump-patch small, white; mane thick, light-colored; facial chevron absent or vague. Metatarsal glands small. Inside of the ears white. Males darker than the females.

Vietnam and S China.

As noted by Geist (1998), the type of *C. mandarinus* resembles these S populations, despite its having been obtained from the Summer Palace.

Addendum to the *Cervus elaphus* and *Cervus nippon* Groups

A problem remains: what are the interrelationships of these two species-groups? Morphologically, red deer, shou, and wapiti share several evidently derived character states, compared with other deer (including sika): the more complex antlers, including (except in some individuals of the dwarfed *C. corsicanus*) the presence of a bez tine, and the brow tine being at a right angle or more to the beam; the much larger rump-patch; the shorter tail; the smaller rhinarium; and the uniform body color of the adults (with one exception—*C. maral* retains not very conspicuous spotting into adulthood). By contrast, sika have primitive states: three-point-plan antlers, without a bez tine and with the brow tine being at an acute angle to the beam; the rump-patch restricted to medial aspect of the buttocks; the tail long; the rhinarium larger; and the adults conspicuously white-spotted. Yet the mtDNA of *C. wallichii*, *C. macneilli*, *C. alashanicus*, *C. xanthopygus*, and *C. canadensis* is close to that of sika, forming merely a somewhat separate subclade. It is particularly noticeable that, morphologically, *C. hanglu* closely resembles *C. wallichii* and *C. macneilli*, to the extent that Flerov (1952) even synonymized them, yet the mtDNA of *C. hanglu* is that of the W (red deer) group. What is the cause of this anomaly?

Our explanation is that hybridization has taken place. We propose that shou related to *C. hanglu* spread E and then NE through the mountain forests of the Himalayas and the E edges of the Tibetan plateau, meeting indigenous sika populations and interbreeding with them asymmetrically—shou stags with sika hinds, and then with backcross hinds, and so on—until the local sika populations had been eliminated by nuclear swamping, but had left their mtDNA in the transformed population. Only a single sika taxon, *C. sichuanicus*, remains in these mountain forests. The shou populations continued to spread (and to acquire successively more apomorphic character states, becoming "wapiti") along the E edge of the plateau, bearing the sika mtDNA, and then spread through similar country farther N and NW, and to similar forest types to the NE, ultimately crossing the Bering Land Bridge into North America.

Cervus unicolor Group

rusa and sambar

On the general biology of this species-group, see Banwell (2006).

Antlers with only three points; rhinarium extending well below the nostrils; tail long and bushy; pelage coarse and shaggy.

According to the molecular clock of Pitra et al. (2004), this group, the *albirostris-canadensis-nippon* group, and the *C. elaphus* group form a three-way split, dating from about 3.5 Ma. The authors dated the separation between *C. unicolor* and *C. timorensis* at about

1 Ma. These dates, as noted earlier, should be pushed back somewhat, because the muntjac-cervine split is now known, from fossil evidence, to be earlier than was formerly thought.

Cervus alfredi Sclater, 1870
Prince Alfred's deer

1870 *Cervus alfredi* Sclater. "Manila."
For the full synonymy of this species, see Grubb & Groves (1983).

Color blackish brown, light-spotted on the flanks; dorsal band dark, bordered with faint ochre spots. Underside, extending to the inner aspects of the limbs and the chin, a sharply contrasting buffy. Lower segments of the legs paler than the body. Head somewhat paler than the body. Pelage thick and soft.

Leyte, Cebu, Guimaras, Negros, and neighboring islands in the C Philippines.

According to Randi et al. (2001), this species is sister to the rest of the genus *Cervus*; morphologically, as noted by Grubb & Groves (1983), it sits very uneasily amid the rusa/sambar group.

Cervus unicolor Kerr, 1792
Indian sambar

$2n = 60$.

Long-legged; color brown, varying from grayish to very dark brown in winter, more yellowish red-brown in summer; belly darker than the back; tail long-haired, very bushy, contrastingly darker than the body. Forehead flat; preorbital fossa very deep. Antlers robust, relatively short, with a short brow tine; the two terminal tines tending to form an equal fork, with the posteromedial tine usually being longer. (Following van Bemmel, 1949). Fawns spotted.

Sri Lanka and most of mainland of S Asia.

Sri Lankan specimens average slightly smaller than the Indian ones.

Cervus equinus (G. Cuvier, 1823)
Southeast Asian sambar

Synonymy includes numerous names by Heude!

$2n = 62$ (Indochinese sample).

Smaller than *C. unicolor*; brow tine longer; anterolateral tine always continuing the line of the beam, and longer than the posteromedial tine, the latter turning inward and backward. Tail somewhat more bushy than in *C. unicolor*. Dark brown in winter; lighter, more yellowish or reddish, in summer. Fawns spotted at first, the spots disappearing rapidly with age.

Bengal and Assam, NE into S China (including Taiwan) and SE to the Malay Peninsula, Sumatra, and Borneo.

We have relatively little information on geographic variation in this species, although it is clear that the mainland specimens are of relatively larger size, are somewhat more sexually dimorphic, and have antlers that are often much longer than those of Sumatra, Borneo, and other islands; but we cannot find any absolute differences. Groves (2006a) drew attention to the apparent differences of the Chinese animals from S Asian and Southeast Asian sambar.

Cervus mariannus Desmarest, 1822
Philippine deer

1822 *Cervus mariannus* Desmarest. Guam.
And numerous synonyms by Heude!
For the full synonymy of this species, see Grubb & Groves (1983).

$2n = 65$.

Deep brown, with a blackish tinge on the back, paler on the cheeks, slightly paler on the underside. Tail white beneath. Muzzle dark; rest of the head as in the body. Legs paler than the body, especially on the lower portions. Spots present in the fawns, but not in the adults.

Luzon, and introduced to Guam; Mindanao; Basilan.

There is a possibility of a second species in the lowlands of Mindanao, around Sarangati Bay, as described by Grubb & Groves (1983), and this needs further investigation.

Cervus barandanus (Heude, 1888)
Mindoro deer

1888 *Ussa barandanus* Heude. Mindoro.

C. barandanus, for the first time, is regarded here as a distinct species, because the width of the braincase, the mastoids, and the nasals, and the length of the toothrow, are consistently and distinctively less than in *C. mariannus*, and the palate is wider.

Cervus nigellus (Hollister, 1913)
Mindanao mountain deer

1913 *Rusa nigellus* Hollister. Mt. Malindang, Mindanao.
? 1952 *Rusa nigellus apoensis* Sanborn. Mt. Apo, Mindanao.

Size very small; pelage soft; higher frequency of neck hair reversal than in lowland deer.

The question of the separation of the mountain deer from the lowland ones—and, indeed, of those on the two mountains from each other—was discussed by Grubb & Groves (1983), who concluded that the mountain form is "very unlikely to be a distinct species"; but here we are inclined to reverse that decision, despite the fact that deer from the lower slopes are larger than those from the upper slopes. Nonetheless, we recommend further study.

Cervus timorensis

rusa

$2n = 60$.

Smaller than *C. unicolor* and *C. equinus*, but larger than the Philippine forms. Relatively short legged and long bodied. Tail less bushy than the other species, relatively long and narrow, not contrasting in color with the body. Forehead concave; preorbital fossa less deep. Antlers relatively long and slender; posterior tine continuing the direction of the beam, and generally longer. (Following van Bemmel, 1949.)

Java and Bali; widely introduced by human agency into the E Indonesian islands as far as Timor and the Moluccas.

There appear to have been changes in pelage and size over the several centuries (up to 3 millennia) since their introductions, witnessed by the description of several subspecies—conspicuous enough to have been recognized by van Bemmel (1949)—from the deep-water islands of E Indonesia.

II

Bovidae

The Bovidae: the largest, most diverse family of ungulates. Particular themes include the role of sexual selection; African biogeography; the recency of speciation in some cases; gene flow between related but distinct species; why many species simply must be split up. The question of a hybrid origin for the genus *Capra* (as proposed by Alexandre Hassanin); are there any other taxa of hybrid origin lurking within the Bovidae?

BOVIDAE GRAY 1821

We here use the nomenclature established by Grubb (2001a).

Usually there is one single, circular lacrimal orifice, inside the orbital rim; there is, however, some variation to this pattern in the Bovinae (Leinders & Heintz, 1980).

The anterior sagittal gulley of the metatarsus remains open over its entire length.

The sagittal crests on the articular surfaces of the distal condyles of the metatarsus are more protruding, more sharp edged, and more proximately extended than in the Cervidae; there is no "plateau postarticulaire" on the medial phalanx; and the proximal articular surface of the distal phalanx is more concave and posteriorly extended, with an "éminence pyramidale," that is, an extensor process on its anterior edge (Leinders, 1979). As that author notes, these features give the articular surfaces a wider range of movement (especially of hyperextension) than in the Cervidae and greater shock-absorbing capability, allowing for what he dubs "greater pogostick effect." According to Leinders (1979), there is some reversal of these characters in the swamp-loving sitatunga (*Tragelaphus spekii*).

A gall bladder is present.

Janis & Scott have fully discussed the relationships of the Bovidae among the ruminants. While undoubtedly the family forms a monophyletic clade, "it appears that no apomorphic characters, apart from the presence of horn cores, can be used to characterize the Bovidae" (Janis & Scott, 1987:47).

The overriding theme in bovid evolution is surely sexual dimorphism.

A widely accepted model of social organization (Jarman, 1974) relates feeding style to group size via the dispersion of food items, habitat choice, and antipredator behavior, which, taken together, help to determine the females' tendency to aggregate; and, therefore, the males' best strategy to secure mating. This, in turn, relates back to sexual dimorphism.

Lundrigan (1996) associated horn form with male strategy, specifically, fighting behavior. Short, simple horns would be associated with stabbing in fighting (e.g., nilgai, *Oreamnos*, most dwarf antelope); long horns, with wrestling (the larger Tragelaphini) and fencing (the Hippotragini); a long "catching arch" (the anteriorly concave midsection of a sigmoid or twisted horn) is especially associated with wrestling (the Tragelaphini, *Nanger*, the Reduncini, impala); and thick horn bases, with ramming during fighting (Cape buffalo, thick-horned duiker, *Ovis*).

Returning again to the sexual dimorphism theme much later, Jarman (2000) plotted sexual dimorphism (in body weight) against the log of the female's body weight, showing that, on the whole, it is the medium-sized bovids (peaking at a female weight of about 55 kg), not the very large ones, that have the greatest sexual dimorphism. He concluded that, just as a larger male can monopolize mating, a larger female can more effectively monopolize resources to the benefit of her philopatric daughters. Estes (2000) likewise drew attention to the importance of female fitness, especially in sexually integrated social groups, where females may mimic males—for example, in color.

This is not to say that natural selection itself has not been a powerful force in bovid evolution (as one, of course, predicts). Stoner et al. (2003) examined color and possibly associated characters by optimizing character states on a phylogenetic tree. They examined all artiodactyls, although obviously bovids constituted the largest component of their dataset. They concluded that stripes and spots really do aid in concealment, whereas flank bands, facial coloration patterns, and conspicuous, contrastingly dark or light legs and tails are all communicatively significant.

An analysis of morphological and behavioral changes in the Bovidae is in Vrba & Schaller (2000). A concise account of the evolution of the Bovidae in Africa, with special attention to the fossil record, has been given by A. W. Gentry (1990).

Patterns of diversification within restricted taxonomic groups follow trends that are, to a considerable extent, predictable from ecology. Vrba (1980), studying fossil bovids—in particular, comparing the Alcelaphini with the Aepycerotini (which at that time were considered probable sister groups)—proposed that stenotopic lineages (i.e., those specialized in particular ecological niches) speciate more than do eurytopic lineages (those of more diverse ecological tolerance), and this provides a potential background against which to analyze diversification patterns in living species, including (in this case) bovids. We have found that stenotopic taxa, such as the Alcelaphini in general, indeed do speciate readily (the genus *Damaliscus* is unexpectedly speciose). Examining ecological parameters, as given by Estes (1991) and by Kingdon (1982a, b), suggests that grysbok are more stenotopic than steenbok; lechwe than waterbuck; sable antelope than roan antelope; and lesser kudu than greater kudu or, especially, eland, which is eurytopic in the extreme. If Vrba's (1980) model is correct, one should therefore expect greater taxonomic diversity at the stenotopic end of the scale; and this, we have found, turns out to be the case.

Grubb (2000a) drew attention to the existence of striking morphoclines among African bovids, and also giraffes. Examples include the genera *Oryx* (running from Arabia to S Africa); *Tragelaphus* and *Syncerus* (both running from closed habitats in W Africa to open habitats in E and S Africa); the *Nanger granti* group (from SE Kenya, in a ring via Ethiopia and Sudan to Maasailand); and *Ovis* and, perhaps, *Capra*.

Morphological analyses have always supported a basic division of the family Bovidae into two major clades: (1) the Bovinae, and (2) the rest, for which the name Antilopinae has been generally adopted (see, for example, A. W. Gentry, 1992). Molecular studies are universal in supporting such a division (see, for example, Allard et al., 1992; Gatesy et al, 1997; Hassanin & Douzery, 1999a; Matthee & Davis, 2001; Kuznetsova et al., 2002; Ropiquet, 2006).

The most complete DNA phylogeny of the family, fully compatible with most or all other published trees, is that of Ropiquet (2006), who proposed a phylogenetically based taxonomic arrangement, which is here accepted, as follows:

Subfamily Bovinae Gray, 1821
- Tribe Bovini Gray, 1821
  - *Bos, Bubalus, Syncerus, Pseudoryx*
- Tribe Boselaphini Knottnerus-Meyer, 1907
  - *Boselaphus, Tetracerus*
- Tribe Tragelaphini Blyth, 1863
  - *Ammelaphus, Nyala, Tragelaphus, Taurotragus, Strepsiceros*

Subfamily Antilopinae Gray, 1821
- Tribe Neotragini Sclater & Thomas, 1894
  - *Neotragus* (presumably including *Hylarnus, Nesotragus*)
- Tribe Aepycerotini Gray, 1872
  - *Aepyceros*
- Tribe Antilopini Gray, 1821
  - *Raphicerus, Antidorcas, Ammodorcas, Litocranius, Saiga, Antilope, Nanger, Gazella, Eudorcas, Dorcatragus, Madoqua, Ourebia, Procapra*
- Tribe Reduncini Knottnerus-Meyer, 1907
  - *Redunca, Kobus* (probably including *Adenota, Hydrotragus*), *Pelea*
- Tribe Hippotragini Sundevall, 1845
  - *Hippotragus, Addax, Oryx*
- Tribe Alcelaphini Brooke, in Wallace, 1876
  - *Alcelaphus, Beatragus, Damaliscus, Connochaetes*
- Tribe Caprini Gray, 1821
  - *Pantholops, Oreamnos, Budorcas, Ammotragus, Arabitragus, Hemitragus, Pseudois, Capra, Nilgiritragus, Ovis, Rupicapra, Naemorhaedus, Capricornis, Ovibos*
- Tribe Cephalophini Blyth, 1863
  - *Philantomba, Cephalophus, Sylvicapra*
- Tribe Oreotragini Pocock, 1910
  - *Oreotragus*

Subfamily Bovinae Gray, 1821

The horns are not regularly ridged, but they are usually strongly keeled and spirally twisted.

The mastoid is narrow, and, in the last two mandibular premolars, the paraconid is differentiated from the parastylid (A. W. Gentry, 1992).

Leinders & Heintz (1980) described some variations in the pattern of the lacrimal orifice in the Bovini and the Tragelaphini (these are not seen in the Boselaphini). There may be a single orifice, as in all other Bovidae, or there may be two, and their positions are slightly variable, usually inside the orbital rim (as in most Bovidae), but sometimes on the rim itself.

There is little disagreement that there are three tribes, which form almost a three-way phylogenetic split. In analyzing the COII gene, however, Janeck et al. (1996) found that *Boselaphus* (representing the Boselaphini) and *Ammelaphus* (representing the Tragelaphini) formed a clade, but with only 51% bootstrap support, although the association of the buffaloes to *Bos* also had only 51% bootstrap support.

Tribe Bovini Gray, 1821

As elucidated by A. W. Gentry (1992), the Bovini have low, wide skulls; horn cores emerging transversely; internal sinuses in the frontals extending into the horn cores; a short braincase and widened occiput; molars with large basal pillars and complicated central cavities; and upper molars with prominent ribs between the styles.

Groves (1981b) attempted a phylogenetic arrangement based on cranial characters: in the genus *Bos*, *B. gaurus* was conceived as sister to other species, followed by *B. javanicus*, and there was then a dichotomy between a *B. sauveli* / *B. primigenius* clade and a *B. bison* / *B. mutus* clade. Aspects of this have since been shown to be incorrect, but the association of bison and yak has proved justified. In the main, the phylogeny of the Bovini is depicted in DNA trees as an initial separation between buffaloes and the rest, followed among "the rest" by a more or less three-way split: true cattle, gaur-banteng-kouprey (the erstwhile subgenus *Bibos*), and bison-yak (the erstwhile subgenera *Bison* and *Peophagus*).

The dendrogram generated by Hartl et al. (1988), on the basis of 15 protein loci studied by electrophoresis, gave buffaloes as a clade separate from *Bos* (including the bison). European and American bison formed a clade with almost no difference between the two, and this was sister to a clade containing the following:

((*gaurus*, *javanicus*) (*mutus*, *primigenius*)).

In all the phylogenetic trees generated by Janeck et al. (1996), *Bos sensu stricto* and the bison formed a clade with respect to the buffaloes; while the precise interrelationships of the species varied, *B. bonasus* was always sister to the other species, with strong bootstrap support, and *B. taurus* and *B. grunniens* always formed a clade—rather unexpected findings that do not cohere with other phylogenetic results.

On the basis of genetic distances calculated from 20 microsatellites, Ritz et al. (2000) constructed two alternative neighbor-joining trees for the Bovini, using two different methodologies. In one, *B. indicus* and *B. taurus* formed a sister clade to *B. frontalis* plus *B. grunniens*, and the two together were the sister clade to American bison. The African and Asian buffaloes constituted separate branches, which, together with the bison/cattle clade, formed a basal trifurcation. The other methodology was similar, except that *B. grunniens* was no longer joined to *B. frontalis*, but was sister to a *B. frontalis* / *B. taurus* plus *B. indicus* clade. Divergence times were calculated, and compared with those of Hartl et al. (1988), which appeared slightly older: in the order of 0.5–1.5 Ma for *B. taurus* / *B. indicus* versus either mithan, yak, or bison; 1–2.61 Ma versus African buffalo; and 1.9–4.9 Ma versus Asian buffalo.

Using multilocus DNA fingerprinting, Vasil'ev et al. (2002) arrived at the following cladogram, based on genetic similarity coefficients:

(*bonasus* (*bison* ((*gaurus*, *indicus*) (*javanicus*, *taurus*)))).

While it is possible that the inclusion of interspecific hybrids in these calculations may have artificially brought some species pairs together (*B. gaurus* and *B. indicus*, in particular), inspection of the individual values in these authors' table 3 shows that the genetic similarities are more or less as in their dendrogram—for example, the genetic similarity between *B. bonasus* and other taxa is lower than that for *B. bison*.

The Y-chromosome phylogeny of Nijman et al. (2008) found a rather different tree: the two genera of buffaloes were linked (which they often are not in mtDNA phylogeny), and the first (short, but significant) branch from the non-buffalo lineage was the yak; then there is a split between bison (both *B. bison* and *B. bonasus*) and other cattle. This supports the hypothesis of Verkaar et al. (2004) that *Bos bonasus* is of hybrid origin; and Nijman et al. (2008) suggested that the genus *Peophagus*, for the yak, should be revived.

Cattle Group

Bos primigenius Bojanus, 1827

aurochs

The name *Bos primigenius* Bojanus, 1827 was placed on the Official List of Specific Names in Zoology by ICZN, Opinion 2027 (2003).

A study of mtDNA sequences from modern domestic cattle (Loftus et al., 1994) separated taurine and zebu breed-groups very strongly, with an average pairwise distance of 7.86%, not very much less than that between cattle and bison at 10.62%. The authors argued that, if bison and cattle lineages separated 1–1.4 Ma, then the two cattle lineages separated 740,000 to 1.04 Ma.

Cattle horns were used in the early 20th century in Cambodia to create a mysterious animal, known only by horns and frontlets, called the "linh duong" or "khting vor" (*Pseudonovibos spiralis*). Both its mtDNA and nDNA are identical to those of domestic cattle, specifically Vietnamese cattle (Hassanin et al., 2001), and the method of faking was discovered by examining longitudinal sections of the horns (H. Thomas et al., 2001). A brief history of this episode—which led to a spurious species being listed on conservation databases, and even to frantic searches for it in the field—was recounted by Galbreath & Melville (2003).

Gaur-Banteng-Kouprey Group

Bos gaurus Hamilton Smith, 1827

gaur, seladang

1827 *Bos gaurus* Hamilton Smith. Mainpat, Sarguja Tributary States, approximately 23° N, 83° E (see Harper, 1940).

The name *Bos gaurus* Hamilton Smith, 1827, was placed on the Official List of Specific Names in Zoology by ICZN, Opinion 2027 (2003). It had previously been conserved (because it was threatened by an earlier name; see Groves, 1982c) under ICZN, Opinion 1348 (1985).

We found no differences among samples of gaur from different parts of India (S, N, Sikkim). Within Southeast Asia, 83% of the samples from Thailand and 63% from the Indochinese region are correctly sorted.

Using six cranial variables (greatest skull length, nasal length, postorbital breadth, breadth between horn bases, and greatest and least occipital breadths), however, showed that there is complete separation between the Indian and the Southeast Asian samples ($N=7$, 10, respectively). A skull from Tripura assorts with Southeast Asia. Absolute size is the main discriminator, with greatest skull length, nasal length, and postorbital breadth contrasting with least occipital breadth. A more complete definition of the boundary between them is needed, but on this cranial basis, probably the S and Southeast Asian gaur would better be classed as two distinct species.

The horn variables give a much less complete separation: 80% of the Indian specimens are correctly identified, but only 67% of the Southeast Asian ones.

Table 21 Univariate statistics for *Bos gaurus*: males

	Gt l	Gt occ br	Teeth	Tip–tip	Span
Southeast Asia					
Mean	581.13	298.72	147.56	569.50	874.76
N	15	18	16	20	21
Std dev	24.118	36.950	7.831	153.207	110.226
Min	537	202	127	313	528
Max	626	334	160	941	1088
India					
Mean	544.89	293.05	145.30	473.07	827.68
N	45	38	40	41	41
Std dev	20.589	18.271	10.634	135.001	99.261
Min	506	233	130	248	668
Max	586	318	173	964	1180

Table 22 Univariate statistics for *Bos gaurus*: females

	Gt l	Gt occ br	Teeth	Tip–tip	Span
Southeast Asia					
Mean	524.40	237.75	142.40	374.71	643.57
N	5	4	5	7	7
Std dev	6.877	8.539	10.479	273.229	149.203
Min	516	230	127	90	510
Max	531	247	155	765	867
India					
Mean	539.88	270.38	145.50	417.50	718.50
N	8	8	8	8	8
Std dev	17.956	23.323	8.536	158.073	121.959
Min	522	237	135	64	535
Max	579	306	157	593	870

In this case, the Tripura specimen assorts with India, as do specimens from the Upper Chindwin and Mogok; one specimen from Chittagong assorts with India, and the other with Southeast Asia.

Tables 21 and 22 show that Southeast Asian gaur tend to be considerably larger than Indian guar, though there is much overlap. Horns continue this tendency. In the females the relationship is reversed: both span and tip-to-tip distance are greater for the Indian than for the Southeast Asian females.

Gaur, of unknown affinity, formerly occurred in Sri Lanka (Groves, 1982c).

Bos gaurus gaurus Hamilton Smith, 1827

$2n=58$.

Rather smaller in size than *B. gaurus* but with relatively longer nasals and a wider occiput.

Bos gaurus laosiensis (Heude, 1901)

For the full synonymy of this subspecies, see Braun et al. (2001).

Vadhanakul et al. (2004) found that a male from Thailand had a chromosome number of $2n=56$, with a Robertsonian fusion; but this is not a consistent difference between Southeast Asian and Indian gaur, because Chaveerach et al. (2007) later found $2n=57$ in another Thai individual.

Larger, with relatively short nasals and a narrow occiput; horn tips less turned in relative to the span.

Bos javanicus d'Alton, 1823
banteng

1830 *Bos leucoprymnus* Quoy & Gaimard. Supposed by Blyth (1842) to be based on a hybrid, but there is no evidence for this.

1840 *Bos sondaicus* Schlegel & Müller. Java.

The three banteng "subspecies" do not separate well on multivariate analysis, but one problem is that many of the available specimens originated as hunting trophies; thus they are very incomplete, and it is not possible to build up a large dataset in which each specimen has all available variables. Additional specimens might well enable complete separation to be made, but to do so at the moment would be unsafe. Table 23 gives univariate statistics. The Bornean *B. j. lowi* is considerably smaller than the other two taxa, with a smaller horn span, but the ranges of variation overlap. Mainland *B. j. birmanicus* is not larger in cranial dimensions than *B. j. javanicus*, but the horns, on average, are larger in all dimensions.

Bos javanicus javanicus d'Alton, 1823
$2n=56$.

Mature bulls black, with horns spreading out and then curving up, connected by a cornified band across the intercornual ridge; cows and calves ochery-colored, with smaller, crescentic, upstanding horns.

Many populations of the domesticated form of this species, known as Bali cattle, turn out to have introgressed genes from ordinary (humped) domestic cattle, and the same cannot be excluded for remnant wild populations (Bradshaw et al., 2006). In the opinion of those authors, this makes the feral population in the Cobourg Peninsula, Northern Territory, Australia, especially valuable, as there is no trace of any non-banteng DNA in 54 sampled individuals. Nonetheless, the Australian banteng do have domestic (Bali cattle) ancestors, and they have not redeveloped the full wild phenotype; for example, the horn tips of bulls point outward, not inward as in wild banteng bulls.

Bos javanicus lowi Lydekker, 1912.
Smaller, with less outcurved horns.

There are too few specimens to decide whether this is a distinct species.

Table 23 Univariate statistics for *Bos javanicus*: males

	Gt l	Biorb	Teeth	Tip–tip	Span	Horn l curve
birmanicus						
Mean	513.75	228.79	139.78	555.43	829.40	314.58
N	28	38	32	40	42	24
Std dev	23.609	20.287	8.178	142.579	121.935	40.239
Min	465	202	117	205	580	235
Max	566	331	157	855	1080	382
javanicus						
Mean	511.55	222.10	137.28	482.31	733.57	295.23
N	47	48	47	42	44	22
Std dev	19.053	9.825	4.610	112.601	88.222	40.959
Min	461	203	128	310	577	226
Max	557	247	148	865	896	362
lowi						
Mean	472.00	208.24	139.27	319.85	546.70	269.10
N	3	25	11	33	33	10
Std dev	19.079	11.505	18.639	100.856	85.753	41.816
Min	450	184	119	113	381	204
Max	484	237	191	497	697	318

Bos javanicus birmanicus Lydekker, 1898.

1898 *Bos sondaicus birmanicus* Lydekker. Burma.
1909 *Bos sondaicus porteri* Lydekker. MeWong district, C Thailand.
For the full synonymy of this subspecies, see Braun et al. (2001).

A study of two males and two females from Cambodia showed a difference in karyotype from Javan banteng: $2n = 60$ in the Cambodian banteng, contrasting with $2n = 56$ in the Javan ones (Ropiquet et al., 2008). We need to know the consistency of this difference, as well as the karyotype of the Bornean taxon, before we can use this information as the basis for a taxonomic regrading.

In Burma, N Thailand, and the Indochinese parts of the range, the bulls are ochery-colored, like the cows, but in the peninsula many bulls turn black. As this is the external difference that is commonly cited between mainland and Javanese banteng, it appears that no firm differentiation can be made on the species level.

The skull of the type specimen of *Bos sondaicus butleri* Lydekker, 1905, often considered a synonym of this taxon, appears to us to be an ordinary domestic cow.

***Bos sauveli* Urbain, 1937**
kouprey

Hassanin & Ropiquet (2004), using both mitochondrial and nuclear sequences, placed kouprey, gaur, and banteng as three approximately equal splits from a common clade.

A brief flurry was caused when Galbreath et al. (2006) found that the kouprey and Cambodian banteng contained very similar mtDNA, and suggested that the kouprey was, in fact, a feral hybrid between banteng and humped cattle, as had been first proposed long before. This, however, was laid to rest by Hassanin & Ropiquet's (2007) study of nDNA, which showed that, on the contrary, it was the Cambodian banteng which had incorporated mtDNA from the kouprey, not vice versa. This makes us wonder whether the distribution of the kouprey was at one time much wider, and has been progressively reduced by peripheral populations undergoing nuclear swamping as the banteng spread into their range, with banteng males mating with kouprey females and then with the female hybrids and backcrosses.

Moreover, Hassanin et al. (2006) obtained DNA from a mounted specimen in the Natural History Museum of Bourges, deemed to have been a Cambodian domestic ox but physically resembling a kouprey. Indeed, the specimen clearly was a kouprey, resurrecting the possibility, which had been suggested half a century before, that domestic kouprey exist.

Bison-Yak Group

Flerov (1965) and Bohlken (1967) made craniological comparisons between the taxa of bison. Flerov described the skull of the European bison in some detail: smaller than that of the American bison, with a depression on the forehead behind the nasals and between the orbits; the premaxillae narrower; the horn cores projecting less backward; the facial skeleton low; the choana relatively low; the occipital condyles less recurved forward at their lateral tips; the basioccipital shortened and narrowing anteriorly; the mandibular condyles short; the external auditory meatus larger; and the (mandibular) incisor-canine row relatively narrow. These are the major points of his extremely long and detailed description, but their degree of consistency, and even the number of specimens studied, is unclear (of bison in total, living and fossil, he mentioned "more than 300 specimens of different age and sex"). For the moment, we list these characters in summary form, in the hope that they can be tested in the future.

Bohlken (1967), on the other hand, using mostly metrical characters, found very few major differences between American and European bison, which, in all allometric comparisons, tended to fall on the same or parallel slopes; the main difference was the less robust horns of American bison, with a correlated lower occipital height.

***Bos bonasus* Linnaeus, 1758**
wisent, European bison (Polish/lowland bison)

Flerov (1965) opined that Linnaeus described the species "on the basis of specimens from the Bialowieza Forest."

Distinguishing the lowland bison from the Caucasian bison, Flerov (1965) described the former as having a convex forehead; the preorbital depression is less clearly defined; the occipital condyles are simpler in form; and the nasal cavity is relatively more voluminous, so that the skull height is slightly greater—he stated that the distance from the lower orbital margin to the alveolus of the posterior molar is about a quarter of the toothrow length. The nasal process of the premaxilla ends sharply, falling short of the nasal by more than the length of the first upper molar. The horns of cows are comparatively slightly bent inward, whereas those of bulls are "spiral: they point out, in, and slightly back," and the horn cores are directed laterally and slightly back, their tips bent upward.

Verkaar et al. (2004) pointed out the anomaly that, whereas the Y-chromosome DNA of this species is close to that of *Bos bison*, mitochondrial sequences align it, if distantly, with the *Bos taurus* group. After considering lineage sorting as a possible (if not particularly plausible) explanation, they proposed instead that the European bison is a hybrid species, in which bulls of one of the European Pleistocene bison species (probably *Bos priscus*) interbred with cows of an ancient *Bos primigenius*–like species, then with the hybrid cows, and so on, until nuclear swamping of the latter had occurred, while the original mtDNA was retained in the new hybridogenetic species. We agree that this is the most plausible explanation.

Two well-known breeding lines of this species are preserved in captivity: the lowland and the lowland-Caucasian lines. The former are descended solely from bison ultimately from the Bialowieza region, so they are the surviving purebred representative of the present species. The latter, the lowland-Caucasian line, are descended from Bialowieza bison plus a single Caucasian male; that is to say, they represent *Bos bonasus* with some *Bos caucasicus* genes. As described by Rautian et al. (2000), there is, in fact, a third line, the highland line, descended from a mixture between the other two lines plus three American bison: in 1940, five hybrid bison were brought to the Caucasian State Nature Reserve and bred there in an enclosure; in 1950 the females were mated with males from the lowland-Caucasian line; and finally, in 1968, they were all released into the reserve, where they now live wild. Rautian et al. (2000) studied 15 male and 19 female skulls of highland bison, collected between 1968 and 1994, as well as plentiful material of what we here call *B. bonasus* and *B. caucasicus*. They found, interestingly, that the highland bison differ substantially from the other two, mostly by being intermediate, but in some cases tending in the direction of *B. bison*, and—in a few cases—deviating in an unexpected direction from all putative parents. This microevolution clearly deserves fuller study, but what is not so clear is that those authors were justified in describing a new subspecies, *Bison bonasus montanus*, on this mixed population!

Bos caucasicus Satunin, 1904

Caucasian bison

Differences between this species and *Bos bonasus* have been listed as follows: the pelage more clearly extends down to the fetlocks; the color is more of a warm sepia, as opposed to "fahlbraun mit ockerbrauner Schattierung"; the height of the withers of the adult bulls is about 160 cm, as opposed to 185 cm or more for *B. bonasus*; the hoofs are shorter, the forehoof <85 cm, as opposed to >90 cm, and the hindhoof <90 cm, as opposed to >100 cm (Mohr 1952; Heptner et al., 1961). Bohlken (1967) found that the tuber malar is less developed than in *B. bonasus*, giving a smaller relative breadth across the facial skeleton; the frontal is nearly flat, as opposed to convex, behind the orbits; and, especially, the occiput is strongly constricted below the occipital crest. Although all the cited authors regarded Caucasian bison as only subspecificially distinct from lowland bison, on the evidence they are consistently (diagnosably) different. The difference in the occiput is not only relative to overall skull size, but is an absolute univariate difference (occipital constriction breadth in *B. caucasicus* 101–130 mm ($N = 7$), compared with *B. bonasus* 149–204 mm ($N = 9$). The horns are also somewhat smaller and more slender, on average (Bohlken, 1967).

Meanwhile, Flerov (1965) distinguished *B. caucasicus* from the European lowland bison as having "always two distensions" at the base of the horn cores; the preorbital depression is clearly defined; the nasal cavity is smaller, so that the facial skeleton is low; the distance from the lower margin of the orbit to the alveolus of third molar is about "one-third of the toothrow length" ([*sic*]; this would seem to make it higher than in *B. bonasus*!). The nasal process of the premaxilla ends bluntly in most cases and may contact the nasals, or, if it does not, the distance between them is less than the length of the first molar. Flerov, like Bohlken (1967), found that the Caucasian bison has a much less convex (usually almost flat) forehead, and the tuber malar is less developed. Flerov described the horns thus: those of the females are noticeably bent inward; those of the males are more strongly spiral, their cores pointing straight out to the side, or even slightly forward, and their tips pointed upward. As before, we are uncertain what to make of Flerov's descriptions, since it is unclear whether they are consistent, and on what samples they are based.

Kretzoi (1946) described a new (subfossil) subspecies, *Bison bonasus hungarorum*, based on a complete skeleton of a female and a cranial fragment of a male. The differences would be first, that the horn cores initially diverge horizontally, and are then slightly curved upward, with the tips curved backward; second, that the withers are much lower, compared with the pelvis; and third, that the size is small. He stated that he had examined drawings and paintings from the 17th and 18th centuries depicting Hungarian bison, and these were consistent. On the other hand, he admitted that some living bison apparently have the same horn form, which he took as meaning that some ancestry from the old Hungarian bison persists in them (!).

The DNA of this species is not known. Extracting both mtDNA and Y-chromosome DNA from preserved specimens will help to show whether this species, like *B. bonasus*, is of hybrid origin, or whether it is a separate lineage.

Bos bison Linnaeus, 1758

American bison ("buffalo" in North America)

1758 *Bos bison* Linnaeus. C Kansas (fixed by Hershkovitz, 1957).

? 1898 *Bison bison athabascae* Rhoads.

1899 *Bison occidentalis* Lucas. St. Michael, Alaska.

? 1915 *Bison americanus pennsylvanicus* Shoemaker, nom. nud.

1932 *Bison bison oregonus* Bailey. Malheur Lake, 43° 5' N, 119° W, Oregon.

1933 *Bison bison haningtoni* Figgins. Rock Creek, NE of South Park, Colorado.

1933 *Bison bison septentrionalis* Figgins. Near Palmer Merrick, Nebraska.

1980 *Bison bison montanae* Krumbiegel. Montana.

Krumbiegel (1980) gave the following table of the three subspecies of North American bison that he recognized (here somewhat modified):

1. *Bison bison athabascae*—wood bison. Large; long-bodied, long-legged. Horns long, more curved, tips often pointing forward, horn base narrow. Hoof broader. Hair shorter, softer, not curly. Color darker, especially the head, legs, and tail.
2. *Bison bison montanae*—northern plains bison. Smaller; shorter legs and a lower, longer body. Horns short, thick, tending to be covered by the thick, upwardly directed hair on the top of the head, the tips not pointing forward. Hoof narrower. Hair longer, denser, more curly; "chaps" (hair on the upper forelegs) longer, thick. Color lighter.
3. *Bison bison bison*—southern plains bison. Small; short-legged, like the previous taxon, but the body higher and shorter. Horns longer, less covered by long hair on top of the head. Chaps longer and less thick, grading more into the hair of the foreparts.

The evidence for the differentiation of Krumbiegel's new subspecies *B. b. montanae* is not all that abundant, but his two photos of southern plains bison (*B. b. bison*, in his classification) are most striking, and quite unlike most plains bison one would see today, which would be *B. b. montanae*. Clearly much is to be learned about the former geographic variation of this species, as Krumbiegel indeed admitted in his paper.

The wood bison, *B. b. athabascae*, is generally recognized as a subspecies of American bison, but there are complications, which must now be faced.

For Flerov (1965), *B. b. athabascae* is a large-sized and primitive form that retains a great many characters of the extinct *Bos* ("*Bison*") *priscus*, such as the more rounded, less widened muzzle; the anterior ends of the premaxillae lacking the bladelike expansion of *B. b. bison*; the nasals straight and not narrowing toward the front; the presence of a depression on the forehead behind the nasals; the long, thick, less downcurved horn cores; and the narrow incisor-canine row.

Collecting data from the literature, Bohlken (1967) concurred that the skull of *B. b. athabascae* is a great deal larger, on average, than that of nominotypical *B. b. bison*, with longer, more spreading horns.

Any study of the differences between *B. b. athabascae*, the wood bison, and other American bison is, of necessity, somewhat constrained, because the only known population of *B. b. athabascae*, living in Wood Buffalo National Park in N Canada (spanning the Alberta–Northwest Territories border), was forever destroyed by the 1925 introduction by the Canadian government (apparently in the teeth of much opposition from the public) of a large number of plains bison from Montana. The later discovery of a presumed isolated population in the park, at Nyarling River, was supposed to have saved the integrity of *B. b. athabascae*, but it is now recognized that even this population had not escaped gene flow from the introduced plains bison, though how much would be in question. In any event, bison from the Nyarling River herd, selected for their *athabascae* phenotype, were subsequently translocated to the Mackenzie Bison Sanctuary (Northwest Territories) and to Elk Island National Park (N Alberta). (There is also a herd of plains bison, also originating from Montana, in Elk Island National Park, kept separately from the putative wood bison).

It is in this context that genetic studies must be interpreted. Frequencies of red cell carbonic anhydrase alleles were different in *B. b. athabascae* (from the Elk Island National Park herd) from plains bison, but not more than the frequencies of plains bison in

Wichita Mountains Wildlife Refuge (Oklahoma) differed from those in Santa Catalina Island Conservancy in California and the National Bison Range in Montana (these two herds have a common origin). Interestingly, the presumed hybrids in Wood Buffalo National Park were quite close to the plains bison herd in Elk Island National Park, and these, in turn, were rather closer to *B. b. athabascae* than to other plains bison.

In another study, a genetic comparison of the two groups of bison in Elk Island National Park showed very low differentiation (Bork et al., 1991). Recall that the plains bison there originate ultimately from Montana, as did a component of the ancestry of the wood buffalo population.

A very thorough study on the differentiation between plains bison (nominotypical *B. b. bison*) and wood bison (*B. b. athabascae*) was conducted by van Zyll de Jong (1986), using both observational data on living animals and morphometrics on crania. The study was again compromised by the hybridization that had taken place, due to the introduction in the 1920s of plains bison to Wood Buffalo National Park. Van Zyll de Jong found that many Nyarling River bison skulls indeed were very like old specimens of *B. b. athabascae*, though not completely so, and he was constrained to selecting specimens. Nonetheless, in many of his craniometric analyses, the two putative taxa were completely separated (see, for example, his figure 6, based on six horn measurements and five cranial ones). The differences he found were, essentially, that *B. b. athabascae* were larger than *B. b. bison* in the length and transverse diameter of the horn cores, posterior cranial width, and basal skull length; but that the former were smaller in the width of the skull across the masseteric processes and, curiously, the occipital condyles. Moreover, although based on rather small sample sizes, *B. b. athabascae* was distinguished by the greater length of the long bones and the pelvis, and the much greater width of the pelvis across the ischia. *B. b. athabascae* has/had a "suddenly tall" hump, which continues farther back along the spine as well; whereas *B. b. bison* has a deeper neck, from which the hump rises slowly, and its hump does not decline as much behind. There were also external differences in *B. b. athabascae*: the color of the cape; the form of the frontal display hair and the length of the display hair on the upper forelegs ("chaps"); and the form of the ventral neck mane, making the beard stand out much more and appear more pointed.

Geist (1991b) took great exception to the idea that *B. b. athabascae* is distinct at all. He applied his findings on deer (Geist, 1989) to bison, claiming that *B. b. athabascae* is, in fact, only an ecotype—that is to say, its differences from plains bison are not genetic at all. Reviewing many of the features deemed distinctive by van Zyll de Jong (1986), he found that they are not consistent, not only as concerns the Nyarling River herd but even, to some extent, when old pre-1925 photos are studied. The photos of extinct southern plains bison from Krumbiegel (1980) and other sources do, indeed, also bear some similarity to wood bison. In addition, Geist (1991b) maintained that the Nyarling River bison that had been translocated to Elk Island National Park, and those of their descendants that were translocated elsewhere, were more reminiscent of plains bison and appeared to have converged on them; that is, the differences in size, body shape, and features of the pelage were actually not hereditary at all, but resulted from environmental plasticity. As for the skull differences, he attacked morphometrics as being completely invalid, because it did not take phenotypic plasticity into account.

Van Zyll de Jong et al. (1995) responded by reexamining some of the characteristics that had been previously used by van Zyll de Jong (1986) to differentiate *B. b. athabascae* from *B. b. bison*, and they scored various populations of known origin for the development of the hump, the cape, the chaps, the frontal display hair, the ventral neck hair, and the beard. Strong differences between presumably purebred plains bison and predominately wood bison remained, casting doubt on the view of Geist (1991) that these were all subject to environmental modification.

Given that there are no remaining pure *B. b. athabascae* populations, with even the Nyarling River herd being mixed to some degree, these positive results do suggest that there really was such a taxon as *Bos bison athabascae*, with its anteriorly high hump, reduced foreleg chaps, and so on. We, very clearly, do not support the disparagement of morphometrics/craniometrics, since we have used this method throughout this book. We can find no evidence that (nonpathological) specimens, let alone whole populations, show such all-pervasive environmental modification as to completely override the genetic basis of cranial (or other) shape differences.

When all is said and done, the subspecies (or species?) *B. b. athabascae* no longer exists. The case is somewhat analogous to that of *Bos caucasicus*; in both cases, all we have now are some of their genes. The difference is that in the *athabascae* case we do not know how much of its genome we possess, and how much the selected lookalikes really are, in any real sense, wood bison.

Bos mutus (Przewalski, 1883)

yak

1883 *Poephagus mutus* Przewalski. Nan Shan, at approximately 32° 20' N, 95° E.

The name *mutus* Przewalski, 1883, was placed on the Official List of Specific Names in Zoology by ICZN, Opinion 2027 (2003).

The complete mitochondrial sequence of domestic yak was determined by Gu et al. (2007), who dated the divergence between yak and domestic cattle at 4.38–5.3 2 Ma. Schaller (1998) found that, in contrast to wild and domestic Bactrian camels, wild and domestic yak are really not very different genetically.

Bubalus Hamilton Smith 1827

Asian buffalo

This genus contains at least four living species: wild buffalo, tamaraw, and two species of anoa. There may even be more.

Pitra et al. (1997) provided an alternative view of the relationships of the anoa. They sequenced two nuclear genes (*Cyp19*, and the promoter region of *Lf*) and found that, when *Capreolus* was taken as an outgroup, the anoa formed a clade with *Boselaphus*, whereas other buffaloes (both Asian and African) were, as expected, a sister group to *Bos*.

Bubalus arnee (Kerr, 1792)

Asian wild buffalo

1792 *Bos arnee* Kerr. Kuch Behar (as restricted by Harper, 1940).

The name *arnee* Kerr, 1792, was placed on the Official List of Specific Names in Zoology by ICZN, Opinion 2027 (2003).

The karyotype of wild Asian buffalo is unknown, but those of its putative descendants, the river and swamp breed-groups of domestic buffalo, show differences: the river buffalo has $2n=50$, the swamp buffalo $2n=48$ (Bongso & Hilmi, 1982).

The taxonomy follows Groves (1996b); in figure 1 of his paper, the three mainland subspecies separate cleanly on multivariate analysis, so there is a possibility that they actually should rank as distinct species, although the analysis was based on 10 skull measurements and, as only one of the three samples had more than 10 specimens, there is a reasonable possibility of a type I error (a false positive). Therefore, a conservative course is provisionally followed here.

Analysis of the mitochondrial D-loop and cytochrome *b* gene indicated that river and swamp buffaloes formed two distinct clades, separated by up to 270,000 years, indicating that they would have been independently domesticated (Kumar et al., 2007). But what their respective wild ancestors are is not clear.

There is, of course, a constant worry over the genetic purity of wild Asian buffalo. Flamand et al. (2003) used microsatellites to examine the status of the wild buffalo of Kosi Tappu Wildlife Reserve (Nepal), finding evidence for a little (but very little) gene flow from domestic to wild animals, and rather more gene flow the other way.

Bubalus arnee arnee (Kerr, 1792)

1792 *Bos bubalus arnee* Kerr.

1852 *Bubalus buffelus macrocerus* Gray. Nepal.

1912 *Bubalus bubalis septentrionalis* Matschie. Kukri-Mukri, an island on the seaward side of the Sundarbans.

Relatively small in size; skull length usually <570 mm; relatively large teeth, >27% of the skull length; horn span usually 1000–1200 mm; tip-to-tip distance nearly always <80% of the span. Color black, with contrastingly white lower limbs below the knees and the hocks, and a white muzzle; tail reaching to the hocks.

Formerly from the Sundarbans SW into Madhya Pradesh and Andra Pradesh, and NW into Nepal; now apparently occurs only in the Raipur and Bastar districts of E Madhya Pradesh and the Kosi Tappu Reserve, SE Nepal.

Bubalus arnee fulvus (Blanford, 1891)

1891 *Bos bubalus* var. *fulvus* Blanford. Mishmi Hills.

Very large size; greatest skull length usually >570 mm; toothrow length approximately 26%–28% of the skull length; horn span >1100 mm in most specimens; tip-to-tip distance as in nominotypical *B. a. arnee*. Color a lighter gray or brownish gray, and the limbs less contrastingly white; tail falling short of the hocks.

Brahmaputra Valley; formerly to the Mishmi Hills, and S to the Chittagong Hills. Still occurs sporadically over most of the Indian portion of the range, and marginally in Bhutan.

Bubalus arnee theerapati Groves, 1996

1996 *Bubalus arnee theerapati* Groves. Mae Wong, Nakhon Sawan, Thailand.

Relatively small in size; skull length <570 mm; toothrow length 24%–27% of the skull length; horn span usually >1200 mm, the tip-to-tip distance >80% of the span.

Formerly from the Irrawaddy Delta through Thailand to Cambodia and probably Vietnam. Still occurs in the Huai Kha Khaeng Wildlife Sanctuary, Thailand, and in various places in Cambodia.

Bubalus arnee migona Deraniyagala, 1952

1952 *Bubalus bubalis migona* Deraniyagala. Yala National Park, Sri Lanka.

Skull relatively small and not sexually dimorphic, <570 mm in both sexes; occipital breadth especially small, usually <260 mm in both sexes (compared with >260 mm in the females, 280 mm in the males, in the mainland taxa). Horns, on average, smaller in all dimensions than in the mainland taxa; span <1000 mm. Apparently almost lacking white markings (though in Groves & Jayawardene's [2009] photos, mud obscures much of the lower part of the limbs).

Sri Lanka; now threatened by interbreeding with domestic buffalo (Groves & Jayewardene, 2009).

Bubalus mindorensis Heude, 1888

tamaraw

Braun et al. (2002) described their rediscovery of the type specimen and gave some historical notes.

Custodio et al. (1996) gave a full account of this species.

Fossil remains (but from a historic period) of a second, even smaller, Philippine dwarf buffalo have been described from Cebu (Croft et al., 2006).

Bubalus depressicornis (Hamilton Smith, 1827)

lowland anoa

1827 *Antilope (Anoa) depressicornis* Hamilton Smith. Celebes.
1853 *Oreas platyceros* Temminck.
1865 *Probubalus celebensis* Rütimeyer.
1905 *Bos depressicornis fergusoni* Lydekker. No locality.

Groves (1969b) defined the species as being large in size (skull length 290–300 mm in the females, 298–322 in the males; toothrow length 82–98 mm); horns triangular in cross section in the adults, with marked transverse ridges and an external keel (183–260 mm long in the females, 271–373 mm in the males); color black in the adults, the hair sparse, the woolly brown juvenile coat lost before maturity; lower segments of the limbs white or yellowish white, except for a black line down the front and across the patterns; generally a white crescent on the throat; groin light to white. Tail length <18% of the head plus body length.

Groves listed localities for this species throughout Sulawesi, in more lowland areas.

Bubalus quarlesi (Ouwens, 1910)

mountain anoa

? 1792 *Bos bubalus anoa* Kerr. Suppressed by ICZN, Opinion 1349 (1985).
1910 *Anoa quarlesi* Ouwens. Mountains of C Toraja, Sulawesi.
The name *quarlesi* was apparently threatened by *Bos bubalus anoa* Kerr, 1792 (see Groves, 1982d), but it was conserved by ICZN, Opinion 1349 (1985).

Groves defined the species as being small in size (skull length 244–290 mm, toothrow length 65–80 mm); horns short, conical, cylindrical, 146–199 mm long; color dark brown to black in the adults, the hair tending to be thick and woolly; legs dark, like the body, with whitish or yellowish spots above the hoofs, or none at all; no white on the throat; groin light, but not white. Tail length <18% of the head plus body length.

Groves listed localities for this species throughout Sulawesi, in mountainous areas.

The relationship between the two species of anoa, and whether there really are only two species, has been discussed by Schreiber & Nötzold (1995). It is clear, at any rate, that both the two putative species, especially *B. quarlesi*, are variable in color pattern, size, length of the horns, and so on; there could well be more taxonomic variation than hitherto recognized.

Kikkawa et al. (1997), on the basis of mitochondrial cytochrome *b*, found differences between what they called lowland, mountain, and quarles anoas (1.2% between "mountain" and "quarles," 3.3%–3.6% between lowland and either of the other two); the sequence divergence between anoas and water buffalo was itself only 3.33%. The meanings of "mountain" and "quarles" are unclear, but it is interesting that the major differentiation among anoas seemed to be equivalent to the difference between anoas in general and other Asian buffalo.

Schreiber & Nötzold (1995) took a more cautious view of taxonomy. They pointed out that European zoos at the time had anoas deriving from four different imports: the Berlin/Krefeld, Antwerp, Rotterdam, and Leipzig groups. The first of these correspond to "mountain" anoas, the other three to "lowland" ones, although they also differ somewhat from each other. The Berlin/Krefeld stock have

$2n=44$ or 45; the others, $2n=48$ (in one individual, $2n=47$). The authors had visited Sulawesi and elsewhere in Indonesia. They reported seeing mountain anoa trophies for sale, and also a living captive animal, in the Toraja highlands; probable lowland anoas in the Gunung Tangkoko National Park on the N tip of Sulawesi, and evidence of both kinds of anoas in or near the Dumoga-Bone National Park. They cited a personal communication from Prof. Nawangsari Sugiri that a Lore Lindu anoa, physically resembling the Berlin/Krefeld stock, had a similar chromosome number ($2n=44$), while a much more pale brown one from the Pompangeo Mountains, E of Lake Poso, had $2n=46$, which is not recorded in any of the European anoas. In Minahassa, Prof. Sugiri found anoas physically resembling the Leipzig stock and with the same $2n=48$, and Schreiber & Nötzold (1995) also described having seen a film of anoas in the Gorontalo district that resembled the Leipzig stock.

In a later publication, Schreiber et al. (1999) found that the two phenotypes of anoas in the European zoos also have extremely divergent cytochrome *b* haplotypes, suggesting a divergence time of 1.25 Ma. The authors also made distance trees (including some previously published data), and used their own previous work on MHC class II genes, and these gave the same results. Given that some of the studies that they incorporated into their analysis had given no evidence that the species of anoa studied were as claimed, they recommended that attention to an external phenotype be given by all future workers—a recommendation that we fully endorse.

Syncerus Hodgson 1847

African buffalo

African buffalo were historically distributed from Senegal in the W to Eritrea and Somalia in the E, and S to the Cape of Good Hope. They seem to have been absent only from extremely arid areas, such as Namibia (except the extreme N) and much of Somalia (except along the Shebeyli and Juba rivers). They have been exterminated from parts of this range, especially around the edges, including Bioko (Butynski et al., 1997). They live in savanna, thornbush, light forest, rainforest, and mountain habitats. One expects, at any rate, only limited taxonomic diversity in such a eurytopic taxon. Nonetheless, there has always been admitted to be a rather dramatic contrast between "savanna" buffalo, called Cape buffalo, which are large, long horned, and black, and rainforest buffalo, which are small, short horned, and red.

After a period in which every slightly (or not at all!) differentiated population was given specific (later subspecific) rank, especially by Matschie (1906, 1918a, b), Lydekker (1913) united them all as subspecies of a single species, with a certain amount of cutting down of subspecies. The forest buffalo appeared to such authors as Lydekker to be considerably more diverse than the big savanna buffalo (Cape buffalo). Christy (1924) at first divided up forest buffalo into six distinct geographic forms, but later lumped them all (Christy, 1929), including W African *Syncerus brachyceros*, into a single catchall *S. nanus*, which he did, however, keep separate from *S. caffer*. Even between forest and Cape buffaloes, however, a forced rapprochement was in the air. The single-species model seemed inexorably to find more favor, for whatever reason, and tended to be adopted by later authors. It was only in the 1970s that Grubb (1972) began to query the basis of Christy's (1929) arrangement.

The taxa of this group can be arranged in a ring-shaped morphocline, from the C African rainforest via W Africa and Sudan to the E and S African savanna (Grubb, 2000a), but within this morphocline there were nonetheless some sharp breaks.

A study by van Hooft et al. (2002) found that, in mtDNA D-loop hypervariable region I sequences, African buffalo fall into two extremely distinct clades, separated by a 6.6% sequence divergence. The first clade consists entirely of buffaloes from Kenya, Tanzania, Uganda, and S Africa (i.e., Cape buffalo); the second, mainly of forest buffaloes from Cameroon, Gabon, Angola, and (supposedly) Namibia, plus two buffaloes from Uganda, one from S Africa, and one from Kenya or Tanzania. They also studied the Y-chromosome microsatellite INRA008; in this, all buffaloes from Kenya, Tanzania, and South Africa ($N=66$) had haplotype a, while two from Cameroon had haplotype b, two from Gabon had c, and one of the two from "Namibia" had d; of two "forest buffalo" of unknown origin, one had d, and the other had a different haplotype, e. On the basis of mismatch distributions, the authors postulated a Middle to Late Pleistocene population expansion of Cape buffalo, expanding from E to S Africa.

We would like to comment on this. A few Cape buffalo had forest mtDNA haplotypes, but there were no "wrong" Y-chromosome haplotypes; this suggests strongly that the distribution of forest buffalo—and, presumably, of the rainforest itself—was much wider in the past, and that vegetation changes (as van Hooft et al. [2002] themselves mention) have enabled Cape buffalo to expand, replacing forest buffalo populations by nuclear swamping. This is what one would expect, given the huge advantage that Cape buffalo bulls would surely have over forest buffalo bulls.

The association of two buffaloes from Namibia with forest buffalo rather than with Cape buffalo, in both mtDNA and Y-chromosome DNA, was raised several times in van Hooft et al.'s (2002) paper. The specimens are identified in their table 1 as being from the Antwerp Zoo, "one with father from Windhoek and one with mother from Okahandja"; on the grounds of relevance, presumably the one with the father from Windhoek was a male used for the Y-chromosome study, while both were used for the mtDNA study. Actually, buffalo occur in Namibia only in the Caprivi Strip and W along the Angolan border just into Ovamboland—not nearly as far S as Windhoek or Okahandja (Shortridge, 1934:440, and see the map opposite his p. 578). Under these circumstances, it is evident that the parents of the two Antwerp buffaloes in the study must have been obtained from dealers or from a Namibian zoo; one could not possibly say without inspecting them personally whether they were phenotypically Cape or forest buffaloes.

We now report on our craniometric analysis, in which we divided the very large sample into 19 subsamples. We found that discrimination on some measurements, especially horn span, was extreme, and in the end we concentrated on these relatively few measurements, which enabled us to incorporate more individuals per sample.

The first important conclusion is that Cape buffalo from S Africa, E Africa, and the W Rift Valley (or W Rift) are all the same—we could not separate them on discriminant analysis. Those from South Africa average somewhat greater in the curved length of the horn, and less in horn span, but these differences are extremely small, and on average only. We consider them a homogeneous taxon, for which the prior available name is *Syncerus caffer*.

All samples that might be classed as *S. brachyceros* and *S. aequinoctialis* (Sudan; Uele; N Nigeria / Chad) are more or less the same. The Uele localities include Ango, Gangala na Bodio, Aku, and Faradje. The type of *S. neumanni* Matschie, 1906 (from Chagwe [= Kyagwe], Uganda, just W of the Nile where it leaves Lake Victoria), identifies itself with this group in the craniometric analyses.

Forest buffalo from Cameroon (and other W-C African areas, especially Gabon), S-C DRC, and SW DRC plus N Angola form another homogeneous group, although the SW DRC / Angola ones tend to be somewhat larger, with a greater horn span but a lesser curvature than the others. W Rift forest buffalo are not different from these either. We consider this group, too, as a homogeneous taxon, for which the prior available name is *Syncerus nanus*.

S. caffer and the *S. aequinoctialis* / *S. brachyceros* group have a greater span and palm width than *S. nanus*, and a somewhat greater curve length; while the *S. aequinoctialis* / *S. brachyceros* group, when compared with *S. caffer*, has a smaller horn curve and a somewhat lower span, contrasting with a relatively slightly greater palm width.

Within W Africa, those from SW Nigeria, Togo, Ghana, and Ivory Coast are larger than those from SE and C Nigeria; and of two specimens from Senegal/Gambia, one falls in the same far W group, while the other is rather closer to SE–C Nigeria. A specimen from Garoua (N Cameroon) falls in with SE & C Nigeria (as discussed below). The far W buffalo differ in their wide mastoids and horn span, compared with the relatively shorter skull.

Buffalo from the CAR (Ndele, Bangui, 6° N 21° E, 7° N 19° E) are clearly different from those from the Uele district (Pandu, Ibembo, Buta) and those from the Ubangui (Dubulu, Bosobolo, Molegbwe, Kekongo, Libenge, Duma, Budja, Ubangi River) in being larger, with a greater horn span compared with horn palm width.

The buffalo from the W Rift fall into three cleanly distinct groups: *S. caffer*, "intermediates," and *S. nanus*. The "intermediates" are somewhat closer to *S. caffer* but, strikingly, do not overlap with them, nor is there overlap elsewhere. The differences are all concerned with absolute size, horn span being most emphasized, followed by curve length.

The position of the "intermediates" is perhaps somewhat unexpected. Rather than spanning the morphospace between *S. caffer* and *S. nanus*, it would appear that there is, in fact, a discrete population of intermediates. For these, the name *S. mathewsi* Lydekker, 1904 (type locality: Virunga volcanoes), is available, with, as its synonym, *S. cottoni* Lydekker, 1907 (type locality: Kasindi, W side of Lake Edward). The localities for the three sharply different taxa in the W Rift are as follows:

1. *Syncerus caffer*—Rutshuru (Katsibwe, 0.41° S, 29.16° E, and S of Lake Edward), Rwindi (0.47° S, 29.18° E), Parc National Virunga (including Kamohorora, 0.41° S, 29.40° E), Ishasha (0.31°–1.00° S, 29.45° E), Usumbura (3.23° S,

29.21° E), Semliki (0.45° N, 29.50° E), Gabiro River (1.31° S, 30.24° E), Lake Rugwero and Kataranga (Burundi), and Beni (0.29° N, 29.29° E)

2. *Syncerus mathewsi*—Kasindi (0.03° N, 29.41° E), Semliki, and Virunga volcanoes.
3. *Syncerus nanus*—Watalinga (0.45° N, 29.48° E), Kylia (Kapanatu), Kikuku (Kombo River, Konto, 0.59° S, 29.04° E), and simply "Rutshuru."

In other words, within this relatively restricted region there are three quite distinct taxa. There may or may not be some marginal interbreeding between them, but we have no evidence for it in our dataset.

It is clearly apparent from our data that buffalo from the SE part of the DRC (mainly Katanga) are fully *S. caffer.* Localities for them are Luisa (7.12° S, 23.25° E), Katombe (9.16° S, 24.40° E), Bushimaie River (6.03° S, 23.75° E), and Lukonzolwa (8.45° S, 28.40° E).

The SE Nigeria sample is not different from that from the rainforest of S Cameroon (i.e., *S. nanus*), but both are different from the sample from the CAR. The CAR sample differs from rainforest *S. nanus* in its greater span, longer skull length, and greater palm width. The major part of a sample from Garoua (N Cameroon) belongs with the CAR group: four of the five available skulls from Garoua fall clearly into that group, although the fifth one does fall in the range of SE Nigeria.

W Rift *S. nanus* are closest to those from S-C DRC, and next closest to S Cameroon; while skulls from SW DRC plus N Angola are different, on average. Differences are that those from SW DRC / Angola have a greater span, followed by higher values for greatest skull length, palm width, and mastoid breadth—contrasting slightly with the not much wider horn curve.

When we compare this SW DRC / N Angola group with South African *S. caffer* and S-C DRC *S. nanus*, we find that, even though this group does differ somewhat from rainforest *S. nanus*, it shows no tendency to resemble *S. caffer,* which is much larger in all measurements than either of the *S. nanus* groups (most especially in the horn span). The SW DRC / Angola buffalo are what have been called "forest-edge *nanus,*" and are often assumed to be different from rainforest *nanus* as a result of gene flow from *S. caffer.* On the basis of this analysis, however, there seems to be no evidence that this is what is going on; they are simply forest-edge rather than deep forest animals, and have adapted accordingly.

Localities of the skulls used in this analysis are as follows:

—S-C DRC *nanus*—Itana, Moma, Idembe, Eligampangu, Itoko, Walatunga, Kylia, Kikaku

—SW DRC / Angola—Kinshasa, Banana, Kwamouth, Mpu, Bas-Congo, Bashongo, Temvo, Matadi, Kunungu, Bokala, "opposite Boka," Bokoro, Tsangata, Lemba, Lake Leopold, Bokuma, Franz-Joseph Falls, Quingombe, Ambriz, Novo Redondo, Braentul, Pama, Binga, Bembe

Samples from N Nigeria and Chad (Bedadi, Mugur, Fort Lamy, Lake Chad, Bahr Salamat, "10° N, 20° E"), from the CAR (N Gaoundere, Kentu, Ndele, Bangui, four out of five of the specimens from Garoua, plus "8° N, 21° E," "6° N, 21° E," and "7° N, 19° E"), and from SE and C Nigeria (N Bamusu, Dundo, Mbam Geh, Katsena, Gabim Ba, Washishi, Mamfe, and one out of five of the specimens from Garoua) are rather different from each other; the first two groups, especially. are different from SE and C Nigeria. These latter have a narrower span, and a lesser curve and palm width, than the first two. As shown previously, they are, in effect, a larger version of *S. nanus*, while the other two groups are different, and somewhat (but not completely) different from each other. The prior available name for the Chad buffalo is *S. brachyceros* Gray, 1837 (synonym *S. bornouensis* Hamilton Smith, 1842; both names with Bornu as the type locality).

Univariate statistics for the working groups of African buffaloes can be found in table 24, summarized using the currently available taxonomic names.

The univariate differences mostly overlap with each other. It is on multivariate analysis that we found strong, nonoverlapping distinctions between taxa. We summarize our findings as follows:

—The sample called *Syncerus aequinoctialis* differs only a little from *S. caffer*; the greatest average difference is in the horn length on the curve.

—The sample called *S. brachyceros* overlaps strongly with *S. aequinoctialis*, and is almost certainly the same.

—The sample called *S. mathewsi* mostly overlaps with the two above, so it is "intermediate" in a way, but the span, especially, is relatively small.

—Excluding its forest-edge component, *S. nanus* is absolutely different from any of the above,

Table 24 Summary of univariate statistics for the working groups of African buffaloes

	Gt l	Mastoid br	Palm w	Span	Horn l curve
caffer	448–575	241–316	166–274	725–1340	660–1160
aequinoctialis	449–528	150–317	170–259	562–1030	530–945
brachyceros	433–599	241–266	178–250	740–1110	545–860
Ubangi/Uele/CAR intermediates	422–486	220–252	90–260	370–825	400–700
mathewsi	450–508	259–302	167–227	621–745	530–720
nanus Central	394–468*	137–257†	105–161‡	341–655	410–690
nanus West	411–492	199–276	119–167	345–720	350–635

*Forest-edge *nanus* (S DRC) goes up to 520.

†Forest-edge *nanus* goes up to 303.

‡Forest-edge *nanus* goes up to 295.

especially in palm width (which distinguishes it absolutely from the closest sample, *S. mathewsi*) and in span.

—The Ubangi/Uele/CAR intermediates are not the same as *S. mathewsi*; the mastoid breadth is absolutely larger, and the specimens are very much more variable, as expected for a hybrid swarm. These are the genuine intermediates, and constitute almost the only evidence for intermediacy between any two taxa.

The general impression of African buffaloes as a single species, grading from large black buffaloes in the savanna to small red ones in the forest, is misleading. Forest and savanna buffaloes are sharply distinct, do not grade into one another, and are distributed in a more mosaic fashion than has previously been recognized. For example, red buffaloes (of which, unfortunately, no specimens appear to be available) occur in some montane forests in Ethiopia (Yalden et al., 1984).

There are, therefore, four species within the genus *Syncerus*.

Syncerus caffer (Sparrman, 1789)

Cape buffalo

Size very large; horns widely spread laterally, each horn strongly curving down, below the skull base, and then up, with the curved length of each horn itself nearly equaling the span; bases of the horns greatly expanded, nearly meeting at the midline of the forehead, where each base forms a convex boss. Color jet-black. Ears meagerly fringed.

The whole of the S and E African parts of the range of the genus, extending into the Rutshuru plains of E DRC, and N into Kenya, at least as far as Mt. Elgon and the N Guaso Nyiro.

Syncerus brachyceros (Gray, 1837)

Lake Chad buffalo

1866 *Syncerus aequinoctialis* (Blyth).

1872 *Syncerus centralis* (Gray).

1906 *Syncerus azrakensis* (Matschie).

1906 *Syncerus neumanni* (Matschie).

1911 *Syncerus solvayi* (Matschie).

Size smaller; horns widely spread laterally, but much less curved, with the length of each horn noticeably less than the total span; bases of the horns greatly expanded, nearly meeting in midline, but not forming a convex boss. Color dark brown, usually only slightly reddish, apparently never quite black. Ears meagerly fringed.

Ethiopia and Sudan W through the savanna belt to Senegal. In C Africa, N of the Shari River. In E Africa, this is evidently the common buffalo of most of Uganda: along the Nile to Lake Victoria, and along the E side of the Great Lakes at least as far as the Kazinga Channel. It appears to enter Kenya in the Lake Turkana region.

Grubb (1972) called this taxon *S. aequinoctialis*, and applied the name *S. brachyceros* to what we are here regarding as hybrids between *S. brachyceros* and *S. nanus*; our study of more abundant material has shown that (as noted above) the buffalo found around Lake Chad, including the type locality of *S. brachyceros*, is the present species.

Syncerus mathewsi (Lydekker, 1904)

Virunga buffalo

1907 *Syncerus cottoni* (Lydekker).

Size similar to *S. brachyceros*; horns about the same length, but less widely spread, with the length of each

Table 25 Univariate statistics for working groups of *Syncerus*

	Gt l	Mastoid	Palm w	Span	Horn l curve
neumanni					
Mean	—	—	191.00	875.00	710.00
N	—	—	1	1	1
Sudan *aequinoctialis*					
Mean	502.52	264.09	210.77	855.77	753.10
N	21	22	22	22	21
Std dev	18.029	32.607	15.350	79.950	92.053
Min	463	150	192	693	640
Max	528	317	259	1030	945
Uele *aequinoctialis*					
Mean	491.93	267.00	198.15	757.79	684.23
N	14	10	13	14	13
Std dev	23.950	21.756	20.145	99.091	74.745
Min	449	236	170	562	530
Max	522	295	233	948	760
W Africa *nanus*					
Mean	454.23	229.64	152.00	527.69	519.09
N	13	14	14	13	11
Std dev	24.177	13.675	17.449	68.930	60.449
Min	411	212	125	447	415
Max	492	250	185	720	585
E Africa *caffer*					
Mean	506.92	283.84	222.04	1003.25	906.36
N	25	25	24	24	22
Std dev	23.615	15.283	22.162	123.869	112.981
Min	448	250	187	798	750
Max	558	312	274	1340	1160
S Africa *caffer*					
Mean	497.50	275.00	227.81	913.45	874.08
N	38	25	27	44	37
Std dev	18.176	13.216	18.874	84.392	84.271
Min	459	241	201	725	720
Max	543	300	258	1150	1060
W Rift *caffer*					
Mean	516.52	284.78	222.14	955.11	856.93
N	25	23	28	27	27
Std dev	25.810	16.384	19.529	94.331	75.655
Min	479	251	166	820	660
Max	575	316	249	1150	1020
CAR cf. *brachyceros*					
Mean	467.60	—	184.45	657.89	587.31
N	10	—	22	19	13
Std dev	13.753	—	36.471	70.806	49.859
Min	446	—	96	540	450
Max	486	—	260	774	640
N Nigeria and Chad *brachyceros*					
Mean	495.00	251.00	204.50	892.58	750.17
N	7	4	12	12	12
Std dev	50.484	12.247	21.682	112.353	89.668
Min	433	241	178	740	545
Max	599	266	250	1110	860
DRC *nanus*					
Mean	425.93	227.14	131.24	458.06	503.82
N	14	14	17	17	17
Std dev	11.118	11.935	10.053	53.150	67.489
Min	410	210	119	385	410
Max	448	257	157	592	650
DRC—Angola *nanus*					
Mean	436.72	222.77	145.39	481.85	446.56
N	46	44	54	55	52
Std dev	17.855	13.518	36.589	56.398	142.526
Min	394	196	115	362	122
Max	468	251	295	655	690
W-C Africa *nanus*					
Mean	420.09	211.26	129.30	431.15	459.58
N	23	23	33	33	33
Std dev	14.823	12.509	13.190	55.710	73.300
Min	378	187	105	341	360
Max	441	237	161	580	655
S DRC *nanus*					
Mean	492.04	273.43	213.77	828.39	780.48
N	23	21	22	23	21
Std dev	16.316	17.113	21.928	91.052	102.187
Min	463	244	185	623	615
Max	520	303	261	1000	955
Ubangui intermediates					
Mean	442.67	234.80	161.29	586.21	544.86
N	6	5	34	33	35
Std dev	17.224	11.520	20.366	120.857	82.091
Min	422	220	124	370	400
Max	464	252	207	825	700
Uele intermediates					
Mean	445.00	222.50	151.80	498.60	495.00
N	3	2	5	5	5
Std dev	9.849	.707	15.106	68.039	59.477
Min	437	222	132	405	430
Max	456	223	174	595	590
W Africa *brachyceros*					
Mean	443.72	234.96	142.13	514.03	541.92
N	25	25	32	31	13
Std dev	19.403	20.163	11.530	93.087	78.357
Min	411	199	119	345	350
Max	474	276	167	690	635
W Rift intermediates					
Mean	478.50	273.00	192.00	670.75	638.75
N	4	4	4	4	4
Std dev	26.388	20.281	25.377	56.240	90.035
Min	452	259	167	621	530
Max	508	302	227	745	720

horn only somewhat less than the total span; bases of the horns expanded, but flattened, not forming a convex boss. Color dark brown to black, often with reddish tones, especially in the cows, which are often reddish animals with a dark dorsal stripe and dark shanks. Ears with a fringe of hair.

Restricted to a the forested mountainous area from the Virunga volcanoes N along the W side of Lake Edward.

As noted by Grubb (1972), these buffalo are, in a sense, intermediate between the large savanna buffalo and *S. nanus*, but further evidence, since that date, has indicated that they are not simply intergrades, but form a homogeneous taxon distinct from all others in the region.

Syncerus nanus (Boddaert, 1785)

forest buffalo, red buffalo

The synonymy is complex, depending on whether the W African populations, which are somewhat larger in size, with a slightly greater horn span, on average, are recognized as a distinct subspecies.

Very small in size; horns short, not much curved laterally, but mainly curved upward, the length greater than the total span, with this discrepancy greater in C than in W African populations. Generally light reddish brown, with dark markings developing on the limbs and the shoulders; sometimes entirely black. Ears very large and prominent, with heavy fringes on the lateral (i.e., inferior) edges, the long, thick hair beginning well inside the inner surface of the pinna, typically in two transverse strips.

Forest belt of W and C Africa.

As noted above, there are slight average differences between W and C African populations, and there are intermediate populations between *S. nanus* and *S. brachyceros* in S Nigeria, the CAR S of the Shari River, and the Uele district. Grubb (1972) used the name *S. brachyceros* for these intermediate populations; abundant new evidence suggests that they are not homogeneous, and represent a fairly narrow hybrid zone.

Within each of these species, there is much more homogeneity than had previously been recognized. So far from relations among them being everywhere clinal, they are mostly sharply distinct where their ranges approach each other (even in the W Rift region, to which *Syncerus mathewsi* is restricted); the only true intermediates we have found are along the N edge of the C African forest: in N Cameroon, in NE Nigeria, and farther E in the Uele region. Both of these intermediates are between *Syncerus nanus* and *S. brachyceros*.

Pseudoryx Dung et al., 1994

The discovery of the saola, *Pseudoryx nghetinhensis*, in the early 1990s (Dung et al., 1993) sent shock waves through both mammalogy and the conservation movement, and signaled the start of a series of discoveries of unanticipated new mammals in Vietnam and Laos.

At first there was some discrepancy over the affinities of this new genus. A resolution of this initial-seeming uncertainty over the proposed relationships—does it belong to the Bovini or to the Caprini?—was proposed simultaneously by Gatesy & Arctander (2000) and by Groves & Schaller (2000). In the first instance, three DNA sequences (12S and 16S, k-casein and b-casein) placed *Pseudoryx* in the Bovinae, most convincingly (with 85% bootstrap support) as sister to *Bos* in the k-casein case (although by a different solution, in a nonsense-tree, using a-lectalbumin intron 1). The second paper reinterpreted the morphological data in this context and produced a parsimonious tree incorporating *Pseudoryx* in the Bovini. At again almost the same time, Hassanin & Douzery (1999b) used both mitochondrial and nuclear sequences to show *Pseudoryx* nested well within the Bovini, slightly closer (but with not very strong support values) to *Bos* than to the buffaloes; subsequently, Hassanin & Ropiquet (2004) concluded that it was one of three basic branches (with *Bos* and the buffaloes) in the Bovini.

Pseudoryx nghetinhensis Dung et al., 1994

saola, spindlehorn, Vu Quang ox

Even today, very little is known about this species; see Dung et al. (1994) and Schaller & Rabinowitz (1995). Groves & Schaller (2000) gave a more detailed description, described and mapped its distribution, and considered its phylogenetic position (see above).

Tribe Boselaphini Knottnerus-Meyer, 1907

The two living genera of this tribe do not appear to be very closely related. As far as the skulls are concerned, A. W. Gentry (1992) could find only a few character states in common: the horn cores are keeled, much inclined, and inserted well behind the orbits; the mastoid is narrow; and the supraorbital pits are situated far forward and widely apart above the front of the orbits.

The two genera have very different karyotypes: in *Boselaphus*, there is a Y-autosome 14 translocation, as in the Tragelaphini, whereas this is not the case in *Tetracerus*. Therefore, Benirschke et al. (1980) proposed that *Boselaphus* is sister to the Tragelaphini, or possibly even to the lesser kudu (here, *Ammelaphus*), with which it shares a fusion of the X chromosome with the same autosome.

Boselaphus de Blainville, 1816

$2n=46$ in both sexes; as in the Tragelaphini, the Y chromosome is fused to autosome 14, but also there is a fusion of the X chromosome to autosome 14—a character shared, curiously, with *Ammelaphus* (Benirschke et al., 1980).

Boselaphus tragocamelus (Pallas, 1766)

nilgai

Groves (2003) could find no marked geographic variation in this species, with the possible exception of a single unusual cranium from Bengal.

Tetracerus Leach, 1825

$2n=38$; all chromosomes are acrocentric, and there is no indication that any of them are compound, that is, there are no autosome / sex chromosome translocations (Benirschke et al., 1980).

The karyotype, therefore, is very different from *Boselaphus*.

Tetracerus quadricornis (de Blainville, 1816)

chowsingha, four-horned antelope

For the full synonymy of this species, see Groves (2003).

Groves (2003) recognized three quite distinct subspecies. It should be noted that in the two four-horned subspecies, the smaller, anterior horn pair do not develop until the second year of life.

Tetracerus quadricornis quadricornis (de Blainville, 1816)

Relatively large; mean skull length of the males 192.9 mm. Nasal bones narrow. Four long horns; mean length of the posterior horn 90.7 mm, of the anterior horn 48.6 mm (Rajasthan), 31.7 mm (elsewhere). Yellow-fawn above, creamy or creamy fawn below, this tone often confined to the midline; forelimbs markedly blackened down the anterior surface; median dorsal region darker than rest of the upperside; tail long, with a bushy white tip.

From Rajasthan and Gujarat E to Bengal and S to Karnataka and Andra Pradesh.

Tetracerus quadricornis iodes Hodgson, 1847

Similar in size, but with wider nasals; horns smaller; mean length of the posterior horn 73.5 mm, of the anterior horn 20.7 mm. Fawn above, light fawn below, this tone often confined to a very narrow midline streak; foreleg with (at most) a vague, dark brown line, often interrupted at the knee; median dorsal region not darkened. Nose diffusely darker.

Nepal and Champa.

Tetracerus quadricornis subquadricornis (Gray, 1843)

Size smaller; mean skull length of the males 187 mm. Nasal bones very broad. Only a single pair of horns at any age; horns rather long, averaging 83.5 mm. Red-fawn to more olive; underside whitish at the midline, pale yellow laterally; foreleg line very vague, or absent; median dorsal region slightly darkened; midline of the nose sometimes sharply darker. Tail short.

S India: Palkonda Hills, Madras, Dharwar, Kuckanalla.

Tribe Tragelaphini Blyth, 1863

Grubb (2004a) noted that, strictly speaking, the name Tragelaphini is antedated by Strepsicerotini Gray, 1846 (published as "Strepsiceriae").

As in the Boselaphini, the horn cores are keeled and the cranial roof is horizontal. As in both the Boselaphini and the Bovini, the mastoid is narrow and, in the last two mandibular premolars, the paraconid is differentiated from the parastylid (A. W. Gentry, 1992).

$2n$ is as follows (males/females): greater kudu, $2n=31/32$; lesser kudu, $2n=38/38$; nyala, $2n=55/56$; sitatunga, $2n=30/30$; bushbuck, $2n=33/34$; common eland, $2n=31/32$; all have a Y-autosome translocation (Benirschke et al. 1980).

Benirschke et al. (1980) postulated that the lesser kudu (plus the nilgai) and the nyala separated first, and the original Y chromosome was subacrocentric; then a pericentric inversion occurred in the Y chromosome (making it submetacentric), after the separation of the lesser kudu and the nyala. Benirschke et al. (1980) proposed the following phylogeny:

((nyala) (lesser kudu, nilgai) ((bushbuck) (bongo) (greater kudu, sitatunga, eland))).

Willows-Munro et al. (2005) compared mtDNA and nDNA data (the latter being introns). The differences were as follows:

1. The lesser kudu, sister to all the rest in the mitochondrial tree, was closely associated with the nyala in the nuclear tree.
2. In the mitochondrial tree, the bushbuck are sister to the sitatunga plus the bongo, with the mountain nyala as sister to all three; yet in the nuclear tree, the bushbuck, the mountain nyala, and the sitatunga plus the bongo are three equal branches. Moodley et al. (2009) noted that the bushbuck sequences used by Willows-Munro et al. (2005) were, in effect, *T. sylvaticus*.
3. In the mitochondrial tree, the eland, the greater kudu, and the mountain nyala / bongo/ sitatunga/bushbuck clade are three equal branches; whereas in the nuclear tree, the eland is sister to the rest, then the greater kudu, then the final clade.

As every one of these branches in both trees is highly supported using Bayesian posterior probabilities and both ML and parsimony bootstraps, the branching patterns seem real and presumably denote past episodes of hybridization. In particular, the nuclear (but not the mitochondrial) relationship of the lesser kudu (here, the genus *Ammelaphus*) and the nyala (here, the genus *Nyala*) strongly suggests that the ancestors of the lesser kudu have undergone nuclear introgression from the proto-nyala.

Hybridization is well known between bongos and sitatungas; a less expected hybrid was between an eland and a kudu. This strange-looking animal, a male, was illustrated, and its genetics described, by Jorge et al. (1976). It was sterile.

The genera recognized here follow Ropiquet (2006:figure 9), who proposed the following phylogeny (using common names):

(lesser kudu (nyala (eland, greater kudu)
(bushbuck (mountain nyala (sitatunga, bongo)))).

In her figure 18, she indicated a separation time for the lesser kudu well back in the Late Miocene; and of the greater kudu, the eland, and the bushbuck/ sitatunga, toward the end of the Late Miocene. The nyala is not represented in the figure, but she later specified (Ropiquet, 2006:66) that "le nyala (*N. angasii*) a divergé des autres Tragelaphini juste après *A. imberbis*." She specifically linked this recognition of several genera to their "divergence ancienne (Miocène supérieur)" (Ropiquet, 2006:67), and we agree with her employment of this criterion.

We examined skulls in the Natural History Museum (London) and elsewhere, and found that their morphological relationships cohere very nicely with those of Ropiquet's molecular-based taxonomy.

Nyala Heller, 1912

$2n = 55$ (males), 56 (females); the Y chromosome is translocated onto chromosome 14, as in other Tragelaphini, but there is no inversion of the Y chromosome, unlike all other Tragelaphini, with the exception of *Ammelaphus* (Benirschke et al., 1980).

Anterior median tips of the nasals very long; lateral prongs virtually absent. Several supraorbital foramina, but only the anterior ones large; foramina sitting in an elongated depression. Lacrimal vacuity relatively short, not reaching the maxilla. Malar-maxillary suture as in *Tragelaphus*, but more deeply penetrating anteriorly. Auditory bulla small, angular; postglenoid foramen an elongated oval shape, with a heart-shaped anterior margin. Basioccipital of even width between the tuberosities, which are low, but angular. Mesopterygoid fossa forming a wide U; vomer forming a ridge. Free ends of the paroccipital processes short, slightly curved. Buccal styles of the upper molars very prominent.

According to Ropiquet (2006), *Nyala* has been a separate genus since the Late Miocene.

Nyala angasii Angas, 1849

nyala

Grubb (2004a) concluded that this name must not be attributed to Gray, as previously assigned, but to Angas.

A SE African species that has no geographic variation, as far as is known.

Tragelaphus de Blainville, 1816

Anterior median and lateral nasal tips more or less equal in length, the median ones being somewhat longer. One to several foramina in a shallow, elongated depression in the frontals. Lacrimal vacuity long, usually reaching the maxilla. Malar-maxillary suture straight, more or less oblique. Bulla fairly rounded or somewhat angular, small; postglenoid foramen forming an elongated oval. Basioccipital of even width, or slightly constricted between the tuberosities, forming distinct, often rounded ridges. Mesopterygoid fossa forming a wide V or a narrow U; vomer rounded, not ridged. Free ends of the paroccipital processes long. Buccal styles on the upper cheekteeth very prominent (the least so in *T. scriptus*).

According to Ropiquet (2006:figure 18), this genus separated from its sister genera, *Taurotragus* and *Strepsiceros*, toward the end of the Late Miocene, and the species in the genus began to diverge in the early part of the Late Pliocene. The bushbuck are depicted as separating first, but Ropiquet did not state which taxon of bushbuck was used in the analysis.

Tragelaphus scriptus Group

bushbuck

Grubb (1985a) recognized the following taxa in E Africa, which, at that time he classed as subspecies of *T. scriptus*:

1. *Tragelaphus scriptus bor*
   —Synonyms—*T. cottoni, T. meridionalis, T. dodingae*; perhaps *T. pictus, T. signatus, T. punctatus, T. uellensis*.
   —Lowland savanna and swamps of Sudan and N Uganda, across the Albert Nile, but not S across the Victoria Nile.
   —Forms a species-group with *T. s. scriptus*.
2. *Tragelaphus scriptus decula*
   —Synonyms—*T. multicolor, T. nigrinotatus, T. fulvoochraceus*.
   —Eritrea, Sudan near the Atbara, Ethiopia to about 10° N.
3. *Tragelaphus scriptus meneliki*
   —Synonym—*T. powelli*.
   —Higher altitudes of Ethiopia.
   —Was considered close to *T. s. decula*.
4. *Tragelaphus scriptus fasciatus*
   —Synonyms—*T. olivaceus, T. reidae*.
   —Riverine forest and coastal forest–savanna mosaic, at low altitudes, from Tanzania and Kenya to Somalia and along the Webi Shebeyli, and perhaps also the Juba River, into Ethiopia.
5. *Tragelaphus scriptus dama*
   —Synonyms—*T. dianae, T. simplex, T. sassae*.
   —Ituri, Uganda N to the Victoria Nile, Rwanda, extreme N of Lake Tanganyika.
   —Intergrades with the *T. s. scriptus* group along a narrow and discontinuous belt from Lake Edward to Lake Tanganyika, adjacent to the W Rift Valley, but no intermediates are known from the apparent contact area W of Lake Albert or in S Sudan.
6. *Tragelaphus scriptus delamerei*
   —Synonyms—*T. massaicus, T. haywoodi, T. meruensis, T. tjaderi, T. brunneus, T. eldomae, T. locorinae, T. laticeps, T. heterochrous, T. barkeri*.
   —E Uganda (highlands), Kenya, S Sudan (highlands), N Tanzania.
   —Forms a species-group with *T. s. dama* and *T. sylvaticus*.
7. *Tragelaphus scriptus ornatus*
   —Angola, S DRC, Zambia, N Botswana, W Zimbabwe.
   —Pelage seemed rather more like *T. s. scriptus*; in other respects shown to be close to *T. s. sylvaticus*.

Here, we recognize several species-level taxa within what has previously been called *T. scriptus*. The differences in the *scriptus*-group taxa may be tabulated as follows:

In our craniometric analyses, we can find no differences among *T. scriptus* populations W of the Nile. Nor are there any differences between so-called *T. dodingae* and *T. bor* (i.e., the *scriptus*-like populations E and W of the Nile), but *T. decula* is quite different.

In a comparison of montane populations from Sudan and Ethiopia, *T. decula* is at the edge of the variation of *T. meneliki*; skulls from the distributional area assigned to *T. powelli* are within the range of variation of *T. meneliki*, consistent with the evidence from the skins (Grubb, 1985a) that indicates the two are synonymous. The Sudanese montane population described as *T. barkeri* is quite distinct, differing from the others in its larger size (mainly in greatest length and preorbital length; there are almost no differences in the horns, and very little difference in the bizygomatic breadth).

In populations from Sudan and Uganda, *T. bor* is separate from the others; *T. barkeri* and *T. dama* are the same, and differ from *T. bor* in the longer skull but shorter preorbital length, and the slightly greater horn span. Of two skulls labeled "S Kordofan," one is evidently an example of *T. bor*, while the other is closer to *barkeri/dama*. A skull from Jebel Manda (N Bahr-el-Ghazal) is like *barkeri/dama*; one from Niangara represents *T. bor*.

When *T. barkeri* and *T. dama* are compared with E African "*T. delamerei*," they do not separate—hardly even on average. But there is good separation between *T. meneliki* and the sample here called *T. delamerei*, whereas *T. fasciatus* does not separate as well from *T. delamerei*; these two E African taxa differ from *T. meneliki* by having a longer skull and greater preorbital length, and a slightly narrower horn span.

Table 26 Differences in East African *Tragelaphus scriptus* group

	bor	*decula*	*meneliki*	*fasciatus*	*dama*	*delamerei*	*ornatus*
Neck hair	Short	Very short	Long	Very short	Long but with a gray collar of short hairs	Long but with a gray collar of short hairs	Long but with a gray collar of short hairs
Dorsal stripe	Well marked, dark brown, more or less extending to the neck	Black- brown; often expanded to a blackish saddle in the females	Young males only	Brownish, poorly defined in the females	Dark brown in the males; often narrow in the females	Dark brown in the males; often narrow in the females	Dark brown in the males; often narrow in the females
Median crest of white hairs	Especially in the males	Often absent	Often a few hairs posteriorly	Some hairs anteriorly	Usually prominent	Few	Few
Nuchal crest	Usually not reversed	Usually not reversed	Never reversed	Usually reversed	Reversed	Reversed	Reversed
Adult males							
Body color	Light yellowish brown	Dull sandy ochre	Uniform glossy dark (often grayish) brown	Grayish ochre or brownish gray	Light grizzled gray-ochre in the adults, yellowish chestnut in the subadults	Dark brown in adults, reddish chestnut in the young; grizzled	Bright dark red-brown
Blackish grizzling	Present	Restricted to the saddle in the young, dominant in adults	None	None	Blacker	Blacker	Blacker
Underside and upper limbs	Blackish	Blackish	Darker	Darker brown	Dark gray in the males, buff in the females	Dark blackish brown	Darker
Neck	Paler, grayer	Lighter, often with a dark patch at the base	As in the body	As in the body	As in the body	As in the body	As in the body
Adult females							
Body color	Yellower; grizzling less evident	Light sandy ochre to light rufous	Uniform glossy dark brown	Rich yellow-ochre	Yellowish chestnut, not obviously grizzled	Chestnut, not obviously grizzled	Chestnut
Blackish markings	Absent	Marked	Absent, or reduced to 2 faint spots	None	None (more red-ochre below)	None (more red-ochre below)	None (more red-ochre below)
Neck	Relatively lighter	Lighter	As in the body	As in the body	As in the body	As in the body	As in the body
Upper lateral stripe	15–30 cm (occasionally absent or faint)	Long, or absent	Absent	Absent	Absent	Absent	Absent
Lower lateral stripe	Usually continuous	Absent, or a few spots posteriorly	Absent, or a few very small rump spots	Line of scattered spots, but many spots on the haunches	Reduced to spots; rump spots	Absent, or just a few faint spots, but usually some rump spots	Broken up into spots
Vertical stripes	2–9, rarely faint, or a line of spots	None	None	3–7	Absent, or faint spots	Absent, or faint spots	Absent, or faint spots
White streaks on the lower limbs	May be confluent	Extensive	Often very reduced	Yellowish	Poorly marked, or absent	Poorly marked, or absent	Poorly marked, or absent
Horns	Straight, smooth	Smooth	Smooth	Widespread, rugose, well-marked annuli	Massive, widespread	Massive, widespread	Massive, widespread
Keels	Virtually absent	Present, but weak	Present, but weak	Pronounced, “folded”	Strong	Strong	Strong

A skull from Tunduru in SE Tanzania is *T. fasciatus* (or could perhaps be *T. delamerei*). One from Mikunduni (on the lower Tana) is on the edge of the *delamerei* range of variation.

Bringing in the S African taxa, *T. ornatus* is somewhat—but not absolutely—different from *T. delamerei*, with *T. sylvaticus* being intermediate, but they overlap fairly widely.

We tested the affinities of *T. ornatus*, which has clear white body markings, by bringing in skull measurements of C African forest bushbuck (*T. phaleratus*) of the *T. scriptus* group. The *T. ornatus* sample is not intermediate, but it is clearly close to *T. sylvaticus*, differing from *T. phaleratus* by its larger size but shorter preorbital length, narrower bizygomatic breadth, and longer horns.

We find that a specimen from Butele, 5.02° S, 23.48° E, represents *T. ornatus*.

We cannot split up the C African *scriptus*-like bushbuck by craniometrics. Comparing the W-C African (*T. phaleratus*), C African (*T. johannae*) and Uele Valley (*T. punctatus*) populations, there are different centers of gravity, but no definite divisions.

Table 27 gives univariate statistics for populations of bushbuck.

The study of the mitochondrial control region and cytochrome *b* by Moodley & Bruford (2007) showed considerable genetic diversity among bushbuck, with rather more significant clades than is indicated by the morphological data.

The significant separation in their phylogenetic tree is between the *T. scriptus* and *T. sylvaticus* clusters; interestingly, *T. decula* is well within the former, and part of the *T. bor* subclade. The morphological affinities of *T. decula* are so very clearly with the *T. sylvaticus* group that a hybrid status seems the most plausible solution, conjecturing that formerly the Ethiopian plateau was occupied in suitable areas by forest-living bushbuck of the *T. scriptus* group. As the climate changed, *T. sylvaticus* bushbuck moved northward, replacing the indigenous phenotype by nuclear swamping (the larger, more sturdily horned males dominating the smaller, weaker-horned indigenous males).

Within the *T. scriptus* clade, Moodley & Bruford (2007) found a major split between those from Sierra Leone and Guinea-Bissau (*T. scriptus*) and those from other localities; the support for the non-*scriptus* clade is only 79%, while the two significant subclades within it (Ghana to NW DRC [*T. phaleratus*] and N DRC / Chad / Sudan / NW Uganda [*T. bor*]) have support values of 100% and 90%, respectively.

Within the *T. sylvaticus* clade, the only really significant division is between *T. ornatus* (94% support) and the rest (only 83% support). *T. meneliki* is divided into two apparently significantly different clades, but with only two specimens in each, and is associated nonsignificantly (67% support) with a cluster of southernmost *T. sylvaticus*. Haplotypes of *T. fasciatus* do not form a monophyletic clade but, instead, recur in two places within the major *sylvaticus* clade; we take this to mean that there has been a great deal of gene flow between this (phenotypically well-defined) coastal species and the more inland *T. sylvaticus*.

Further analysis (Moodley et al., 2009) found that not only were there two major clades, but that they seem to occupy different positions among the Tragelaphini. The *T. sylvaticus* clade was preferentially associated with the *T. spekii* / *T. euryceros* clade (although, as the authors admit, not with a very high level of support), whereas the *T. scriptus* clade was either the sister group of *Nyala* (*Tragelaphus angasi* in the "traditional" taxonomic arrangement, which the authors used) or else sister to all other Tragelaphini. Whatever the true affinities of the two bushbuck clades, nonmonophyly (at least of mtDNA) was clearly favored over monophyly.

Tragelaphus scriptus Cluster

harnessed bushbuck

Tragelaphus scriptus (Pallas, 1766)

1766 *Antilope scripta* Pallas. Senegal.

1844 *Antilope leucophaea* Forster; not Pallas, 1766 (= the blaubok).

1882 *Tragelaphus gratus* Rochebrune; not Sclater, 1880 (a sitatunga). Mouth of the Senegal River, between Cayor and Walo.

1898 *Tragelaphus obscurus* Trouessart. To replace *Tragelaphus gratus*.

Specimens available to us are rich dark rufous with a blackish suffusion; between 3 and 10 distinct, transverse white stripes; an upper and a lower white longitudinal flank band; a circle of white haunch-spots. Upper part of the limbs blackish, the inner side white; often a black line right down the front of the forelimbs. Underparts black. Pelage fairly long. Females paler than the males, marked similarly.

The mtDNA data (Moodley & Bruford, 2007) unite specimens from far W Africa, specifically Sierra Leone and Guinea-Bissau, to the exclusion of others. The support value for this clade is 99%.

Table 27 Univariate statistics for putative taxa of the *Tragelaphus scriptus* group: males

	Gt l	Bizyg	Preorb l	Horn l	Span	Horn br
W Africa *scriptus*						
Mean	237.43	93.89	119.14	232.50	121.75	39.50
N	7	9	7	8	8	2
Std dev	6.901	2.369	5.460	20.529	15.341	—
Min	228	90	112	207	106	37
Max	247	97	127	263	155	42
Lower Volta *scriptus*						
Mean	257.00	99.00	124.00	299.60	160.00	43.00
N	1	1	1	5	4	2
Std dev	—	—	—	14.450	14.445	—
Min	—	—	—	279	148	43
Max	—	—	—	317	180	43
phaleratus						
Mean	235.61	93.44	116.91	237.80	125.03	42.00
N	28	39	29	41	37	17
Std dev	7.529	3.331	5.011	25.040	23.410	3.623
Min	217	87	106	185	19	36
Max	245	100	130	290	184	48
bor						
Mean	238.14	94.42	118.91	253.43	134.58	44.75
N	22	39	22	42	40	20
Std dev	7.298	2.701	4.492	26.821	18.832	5.911
Min	227	90	112	197	110	34
Max	253	101	129	320	188	56
dodingae						
Mean	238.00	95.00	115.00	230.00	127.50	—
N	2	2	2	2	2	—
Min	237	93	113	218	107	—
Max	239	97	117	242	148	—
decula						
Mean	223.43	91.44	105.07	250.86	143.67	—
N	7	8	7	7	6	—
Std dev	5.192	2.921	3.297	29.464	14.180	—
Min	219	86	101	212	123	—
Max	234	95	110	305	161	—
powelli						
Mean	221.67	90.75	107.00	226.75	147.33	—
N	3	4	3	4	3	—
Std dev	4.041	5.679	1.000	21.469	15.144	—
Min	218	87	106	200	130	—
Max	226	99	108	249	158	—
meneliki						
Mean	228.05	92.43	109.16	265.58	150.88	—
N	19	20	19	18	17	—
Std dev	7.051	3.836	4.717	25.308	15.029	—
Min	217	83	102	220	130	—
Max	238	98	119	320	191	—
barkeri						
Mean	254.88	101.75	127.13	321.00	178.20	50.67
N	8	8	8	10	10	3
Std dev	4.016	3.151	2.949	43.586	28.177	4.509
Min	248	96	123	276	148	46
Max	262	105	132	404	248	55

Table 27 (continued)

	Gt l	Bizyg	Preorb l	Horn l	Span	Horn br
dama						
Mean	254.81	100.29	125.76	312.19	175.41	50.05
N	40	64	44	77	76	56
Std dev	9.119	3.809	5.874	39.079	20.854	4.518
Min	235	92	115	202	132	40
Max	274	108	137	420	217	59
delamerei						
Mean	253.64	99.59	125.02	322.23	176.05	48.07
N	29	46	32	47	46	27
Std dev	8.612	5.010	5.600	36.756	27.138	4.682
Min	232	89	113	238	135	39
Max	274	110	136	405	263	60
fasciatus						
Mean	253.44	98.85	127.00	344.85	202.90	58.00
N	9	10	9	10	10	2
Std dev	9.700	4.972	6.016	46.357	49.492	16.971
Min	236	93	115	261	150	46
Max	271	106	136	420	332	70
ornatus						
Mean	243.27	94.98	118.61	300.79	168.36	47.03
N	33	45	33	48	45	32
Std dev	8.255	4.251	5.267	39.295	32.343	3.403
Min	229	84	107	205	109	39
Max	268	106	132	372	309	53
sylvaticus						
Mean	243.44	94.32	122.00	295.06	164.59	45.75
N	18	25	10	32	17	8
Std dev	8.016	2.911	3.300	33.069	18.456	1.165
Min	230	89	117	232	130	43
Max	258	102	127	380	219	47

Tragelaphus phaleratus (Hamilton Smith, 1827)

1827 *Antilope phalerata* Hamilton Smith. W bank of Stanley Falls, Congo.

1905 *Tragelaphus scriptus knutsoni* Lönnberg. Upper Manns Valley, SE Nigeria.

1914 *Tragelaphus scriptus pictus* Schwarz. Duguia, lower Shari Valley.

1914 *Tragelaphus scriptus punctatus* Schwarz. Duma, Oubangui Valley.

1914 *Tragelaphus scriptus signatus* Schwarz. Tome Valley, near the Gribinge-Oubangui watershed.

Body color reddish brown, without any blackish suffusion, except on the withers. Upper longitudinal band often absent, especially in the females.

Ghana, Togo, Nigeria, Cameroon, and Congo; the easternmost locality is Bolobo, in the far NW of the DRC.

The support value for this clade in the Moodley & Bruford (2007) analysis is 100%.

Tragelaphus bor Heuglin, 1877

1877 *Tragelaphus bor* Heuglin. Req swamps (restricted by Grubb, 1985a), near Bor, White Nile, S Sudan.

1912 *Tragelaphus cottoni* Matschie. Farajala, W of Lado, on the Koda River, White Nile.

1912 *Tragelaphus cottoni dodingae* Matschie. Kedef Valley, W slope of the Didinga Hills, E of White Nile, Sudan, 4° 20' N, 33° 30' E.

1912 *Tragelaphus cottoni meridionalis* Matschie. Three days' march N of Wadelai, Uganda.

1914 *Tragelaphus scriptus uellensis* Schwarz. Angu, Uele Valley.

Color more ochery, less red, with no dark suffusion on the withers; transverse stripes less distinct, especially in old animals; upper longitudinal band short or absent, the lower one generally broken into spots and streaks; several haunch spots; white mark on the throat.

From Bambio in the CAR, Koumoudjou in Chad, and the Uele district in NE DRC to the Victoria Nile in Uganda and the S Sudan as far E as the Kedef Valley, well to the E of the Nile.

Support for this clade is 90%.

Tragelaphus sylvaticus Cluster
keeled bushbuck

Tragelaphus decula (Rüppell, 1835)

1835 *Antilope decula* Rüppell. "Abyssinia."

1902 *Tragelaphus multicolor* Neumann. Awash Valley, SE of the Sekuala Range, Shoa.

1902 *Tragelaphus nigrinotatus* Neumann. Barsa Valley, Maleland, N of Lake Stephanie.

1912 *Tragelaphus decula fulvo-ochraceus* Matschie. Dungoler, near Lake Tana.

Color ochery to yellow-brown, often with a black suffusion on the back; white markings indistinct, fading completely in old animals, with the exception of some haunch spots; two longitudinal bands often visible in young animals; white throat-patch present, sometimes also a breast-patch. Front of the forelegs black, with white on the front of the knees.

Dry lowlands of Ethiopia.

Evidently a species of hybrid origin, because it possesses the phenotype of the *T. sylvaticus* cluster but the mtDNA of the *T. scriptus* group. Moodley & Bruford's (2007) specimens were from Lake Abaya and Ghimbi.

Tragelaphus meneliki Neumann, 1902

1902 *Tragelaphus meneliki* Neumann. Gara Mulatta, Burka, and Jaffa ranges, the upper Webi Shebeyli watershed.

1912 *Tragelaphus powelli* Matschie. Managasha Forest, W of Shoa.

Dark brown to black in the adult males, with no white except in the axillae, but with the occasional exception of indistinct white flecks on the haunches and above the hoofs; very occasionally, traces of white lines down the forelegs. Females lighter, red, with a tendency for the white marks to be clearer. Coat long, full, shining.

Highlands of Ethiopia.

Tragelaphus fasciatus Pocock, 1900

1900 *Tragelaphus scriptus fasciatus* Pocock. Sen Morettu, Webi Valley, SE Ethiopia, 6° 20' N, 42° 50' E.

1913 *Tragelaphus scriptus olivaceus* Heller. Maji-Ya-Chumbi, SE Kenya.

1949 *Tragelaphus scriptus reidae* Babault. Amboni, NE Tanzania.

Males dark gray-brown above, gray on the sides; females more yellowish; young animals more rufous. About six white stripes, generally distinct; broken longitudinal flank-band; many haunch-spots. No black on the crown and the nose; chevron below the eyes in the females.

Coastal forests of E Africa, from the Webi Shebeyli S at least to S Tanzania.

Moodley & Bruford (2007) found that the mtDNA of two specimens of this species (one from the Tana River, one from Mona Mofa in Somalia) clustered with a *T. sylvaticus* specimen (ascribed to *T. roualeyni*) from Thabazimbi, South Africa, while two others from Mona Mofa clustered with a number of *T. sylvaticus* specimens from localities in Tanzania, Malawi, and Zimbabwe. There has evidently been gene flow between *T. fasciatus* and *T. sylvaticus*, but its history and direction are hard to work out.

Tragelaphus ornatus Pocock, 1900

1900 *Tragelaphus scriptus ornatus* Pocock. Linyanti, Chobe River between Lake Ngami and the Zambezi River.

Males rich dark rufous, becoming black on the withers; 6–8 white stripes; many haunch-spots; longitudinal bands represented only by a row of spots. Outer sides of the legs black above the knees, the inner sides white, with a broad black "garter" above the knee and the hock; white stripe from the knees and the hocks to the pasterns. Females light red-brown, with fewer stripes.

From the Chobe and Okavango rivers E along the Zambezi and its tributaries, at least as far as Malawi.

Moodley & Bruford (2007) found 94% support for this species in general, as represented by their clade containing specimens from Zambia, Zimbabwe, Botswana, and Angola; in turn, the Angola and non-Angola clades also had support values of 97% and 98%, respectively; the affinities of these two geographic groups need to be investigated morphologically.

On both DNA and morphological grounds, this species is a clear member of the *T. sylvaticus* cluster, despite the presence of striping in the adults.

Tragelaphus sylvaticus (Sparrman, 1780)

1780 *Antilope sylvatica* Sparrman. Groot Vatersbosch district, S Cape.

1850 *Antelopus roualeynei* Gordon-Cumming, nom. nud.

1891 *Tragelaphus scriptus roualeyni* Thomas. Bakarikari, near the sources of the Limpopo River.

1900 *Tragelaphus delamerei* Pocock. Saya Valley, 0° 55' N, 36° 55' E.

1902 *Tragelaphus massaicus* Neumann. Upper Bubu Valley, NW of Irangi, Tanzania.

1902 *Tragelaphus scriptus dama* Neumann. Kavirondo Gulf.

1905 *Tragelaphus haywoodi* Thomas. Nyeri, near Mt. Kenya.

1908 *Tragelaphus sylvaticus meruensis* Lönnberg. Between Mt. Kilimanjaro and Mt. Meru, N Tanzania.

1909 *Tragelaphus tjaederi* J. A. Allen. Nakuru.

1912 *Tragelaphus dianae* Matschie. Kabakaba, near Mahagi, N end of Lake Albert, 2° 10' N, 30° 45' E.

1912 *Tragelaphus dianae simplex* Matschie. Ituri Valley, near Kifuku and Irumu.

1912 *Tragelaphus eldomae* Matschie. Eldama Ravine, Mau Forest, Kenya.

1912 *Tragelaphus haywoodi brunneus* Matschie. W of Mt. Kenya.

1912 *Tragelaphus locorinae* Matschie. Mt. Lcorina, 2950 ft., S of the Didinga Hills, and N of the Marangole Mountains, Sudan. Not "near Narringepur," as stated by Grubb (1985a); this is the locality of the juvenile female specimen given as the type in the printed version of the type description, but a note in Matschie's handwriting on the Berlin Museum offprint of the paper deletes the "Typus" against that specimen and writes it against the adult male specimen from Mt. Locorina.

1912 *Tragelaphus locorinae laticeps* Matschie. NW foot of Mt. Kadam, in the Debasien Range, 3950 ft., N of Mt. Elgon.

1912 *Tragelaphus scriptus makalae* Matschie. Makala, S of the Lindi Valley, about 27° 50' N, 0° 35' E.

1912 *Tragelaphus scriptus sassae* Matschie. Ishasha River, DRC–Uganda boundary S of Lake Edward.

1919 *Tragelaphus scriptus heterochrous* Lönnberg. W slope of Mt. Elgon.

1924 *Tragelaphus barkeri* Millais. Imatong Mountains, SE Sudan.

?1961 *Tragelaphus insularis* or *Tragelaphus scriptus insularis* Zukowsky. Islands in the Rufiji delta.

Older males deep brown to blackish brown, with grayish sides, more chestnut above; younger males more red-brown. Forehead black-brown; nose black, with two white suborbital spots. White mark on the throat; white spots (usually indistinct) on the face and the haunches; little or no trace, in the adult males, of transverse or longitudinal stripes. Females dark yellow-brown to more reddish, tending to be lighter on the shoulders and the upper part of the forelegs; often with distinct traces of stripes.

T. sylvaticus has an enormous range, from the Cape N to the Kenya highlands, Uganda, and NE DRC, with outlying populations in the Imatong and Didinga-Marangole mountains, Sudan. It is, in general, impossible to tell a Kenyan from a Cape animal of the same sex and similar age. Samples in Moodley & Bruford's (2007) phylogenetic tree come from as far N as the Imatong Mountains, Lango, and Karamoja (N Uganda) and Irumu, the Ishasha River and the Semliki district (DRC), down through Kenya and Tanzania to South Africa. As with the morphological characters, the haplotypes are all intermixed, and no real geographic divisions can be discerned.

In a few areas, however, some distinctive morphological types predominate, and it may be that a few subspecies ought to be recognized. Mt. Elgon individuals tend to be more of a dark red-brown in the males, with almost no traces, even in the females, of white marks, except for some haunch-spots; some skins can be very similar to the lighter skins of *T. meneliki*. Possibly these should be separated under the name *T. heterochrous* Lönnberg, 1919.

Those from W Uganda and E DRC are clearly *sylvaticus*-like in size, horns, short collar, and so on, but they tend to be lighter, with clear white markings and, generally, a white longitudinal band composed of a line of flecks. It is possible that they represent a population with some gene flow from *T. bor*. This is particularly the case for the type specimen of *T. s. sassae*, from the Sassa (= Ishasha) River, which is ochery brown with a hint of rufous, and faint but distinct white stripes.

The Imatong Mountains population, described as *T. barkeri*, and the Didinga-Marangole Mountains population, described as *T. locorinae*, are of special interest. The Imatongs, or Al Istiwa'iyah, a small, isolated range at 4° 5' N, 32° 51' E, just N of the Uganda border, and the more extensive Didingas farther E (and their S continuation across the border, the Marangoles), both have bushbuck that were described as separate species (*T. barkeri*, *T. locorinae*). Both of these species were thought to be "doubtfully distinct from *T. scriptus bor* Heuglin" by G. M. Allen (1939), but they are, in fact, populations of *T. sylvaticus* in highland forest, surrounded by a sea of harnessed bushbuck, *T. bor* (Grubb, 1985a). This was the first indication that what had been taken to be simple geographic variation within a single species was, in fact, more complicated, involving interdigitation of sharply distinct species of bushbuck, segregated by habitat.

A dwarf bushbuck was described as *T. insularis* by Zukowsky (1961), from islands in the Rufiji delta. It is impossible to say, from the type description, to which species this might be allied—presumably either *T. sylvaticus* or *T. fasciatus*.

Tragelaphus spekii Group

sitatunga

Much larger than bushbuck, with long, robust horns; particularly characterized by extremely elongated hoofs, adapted to their marshy habitat; only those of the dry Sesse Islands, in Lake Victoria, appear to have shorter, more "normal" hoofs.

The results of our craniometric analyses are given below.

Okavango sitatunga are large, but short-faced, with relatively short horns; the skulls seem closer to those from Uganda than to those from Bangweulu or other S African samples.

As far as population samples from Lake Victoria and elsewhere in Uganda are concerned, skulls from Zinga Island, just offshore and SW of Entebbe, and Damba Island, only a little farther offshore, are not essentially different from specimens from the mainland, but one from Nkosi Island, in the Sesse group, well offshore of the W side of the lake, is quite different.

The Cameroon sample is also somewhat different from Okavango and Uganda specimens, and Sudan sample even more so—its large size and long face contrasting with its short horns and narrow skull (Cameroon has these features to a lesser extent).

A skull from the Kouyou River, Congo, is more like the ones from Bangweulu and Uganda than like those from Cameroon (which is the closest sample geographically). A skull from Elephant Bay, Angola, is different from all the others, being especially broad for its size. The type of *T. albonotatus*, from "possibly either upper Guinea or Angola," is clearly close to the Cameroon sample; the Cameroon and *T. albonotatus* skulls are large and long-faced, but they have contrastingly short horns.

For their size, sitatunga from Cameroon, Sudan, and Okavango have comparatively short horns; Bangweulu, Tanzania, and Uganda sitatunga have relatively long ones.

Univariate statistics are given in table 28. The type specimen of *T. albonotatus* (in the Zoologisches Museum Alexander Humboldt in Berlin) is an extremely aged animal with strongly worn horns, their tips being reduced to rounded stubs. Hence the horn length is not comparable to those of other specimens, and is enclosed in parentheses.

Okavango (*T. selousi*) specimens are the same size as those from Bangweulu (to which the name *T. inornatus* presumably applies), but broader and shorter-faced, with shorter horns. Tanzania (*T. ugallae*) skulls

Table 28 Univariate statistics for the *Tragelaphus spekii* group: males

	Gt l	Bizyg	Preorb l	Horn l
albonotatus				
Mean	303.00	122.00	151.00	(366.00)
N	1	1	1	1
Angola				
Mean	327.00	133.00	175.00	665.00
N	1	1	1	1
Bangweulu				
Mean	298.33	119.67	159.33	582.33
N	3	3	3	3
Std dev	5.859	6.351	3.055	51.598
Min	294	116	156	524
Max	305	127	162	622
Cameroon				
Mean	304.00	120.38	160.20	509.44
N	6	8	5	9
Std dev	10.450	2.326	7.014	50.028
Min	290	116	150	425
Max	316	124	167	568
Congo				
Mean	298.00	126.00	152.00	529.50
N	1	1	1	2
Min	—	—	—	503
Max	—	—	—	556
Gabon				
Mean	323.00	120.00	166.00	554.00
N	1	1	1	1
Gambia				
Mean	—	—	—	608.00
N	—	—	—	1
Okavango				
Mean	297.50	126.67	150.50	540.33
N	2	3	2	3
Std dev	—	7.371	—	20.108
Min	295	121	147	518
Max	300	135	154	557
Sesse Island				
Mean	270.00	116.67	131.00	495.50
N	2	3	2	2
Std dev	—	3.786	—	—
Min	265	114	128	481
Max	275	121	134	510
Sudan				
Mean	326.00	121.00	173.00	533.33
N	1	2	1	3
Std dev	—	—	—	56.439
Min	—	117	—	494
Max	—	125	—	598
Tanzania				
Mean	285.00	116.00	141.00	531.00
N	1	1	1	1
Uganda				
Mean	286.20	120.17	151.70	553.16
N	10	18	10	19
Std dev	13.248	3.930	7.258	40.237
Min	262	112	143	467
Max	304	125	166	608

are smaller, compared with Bangweulu ones, and much shorter-faced, with shorter horns, thereby somewhat resembling those from Okavango.

Most skulls from Uganda (*T. spekii*) are also smaller than the Bangweulu ones, and they have similar skull proportions but slightly shorter horns. Those from the Sesse Islands (*T. sylvestris*), though only slightly smaller than other Uganda skulls, have a much shorter face and very short horns.

Sudan specimens (*T. larkenii*) are larger than Uganda ones, the skull being relatively narrower and the horns shorter.

In the Cameroon sample, the size is like the Bangweulu specimens, with the same skull proportions and very short horns. The Gabon skull (presumably representing *T. gratus*) does not fall into the range of the Cameroon specimens, but it is larger than any from Cameroon, is relatively narrower, and has longer horns. The Congo skull is the same size as those from Cameroon, but the former is relatively broader and, again, has slightly longer horns. Finally, the skull from Elephant Bay, Angola, is larger than any from Cameroon—in fact, it is the largest of the entire set—and has a shorter face and extremely long horns.

Sitatunga habitat, in deep-water swampland, is discreet and intermittent. It is no surprise that these animals have differentiated into quite a number of distinct local forms. We propose the following revision, based on our analyses (above) plus our study of skins, but, as noted above, there are hints that the diversity may actually be even greater.

Tragelaphus spekii Speke, 1863

1863 *Tragelaphus spekii* Speke. Karagwe.

The correct spelling of the specific name, the date of the description, and the person to whom it should be attributed have all been much discussed; the correct author should be Speke, the date 1863, the spelling *-ii* (Grubb, 2004a).

Preserved skins seen: 13/4 (males/females).

Sexes dissimilar in color. Males brown, females dull reddish brown; males with prominent preorbital spots, occasionally with cheek spots; females with no or only faint preorbital markings. Males occasionally with some white in the crest. Females sometimes with rump spots or flank spots. Young animals a uniform yellow-brown or dull reddish brown, with faint or no white marks; preorbital spots sometimes distinct in the females; red on the head; neck paler. Skull relatively small; horns long.

We presume that the name *Tragelaphus wilhelmi* (Lönnberg & Gyldenstolpe, 1924)—type locality: Lake Bunyonyi, Kigezi, Uganda—is a synonym of *T. spekii*. If so, then this is a species found in suitable areas almost throughout C and W Uganda.

Tragelaphus sylvestris (Meinertzhagen, 1916)

Skins seen: 3/2.

Males with the neck paler, the top of the head reddish, and preorbital marks sometimes faint; females dull red-brown, with a dark dorsal stripe and no white marks. Young animals more reddish; no markings, or perhaps a flank band; neck paler; top of the head reddish; dark muzzle. Skull small; facial skeleton very short; horns short. According to the describer, *T. sylvestris* lacks the elongated hoofs of all other sitatunga.

Nkosi Island; apparently a dryland taxon.

This is Meinertzhagen's other contribution to mammalian taxonomy, apart from the giant forest hog, and, despite his reputation (Garfield, 2006), we can find no fault with him in this instance.

Tragelaphus larkenii (St. Leger, 1931)

Skins seen: 4/4.

Males rich dark brown, 4–8 stripes, usually faint, and a lateral stripe that may be reduced to 3–4 spots; females dark red-brown, 6–7 faint or broken stripes and a lateral stripe or spots; dorsal crest usually with white in both sexes; much gray in the necks of the females. Young animals rich red-brown, with 6–9 stripes and a lateral stripe or spots; dorsal crest sometimes white or all dark; neck pale. Large in size; long face; relatively short horns; relatively narrow skull.

Fifty mi. S of Yambio, Sudan. We have seen specimens from the swamps of the White Nile and the Bahr-el-Ghazal, S Sudan.

?*Tragelaphus ugallae* Matschie, 1913

1913 ?*Tragelaphus ugallae* Matschie. Sindi, Ugalla, Tanzania.

Skins seen: 0/2.

Dull yellow-brown, with only vague white markings (in the skin from Tabora), or reddish (a nice red-yellow-gray) with seven clear but ill-bordered stripes and a flank stripe (in the skin from Kigoma); dark dorsal stripe. Skull relatively small; short-faced; relatively short horns, as in *T. selousi* from the Okavango.

We use this name provisionally for specimens from Tanzania that we have seen, although the two available skins are more variable than seems likely in a single taxon.

Clearly, more needs to be known about these sitatunga: whether they are all the same or, as seems probable, they represent different taxa.

Tragelaphus gratus Sclater, 1880

1880 *Tragelaphus gratus* Sclater. Gabon.

Skins seen: 2/2.

Large; pale pink-brown; white dorsal crest; white lateral stripes and rump spots. A female from Mbi Crater, 15 mi. NE of Bamenda, is dull gray-yellow-brown, with 4 *very* faint stripes, but the white dorsal stripe is very clear. A male from Tiko, W Cameroon, is rich pale rufous, with spots high on the back rather than stripes and a clear black dorsal stripe. The skull, if the Cameroon sample is typical, is somewhat like that of *T. larkenii*, but with very short horns. We are unsure if all of these represent the same taxon.

The type of *Tragelaphus albonotatus* Neumann, 1905—type locality: "possibly either upper Guinea or Angola"—is like *T. gratus*, but with a longer and wider frontal chevron. The skull shows no difference.

Three males from Angola are pale to dark brown, with paler hairs in the dorsal crest, and hints of 1–2 stripes. They evidently represent a different, undescribed taxon, but more evidence is needed before it can possibly be described.

Tragelaphus selousi Rothschild, 1898

1898 *Tragelaphus selousi* Rothschild. Zambezi Valley; exact locality uncertain (Grubb, 1999a).

Skins seen: Okavango region 3/7, Bangweulu region 7/6.

Okovango skins are brown in both sexes; preorbital spots are prominent in the males, faint or obsolete in the females; one female has a faint lateral line and rump spots. Young animals have long fur and are dull brown, with faint preorbital spots, a dark dorsal stripe, a faint lateral line, rump spots, and hints of stripes.

Bangweulu skins are mostly the same, except that one is pale yellow-brown; but, as shown above, they differ in skull form (by being longer-faced and relatively longer-horned) from the Okavango individuals, which may be taken (rightly or wrongly?) as representing *T. selousi*. The type locality of *T. inornatus* Cabrera, 1918, is Lake Young, NE Zambia; this may be the name that would be applicable to the Bangweulu sitatunga if further research finds it to be a distinct species.

Tragelaphus eurycerus (Ogilby, 1837)

bongo

$2n = 33$ (males), 34 (females), there being a Y-autosome translocation; the X chromosome is polymorphic, either a large acrocentric (as in other Tragelaphini), or an apparently uniquely derived submetacentric (Benirschke et al., 1982).

The bongo, despite its more robust build and lack of significant sexual dimorphism (including the presence of horns in the females), is essentially a dryland sister species of the sitatunga. Their skulls are similar, and the horns are very alike.

In our analysis of the skulls of males from W and C Africa (see table 29), the three major geographic groups—W-C African, Uele, and W African—separate nicely. They have a progressively larger size, and horns with a somewhat greater span, in that sequence; but the horn length, the tip-to-tip distance, and relative skull breadth are not quite consistent,; that is to say, the cline is not a geographic one. The picture for females is similar (see table 30): W Africa are again somewhat broader in biorbital breadth than W-C Africa, with a wide tip-to-tip distance; while Uele are smaller, with a broader span, and a somewhat greater tip-to-tip distance than the other two groups.

The univariate statistics for males (table 29) emphasize that these are possibly significant differences, although the very small sample sizes in most cases suggests that the results should be interpreted cautiously. Skulls of W-C African males are absolutely shorter than W African ones, but not broader; the horns, on average, are shorter, with a narrower span, but the tip-to-tip distance is roughly the same. Uele is intermediate in size, with small teeth, and long, divergent horns (like W Africa), but the tip-to-tip distance is greater, on average, and the basal diameter is larger (very slight overlap).

E Kenya is also intermediate in size, but narrow, with short nasals, and fairly long horns (more like W Africa), with a narrower span but divergent tips.

The single specimen from W Kenya appears larger than any other specimen, with long, robust, divergent horns with the tips close together.

The females (table 30) show no great differences in size, implying different degrees of sexual dimorphism, but the two Kenya samples are narrow, with (in W Kenya) short nasals. There are no great differences in horn span, and the sample size is too small to guarantee that differences in the tip-to-tip distance

Table 29 Univariate statistics for bongo: males

	Gt l	Biorb	Nas l	Max teeth	Horn l str	Span	Tip–tip	Bas diam
W Africa								
Mean	417.40	156.45	157.86	122.00	637.08	416.00	264.09	84.00
N	5	11	7	2	12	11	11	13
Std dev	12.582	6.346	13.692	.000	47.354	71.208	118.875	10.654
Min	405	150	134	122	558	283	70	66
Max	437	168	172	122	728	525	482	100
W-C Africa								
Mean	384.25	156.00	146.50	—	550.00	351.00	260.50	88.00
N	4	6	2	—	6	4	4	4
Std dev	15.586	6.782	16.263	—	32.398	15.979	56.122	5.598
Min	364	150	135	—	505	335	181	80
Max	402	164	158	—	590	370	312	93
Uele								
Mean	406.00	159.50	142.00	113.00	653.33	405.50	339.17	99.00
N	4	4	1	2	6	6	6	2
Std dev	28.012	10.279	—	9.899	57.270	50.469	56.782	1.414
Min	365	146	142	106	600	340	260	98
Max	427	171	142	120	760	472	430	100
W Kenya								
Mean	—	170.00	—	—	712.00	455.00	194.00	95.00
N	—	1	—	—	1	1	1	1
Std dev	—	—	—	—	—	—	—	—
Min	—	170	—	—	712	455	194	95
Max	—	170	—	—	712	455	194	95
E Kenya								
Mean	394.00	147.67	145.00	117.50	616.00	385.00	316.00	85.33
N	1	3	2	2	1	3	3	3
Std dev	—	5.132	7.071	.707	—	28.583	60.506	7.572
Min	394	142	140	117	616	352	272	80
Max	394	152	150	118	616	402	385	94

are valid. The skulls in the W Kenya sample have very robust horns. The tips sometimes cross in the W Kenya ones, and they always tend to be closer together than in the males.

There is a strong impression of different amounts of sexual dimorphism, although the very small sample sizes necessitate great caution in interpretation. On the face of it, however, it appears that the size difference between the sexes is greatest in the Uele region, followed by W Africa, and is least in W-C Africa. This trend, if valid, seems correlated with habitat: considerable sexual dimorphism in more open forest, and hardly any in deep rainforest.

Therefore, we cannot make any secure conclusions about taxonomic variation in this species (or species-group).

Tragelaphus buxtoni Lydekker, 1910

mountain nyala

T. buxtoni is sister to the sitatunga/bongo clade. The horns (males only) are much more open-spiraled, with a very marked, raised anterior keel, making them look superficially like those of greater kudu, although there are fewer turns and the spiral is even more open in *T. buxtoni*. Body stripes are less well marked than for any other member of the Tragelaphini (except for some bushbuck), the main body markings being the white facial chevrons, the white jaw, the two white throat bands (one on the upper throat, one near the brisket), the white on the inner surfaces of the forelegs, and an upwardly curved row of white spots on the haunch.

Restricted to the forests of the Bale Mountains of Ethiopia.

Table 30 Univariate statistics for bongo: females

	Gt l	Biorb	Nas l	Max teeth	Horn l str	Span	Tip–tip	Bas diam
W Africa								
Mean	395.75	152.83	153.67	—	580.13	256.13	93.43	60.50
N	4	6	3	—	8	8	7	8
Std dev	10.243	6.911	4.041	—	72.878	25.396	71.666	8.211
Min	383	141	149	—	514	208	0	48
Max	405	161	156	—	740	280	173	70
W-C Africa								
Mean	386.67	150.50	151.00	—	535.50	240.25	89.00	60.75
N	3	6	1	—	4	4	4	4
Std dev	6.351	7.635	—	—	55.296	7.411	35.449	6.292
Min	383	140	151	—	495	231	47	55
Max	394	162	151	—	617	248	131	68
Uele								
Mean	381.60	151.40	—	—	568.29	261.57	96.86	—
N	5	5	—	—	7	7	7	—
Std dev	3.209	1.949	—	—	33.470	32.382	65.226	—
Min	377	150	—	—	518	206	23	—
Max	386	154	—	—	621	300	207	—
W Kenya								
Mean	392.67	144.33	134.67	108.00	564.00	233.33	88.00	67.67
N	3	3	3	1	3	3	3	3
Std dev	8.327	10.017	6.658	—	65.506	4.163	89.370	3.786
Min	386	133	127	108	489	230	11	65
Max	402	152	139	108	610	238	186	72
E Kenya								
Mean	384.00	138.00	—	113.00	—	253.00	177.00	58.00
N	1	1	—	1	—	1	1	1
Std dev	—	—	—	—	—	—	—	—
Min	384	138	—	113	—	253	177	58
Max	384	138	—	113	—	253	177	58

Ammelaphus Heller, 1912

lesser kudu

$2n = 38$ in both sexes for one female and four male lesser kudu, of unrecorded origin, that were karyotyped by Benirschke et al. (1980). Unlike other Tragelaphini, both X and Y chromosomes were fused with autosomes.

Benirschke et al. (1980) proposed that lesser kudu separated early from other taxa, possibly along the lineage leading to *Boselaphus*. DNA data (see, for example Ropiquet, 2006) confirm that *Ammelaphus* was the earliest lineage within the Tragelaphini to separate, although not that it was part of the *Boselaphus* lineage.

Anterior median and lateral nasal tips more or less equal in length, with a wide cleft between them. Several supraorbital foramina, in a deep channel, but only the anterior one large. Lacrimal vacuity short, not reaching the maxilla. Superior part of the malar-maxillary suture oblique, the suture then angling forward, then turning and running backward. Auditory bulla rounded; postglenoid foramen huge, an elongated oval. Basioccipital constricted between the mesopterygoid plates; mesopterygoid fossa forming a narrow U; vomer forming a sharp ridge; basioccipital tuberosities high and angular. Free ends of the paroccipital processes long, straight. Buccal styles on the upper cheekteeth not as prominent as in other tragelaphines. Horn spiral comparatively open; anterior keel well marked, somewhat raised from the plane of the rest of the horn.

The two sexes very alike in external features, with the only noticeable difference being in the overall tone, redder in the females and grayer in the males. (The skins described below are males, unless otherwise indicated.)

Two sets of horns of lesser kudu are known from the Arabian Peninsula: one from Yemen, and one from Saudi Arabia. It is unknown whether this animal is, or was, indigenous, or whether the horns were ultimately brought from Africa (D. Harrison & Bates, 1991).

Ammelaphus imberbis (Blyth, 1869)

1869 *Ammelaphus imberbis* (Blyth). Somalia.

Skins seen: 10.

Males paler, grayer than the more S specimens, with much fewer contrasting white markings. Whole tail with a dark dorsal surface, whereas S specimens with the tail dark only on the tip and at the base (and along the sides). Upper throat spot fairly small; lower one widened to a chevron (usually not widened in S skins). Stripes beginning farther back on the shoulders. Body stripes any quantity between 8 and 14, usually more than 10, the last 2 in each case being short, somewhat converging below, and going somewhat obliquely onto the haunch. Striping very characteristic of *A. imberbis*: the stripes more spaced than in *A. australis*, and rather less marked. Hind pasterns generally more fully black, but the anterior ones white.

N Somalia. We have seen specimens from Orthar, Gurgura, Ginoble, Ulfula, Dire Dawa, and Berbera.

The type specimen of *A. imberbis* resembles other, better-preserved skins of this species, although the skin of the type is much too yellowed to detect its original color.

Ammelaphus australis Heller, 1913

1913 *Ammelaphus australis* Heller. Longaya, Marsabit.

A. australis was said to differ from *A. imberbis* by being darker (bright ochery tawny), with no white spot on the front of the pasterns; horns shorter.

In our study, the skins differed slightly according to different localities, and there may be more taxonomic variation than hitherto recognized, as follows:

1. Longaya Water (2 skins, both females; the type and paratype of *A. australis*)—Bright ochraceous-tawny; 12 stripes; no spots on the front pasterns, but a white spot on the hind pasterns. Tail tawny above; tip seal-brown. Lower throat-spot larger than the upper one. This description is taken from Heller (1913b), because we have seen no specimens from this region ourselves.
2. Turkana area (10)—Gray-white on the neck; body chocolate-brown; black band on the inner side of the upper arm; striking white chevron; 1–2 white spots on each side of the jaw angle; ears dark, with white rims; one upper and one lower white spot, usually fairly small, on the throat; 7–13, but usually 10–12, contrasting white stripes on each side, and no oblique stripes on the haunch. Black, round posterior false hoofs; no (or very little) white on the pasterns. Brown mane shortening to a line, then becoming a mane again on the withers and the anterior back, with white hairs in it; mane continuing as a thin white dorsal stripe, and becoming lengthened a bit, with more brown on the rump. Tail black at the base and on the tuft, white in between and on the sides. The localities for the skins are Lotholier, Turkana, Kedef Valley, Karamoja, Engaruka, an Omo tributary, Lake Abaya, and Lake Zwai.
3. Galana (1)—Dark gray–chocolate-brown; 14 stripes; big white spots on the front feet; large white area on the back feet.
4. Iringa (1)—More reddish brown than the previous skins; big white throat spots; 12 stripes; big white spots on the feet. The locality for the skin is Mtandika, Iringa.
5. Jubaland—There are, curiously, no skins available from Jubaland.

The skulls fall into three completely separate clusters: N Somalia / Ogaden, Jubaland, and Turkana/ Abaya. These are 100% distinct. The N Somalia / Ogaden sample represents *A. imberbis*; the Jubaland sample is probably *A. australis* (although, having no available skins from Jubaland, we cannot be certain of this); the third sample would therefore seem to represent an undescribed taxon (and, as implied above, the skins we have seen from this region are different from the descriptions of the type and paratype of *A. australis*). Table 31 gives the univariate statistics. The number of female skulls is too small to break the sample down by locality; the figures for these skulls are listed merely to give an idea of the amount of sexual dimorphism.

The described differences in horn length (*A. imberbis* was stated to be longer than *A. australis*) do not hold; the horns, on average, are considerably longer, and more spreading, in the southernmost sample (Lake Turkana, Lake Abaya)!

The number of horn turns varies by locality:

N Somalia—5 have 2 turns, 3 have 2.5 turns
Jubaland—6 have 2.5 turns, 1 has 3 turns

Table 31 Univariate statistics for lesser kudu

	Gt l	Cb l	Gt br	Preorb	Horn l str	Tip–tip	Span
Males							
N Somalia, Ogaden							
Mean	319.80	314.40	122.71	176.00	526.62	261.29	290.50
N	5	5	7	6	8	7	6
Std dev	7.014	9.099	3.684	5.254	52.236	62.147	79.490
Min	310	303	118	168	433	187	187
Max	328	327	129	184	588	360	403
Jubaland							
Mean	308.67	299.33	119.86	170.17	532.71	257.40	274.80
N	6	6	7	6	7	5	5
Std dev	7.633	6.683	4.525	4.579	42.976	65.653	29.811
Min	296	292	114	166	469	168	238
Max	318	310	128	178	577	315	305
Lake Turkana region							
Mean	319.37	310.86	119.45	174.25	559.60	303.50	328.30
N	8	7	11	8	10	8	10
Std dev	7.763	5.640	6.532	5.800	45.957	77.371	52.603
Min	309	303	104	166	508	222	263
Max	332	319	127	181	655	415	415
Females							
Total for all regions							
Mean	290.67	277.33	110.67	157.33	—	—	—
N	3	3	3	3	—	—	—
Std dev	.577	3.786	4.041	.577	—	—	—
Min	290	273	107	157	—	—	—
Max	291	280	115	158	—	—	—

Lake Turkana—4 have 2 turns, 1 has 2.25 turns, 4 have 2.5 turns, 2 have 3 turns

Strepsiceros Hamilton Smith, 1827

greater kudu

Anterior lateral nasal tips much longer than the central one, with a deep cleft in between. Supraorbital foramina large, sometimes several, or sometimes all fused into one, in an elongated depression. Lacrimal vacuity short, not reaching the maxilla. Superior part of the malar-maxillary suture oblique backward. then the suture arching forward. Auditory bulla deep, rounded in lateral view, the inferior surface angular; postglenoid foramen small and round. Basioccipital of even width, the tuberosities low, rounded; mesopterygoid fossa forming a wide V; vomer sharply ridged. Free ends of the paroccipital processes long, curved forward. Buccal styles on the upper cheekteeth very prominent.

Nersting and Arctander (2001) studied the mitochondrial control region of kudu from Namibia NE to Samburu (N Kenya); that is to say, from much of their range. In the neighbor-joining haplotype tree, the single specimen from Samburu was sister to all of the others (the range of these "others" extending as far NE as Maasailand in N Tanzania); the bootstrap value for the non-Samburu clade was 100%. In this other clade, most of the SW individuals clustered separately from the rest, but the bootstrap values were not high. Although the specimens in the Tanzanian sample clustered together, it was a subclade nested well within the SW–S African sample.

There is clearly need for a wider geographic study of the greater kudu; for the moment, it is notable that, as in our morphological sample, there appears to be a deep split between those from NE Africa and those from farther S.

Strepsiceros strepsiceros (Pallas, 1766)

1766 *Strepsiceros strepsiceros* (Pallas). SE Cape (Grubb, 1999a).

Grubb (1999a) noted that there has historically been discontinuity within South Africa for *Strepsiceros*; greater kudu occur in the S Cape, and not again until the N Cape / Transvaal/Botswana/Namibia population. Shortridge (1934) found that a skin from Keiskama, E Cape, was considerably darker than those from Namibia; so did A. Roberts (1951:308), who wrote: "Flat skins which I have seen in shops in the Great Karas Mountains area of Great Namaqualand appeared to me to be darker than those from the East." We ourselves have seen no skins from the S Cape, apart from the skin of a young female from "Steytherville, Cape Colony," that is dark red-brown, with 12 stripes, a good mane, and very little black on the feet.

Strepsiceros zambesiensis Lorenz, 1894

1894 *Strepsiceros zambesiensis* Lorenz. Lesuma Forest, on the Zimbabwe–Botswana border.

1914 *Strepsiceros frommi* Matschie. 7° 20' S in Tanzania.

Males (17 skins seen): grayer to light ochery to reddish gray-brown, some white hairs mixed in. Ear-tips occasionally white; ears otherwise a darker shade of the body color. Very clear white chevron on the face, often rather short. Long throat mane, straw-colored or mixed pale red-brown and black, often the whole hair tip blackish. Short red-brown nuchal mane, becoming a clear white dorsal stripe (with slightly lengthened hairs) at the commencement of the first stripe, but with black hairs sometimes mixed in. Stripes 8–11, variable, but tending to be contrasty. Pasterns brown, usually not black, on the backs. Lower legs more cinnamon than the color of the body and the upper segments.

Localities from which we have seen male specimens: Hillingdale, Kangongo, Koreis, Bambi, Etosha (Namibia), Angola, Olifants River (Mpumalanga), Mahura, Katumbas (E Zambia), Lengwe (Malawi), "Malawi."

Females, young males (17 skins seen): ochery to bright grayish brown; longer, redder hair than the adult males. Chevron on the face more poorly developed than in the adult males. Very short, straw-colored throat mane. No white tips to the ears. Stripes 5–10. Nuchal mane longer, reddish, with more black hairs mixed in on the withers than in the adult males; dorsal stripe vaguely marked in parts, white, except where sporadically interrupted with black in the middle of the back (after the fourth stripe). Pasterns usually black (sometimes not much—they may be narrowly brown).

Localities of female and young male specimens we have seen are Nubis, Hillingdale, Struispan, Omataka (Namibia), "Namibia," Okavango, Angola, Banga (Zimbabwe), Kwa Kzembe (DRC), Kruger, Rupias.

Strepsiceros chora (Cretzschmar, 1826)

1826 *Strepsiceros chora* (Cretzschmar). "Eastern Sudan."

1913 *Strepsiceros bea* Heller. E of Lake Baringo, Kenya.

Males (20 skins seen): Pale, light gray-brown; big facial chevron; long, pale nuchal mane, with dark brown tips, on both the throat and the nape (in Andal and Eireirib specimens), but only a very short, dark, excuse for a throat mane, restricted to the lower throat, in the others (and this is pale in the Blue Nile specimen). Only 3–7 stripes. Feet with black or brown pasterns, and a white band above the hoofs; this region always whitish above the dark zone. Ear-backs wholly black or dark gray (except the rims), or becoming black distally (in Andal specimens).

Localities of male specimens we have seen: Andal (Sudan); Eireirib; "Somalia"; Gadabusi (Somalia); Sennar; Blue Nile; Bogos; Arroweina (S Ethiopia); Laikipia (Kenya).

Females (3 skins seen): Dull to light brown; full facial chevron. In one skin, 5 clear stripes, plus a sixth on the haunch; in another skin, 3 stripes on the haunches, 3 others well spaced farther forward, a vague stripe between these two groups. Dark ear-backs; dark ring all around the hoofs; no mane, just very slightly lengthened hair on the nape and the throat.

Localities of female specimens we have seen: Chuowkan (Abyssinia); Andal; "Kenya."

Strepsiceros cottoni Dollman & Burlace, 1928

Males (3 skins seen): Pale brown; facial chevron poorly expressed. Stripes 4–8, the dorsal stripe white or light brown, the hair somewhat lengthened, with more black hair toward the croup. Ear-backs dark gray. Hoofs black, round, very small; this region is always whitish above the dark zone. Mane pale mane, very short on the nape, and long on the throat.

Females (1 skin seen, the type): Like *S. chora*. Light brown; full facial chevron. Stripes 3 on the haunches, 3 others well spaced farther forward, with a space between the two groups. Dark ear-backs; dark ring all around the hoofs; no mane, just very slightly lengthened hair on the nape and the throat.

Aya.

The two N species (*S. chora* and *S. cottoni*) are very alike in many respects, but, on the limited evidence, the skins from Chad fall outside the range of variation of the more numerous specimens in the sample from the Horn of Africa.

Table 32 Regional differences in the number of turns in the horns of greater kudu

	<2.5	2.5	>2.5
S Africa	5	13	5
Kenya, Somalia, Chad	12	7	—

Craniometric analysis is restricted, because of comparatively few complete skulls; most available specimens are trophies, and consist only of frontlets and horns.

Skulls from Namibia are larger all around, compared with those from Malawi and Somalia. Somali skulls are generally small, especially their horn length, but the skull breadth is not small.

When the small Chad sample ($N=4$) is compared with the Somali sample ($N=8$), they are completely separated, most strikingly in the horn measurements.

Greater kudu show regional differences in the number of turns in the horns.

Table 33 gives the univariate statistics for both the males and the females. The Kenya specimens, which all come from N Kenya, clearly belong with the Somali sample. The three S samples probably belong together. Thus, three species are recognizable in the main series, and the poorly known S Cape taxon surely represents a fourth.

Taurotragus Wagner, 1855

eland

The eland skull, in many respects, is very like that of *Strepsiceros*: anterior lateral nasal tips much longer than the median ones, with a deep cleft in between; lacrimal vacuity short; malar-maxillary suture sinuous, forming an anterior-pointing W; auditory bulla large, its inferior surface slightly angular, with a small postglenoid foramen anterior to it; basioccipital of even width, with the basioccipital ridges relatively low; buccal styles very prominent. Differences: supraorbital foramina in *Taurotragus* 1–2, very large, in a very deep, wide, and elongated depression; mesopterygoid fossa forming a wide U; the free ends of the paroccipital processes short.

Taurotragus oryx (Pallas, 1767)

common eland

$2n=31$ (males), with the Y/14 translocation that is usual in the Tragelaphini. The chromosomes are almost identical to those of the greater kudu, with small differences in four autosomes (Jorge & Benirschke, 1976).

Taurotragus oryx oryx (Pallas, 1767)
No stripes in the adults.

Taurotragus oryx livingstonii Sclater, 1864
About 8 white stripes. As far as we can tell, mainly from photographs taken in the wild, eland N of the Zambezi tend to retain their stripes until old age.

We are utterly unable to distinguish any geographic samples of *T. oryx* by multivariate analysis.

Univariate statistics are presented in table 34. For the males, E African eland average smaller than other samples, with shorter, more slender, less divergent horns. The W Zambian sample has large teeth, and is also fairly small. South Africa, Botswana, and Angola specimens are large, with large teeth and slender horns; the single pair of South African horns are extremely short and not very divergent. Except perhaps for the E and the South African samples, these differences are not consistent enough for subspecific recognition.

For the females, the skulls in the E Africa sample are not smaller than those in other samples, but the shorter, more slender, less divergent horns are similar to the horns of the males. The South African (Cape) specimen is large, and the horns ($N=1$!) are again very short and not very divergent. The Namibia skull (Grootfontein) has large teeth and slender horns (as does the type specimen for *Antilope triangularis* Günther, 1889, of unknown origin). Except perhaps for the E African and the South African samples, these differences, again, are not consistent enough for subspecific recognition.

Taurotragus derbianus (Gray, 1847)

giant eland

There are no morphometric differences among geographic samples (table 35), and our inspection of photos shows no consistent external differences, either. Thus all giant eland, from Gambia to the Nile, would be taxonomically the same.

Comparing *T. derbianus* and *T. oryx*, and combining all samples into one for each species (table 36), *T. derbianus* averages slightly the smaller of the two species in the males (but is very slightly larger in the females), while the horns are very much larger—thicker at the base, longer (absolutely so in the case of the males), and more divergent.

Table 33 Univariate statistics for greater kudu

	Gt l	Cb l	Gt br	Preorb	Horn l str	Tip–tip
Males						
Cape						
Mean	421.67	408.33	166.75	258.00	895.00	708.67
N	3	3	4	2	3	3
Std dev	20.232	13.051	2.500	—	110.340	98.592
Min	409	398	164	221	780	595
Max	445	423	170	295	1000	771
Zambia, Namibia, Zimbabwe						
Mean	415.10	398.11	171.14	226.67	951.33	747.77
N	10	9	14	9	15	13
Std dev	11.308	10.374	4.737	8.337	67.969	182.179
Min	394	379	164	210	850	423
Max	428	413	181	241	1110	1034
Malawi, S Tanzania						
Mean	415.00	402.00	166.00	227.50	1067.50	803.75
N	4	3	5	4	4	4
Std dev	17.720	15.620	3.317	9.434	47.170	84.200
Min	396	384	161	214	1000	685
Max	435	412	170	234	1100	880
N Kenya						
Mean	397.00	—	168.00	212.00	903.00	594.00
N	1	—	2	1	2	1
Min	—	—	166	—	883	—
Max	—	—	170	—	923	—
Somali, Eritrea, NE Sudan						
Mean	393.44	380.86	165.25	216.44	831.73	630.36
N	9	7	12	9	11	11
Std dev	14.917	15.432	3.769	8.278	114.990	119.576
Min	373	363	160	206	690	426
Max	419	403	170	230	1110	810
Chad, W Sudan						
Mean	375.17	358.25	159.86	205.67	780.50	614.83
N	6	4	7	6	6	6
Std dev	12.352	8.221	5.146	11.843	84.033	139.481
Min	358	351	152	188	636	510
Max	393	370	167	221	850	877
Females						
Cape						
Mean	391.00	372.00	155.50	—	—	—
N	2	1	2	—	—	—
Min	387	—	151	—	—	—
Max	395	—	160	—	—	—
Zambia, Namibia, Zimbabwe						
Mean	375.89	362.33	145.70	210.22	—	—
N	9	9	10	9	—	—
Std dev	12.624	11.269	6.038	9.550	—	—
Min	354	345	134	196	—	—
Max	391	378	156	223	—	—
Mean	375.89	362.33	145.70	210.22	—	—
Malawi, S Tanzania						
Mean	378.00	357.00	146.00	209.00	—	—
N	1	1	1	1	—	—

(continued)

Table 33 (continued)

	Gt l	Cb l	Gt br	Preorb	Horn l str	Tip–tip
N Kenya, Somali, Eritrea, NE Sudan						
Mean	367.00	551.67	146.33	202.00	—	—
N	3	3	3	3	—	—
Std dev	11.533	353.670	4.041	1.732	—	—
Min	358	342	142	200	—	—
Max	380	960	150	203	—	—
Chad, W Sudan						
Mean	362.00	350.00	146.00	200.50	—	—
N	2	2	2	2	—	—
Min	354	346	145	197	—	—
Max	370	354	147	204	—	—

The name "giant" eland is a bit of a misnomer; presumably it is the enormous horns that have led to it being awarded this name.

Subfamily Antilopinae Gray, 1821

In his morphologically based cladogram, A. W. Gentry (1992) found the Cephalophini to be sister to all other groups, followed by *Neotragus* (as represented by *N. moschatus*). There was then a split between the Antilopini and others; he was thus probably the earliest to recognize the affinity of "antelopos" (such as the Alcelaphini and the Hippotragini) with the Caprini.

On the basis of partial mitochondrial sequences (12S and 16S rRNA), and with an unavoidably limited Chinese-centered dataset, Lei, Jiang, et al.'s (2003) neighbor-joining tree of this subfamily was as follows:

(*Neotragus* (*Oreotragus* (*Ourebia*, *Procapra*, *Saiga*, *Gazella*/*Eudorcas*, *Antidorcas*/*Madoqua*, *Antilope*, *Raphicerus*)) ((*Damaliscus*, *Oryx*/*Hippotragus*) (*Ovis*, *Pantholops*, *Pseudois*/*Capra*/*Hemitragus*))).

Kuznetsova & Kholodova (2003), however, also using 16S as well as a nuclear fragment (b-spectrin), found that the data supported, at most, a weak association of the Neotragini with the Antilopini, so these authors instead placed the Neotragini in a clade with the Aepycerotini, which is also the conclusion of Ropiquet (2006) and others.

Tribe Neotragini Sclater & Thomas, 1894

A. W. Gentry (1992) noted many apparent primitive retentions in this tribe, such as the low infraorbital foramen, the posterior level of the median indent at the back of palate, and the lack of differentiation of the parastylid from the paraconid on the two posterior lower premolars.

Neotragus Hamilton Smith, 1827

The distribution and systematics of the dwarf antelope (*Neotragus* species) are not well known. Although we here continue to unite them all into a single genus, as is conventional, whether they are all truly closely related to one another has never been tested, just merely assumed.

The dwarf antelope are among the smallest ungulates. They have relatively prognathous skulls, with elongated premaxillae; very brachyodont dentition, with relatively small premolars; and the horns are not erect, but directed back in the plane of the face. The genotypic species, *N. pygmaeus*, differs from all other bovids in the frequency with which it retains milk canines.

Neotragus pygmaeus (Linnaeus, 1758)

royal antelope

1758 *Capra pygmaea* Linnaeus. "Guinea, India" (W Africa).

1777 *Antilope regia* Erxleben. Senegal.

1827 *Antilope opinigera* Lesson. Coasts of Guinea and Loango.

1851 *Nanotragus perpusillus* Gray. No locality.

Size small, shoulder height to 25 cm (Lydekker 1913; Ansell 1971); weight 1.8–2.3 kg (five males), 3.1 kg (two adult females).

Lateral hoofs said to be present, but rudimentary (Urbain & Friant 1942; Dragesco et al, 1979); we found no trace in the specimens we examined. Horns smooth and short (to about 38 mm, according to Ansell, 1971).

Table 34 Univariate statistics for common eland (*Taurotragus oryx*)

	Gt l	Cb l	Biorb	Max teeth	Bas diam	Horn l	Tip–tip
Males							
S Africa							
Mean	511.00	484.00	213.00	152.00	79.00	433.00	303.00
N	1	1	1	1	1	1	1
Bakwena							
Mean	508.00	490.00	198.00	148.00	77.00	676.00	436.00
N	1	1	1	1	1	1	1
Angola							
Mean	490.00	471.00	198.00	148.00	78.00	580.00	402.00
N	1	1	1	1	1	1	1
S Mozambique							
Mean	488.29	467.60	203.14	139.00	87.00	589.00	316.86
N	7	5	7	7	7	7	7
Std dev	13.937	9.017	2.854	4.435	4.690	37.216	14.622
Min	463	457	200	130	80	519	300
Max	506	479	207	144	94	638	337
Zimbabwe							
Mean	496.00	459.00	204.50	141.00	91.50	654.00	312.50
N	2	1	2	2	2	2	2
Min	491	—	203	132	89	650	280
Max	501	—	206	150	94	658	345
W Zambia							
Mean	480.33	468.00	200.29	148.14	85.83	645.83	385.33
N	6	3	7	7	6	6	6
Std dev	14.597	15.588	17.689	9.805	6.824	45.429	94.299
Min	459	450	182	135	73	595	257
Max	500	477	236	161	93	700	515
Malawi, S Tanzania							
Mean	491.14	467.40	201.78	139.43	83.56	664.33	353.22
N	7	5	9	7	9	9	9
Std dev	11.611	8.849	6.741	7.458	2.920	43.629	85.130
Min	474	459	191	126	80	603	176
Max	508	477	210	151	88	740	480
E Africa							
Mean	477.38	457.43	201.11	138.22	80.89	565.78	272.33
N	8	7	9	9	9	9	9
Std dev	20.311	16.481	8.638	7.412	7.356	53.399	43.709
Min	450	439	190	126	69	512	181
Max	515	485	214	146	88	661	322
Females							
Cape							
Mean	480.00	456.00	192.00	144.00	74.00	542.00	341.00
N	1	1	1	1	1	1	1
Type of *triangularis*							
Mean	—	—	155.00	—	56.00	795.00	563.00
N	—	—	1	—	1	1	1
Grootfontein							
Mean	453.00	—	184.00	150.00	58.00	648.00	233.00
N	1	—	1	1	1	1	1
S Mozambique							
Mean	445.20	442.75	183.80	132.00	64.40	628.20	349.00
N	5	4	5	5	5	5	5
Std dev	34.981	7.136	6.760	9.592	7.503	93.919	149.355
Min	384	434	177	123	54	533	212
Max	470	451	193	145	73	773	570

(*continued*)

Table 34 (continued)

	Gt l	Cb l	Biorb	Max teeth	Bas diam	Horn l	Tip–tip
Zimbabwe							
Mean	461.00	440.00	189.00	138.00	63.00	644.00	361.00
N	1	1	1	1	1	1	1
W Zambia							
Mean	456.67	442.00	185.29	135.83	61.33	642.00	385.33
N	6	5	7	6	6	6	6
Std dev	6.713	16.016	10.452	4.262	5.007	41.323	97.867
Min	444	427	165	130	56	598	230
Max	463	469	196	143	68	704	506
Malawi, S Tanzania							
Mean	439.00	418.50	176.80	133.75	57.20	673.00	278.00
N	5	2	5	4	5	5	5
Std dev	13.096	—	6.458	4.272	4.324	114.645	132.293
Min	427	417	170	128	51	535	175
Max	460	420	185	137	63	823	500
E Africa							
Mean	441.90	427.00	179.80	131.00	60.78	601.33	305.50
N	10	6	10	9	9	9	8
Std dev	22.218	19.411	12.674	8.703	7.839	47.326	46.448
Min	402	404	153	114	54	524	246
Max	473	455	201	140	79	646	376

Table 35 Univariate statistics for *Taurotragus derbianus*

	Gt l	Cb l	Biorb	Max teeth	Bas diam	Horn l	Tip–tip
Males							
Sudan							
Mean	471.00	455.50	195.75	140.67	100.67	866.00	691.25
N	3	2	4	3	3	4	4
Std dev	15.100	13.435	.957	2.309	10.263	54.437	91.587
Min	455	446	195	138	92	814	623
Max	485	465	197	142	112	922	822
Central Africa							
Mean	477.75	464.75	200.00	140.00	94.71	878.71	665.14
N	4	4	5	4	7	7	7
Std dev	18.337	14.245	7.810	5.477	4.751	47.419	188.552
Min	460	452	187	134	89	800	483
Max	502	479	207	147	100	933	985
West Africa							
Mean	—	—	—	—	91.00	810.00	737.00
N	—	—	—	—	2	2	2
Min	—	—	—	—	91	800	724
Max	—	—	—	—	91	820	750
Females							
Central Africa							
Mean	450.71	439.57	182.88	138.00	69.63	718.63	431.75
N	7	7	8	8	8	8	8
Std dev	16.101	11.013	5.489	4.781	7.425	81.367	131.947
Min	418	419	171	133	55	579	211
Max	471	455	189	145	77	796	575

Table 36 Univariate statistics for the two species of *Taurotragus*

	Gt l	Cb l	Biorb	Max teeth	Bas diam	Horn l	Tip–tip
Males							
oryx							
Mean	486.61	466.04	202.26	140.89	84.27	611.46	329.89
N	33	24	38	36	37	37	37
Std dev	16.105	13.637	9.512	9.504	6.323	65.562	73.738
Min	450	439	182	112	69	433	176
Max	515	490	236	161	97	740	515
derbianus							
Mean	474.86	461.67	198.11	140.29	95.58	864.23	684.23
N	7	6	9	7	12	13	13
Std dev	16.036	13.441	5.988	4.112	6.543	49.954	143.507
Min	455	446	187	134	89	800	483
Max	502	479	207	147	112	933	985
Females							
oryx							
Mean	447.38	435.58	181.23	134.11	61.17	634.45	335.32
N	29	19	31	27	29	29	28
Std dev	21.316	16.567	10.938	7.876	6.693	77.171	109.664
Min	384	404	153	114	51	524	175
Max	480	469	201	150	79	823	570
derbianus							
Mean	450.71	439.57	182.88	138.00	69.63	718.63	431.75
N	7	7	8	8	8	8	8
Std dev	16.101	11.013	5.489	4.781	7.425	81.367	131.947
Min	418	419	171	133	55	579	211
Max	471	455	189	145	77	796	575

General color dull grayish brown, finely speckled with golden-brown (speckling absent or weak on the neck and the blaze), shading quite sharply to bright orange-fulvous, without hair-banding, on the rump, the tail, the margins of the white belly-patch, the forelegs, the side of head (well below the eye), and the underside of the neck; dark brown on the pasterns and the digits, and the fronts of the forelegs from the carpus downward; pale patches on the digits (almost white in the juveniles); white belly-patch narrow, extending down to the hock and the carpus on the insides of the limbs; separate white throat-patch; tail colored like the upperparts above, but white below; ears with whitish marks along the margin near the base, and white inside.

N. pygmaeus differs in cranial features from *N. batesi* and *N. moschatus* in having the free anterior margin of the premaxilla bent and proximally more vertical, as well as in having a median spinous process; no maxillary-premaxillary fissure; nasals narrow at the base, not reaching back to the point of the lacrimofrontal suture on the orbital rim; minute maxillary canines frequently present in the juveniles, more rarely so in the adults; palatine foramina long; lateral palatine notches extending farther forward, to the front edge of the third upper molar. *N. pygmaeus* resembles *N. batesi* and differs from *N. moschatus* in being less prognathous; and in having both the anterior rim of the orbit farther forward, above the last premolar, and the auditory bullae large and inflated.

Known only from the forests of upper Guinea W of the Volta River, from Sierra Leone to Ghana.

Neotragus batesi (De Winton, 1903)

dwarf antelope

1903 *Neotragus batesi* De Winton. Efulen, Cameroon.
1906 *Hylarnus harrisoni* Thomas. Semliki Forest, DRC.

Said to differ from *N. pygmaeus* in size: nearly half as large again as in the royal antelope, according to Lydekker (1915a), and shoulder height 30 cm, according to Ansell (1971), but mean measurements for the males and the females, respectively, are 109 cm and 111 cm (total body length), compared with 120 cm and 114 cm for *N. pygmaeus*, suggesting that Lydekker (1915a) exaggerated the size differences.

Cranial measurements are very close in the two species. Dragesco et al. (1979) have provided measurements from 138 adult animals: shoulder height (with standard deviation) was 303 ±27.0 mm in 62 males, 310 ±22.6 mm in 75 females; weights were 2.26 ±0.46 kg in 54 males, 2.53 ±0.45 kg in 66 females, but not necessarily heavier than *N. pygmaeus*, for which the data are scanty.

Said to differ from *N. moschatus* in the presence of rudimentary lateral hoofs (Rode 1943), but these are not present in the specimens we have seen, nor were they in those examined by Dragesco et al. (1979).

General color superficially like *N. pygmaeus*, but with less contrast between the dorsum and the lighter flanks; speckling of the hairs coarser and more contrasting; speckling extending onto the blaze and the neck and, very finely, onto the ears; speckling more olive-ochre than orange-ochre; blaze narrower, with the transition from the dark blaze to the lighter face color occurring above the eye; anterior surfaces of the forelegs and the metatarsi above the pasterns usually darker than the body.

N. batesi differs cranially from *N. pygmaeus*, and resembles *N. moschatus*, in having the premaxilla straight along its anterior margin and at a greater angle from the vertical, with the median process usually absent (present as a tiny spicule in a few specimens); a vacuity present between the maxilla and the premaxilla; nasals broad at the base, extending back to the level of the lacrimofrontal suture on the orbit rim; canines absent at all ages; palatine foramina short; lateral palatine notches not extending so far forward, only reaching the level of the hind edge of the third upper molar. *N. batesi* differs from both *N. pygmaeus* and *N. moschatus* in the absence of an ethmoid fissure (Ansell 1971; Dragesco et al. 1979).

Dragesco et al. (1979) detected significant differences between the sexes in several measurements, including mass and limb length. Mean head plus body length for the males was 98% of that for the females; mean body width, 89%; and mean skull length (the precise nature of the measurement was not given) 97%, compared with 98% for our data from the DRC specimens.

Geographical variation was implied by the recognition of the species (or subspecies) *harrisoni*. This form was described as being less contrasting in coloration than typical *N. batesi*, but the holotype skin is not accessible, as it was returned to the donor after it had been described. A more extensive series now available indicates that *harrisoni* cannot be distinguished from *N. batesi*. Differences in measurements are not significant, and the cranial characters of the holotype, described by O. Thomas (1906), are of an individual nature: the maxillo-premaxillary vacuity is very large in the holotype, but values for the series of specimens overlap quite considerably: 6.5–13 mm long in 20 W-C African *N. batesi* (mean 9.45 mm), and 5.5–15 mm in 23 E-C African *harrisoni* (mean 9.76 mm). The nasals in the holotype are broad at the base, but nasals tend to broaden with age, as is also the case with *N. moschatus*.

N. batesi has a discontinuous distribution, having been recorded from three areas: (1) in the Owerri region of Nigeria, between the Niger and Cross rivers; (2) in Cameroon S of the Sanaga River and E of 10° E, extending into N Gabon and the Congo Republic (absent in Mimongo, MBigou, Koula Mouton, Mouila, Fougamou, Fernan Vaz, and neighboring areas of S Gabon [Malbrant & Maclatchy]); and (3) in the DRC between the Congo River and the Rift Valley highlands E of 24° E. (Beni, Epulu River, and Semliki Forest, S to Shabunda; Kibale, Kalinzu, and Maramagambo forests in Uganda [Kingdon 1982a]).

Neotragus moschatus Group

suni

The suni is larger than the other species of *Neotragus*: skull more than 110 mm in greatest length, compared with less than 110 mm in the other species; shoulder height 1314 in. (= 3335 cm) in *N. moschatus* and 1415 in. (= 3538 cm) in *N. livingstonianus*, according to Lydekker (1915a); horns relatively longer, not below 60 mm in length, and ringed.

Suni differ from *N. pygmaeus* and *N. batesi* in having a more prognathous skull, with the front border of the orbit farther back, above the first or second upper molar; preorbital fossa relatively larger; auditory bullae small, not inflated. Resemblances to *N. batesi* have already been discussed.

Sexual dimorphism in suni is possibly less than in *N. batesi* (at least); skull length is virtually identical in the two sexes.

Two main groups of taxa may be recognized: the smaller animals from N of the Zambezi (*N. kirchenpaueri* and *N. moschatus*) and the larger ones from S of that river (*N. livingstonianus* and *N. zuluensis*). *N. livingstonianus* differs from the nominate group not only in size, but also in the simplification of the banding pattern of the hairs, so that the pelage is streaked rather than speckled. These two species, from N and S of the Zambezi River, have been confused, yet they were generally treated as separate species until the intervention of Ellerman et al. (1953).

Table 37 Skull measurements for taxa referred to *Neotragus*

	Skull l			Biorb br			Horn l		
	Mean	Range	*N*	Mean	Range	*N*	Mean	Range	*N*
pygmaeus	103.6	99–108	19	46.8	47–50	21	25.5	18–35	5
batesi	104.4	99–109	35	47.6	47–51	36	31.4	26–38	7
moschatus	115.6	113–114	5	56.5	55–58	10	57.0	57–57	3
kirchenpaueri	118.0	113–120	22	58.3	50–57	22	67.2	63–82	14
livingstonianus	124.6	122–129	15	57.7	55–61	14	73.5	65–87	6

There is considerable variation in color in the *N. moschatus* group, with the darkest specimens coming from montane localities (inland in Kenya and N Tanzania) and paler animals from lower altitudes (coastal Kenya and Tanzania, Zanzibar, and possibly also in Mozambique). These are clearly distinct, with no intermediates, although no cranial measurements differentiate them (there may be a difference in horn length, however; see table 37). A few skins from Malawi present a zoogeographic problem.

Neotragus kirchenpaueri (Pagenstecher, 1885)

1885 *Nesotragus kirchenpaueri* Pagenstecher. Great Arusha, Tanzania.

1913 *Nesotragus moschatus akeleyi* Heller. SE slope of Mt. Kenya, 7000 ft.

$2n$ = probably 52 (Kingswood, Kumamoto, Charter & Jones, 1998). A captive group of suni, whose origin was from stock caught both in KwaZulu-Natal and on Mt. Kenya, had $2n$ = 52–56; certain indications, including the relatively high perinatal mortality of cytotypes 53, 54, and 55, suggested to the authors that these latter numbers were due to hybridization between the two original stocks.

Coloration dark, dull brown, speckled dorsally with browny buff, shading to dull sandy along the white belly-patch, and warmer sandy on the lower neck, in both cases with loss of the dark bands on the hairs; light fawn patch at the base of the ears; tail not contrasting with the body, colored like the back, but darkening toward the tip, with its unbanded hairs, and white below; legs from the hock and the top of the forelimb light sandy brown, like the lower neck; pasterns and digits dark brown, with a dark line running up to the carpus in some specimens; ears gray-brown; blaze chestnut, with a dark brown to black streak on the muzzle; top of the head dark brown, with chestnut speckling, or wholly deep chestnut.

Confined to forests in the highlands of Kenya E of the Gregory Rift Valley, and in N Tanzania: Aberdares Escarpment, Mt. Kenya, Kikuyu and Langata near Nairobi, Mt. Kiliminjaro, Mt. Meru, Ngorongoro (Swynnerton & Hayman, 1951), Phillipshof (= Magamba) (G. M. Allen & Loveridge, 1927), and the Olmoti and Empakaai craters (Frame, 1982). *N. kirchenpaueri* is not positively recorded from isolated mountain forests in S Tanzania (Kungwe, Ufipa, Poroto, Rungwe, Uluguru or Udzungwa), though the species is said to occur in the Iringa district (Swynnerton & Hayman, 1951). It also occurs in the highlands of Malawi.

Neotragus moschatus (Von Dueben, 1846)

1846 *Nesotragus moschatus* Von Dueben in Sundevall. Chapani Islet, Zanzibar, Tanzania.

1913 *Nesotragus moschatus deserticola* Heller. Maji-Ya-Chumvi, Kenya.

Very much lighter in color, and somewhat less strongly speckled, than *N. kirchenpaueri*; blaze not blackish; ventral tone more sandy.

Thirty skins from Malawi have been examined by us, and most cannot be separated from the E African material. But four British Museum skins (Mpezo; hills near Blantyre; two without exact locality) are very dark, indistinguishable from *N. kirchenpaueri*. Apparently some show as wide a range of color within Malawi as they do in Kenya and Tanzania, but over a much smaller area (these enigmatic specimens were discussed by Grubb, 1989). As the localization of the dark specimens is poorly known, at present it is not possible to relate color variation to habitat, but it may be that these dark skins could be representatives of a further taxon, if they are not identical to *N. kirchenpaueri*. This latter species, then, appears to be polytopic, with an exceptionally discontinuous distribution.

Coastal Kenya and Tanzania (Maji ya Chumvi; Mombasa Island; Sokoke Forest; Takaungu); Zanzibar (including Chwaka and Dimani, per Rutzmoser, pers. comm. to PG), Chapani Islet (Von Dueben, 1846); Bawani Islet (Sclater & Thomas, 1895); Boydu Island, Rufiji River delta; Sambala; Kilwa, Lindi, Liwale, and Mikindani districts (Rushby & Swynnerton, 1946); Kidenge, Uzaramo; Mpwapa; Tununguo, Ukami (Matschie, 1895); Mafia Island; Mbulu and Iringa districts (Swynnerton & Hayman, 1950); inland from Kilwa; Mikindani; Morogoro; S presumably to Mozambique (from which the species is recorded from N of the Zambezi by Smithers & Tello [1976], but no specimens from there are available for study), and from there into S Malawi. Believed to have been introduced to the Grave Islets off Zanzibar (Sclater & Thomas, 1895), namely Bawani and the type locality of the species, Chapani; not present on Pemba.

The Zanzibar population cannot be distinguished in color or in size from some mainland populations, so it cannot be regarded as a separate insular subspecies.

***Neotragus livingstonianus* (Kirk, 1865)**
Livingston's suni

Pelage not obviously speckled; pale band to the hair just below the long dark tip, but much narrower than in *N. kirchenpaueri* and *N. moschatus*, grading evenly into the somewhat darker shaft; general color much browner than in *N. moschatus*, but with greater contrast between the dorsum and the haunches; tail dark above, white below; flanks shading to buffy, without dark hair tips, in the vicinity of the white belly-patch; muzzle paler, black on the top; forehead deeper toned.

Known only from S of the Zambezi. Consistently larger than the two E. African species.

Neotragus livingstonianus livingstonianus (Kirk, 1865)

1865 *Nesotragus livingstonianus* Kirk. Shupanga, Mozambique.

In coloration, resembles *N. l. zuluensis* through being superficially nonspeckled, but *N. l. livingstonianus* is less streaked with black, the thighs less contrastingly grayish, and the tail not so dark. being speckled gray above, and white below; general color light fulvous-brown, with very fine, light ticking, particularly on the hindquarters, formed by the pale pigment of the hair shaft lightening just below the dark tip; muzzle paler, black on the top; forehead deeper toned; hairs on the whole body faintly tipped with dark brown; throat, belly, and groin to below the hock white; axilla to the carpus white.

Presumably intergrades with *N. l. zuluensis* in Mozambique; said to range S to about the Vila Pery and Beira districts (Smithers & Tello, 1976); a specimen from Masangena, Save River, on the S border of these districts, is *N. l. livingstonianus*, but so is one from Gazaland, still farther to the S (Smithers & Tello, 1976). Suni occur in Zimbabwe (Chorley, 1956; Child & Savory, 1964), although the taxonomic status for the Zimbabwe suni is yet to be confirmed.

Neotragus livingstonianus zuluensis (Thomas, 1898)

1898 *Nesotragus livingstonianus zuluensis* Thomas. Umkuja (= Mkuzi) Valley, Zululand.

$2n = 56$ (Kingswood, Kumamoto, Charter & Jones, 1998).

Compared with *N. l. livingstonianus*, greater contrast between the dorsum and the haunches, which are gray; black hair tips overlap, forming dark streams on the rump; dark rump streams flowing together into the blackish brown color of the tail; tail white below; white belly-patch extending up the throat (Zululand), or a separate white patch on the throat (Coguno); lower throat and hind shanks sandy buff, less rich in tone than in *N. moschatus*; back of the ears dark gray. These are strong tendencies, but thus far it is not possible to draw absolutely sharp distinctions between *N. l. zuluensis* and *N. l. livingstonianus*.

Mozambique S to the Hluhluwe River, Zululand (Ansell, 1971); said to range N toward the Vila Pery and Beira districts (Smithers & Tello, 1976), but may not occur N of the Save River (A. Roberts, 1915; Pienaar, 1963; Rautenbach, 1982); other localities in Natal mapped by Mentis (1974).

Tribe Aepycerotini Gray, 1872

Like the Alcelaphini, the Aepycerotini have transverse ridges on the horn cores, the face is elongated, the braincase is inclined, the zygomatic arch is expanded below the orbit, there is no ethmoid fissure, there are strong longitudinal ridges on the basioccipital, the central incisors are expanded, and the paraconid and metaconid are fused on the posterior lower premolar. For these reasons, A. W. Gentry (1985) placed *Aepyceros* in the Alcelaphini, although he later (1992) reversed his opinion.

On the basis of both a mitochondrial (16S) and a nuclear (b-spectrin) sequence, Kuznetsova & Kholodova (2003) supported an association between *Aepyceros* and *Neotragus*.

Aepyceros Sundevall, 1847

impala

Impala have a gland on the forehead in both sexes; in the males, these swell and increase in activity during the rut, whereas in the females they do not change seasonally; metatarsal glands likewise show no seasonal changes (Welsch et al., 1998). The males have a thickened dermal shield, covering at least the shoulders, presumably related to their fighting style (Jarman, 1972).

The phylogeny by Nersting and Arctander (2001), using the mitochondrial control region, separated most samples of the two species; in pairwise comparisons, the two species were separated by an average of 9.8 substitutions, whereas only an average of 2.6 substitutions was found among *A. melampus*. One haplotype from *A. petersi*, however, was nested very deeply within the *A. melampus* clade, presumably a result of recent interbreeding.

Aepyceros petersi Bocage, 1879

black-faced impala

Although there has been widespread translocation of *A. melampus* into game farms in Namibia, on the borders of Etosha National Park, a study of eight microsatellite markers found no evidence for introgression (Lorenzen & Siegismund, 2004), so any interbreeding, as inferred from the finding of Nersting & Arctander (2001—see above), must be very infrequent. The differentiation between *A. petersi* and *A. melampus*, in both mtDNA and microsatellites, is very high.

Aepyceros melampus (Lichtenstein, 1812)

common impala

Lorenzen, Arctander, et al. (2006), using mtDNA sequences and microsatellites, found some evidence of genetic differentiation between S and E African samples of *A. melampus*.

Bastos-Silveira & Lister (2007) made a metrical analysis of skulls and horns, with an assessment of pelage color and patterning. They found that *A. petersi* is thoroughly distinct, and they were also inclined to recognize a South African, an E African, and a Zambian/Malawi subspecies. Their maps of coat color and facial patterns showed that, among *A. melampus*, there was no significant geographic variation, except that the Zambian sample did tend to stand out somewhat. Again, in their principal components analysis, the Zambia sample (the one in the middle) was differentiated much more from the S and E African samples than the latter were from each other. This is a very interesting result, as far as evolutionary theory goes—centrifugal speciation is strongly indicated—but it seems impossible to make a taxonomic arrangement from it. At the moment, therefore, no subspecies are recognized in the common impala.

Tribe Antilopini Gray, 1821

A. W. Gentry (1992) noted character states shared by all members of this tribe, including those "dwarf antelope" that he included in it for the first time: an expanded lacrimal bone; absence of basal pillars on the molars; very wide temporal ridges; palatal ridges touching one another; fusion of the metaconid and the entoconid on the posterior lower premolar; a laterally swollen femoral head; and an expanded hypoconulid on the posterior lower molar that is not offset labially and has a central cavity.

What was perhaps the first character-based phylogeny of the Antilopini (as the group was restricted at that time, i.e., without the "dwarf antelope" and the saiga) was that of Effron et al. (1976), based on chromosomes. Their proposed phylogeny was as follows:

> ((*Antidorcas*) (*Litocranius*) (*Antidorcas, Eudorcas*) ((*Antilope, Nanger*) *Gazella*)).

Notably, the *Antilope-Nanger-Gazella* clade was characterized by an X-autosome translocation and many Robertsonian fusions among the autosomes.

Rebholz & Harley (1999) sequenced the mitochondrial cytochrome *b* and cytochrome *c* oxidase III genes, producing the following neighbor-joining phylogeny:

> (*Raphicerus* (*Ourebia, Madoqua* (*Antidorcas* ((*Eudorcas, Nanger*) (*Antilope, Litocranius, Gazella*))))).

The parsimony cladogram was slightly different:

> ((*Raphicerus, Ourebia, Madoqua*) ((*Litocranius, Antidorcas*) (*Eudorcas* (*Nanger* (*Antilope* (*Gazella*)))))).

It is fair to say, however, that most of the nodes in the parsimony cladogram that were unresolved in the neighbor-joining cladogram had very low support.

Raphicerus Hamilton Smith, 1827

The early literature on South African members of this genus was surveyed by Rookmaaker (1988), who pointed out that the presently used name *Raphicerus campestris* Thunberg, 1811, is actually antedated by *Antelope dama* var. *rupestris* Forster, 1796, and *Raphicerus melanotis* Thunberg, 1811, is antedated by *Antelope dama* var. *melanotis* Forster, 1790. He argued that although Forster's two names are not clearly available in the sense of the International Code of Zoological Nomenclature, there is room for argument; he therefore recommended continued use of Thunberg's names.

The two species-groups are distinguished by body build (relative length of the legs), pelage, size, horn size, and skull characters. Craniometrically, size, especially the large greatest skull length compared with condylobasal length (indicating greater development of the occipital crest), is the main discriminator. The two groups are very different, and it may well be that they will be found to be generically distinct.

Raphicerus melanotis Group

grysbok

Relatively short legs; small in size, with the rump higher than the withers; short horns; preorbital fossa wider; nasals posteriorly broader, nearly crowding out the lacrimal vacuity; bullae little inflated. We can divide the species in the *R. melanotis* group into two subgroups: the *R. sharpei* subgroup, lacking lateral hoofs, and *R. melanotis* itself, which possesses them.

Raphicerus melanotis (Thunberg, 1811)

Cape grysbok

1790 *Antelope dama* var. *melanotis* Forster, nom. nud. (see Rookmaaker, 1988).

1796 *Antelope dama* var. *grisea* Forster; not Boddaert, 1785.

1811 *Antilope melanotis* Thunberg. Cape of Good Hope.

1816 *Antilope grisea* Cuvier.

1822 *Antilope rubro-albescens* Desmoulins.

Color reddish brown, with white hairs intermingled, becoming yellower on the cheeks, the legs and the sides of the neck and the body; a white patch on the throat. Muzzle brown above, darkest in the midline, but without a defined nose-patch. Sides of the muzzle dirty white.

Lateral hoofs present, unlike the other species of grysbok.

S of the Karroo, as far as S KwaZulu-Natal.

Raphicerus sharpei Thomas, 1897

Sharpe's grysbok

Color like *R. melanotis*, and similar to *R. melanotis* in most other respects, but the ears large and thinly haired at the back, mostly white, but black along the margins; underparts whiter; legs lighter reddish. A striking difference is that there are no lateral hoofs. Tail short, colored above like the back, but white below. Horns much shorter than in *R. melanotis*; teeth smaller.

Zambia E of the Luangwa; Malawi; presumably N Mozambique?

Raphicerus colonicus Thomas & Schwann, 1906

Limpopo grysbok

Size larger than *R. melanotis* and *R. sharpei*, but the skull relatively narrower; preorbital fossa wider; nasals broaden posteriorly, nearly crowding out the lacrimal vacuity; bullae little inflated. Numerous white hairs interspersed among the browner ones, like the other two species, but the overall color richer, darker; legs lighter, more reddish; underparts, including the throat, more buff, with the boundary line more indistinct on the sides of the belly. Horns similar in size to *R. sharpei*. No lateral hoofs. Hindfeet elongated.

South Africa (N of the S coastal strip) and Zimbabwe; probably also S Mozambique, S of the Zambezi.

Possible Other *Raphicerus* Species

A fourth, unnamed, species, from W and C Zambia, included in table 38, is easily separable craniometrically from all other grysbok, although the sample sizes are small. Skull length is somewhat greater than in *R. sharpei*, nasal breadth is much greater than either *R. sharpei* or *R. colonicus*, and the teeth are smaller than in either of them.

Raphicerus campestris (Thunberg, 1811)

steenbok

More lightly built, taller, and longer-bodied than the three (four?) grysbok species, but with the same high rump and short face. Pelage not noticeably white-ticked; broad white eye-rings. Horns longer, sharply upright. Lateral hoofs present.

Craniometrically, there are no absolute divisions between any of the geographic samples within this widespread species (table 39), and those differences in external features are slight. Therefore, the described taxa are subspecifically distinct, at best.

Raphicerus campestris campestris (Thunberg, 1811)

1783 *Antilope grimmia* Sparrman; not *Capra grimmia* Linnaeus, 1758.

1790 *Antelope dama* Forster; not *Antilope dama* Pallas, 1766.

1796 *Antelope dama* var. *rupestris* Forster, nom. nud. (see Rookmaaker, 1988).

1811 *Antilope campestris* Thunberg.

1811 *Antilope capensis* Thunberg; not P. L. S. Müller, 1766.

General color orange-rufous; cheeks and legs ochraceous-tawny. Large white area on the underparts, not extending down the inside of the limbs; white patches on the throat less extensive than in the other sub-

Table 38 Univariate statistics for the *Raphicerus melanotis* group

	Gt l	Cb l	Gt br	Nas l	Nas br	Max teeth	Horn l
melanotis							
Mean	137.79	129.58	67.86	37.29	17.21	44.14	75.20
N	7	6	7	7	7	7	5
Std dev	4.222	5.886	1.345	1.976	3.026	.900	8.927
Min	132	121	66	35	14	43	65
Max	143	136	70	40	23	45	86
sharpei group							
colonicus							
Mean	132.40	126.00	64.00	35.80	14.50	42.17	46.00
N	5	5	6	5	5	6	2
Std dev	2.793	2.828	2.449	1.924	1.658	1.835	
Min	129	123	60	34	13	40	44
Max	135	129	67	39	16	45	48
sharpei							
Mean	127.25	120.75	63.75	37.25	14.25	42.50	—
N	4	4	4	4	4	4	—
Std dev	1.500	1.708	3.775	1.500	1.500	1.291	—
Min	126	119	60	36	13	41	—
Max	129	123	69	39	16	44	—
W + C Zambia							
Mean	128.80	121.60	63.60	38.40	16.80	40.80	43.67
N	5	5	5	5	5	5	3
Std dev	2.683	2.608	2.302	3.209	.837	1.643	5.686
Min	126	119	61	34	16	39	39
Max	132	125	66	42	18	43	50

species. Muzzle dark brown. Dark brown horseshoe-like patch on the crown. Tail short, colored above like the back, and white below.

The limited evidence suggests that this subspecies may be smaller, but with larger teeth, than others ascribed to this species.

S and SW Cape Province, according to A. Roberts (1951).

Raphicerus campestris fulvorubescens (Desmoulins, 1822)

Rather lighter than *R. c. campestris*, near "amber brown." White of the throat extending almost to the chin; lower lip whitish; white on the underparts extending down inside of the upper part of the thighs and the forelegs.

Albany district, according to A. Roberts (1951).

Raphicerus campestris natalensis (Rothschild, 1907)

1914 *Raphicerus campestris zuluensis* Roberts. Umfolosi.

Color much the same as in *R. c. capricornis*; particularly distinguished by the much longer horns.

Apparently from KwaZulu-Natal into the E parts of the former Transvaal.

Raphicerus campestris capricornis Thomas & Schwann, 1906

Amber-brown, more rufous than *R. c. fulvorubescens*, and with the face much lighter. White extends continuously from the throat to the chin, and more extensively elsewhere below.

According to the limited dataset, specimens from the topotypical region seem to have shorter horns.

Between the Limpopo and the Zambezi rivers, according to A. Roberts (1951).

Raphicerus campestris neumanni (Matschie, 1894)

1908 *Raphicerus neumanni stigmaticus* Lönnberg. Lake Natron.

No dark, crescentic coronal mark. General color paler. White eye-rings large, complete. More white on the margins of the ears. White more clearly defined on the underside, also on the lips.

Occupies the species range N of the Zambezi, into S Kenya.

Table 39 Univariate statistics for *Raphicerus campestris* subspecies

	Gt l	Cb l	Gt br	Nas l	Nas br	Max teeth	Horn l
campestris							
Mean	137.00	—	66.75	36.50	13.00	50.80	95.00
N	2	—	2	1	1	2	1
Min	135	—	65	—	—	50	—
Max	139	—	69	—	—	52	—
fulvorubescens							
Mean	145.50	137.00	73.00	45.00	20.00	47.00	101.50
N	1	1	1	1	1	1	1
zuluensis							
Mean	143.44	135.89	68.72	39.78	14.56	46.50	115.00
N	9	9	9	9	9	9	3
Std dev	3.005	2.607	2.623	4.644	2.800	2.179	10.817
Min	139	131	66	34	11	43	103
Max	147	139	74	47	20	50	124
natalensis							
Mean	141.75	135.25	67.50	40.00	13.25	46.00	113.50
N	2	2	2	2	2	2	1
Min	141	133	66	36	13	43	—
Max	143	138	70	44	14	49	—
capricornis							
Mean	144.42	136.50	70.33	42.58	14.42	46.58	79.38
N	6	6	6	6	6	6	4
Std dev	6.583	5.941	1.602	4.576	1.594	1.744	11.309
Min	134	126	68	37	13	45	66
Max	152	143	72	50	17	49	92
steinhardti							
Mean	142.07	135.00	66.71	44.42	13.00	45.64	93.00
N	7	7	7	6	6	7	1
Std dev	2.950	3.096	1.997	4.005	.894	1.994	—
Min	138	131	64	38	12	43	—
Max	147	140	69	50	14	49	—
kelleni							
Mean	142.91	134.82	68.09	42.00	15.91	45.64	86.00
N	11	11	11	11	11	11	6
Std dev	3.390	3.920	2.427	2.898	2.256	1.567	9.143
Min	138	129	63	38	13	43	75
Max	148	142	71	46	21	47	98
neumanni							
Mean	141.82	133.91	69.36	43.36	16.36	46.00	100.22
N	11	11	11	11	11	11	9
Std dev	2.089	2.071	2.014	4.388	2.157	2.490	9.615
Min	139	131	66	35	14	41	90
Max	146	139	72	49	21	50	116

Over the whole of this wide area, we can find no noticeable geographic variation.

Raphicerus campestris steinhardti (Zukowsky, 1924)

A. Roberts (1951) gives the full synonymy for this subspecies.

Paler than other subspecies; white parts extensive, like *R. c. capricornis.*

Orange River N to Ovamboland, throughout the Kalahari and Botswana, and into Angola.

This is certainly the most distinctive of the described subspecies, both in the pelage and the skull.

Antidorcas Sundevall, 1847

springbok

$2n = 56$ in all three taxa of this genus; the X chromosome is a large acrocentric, and the Y chromosome is a small metacentric, the only metacentric chromosome in the genome (Robinson & Skinner, 1976). The karyotype is close to that of the reconstructed ancestral caprine type, and quite unlike that of any "gazelle" (Vassart, Séguéla, et al., 1995).

Horn rings of the males very prominent, widely spaced, circular (all around the horn), connected by longitudinal ridges. Horn cores circular in cross section, with traces of the rings on them, and with broad, deep grooves up the posterior surface. Anterior premolar (P2) very reduced, generally absent in the mandible; styles on the other premolars small, but well developed on the molars; molars elongated, especially the distal selene, with its flat buccal wall. Mandibular P^4 elongate, with the lingual wall open mesially, nearly closed distally; mandibular molars elongate, their buccal midwall complexities developed into stylids, their buccal valleys deep, but their lingual valleys hardly developed; distal lobe on the posterior lower molar well individualized. Supraorbital foramina flush with the forehead; lacrimal bone extending forward and downward onto the face, and into the preorbital fossa; ethmoid fissure short; lacrimal bone with fenestrae; median nasal tips very long, the lateral prongs very short. Basioccipital short and broad; only a narrow channel between the anterior basioccipital tuberosities; anterior tuberosities small, the posterior ones represented only by low ridges, and the longitudinal ridges between the anterior and the posterior pairs barely indicated; pterygoid plates evenly sloping. Cranial profile short, high; facial profile straight; horns placed slightly behind the posterior margin of the orbit, sloping back at a 45° angle in the males; tips of the premaxillae well above palatal level; preorbital foramen rounded, above the posterior premolars; palatal surface of the diastema raised well above the plane of the palate; auditory meatus oblique, fairly rounded. No glandular knee-tufts; preorbital glands present; foot glands present; inguinal glands absent; rhinarium restricted to a median line; one pair of mammae; tail relatively long. (Following Groves, 2000.)

In a survey of this genus, Lange (1970) pointed out the reduced sexual dimorphism of *Antidorcas* in comparison with *Gazella*.

Robinson (1979) found strong and significant size differences between the two N taxa (designated *A. m. angolensis* and *A. m. hofmeyri*; note that his Kaokoveld sample actually represents *A. m. hofmeyri*) and the S nominotypical *A. m. marsupialis*: the mean body mass of the males of the first two taxa was 42.0 and 41.6 kg, respectively, and of *A. m. marsupialis*, 31.2 kg. Body mass was found to be highly correlated with the availability of winter dietary protein, raising the question of the degree to which this difference in body mass is due to genetic factors or to phenotypic plasticity.

The general taxonomy here follows Groves (1981c), except that the three subspecies recognized therein are raised to species rank, as they differ 100%.

Antidorcas marsupialis Zimmermann, 1780

Rich chestnut-brown; flank-stripe deep brown; pygal-stripe well marked. Face-stripes very thin, not dark; forehead brown, fawn, or white, this color not extending in front of the level of the eyes, and (if brown or fawn) not sharply bordered anteriorly. Nose white, or with a brown smudge. Horns exceptionally short in the females, only 60%–70% of the length of those of the males.

S of the Orange River, from the NE Cape to the Free State and the Lombard and Kimberley districts.

Antidorcas hofmeyri Thomas, 1926

Light fawn; lateral band dark brown to nearly black; pygal band medium brown, thin. Face-stripes thin, dark brown. Forehead anteriorly extending in front of the level of the eyes, brown or pale fawn, with no anterior border, so the transition to the white of the face not sharply demarcated. Dark smudge on the nose varying from dark to pale to absent altogether. Larger than the other two species, to judge from the skulls. Horns of the females clearly annulated, large, 77%–85% the length of those of the males.

N of the Orange River, from the Upington district (N Cape) via the Sandfontein district through Botswana to Namibia.

Antidorcas angolensis Blaine, 1922

Brown-tawny; lateral band nearly black; pygal band very dark brown. Face-stripes thick, rich dark brown, extending two-thirds of the way to the muzzle. Forehead, down to in front of the eye level, medium brown, bordered anteriorly by a dark brown edge; rest of the face sharply white, except for a brown spot on the nose. Horns of the females angulated, relatively large, 87% of the length of those of the males.

Angola, N to the latitude of Benguela.

Ammodorcas Thomas, 1891

Females without horns. Horn rings of the males very prominent, widely spaced, angulated in front, much flatter behind than in front, connected by longitudinal ridges. Horn cores rather compressed, with traces of rings on them, and with broad, deep grooves up both the anterior and the posterior surfaces. Anterior premolar (P^2) very small, oval in both jaws; styles on the other premolars small, but well developed on the molars; molars elongated, especially the distal selene, with its flat buccal wall. Mandibular P^4 elongate, with the lingual wall open both mesially and distally; mandibular molars elongate, their buccal midwall convexities developed into stylids, their buccal valleys deep, but their lingual valleys hardly developed; distal lobe on the posterior lower molar simple. Preorbital fossa shallow. Supraorbital foramina flush with the forehead; lacrimal bone extending onto the face, but not into the preorbital fossa; ethmoid fissure short; lacrimal bone with numerous fenestrae; median nasal tips very long, the lateral prongs fairly long. Pterygoid plates evenly sloping. Cranial profile very long and low; facial profile straight; horns placed slightly behind the posterior margin of the orbit, sloping at a 60° angle in the males; tips of the premaxillae at or below the occlusal level; preorbital foramen slitlike, above the anterior premolar; palatal surface of the diastema in the plane of the palate; auditory meatus vertically oval. Glandular knee-tufts and preorbital glands present; foot and inguinal glands absent; rhinarium restricted to the median line; two pairs of mammae; tail very long. (Following Groves, 2000).

Ammodorcas clarkei (Thomas, 1891)

dibatag

On the general biology of this species, see Schomber (2007); on its distribution and advances in its conservation, see Wilhelmi et al. (2007).

A. clarkei is essentially a species of restricted distribution in the Somali arid zone.

Litocranius Kohl, 1886

gerenuk

Horns absent in the females. Horn rings flattened, very close set, circular (all around the horn), connected only by very poor longitudinal ridges. Horn cores very compressed, with only irregular, shallow, short grooves, but with deep grooves up both the anterior and the posterior surfaces. Anterior premolar (P^2) not reduced; styles on the other premolars very prominent, producing a "winged" appearance, but less well developed on the molars; molars not elongated, the selenes with concave buccal walls. Mandibular premolars short, with the lingual wall open both mesially and distally; mandibular molars short, their buccal midwall convexities not developed into stylids, their lingual valleys well developed; distal lobe on the lower molar simple, very small. Supraorbital foramina in a shallow pit; lacrimal bone short, not extending into the preorbital fossa; ethmoid fissure long, extending some way down the lateral edge of the nasal; lacrimal bone with numerous fenestrae; median nasal tips very long, the lateral prongs well developed. Basioccipital long and narrow; only a narrow channel between the anterior basioccipital tuberosities; anterior tuberosities small, the posterior ones represented by bulky ridges, and longitudinal ridges between the anterior and the posterior pairs barely indicated; pterygoid plates with a large hamuli. Cranial profile very long and low; facial profile convex; horns placed directly above the orbit, rising at an angle of nearly 90°; tips of the premaxillae at or below the occlusal level; preorbital foramen slitlike, above the anterior premolar; palatal surface of the diastema in the plane of the palate; auditory meatus vertical-oval. Glandular knee-tufts present; preorbital glands present; foot glands present; inguinal glands absent; rhinarium extending slightly along the dorsal rims of the nostrils; two pairs of mammae; tail relatively long. (Following Groves, 2000).

Taxonomic characters follow Grubb (2002), but the two taxa recognized therein as subspecies are here recognized as full species, because their ranges of variation are discrete.

Litocranius walleri (Brooke, 1878)

1878 *Gazella walleri* Brooke. Grubb (2002) designated Chisimaio (0° 23' S, 42° 34' E) as the restricted type locality.

Size small; skull length of the males 211–241 mm (mean 231.6 mm, $N=25$), of the females 207–212 mm (mean 210.9 mm, $N=5$) (Grubb, 2002). A band of reversed hairs, directed toward the head, on the dorsal midline of the neck.

NE Tanzania through Kenya to Galcaio in E-C Somalia, N of the Webi Shebeyli.

Litocranius sclateri Neumann, 1899

1899 *Lithocranius* [*sic*] *sclateri* Neumann. Berbera, Somalia.

Size large; mean skull length of the males 248–259 mm (mean 251.2 mm, $N=10$), of the females 226–231 mm (mean 227.3 mm, $N=4$). Other cranial measurements, especially of the facial skeleton, much larger than in *L. walleri*, with their ranges not overlapping. No hair reversal on the neck.

No color differences can be detected between *L. sclateri* and *L. walleri*.

NW Somalia: Berbera district and W, to just over the Ethiopian border; Djibouti.

Saiga Gray, 1843

saiga

Females hornless. Horns of the males amber-colored, not blackish like other antilopins, their rings very prominent, widely spaced, circular (all around the horn), connected only by very poorly expressed longitudinal ridges. Horn cores circular in cross section. Anterior premolar (P^2) very reduced, generally absent in the mandible; styles on the other premolars small, but well developed on the molars; molars short. Mandibular P^4 elongate, with the lingual wall closed both mesially and distally; mandibular molars elongate, their buccal valleys deep, but their lingual valleys hardly developed; distal lobe on the lower molar well individualized. Nasal bones very reduced, above a very deep nasal cavity. Supraorbital foramina flush with the forehead; lacrimal bone extending forward and downward onto the face, but not into the preorbital fossa; ethmoid fissure absent, probably correlated with the extremely short nasal bones; lacrimal bone with no fenestrae; median nasal tips very long, the lateral prongs very short. Basioccipital short and broad, with a broad smooth channel between the anterior basioccipital tuberosities; anterior tuberosities small, ridged, the posterior ones represented only by low ridges, and the longitudinal ridges between the anterior and the posterior pairs barely indicated; pterygoid plates with large hamuli. Cranial profile very short and high; facial profile convex, the facial skeleton very deep; horns placed slightly behind the posterior margin of the orbit, sloping back at a 60° angle; tips of the premaxillae at or below the occlusal level; preorbital foramen slitlike, above the posterior premolars; palatal surface of the diastema in the plane of the palate; auditory meatus obliquely oval. Glandular knee-tufts present; preorbital, foot, and inguinal glands all present; rhinarium absent; two pairs of mammae; tail very short. (Following Groves, 2000).

Saiga was formerly included in the Caprini, but A. W. Gentry (1992) pointed out that it shares some very cogent character states with acknowledged Antilopini: the horn core diameter is enlarged, the interfrontal suture is complicated, the anterior basioccipital tuberosities are strong and localized (as in some of the Caprini), there is a central cavity on the hypoconulid of the posterior lower molar, the digital flexor ridge on the tibia is laterally placed, and the lateral tubercle on the radius is high-placed. The cranial roof, however, is not strongly inclined. Gentry's insight has been vindicated by molecular phylogenetic study (Hassanin & Douzery, 1999a; and other sources).

The extreme specializations of the nasal cavity have been studied by Clifford & Witmer (2004). The cavity is enormously enlarged and inflatable, with a lateral vestibular recess, apparently entirely neomorphic, and a largely membranous nasal septum, in which there is a large patch of tissue acting as a baffle, controlling air flow.

Characters of the two very distinctive taxa largely follow Bannikov (1963).

Saiga tartarica (Linnaeus, 1766)

1766 *Antilope saiga* Pallas.

1766 *Capra tartarica* Linnaeus. Ural steppes of Russia.

1767 *Antilope scythica* Pallas.

1827 *Antilope colus* Hamilton Smith.

Horns thick, waxy-colored, semitranslucent, with 12–20 strongly marked rings; horns 25–33 mm thick at the base, 280–380 mm long. Nasal opening enormously enlarged, raised. Summer coat yellowish red, paler on the flanks, the underside white, with darker zones on the shoulders and the loins; winter coat very light, gray-colored. Condylobasal length of the males 222–250 mm, of the females 205–209 mm.

Historically found from the borders of the Carpathians E to Dzungaria; now reduced to small fragments, mainly in Kazakhstan, but as far W as the right bank of the Volga River.

Saiga mongolica Bannikov, 1946

1946 *Saiga mongolica* Bannikov. Tshargin Gobi, Mongolia.

Horns relatively short, thin, with weakly expressed rings; bases of the horns up to 28 mm thick; horn length at most 220 mm; frontals more sloping anteriorly, with the nasal opening less raised than in *S. tartarica*. Summer coat sandy gray; dorsal region not darkened, but the brown spot in the lumbar region large and sharply bordered. Skull length 203–237 mm.

E of the Mongolian Altai, in the basins of the W lakes in Mongolia, from nearly 50° N to 48° N: Kirghiz Nor SE to the Tshargin Gobi, and periodically to Orok Nor. *S. mongolica* was included as a subspecies of the fossil *S. borealis* by Baryshnikov & Tikhonov (1994).

Antilope Pallas, 1766

$2n = 30$ (females), 31 (males).

Antilope emerges as sister to *Gazella* (that is to say, it is closer to *Gazella* than are *Eudorcas* and *Nanger*) in most phylogenies (Rebholz & Harley, 1999), although the earliest proposed phylogeny placed it as sister to *Nanger* (Effron et al., 1976). Its detailed karyotype, in fact, more closely resembles that of the *Gazella subgutturosa* group than that of other gazelles (Vassart, Séguéla, et al., 1995).

Horn of the males fairly prominent, widely spaced; rings circular (all around the horn), connected by poorly expressed longitudinal ridges. Horn cores nearly circular in cross section, with no traces of rings on them, and with one groove up the posterior surface and one up the anterior surface. Anterior premolar (P^2) reduced in the mandible, not in the maxilla; styles prominent on the other premolars and on the molars; molars elongated, especially the distal selene, with its concave buccal wall. Mandibular P^4 short, with the lingual wall open mesially, nearly closed distally; mandibular molars elongate, their buccal midwall convexities not developed into stylids, their buccal valleys shallow, but their lingual valleys more developed; distal lobe on the lower molar well individualized. Supraorbital foramina in shallow pits; lacrimal bone extending forward and downward onto the face, and into the preorbital fossa; ethmoid fissure long, extending some way down the lateral edge of the nasal; lacrimal bone with a single fenestra; median nasal tips shorter than the lateral prongs. Basioccipital somewhat elongated; only a narrow channel between the anterior basioccipital tuberosities; anterior tuberosities large, ridged, the posterior ones more bulky, and the longitudinal ridges between the anterior and posterior pairs barely indicated; pterygoid plates evenly sloping. Cranial profile short, high; facial profile straight; horns placed slightly behind the posterior margin of the orbit, sloping back at a 45° angle in the males; tips of the premaxillae well above palatal level; preorbital foramen rounded, above the posterior premolars; palatal surface of the diastema raised somewhat above the plane of the palate; auditory meatus oblique, fairly rounded. Glandular knee-tufts present; preorbital glands present, even hypertrophied; foot and inguinal glands absent; rhinarium extends all along the dorsal rims of the nostrils; one pair of mammae; tail short. (Following Groves, 2000).

Schreiber et al. (1997) found low protein polymorphism in this genus compared with *Gazella*, *Eudorcas*, and *Antidorcas*, and they ascribed this to a history of rapid evolution of its highly autapomorphic phenotype, aided by strong selection of a few dominant males, due to its lekking behavior, since its common ancestor was with *Gazella* (since the authors were, at the time, including *Eudorcas* as a species of *Gazella*, they regarded blackbuck as being actually nested within *Gazella*).

Antilope cervicapra (Linnaeus, 1758)
blackbuck

The full synonymy of this species will be found in Groves (1982b, 2003).

The taxonomy follows Groves (1982b, 2003). There is a possibility that the two recognized subspecies may actually be distinct species, but more needs to be known about their individual variations.

Antilope cervicapra cervicapra (Linnaeus, 1758)
Smaller, with short, fine hair; dark color of the upperside running all down the limbs to the hoofs; white eye-ring narrowed above the eye. Horns relatively short, not very divergent, with a relatively open spiral.

E and S of the Delhi region, as far S as Madras and Karnataka, and E to Bengal.

Antilope cervicapra rajputanae Zukowsky, 1927
Larger, with longer, roughened pelage; males, in the breeding season, with a gray sheen; shanks largely white, with little or no extension of the dark color from the upper limb segments; white eye-ring broad all around the eye. Horns tending to be longer, more divergent, and more closely spiraled.

W of the Delhi region, to Saurashtra and Vadodara, Amritsar, and into Pakistan.

Nanger Lataste, 1885
$2n = 40$.

There is little similarity between this genus and any other in chromosomes, except that, like *Eudorcas*, *Nanger* has a Y-autosome translocation, as well as the X-autosome translocation characteristic of *Gazella* and *Antilope* (Vassart, Séguéla, et al., 1995).

Differs from *Antilope* as follows: horn cores more compressed; several shallow grooves up the posterior surface of the core, instead of long, deep, broad ones; molar styles less well developed; distal selenes of the molars with a broad, flat buccal wall, instead of being concave; anterior lower premolar less reduced; lingual stylids of the lower molars less well developed; buccal valleys of the lower molars deeper; lacrimal bone usually with no fenestra; pterygoid plates with large hamuli; cranial profile more elongated; horns directly above the orbits, sloping back less; premaxilla tips at or below the occlusal level; preorbital fora-

Table 40 Univariate statistics for the *Gazella subgutturosa* group: males

	Horn l	Tip–tip	Span	Base br	Nas br ant	Nas br post	Gt l	Braincase l
marica								
Mean	270.188	101.50	168.118	56.278	22.682	22.500	181.333	100.200
N	16	14	17	18	11	12	9	15
Std dev	22.0884	41.210	39.6562	2.4448	3.1167	2.2764	6.9101	3.8210
Min	234.0	50	97.0	50.0	17.0	20.0	172.0	92.0
Max	312.0	178	228.0	61.0	28.0	28.0	193.0	107.0
Iraq								
Mean	271.875	108.00	181.375	59.700	23.333	22.800	185.500	104.500
N	8	7	8	10	3	5	2	6
Std dev	16.7369	38.066	25.1733	1.7029	1.5275	2.2804	6.3640	4.0373
Min	251.0	46	152.0	57.0	22.0	20.0	181.0	100.0
Max	296.0	148	228.0	63.0	25.0	25.0	190.0	112.0
Caucasus *subgutturosa*								
Mean	282.000	107.00	196.000	67.500	23.500	26.000	215.000	116.500
N	1	1	1	2	2	2	2	2
Std dev	—	—	—	.7071	.7071	1.4142	5.6569	3.5355
Min	282.0	107	196.0	67.0	23.0	25.0	211.0	114.0
Max	282.0	107	196.0	68.0	24.0	27.0	219.0	119.0
Iran *seistanica*								
Mean	297.333	129.56	200.778	63.950	25.125	26.313	208.500	113.000
N	9	9	9	10	8	8	6	8
Std dev	21.4418	71.544	57.2904	2.5653	1.4577	1.6677	6.5345	3.9641
Min	246.0	56	151.0	60.0	22.0	24.0	199.0	106.0
Max	318.0	245	298.0	69.0	27.0	28.0	219.0	118.0
diversicornis								
Mean	308.835	142.68	219.218	74.400	25.664	27.125	203.779	112.035
N	17	17	17	17	14	12	14	17
Std dev	26.4785	53.170	37.4524	1.8894	1.4302	1.4169	4.2392	3.3429
Min	229.7	78	151.7	70.5	24.0	25.0	195.8	104.7
Max	331.0	250	302.0	77.4	29.5	29.5	210.4	117.3
yarkandensis								
Mean	280.778	92.93	175.875	69.000	27.310	28.840	212.375	112.230
N	9	7	8	8	10	10	8	10
Std dev	32.9802	36.788	26.9997	5.9761	1.4364	1.2607	5.7056	4.8369
Min	213.0	41	128.0	64.0	25.0	27.0	201.0	106.0
Max	318.0	137	214.0	79.0	29.0	31.0	220.0	122.0
hillieriana								
Mean	258.252	109.47	163.737	64.380	27.284	29.332	203.763	107.391
N	21	18	19	25	25	25	19	23
Std dev	16.4623	35.398	28.0888	4.3091	1.5491	1.8443	5.5587	4.3142
Min	233.0	39	120.0	58.0	25.0	26.0	194.0	99.0
Max	310.0	176	213.0	79.0	31.0	33.0	218.5	116.0

men farther forward; glandular knee-tufts hypertrophied; preorbital glands less enlarged; rhinarium extending only slightly along the dorsal rims of the nostrils; females with horns. (Following Groves, 2000). The anterior basioccipital tuberosities tend to be small, as in *Gazella* (A. W. Gentry, 1964).

Lange (1971) argued that the pelage pattern of this genus (which he deemed a subgenus of *Gazella*) is specialized, with strong and progressive pattern reduction, the pattern tending to remain more strongly expressed in the females than in the males. He pointed out a relationship between specialization in horn form and pattern reduction.

In the phylogenies of Rebholz & Harley (1999), *N. soemmerringi* and *N. dama* always formed a sister clade to *N. granti*.

Nanger granti Group

Grant's gazelle

Furley (1986) recorded gestation length as 198–199 days; the first fertile mating of a male at 450 days; and the first fertile mating of a female at 420–450 days, but occasionally as early as 210–270 days.

Grubb (2000a) argued that the species of this group form a ring-shaped morphocline, from *N. petersi* on the Kenya coast via Sudan/Ethiopian *N. notata* (designated *N. brighti* in that paper) to *N. granti* and, finally, *N. robertsi*. Lorenzen et al. (2008a) studied mtDNA in this group and found three reciprocally monophyletic clades, differentiated by numerous substitutions.

Nanger granti (Brooke, 1872)

1872 *Gazella granti* Brooke. W Kinyenye, Ugogo, Tanzania.
1903 *Gazella granti robertsi* Thomas. Mwanza, Speke Gulf, Lake Victoria, Tanzania.
1913 *Gazella granti roosevelti* Heller. Kitanga Farm, Athi Plains.

Skins seen: 12/8, as well as observations of numerous living animals.

White rump-patch large, undivided; whole of the tail base white; lateral extension of the rump-patch intruding largely into the body color, thus extending beyond the dark pygal band. Flank band in the males colored like the dorsum, or somewhat darker, or dark but fading anteriorly, usually present in the females (occasionally fading forward), but sometimes faint, or even absent.

In populations to the W of the Rift Valley (described as a subspecies, *G. g. robertsi*), the horns are strongly outcurved and divergent, with the tips backwardly curved, in about 50% of the adult males, but as there is apparently no other difference, and this unusual horn type does not characterize a preponderance of the population, the subspecies is not recognized here.

We have seen specimens from Lake Elmenteita, Gilgil, Lake Nakuru, Nairobi, Simba Station, Lukenya, Athi River, Sultan Hamud, Taveta, S Guaso Nyiro, Serengeti plains.

Nanger notata (Thomas, 1897)

1897 *Gazella granti notata* Thomas. W slope of the Loroghi Mountains, Kenya.
1901 *Gazella granti brighti* Thomas. 150 mi. E of Lado (5° 20' N, 34° 5' E), Sudan.
1906 *Gazella granti lacuum* Neumann. Suksuk River, S of Lake Zwai, Ethiopia.
1913 *Gazella granti raineyi* Heller. Isiolo, N Guaso Nyiro, Kenya.

Skins seen: 4/6.

Flank band in the males usually somewhat darker, in the females, very faintly darker. Instead of a broad pygal band extending all down the rump (as in *N. granti*), the rump band narrow, extending only halfway down; color a shade darker.

Grubb (1994) found that the type of *Gazella granti notata* is somewhat different from the types and topotypical series of *G. g. brighti* and *G. g. lacuum*, and *G. g. raineyi*; it may represent "a unique localized population, possibly now extinct." For the moment, we retain all these names in synonymy.

All of the specimens we have seen are from N of the Kenya highlands: Marsabit Road; N Guaso Nyiro; Lorian Swamp; Ngare Ndare; Laikipia; Lake Baringo; Rumuruti; Turkwell River. Looking at published maps, in historical records there always seems to have been a geographic discontinuity between *N. notata* and *N. granti* (Grubb, 1994).

Nanger petersi Günther, 1884

1884 *Gazella petersii* Günther. Gelidja, near the mouth of the Osi and Tana rivers.
1887 *Gazella granti* var. *geliadjiensis* Noack. Gelidja.
1913 *Gazella granti serengetae* Heller. Taveta, SE Kenya.

Color darker than others. No dark lateral band in the adult males. White rump-patch relatively small, divided above by a broad band extending onto and along the upper surface of the tail. Lateral prolongation of the rump-patch narrower, and intruding, to a smaller extent, into the body color, thus scarcely, if at all, overhanging the dark pygal band. Horns short, not very divergent, not backward-curved. Skull small, with a relatively narrow nasal opening.

Kenya coast, extending N into southernmost Somalia. There have been noticeable range changes in the recent past.

Gazella granti serengetae Heller, 1913 (type locality: Taveta), was described as being dark cinnamon or fawn on the back; in two males, the fawn color extended backward as a narrow stripe through the middle of the white rump-patch and onto the upper surface of the tail, of which only the basal third is white, the remainder being black. The females lack this division. According to Roosevelt & Heller (1914), *G. g.*

serengetae differs from *N. petersi* in the cinnamon streak being narrow, and in having larger and less parallel horns. Actually, *G. g. serengetae* is clearly a synonym of *N. petersi*—straight, parallel horns; reduced croup-patch, with dark dorsal color bisecting the croup-patch and joining the tail. Yet today, *N. granti* is the species that occurs at Taveta, and, according to Lorenzen et al. (2008a), mtDNA from gazelles at the type locality resembles what they call *Gazella granti robertsi*, (i.e., *N. granti*). Hence, it appears that there has been a shift in distribution since the early 20th century, when *G. g. serengetae* was collected.

Four males in Tervuren—from the Ziwani Swamp, Tsavo—are variable. Three have a slight wedge back toward the tail, and that of the fourth reaches the tail base; this also applies to one from Pika Pika (3° 49' S, 38° 52' E), and one from Ubwegetwe, Tanzania. These suggest a population with some gene flow from *N. granti*. In addition, Lorenzen et al. (2008a) noted that one of 11 samples from Mkomazi, in NE Tanzania near the Kenya border, had mtDNA not like *N. granti* (as in the other 10), but like *N. petersi*, corroborating all of the above indications of the latter's former occurrence far to the SW of its present range.

The study of mitochondrial control region sequences by Arctander et al. (1996) found astonishingly large degrees of separation between these three species, comparable to the distance between any of them and *N. soemmerringi*. There was a tendency for *robertsi* samples to form a subcluster within *N. granti*. A preliminary report on this work appeared in the Antelope Specialist Group's *Gnusletter*, and occasioned some surprise, which was not, in fact, warranted (Grubb, 1994).

Lorenzen et al. (2008a), working on a larger number of samples, likewise separated these three species cleanly; this indicated to those authors that at least *N. petersi* should be regarded as a distinct species, differing from the other two by an amount similar to that differentiating them from *N. soemmerringi*. They drew attention to the opening up of Tsavo East National Park during the late 19th and early 20th centuries by elephant action, bringing into contact *N. petersi* and *N. granti*, previously separated by *Acacia-Commiphora* woodland, as reported by Leuthold (1981), who documented a case of intermale combat between the two species. The range of *N. petersi* has continued to spread; it is now the predominant (or only) species of *Nanger* in Tsavo East National Park (Bruce Patterson, pers. comm. to CPG), thus evidently reestablishing itself in much of its pre-20th-century range.

Nanger soemmerringi (Cretzschmar, 1826)

Soemmerring's gazelle

$2n$ varied, from $2n=34$ to $2n=39$, for 7 male and 20 female individuals in two US zoos (Benirschke et al., 1984). The source of the animals was unclear; perinatal mortality was very high, and whether the stock was actually mixed—perhaps even with a member of the *N. granti* group, as the authors inferred—was not known.

Furley (1986) recorded gestation length as 195–210 days, and the age of a female at earliest fertile mating as 540–570 days.

This (presumed) species is found throughout the lowlands of the Ethiopian region and into Eritrea, Somalia, and far E Sudan. We have little data on the potential taxonomic diversity within the species, but we note that, given the barriers formed by the Ethiopian highlands, the probability of significant differentiation is great.

Nanger dama Pallas, 1766

dama gazelle

1766 *Antilope dama* Pallas. Senegal.

As reviewed by Harper (1940), anywhere in the Lake Chad region cannot be the type locality, because in the time of Buffon (whose descriptions Pallas used), Lake Chad was unknown.

Gestation length is recorded as 174–202 days in *N. d. dama*, 201 days in *N. d. mhorr*, and 210–225 days in *N. d. ruficollis*. The first fertile mating of a male was recorded at 120 days; of a female, 610 days (Furley, 1986).

The revision by Perez (1984), based on study of 50 museum skins and 150 living animals (including the skins of some of the latter on their demise), divided this species into three subspecies:

1. *Nanger dama mhorr* (synonyms *Gazella dama permista*, *Gazella dama lozanoi*)—Ranging from S Morocco to Rio de Oro, Senegal, and the Niger River. Upper parts reddish brown, reaching from the head to the middle of the croup, and down the hindlegs at least to the hocks; the underside, the rest of the legs, and the croup white (sending a V-shaped wedge forward into the haunch), separated from the reddish parts by a clear line of demarcation. The reddish line on the thigh is broadened, restricting the white posteriorly to the

croup-wedge, and anteriorly to the anterior edge of the stifle. The reddish line on the upper segment of the forelegs may or may not be connected with that of the body color. There are a darker and a lighter morph, the latter being the one described as *N. d. lozanoi*.

2. *Nanger dama dama* (synonym *Gazella dama damergouensis*)—Ranging from about 7° E to Chad. Distinguished from *N. d. mhorr* by the absence of the thick reddish thigh-block; instead, a thin reddish line runs back horizontally from the lower part of the body-color block, then turns downward onto the upper thigh. Perez argued that Buffon's "Nanguer," on which the name *Nanger dama* was founded, could not have been from Senegal, as his illustration does not correspond with specimens known to have come from Senegal. Yet, as noted above, a more E place of origin is very unlikely.
3. *Nanger dama ruficollis*—E of about 15° E. The reddish tones are mostly restricted to the neck, sometimes nearly reaching to the croup, but always restricted to the upper part of the body, and the line of demarcation with the (often red-toned) zone on the underparts is diagonal.

The following year, a very similar arrangement was produced by Drüwa (1985), who likewise argued that the name *N. d. dama* could not refer to the Senegal taxon, and saw *N. d. mhorr* as extending E as far as the Niger River. Uniquely, however, he synonymized *N. d. ruficollis* with *N. d. dama*, thus reducing the subspecies to two:

1. *Nanger dama mhorr*—Mountain- and grass-steppes of W Africa. Dark, with a marked brown tone to the colored areas; broad, brown stripes extend from the flank zone down the legs. There is variation in the size of the rump-patch, the face pattern, the extension of the brown tone onto the belly, and of the flank stripes down the legs. Synonyms would include not only *Gazella dama lozanoi* but also *G. d. permista* and *G. d. damergouensis*.
2. *Nanger dama dama*—Dry grass and desert zones of the C and E Sahel. The brown body color is reduced, to a greater or lesser extent, to a saddle; the behind parts and the sides of the body are contrastingly white. A clear flank stripe can extend as a rudiment, from the saddle to the haunch, but the brown stripe on the limbs is only on the lower segments, if at all.

Our data, collected over the years, would tend more to support the arrangement of Perez (1984). There is a decided difference between those W of Lake Chad and those to the E, with an admittedly wide area of intermediacy within the Republic of Chad. But the boundaries in the W between *N. d. dama* and *N. d. mhorr* is incorrect. The specimens we have seen from Senegal are indeed like Buffon's plate—which constitutes the type of *N. d. dama*—and specimens corresponding to *Gazella dama permista* also occur in W Africa, and even in the Chad zone of intermediacy; the dark *N. d. mhorr* and intermixed *Gazella dama lozanoi* types do not occur farther S than the W Sahara. Our arrangement, therefore, is listed below.

Nanger dama dama (Pallas, 1766)

1766 *Antilope dama* Pallas. Probably Senegal (Harper, 1940).
1833 *Antilope nanguer* Bennett. Senegal.
1847 *Antilope dama* var. *occidentalis* Sundevall. Senegal and Morocco.
1906 *Gazella dama permista* Neumann. Senegal.
1907 *Gazella mhorr reducta* K. Heller. Locality unknown.
1921 *Gazella dama damergouensis* Rothschild. Takoukout, Damergou, Niger River.
1935 *Gazella dama weidholtzi* Zimara. Hombori, Mali.

Nanger dama mhorr (Bennett, 1833)

1833 *Antilope mhorr* Bennett. Wednun, near Tafilat, Mogador, Morocco.
1934 *Gazella dama lozanoi* Morales Agacino. Cape Juby, Rio de Oro, W Sahara.

Nanger dama ruficollis (Hamilton Smith, 1827)

1827 *Antilope ruficollis* Hamilton Smith. Dongola, Sudan.
1833 *Antilope addra* Bennett. Nubia and Kordofan.
1847 *Antilope dama* var. *orientalis* Sundevall. Sennar.

Gazella de Blainville, 1816

smaller gazelles

Differs from *Antilope* as follows: horn rings usually somewhat angulated in front, and usually better expressed on the front surfaces of the horns; horn cores of the males slightly more compressed; molar styles less developed; anterior lower premolar less reduced; preorbital fossa deeper; lacrimal bone extending well into preorbital fossa; ethmoid fissure shorter; basioccipital longer and narrower, anterior tuberosities smaller, more triangular, the posterior ones less bulky, the longitudinal ridges between them more expressed; horns of the males more upright; preorbital foramen farther forward; preorbital glands less hypertrophied. (Following Groves, 2000).

Groves (1969a) and Lange (1972) placed *G. pelzelni*, *G. saudiya* and *G. bennetti* in *G. dorcas*, but Lange (172) transferred *G. cuvieri* from *G. rufifrons* (here, *Eudorcas rufifrons*), whereas Groves (1969a,) had placed it, in a moment of mental aberration, as a subspecies of *G. gazella*, thereby returning to the old habitat-dependent model of Ellerman & Morrison-Scott (1951). Lange (1972) also made *G. leptoceros* conspecific with *G. subgutturosa*.

Vassart, Granjon, et al. (1994) calculated an unrooted cladogram on the basis of allele frequencies of 16 genetic loci. *G. dorcas* appeared very different from members of the *G. gazella* group. The use of allele frequencies may not, however, be a satisfactory way of reconstructing phylogeny above the species level (see, for example, J. P. Grobler & van der Bank, 1993).

Rebholz & Harley (1999) found that in their neighbor-joining tree, there were three clades in this genus: the *G. bennetti* group, the *G. subgutturosa* group, and a *G. dorcas* / *G. spekei* / *G. gazella* group. R. Hammond et al.'s (2001) phylogeny of the major taxa of the genus was as follows:

((*subgutturosa* (*bennetti*, *leptoceros*/*marica*)) (*gazella* (*saudiya*, *dorcas*/*pelzelni*))).

The present and former distribution of gazelles in the Middle East was discussed by Uerpmann (1987), with notes on their taxonomy.

Gazella subgutturosa Group

Females often hornless. Rings on the horns of the males somewhat less close-set than in the *G. dorcas* group, but more than in the *G. bennetti* group, and equivalent to the *G. gazella* group, as well as less angulated in front than either of the other two groups and more equal in prominence from back to front, with even poorer longitudinal ridges between them; horn cores more nearly circular in cross section, with a deep groove running up the posterior surface. Upper molars with more convex buccal walls. Preorbital fossa very deeply excavated. Supraorbital foramina 2–3, in deeper pits than the *G. dorcas* or the *G. bennetti* groups. Nasofrontal suture transverse, or slightly W-shaped. Lacrimal bone with a fenestra. Median nasal tips shorter than the lateral prongs. Anterior basioccipital tuberosities smaller than other *Gazella*. Facial skeleton somewhat elongated. Valvular preorbital glands hypertrophied.

Gazella leptoceros (F. Cuvier, 1842)

slender-horned gazelle

Lange (1972) made *G. leptoceros* conspecific with *G. subgutturosa*. Interestingly, in R. Hammond et al.'s (2001) phylogeny, *G. marica* assorted with *G. leptoceros*, rather than with *G. subgutturosa*.

Gestation length is 156–169 days, and the age of a female at first fertile mating can be as early as 152 days (Furley, 1986).

Gazella cuvieri Ogilby 1841

Cuvier's gazelle

$2n=32$ (females) and 33 (males), with an X-autosome translocation like most gazelles; the karyotype strongly resembles *G. leptoceros*, with a slight difference in the banding pattern on the short arm of the X (Kumamoto & Bogart, 1984).

Gestation length is 165 days; the age of a male at first fertile mating is given as 180 days, or alternatively at 360–540 days; the age of a female at first fertile mating, 422 days; the frequency of twinning is about 50% (Furley, 1986).

Gazella subgutturosa and Related Taxa

goitered gazelles

Wurster (1972) reported $2n=30$ (females) and $2n=31$ (males) in "Persian gazelles," Kingswood and Kumamoto (1988) found the same for five males and four females specified as *G. subgutturosa*, and Shi (1987) reported the same for what was identified as *G. subgutturosa* from China.

Kingswood & Kumamoto (1988) also karyotyped 38 males and 33 females of what were identified as *Gazella subgutturosa marica*, finding $2n=31$, 32, and 33 (males), and $2n=30$, 31, and 32 (females). These authors argued that the stock may ultimately have been founded by a mixed stock of *G. marica* and *G. subgutturosa*, and they noted, in addition, that the presumed *G. marica* were also less fertile than captive stocks of *G. subgutturosa*. Granjon et al. (1991) karyotyped 11 males and 19 females of *G. marica* from a captive stock in Saudi Arabia, finding $2n=33$ in all males, but $2n=32$ and $2n=31$ (in 4 out of 20) in the females; they considered this to be support for Kingswood & Kumamoto's (1988) hypothesis, although the presence of the four $2n=31$ females seems to indicate that, even though the hypothesis may be correct, there may have been a little mixture here, as well.

This species complex has been fully described by Kingswood & Blank (1996). It has a characteristic enlargement of the larynx, which is more prominent in the males; horns of the males are close together at the bases, long and lyrate, but (in most taxa) usually absent in the females. Facial markings and the flank band fade with maturity; adults of most taxa have a face that is essentially white. In the skull, the palate and the biorbital breadth tend to be greater than for other taxa; the lacrimal pits are very deep.

In our craniometric comparisons, in the SW members of this group there is a clear distinction between those from Iran and the others, but the samples from Arabia, on the one hand, and from Iraq and Khuzestan, on the other, though they differ in color, do not differ strongly craniometrically. The difference is largely in size; while palate breadth, braincase breadth, and braincase length are hardly different, greatest skull breadth and breadth across the base of the horns account for the size difference (skull length was not included in the analysis, because of the high number of skulls with broken premaxillae).

In the N-C samples, there is (quite unexpectedly) a clear distinction between Iran and Turkmenistan. We have two available "Caucasian" specimens, but they are actually from different areas: one, inside the Iranian range of variation, is from Baku in Azerbaijan, just NW of the Iranian border, whereas the other, from Tbilisi (i.e., the topotypical area for *G. subgutturosa*), seems rather different, distinguished from the Iranian sample by its long, relatively narrow braincase and broad palate. Those in the Turkmenistan group have very high values for the basal breadth of the horns, contrasting with the relatively narrow palate and the short braincase. These findings are unexpected, considering that previously, the Turkmenistan and Tbilisi gazelles had been regarded as consubspecific with those from Iran.

In our four horn variables, we again find a clear difference in Iran versus Turkmenistan. The only Caucasian specimen (Tbilisi) this time assorts with Iran. The Turkmenistan gazelles have a wide tip-to-tip interval, contrasting with a relatively narrow span.

When the comparisons are moved farther to the NE, skulls from the Gobi are quite separate from those from Turkmenistan, and those from Yarkand (in Xinjiang) overlap with the Gobi sample (one specimen, from the Cherchen River, W of Kunlun, approaches Turkmenistan, but does not overlap with it).

Individual skulls from Caidam, Semiretch, and Har Nor (38.20° N, 97.30° E, on the borders of Caidam) are closest to Turkmenistan, although only the last of these is actually within the Turkmen range.

The Turkmenistan group is differentiated from the Gobi sample by the thick horn bases and the long braincase, contrasting with the narrow posterior nasals and the relatively short skull. The Yarkand group tends to be long-skulled but relatively narrow, with a relatively short braincase, but the differences from the Gobi sample are not consistent, and there is no evidence that they are not conspecific.

On horn measurements, we get much the same picture. A specimen from Zaisan Nor falls in the range of the Turkmenistan group.

Accordingly, the arrangement of the taxa of this group is given below.

Gazella marica Thomas, 1897

Arabian sand gazelle

Smallest species, but comparatively robustly built; whitish or pale yellow-brown in color, with only rudimentary face and body markings.

From the Arabian desert, apparently preferably living in sandy areas.

Gazella subgutturosa (Güldenstaedt, 1780)

Persian gazelle

Reddish or grayish sandy above, white below and on the buttocks. Young animals with a white stripe from above the eye to the nostril, and a narrow black line below, but this nearly or entirely disappears in the adults. Females usually (but not invariably) hornless.

Kumerloeve (1969) illustrated Turkish examples of this species and gave a map of the distribution of gazelles in SE Turkey, although probably not all of these refer to *G. subgutturosa*. Karami et al. (2002) have detailed the distribution of this species in Iran.

The gazelles of the Caucasus may be different from those coming from the main Iranian plateau; they sometimes average larger, with relatively short horns, and, according to Heptner et al. (1961), are darker, a brownish gray, with a clear, dark flank stripe and a heavy tail-tuft.

Gazella gracilicornis Stroganov, 1956

Sandy brown to sandy gray in summer; lighter in winter. Very wide across the bases of the horns; relatively narrow palate; short braincase.

Deserts of Kazakhstan, Turkmenistan, and Uzbekistan.

G. gracilicornis was regarded as a synonym of *G subgutturosa* by Groves (1969a), but the new cranio metric information reported here indicates that it is absolutely distinct.

Gazella yarkandensis Blanford, 1875

1875 *Gazella subgutturosa* var. *yarkandensis* Blanford. Yarkand, Xinjiang.

1894 *Gazella hillieriana* and *Gazella mongolica* Heude. Mongolia.

? 1900 *Gazella subgutturosa sairensis* Lydekker. Saiar Mountains, Xinjiang.

? 1931 *Gazella subgutturosa reginae* Adlerberg. W Caidam.

Color much the same as in *G. gracilicornis*. Relatively narrow across the horn bases; long skull; short braincase; very broad posterior nasals.

We have been shown photographs from the wild of gazelles in the Saiar Mountains, and they are conspicuous by their strong coloration and strong, dark body markings; we think it improbable that the name *Gazella subgutturosa sairensis* is, in fact, a synonym of this species. Our material from Caidam, however, is too meager to allow us to make any firm decision on the status of *Gazella subgutturosa reginae*.

Gazella Species (Undescribed)

Gazelles in Iraq, in Anatolia, and in Iran (on the W side of the Zagros) are much smaller than those of the Iranian plateau, with their horns having a narrow span and tips that are very close together, and stronger markings (flank-stripes and face-stripes) on their bodies.

Gazella dorcas Group

Females with comparatively long, but thin, horns. Horn rings less prominent than for other gazelles, but somewhat expressed in the females. Horn cores of the males more or less circular in cross section (the ratio of the transverse to the anteroposterior diameter of the pedicels is given by A. W. Gentry [1964] as 58–66 mm, with a mean of 61.7 mm); no anterior groove on the core, but one running up the posterior surface (these distinctions, first made by Ducos [1968], were confirmed by Tchernov et al. [1987], and they have also been noted by one of us [CPG]). Horns of the adult females usually long, well formed, and slightly ringed. Ascending branch of the premaxilla making a substantial contact with the nasals. Nasals tending to broaden from the posterior to the anterior ends. Posteriorly, the nasals together forming a V, penetrating into the frontals.

Anterior basioccipital tuberosities tending to be small, but larger than the ridges behind them, and lying at the same level as the foramen ovale (A. W. Gentry, 1964).

Alados (1987) took 53 measurements of the skulls and the jaws of 10 OTUs of the *G. dorcas* group, plus *G. saudiya* (group S) and "*G. gazella arabica*" (group Ar; actually a mixture of *G. cora* and *G. erlangeri*). The OTUs within *G. dorcas* were as follows: P I (*G. pelzelni* from Berbera), P II (*G. pelzelni* from Danakil), I (*G. isabella*), L ("*G. littoralis*"), N ("*G. neglecta*"), D I (*G. dorcas* from Darfur and Kordofan), D II (*G. dorcas* "from Hoggar and nearby northern areas"); and A I (*G. dorcas* from Rio de Oro). Unfortunately, the samples were far from clear cut: the samples listed as *G. isabella* and "*G. littoralis*" were nearly coterminous in their geographic scatter; the *G. neglecta* sample was specifically from Hoggar, but so were some in the D II sample; and the D II sample also included specimens from the Maghreb, Egypt, and Sudan. The measurements were subjected to a system of gap-coding and treated cladistically. In the overall cladogram, groups I and L assorted together with Ar, with N as their sister group; on the other branch, successively, were P I, P II, D II, A I, and S. If S was taken as "ancestral," the branching order was A I, P I, D II, P II, N, L, I, and Ar. If Ar was taken as "ancestral," the branching order was I, L, then a three-way split of N, PII plus PI, and DII plus a grouping of S plus A I.

It was most unfortunate that this potentially interesting analysis was marred by the intermixing of samples. In this light, it is unsurprising that *G. isabella* and "*G. littoralis*" always came out close together—because they are, in essence, the same sample. More significant is the finding that they cluster close to the *cora/erlangeri* combination; this rapprochement has been noticed previously. Also significant is the finding that the two *G. pelzelni* samples always came out together, and were not close to *G. isabella* (*pace* Groves, 1981a). The fairly consistent association of A I (*G. dorcas* from Rio de Oro) with *G. saudiya* is puzzling, and perhaps reflects convergence in their extreme desert adaptations.

Gazella dorcas (Linnaeus, 1758)

dorcas gazelle

Gestation is given as 164–174 days; the earliest age for a male at first fertile mating, 589 days; the age of a female at first fertile mating, anything from 192 to 599 days; pregnancy is rare (Furley, 1986).

Tchernov et al. (1987) found that the presence of *G. dorcas* in the Sinai and Israel dates from after the

Pre-Pottery Neolithic B period, when, with increasing aridification, *G. dorcas* largely replaced the *G. gazella* group in that region.

Groves (1981c) recognized five subspecies, apart from *G. pelzelni*. These will be maintained here, but they need restudy, given that different populations within each can be almost as different from each other as the "central" populations of each subspecies.

Gazella dorcas dorcas (Linnaeus, 1758)
Rich fawn; flank band well expressed, pygal band clear; dark lateral face-stripe tending to be suffused with black. Size very small; skull broad, with a short rostrum. Horns long in both sexes, lyrate, with numerous close rings, in the males.

W desert of Egypt, as far W as Tunisia.

Gazella dorcas osiris Blaine, 1913
Color very pale fawn; dark flank band not well marked, but bordered above with a very pale band; pygal stripe poorly marked; face markings paler, with no nose-spot; light face-stripes pure white; ears very pale; little or no darkening on the knee tufts. Skull small; horns long and narrowly ringed in the males, relatively long in the females.

Saharan form, from W of the Nile to the Atlantic.

Gazella dorcas massaesyla Cabrera, 1928
Richer, more ochery color; nose-spot poorly expressed in the adults; skull rather narrow across the horn bases; horn tips less inturned.

Apparently the Maghreb: coastal ranges N of the Atlas Mountains, along the Mediterranean.

Gazella dorcas isabella Gray, 1846
Relatively large; dark brown-gray, with reddish tones; dark markings, including the nose-spot, well expressed. Horns of the males relatively short and stout, with the tips strongly turned inward; horns of the females less elongated than in taxa from W of the Nile.

N Eritrea; Sudan, E of the Nile; up the Red Sea coast to the Sinai, and into the deserts of S Israel.

Gazella dorcas beccarii de Beaux, 1931
Richer and more chestnut than *G. d. isabella*; flank band darker; extremely large in size.

This is an enigmatic taxon, known from only one skull and two skins from the Upper Baraka district in Eritrea, approximately 15° 30' N, 38° E. *G. d. beccarii* needs further examination, preferably including field observations.

***Gazella saudiya* Carruthers & Schwarz, 1935**
Saudi gazelle
A small species, with poorly expressed patterning and long horns—nearly straight, barely turned in at the tips—in both sexes.

G. saudiya is securely known from only two regions: the type locality (Dhalm) and nearby localities (Alam Abyadh, Wadi Markha, Arq Abu Da'ir, Ruwaik, Taraf Al Ain, Wadi Naq'a, Sirr al Yamani), all roughly 150 mi. NE of Mecca, collected during the 1930s; and Kuwait, collected in 1934. The Kuwait skin is darker, grayer than the type series from Dhalm, though the skulls are alike.

It seems unlikely that either population still exists. Nonetheless, there are two captive breeding groups that have been identified as being of this species: one in Al Ain Zoo (Abu Dhabi, United Arab Emirates), the other in Al-Areen (Bahrain). One of us (CPG) saw both of these groups in the early 1990s and he thought that they were both *G. saudiya*, though there were differences between the two (the Al-Areen gazelles are larger and more disruptively colored, for example). Stock from both of these collections have since been shared with the King Khalid Wildlife Research Centre (KKWRC), Saudi Arabia. A third captive group, in Al Wabra, Qatar, has also been identified as "Saudi dorcas," although in this case, CPG identified them as *G. dorcas isabella* and as having probably not been obtained in Saudi Arabia at all.

Rebholz et al. (1991) karyotyped two males and three females from the Al-Areen collection at the KKWRC. $2n = 47$ in the females, and $2n = 51$ and 52 in the two males. There was conspicuous variation in chromosomal morphology. In at least one of the males, two Y chromosomes were present, indicating the presence of the usual X-autosome translocation. The authors noted that the karyotypes appear to be very similar to those of the *G. bennetti*–group gazelles illustrated by Furley et al. (1988).

Kumamoto et al. (1995) karyotyped five female and two male specimens, also from Al-Areen and the KKWRC. $2n = 46$, 48, 49 or 50 in the females, and $2n = 49$ and 53 in the two males; hey had the usual X-autosome translocation. The authors considered it likely that some hybridization with *G. bennetti* might have occurred in this stock. They noted, interestingly, that in two cases the karyotypes were identical with those of two *G. bennetti*–group gazelles from Iran.

R. Hammond et al. (2001) managed to extract samples of the mitochondrial cytochrome *b* gene from museum specimens known or presumed to be of this species, and to compare them with the samples from the "Saudi gazelles" in the Arabian collections (Al Wabra, Al Areen, Al Ain Zoo, and the KKWRC). The authors also sampled other taxa, both from the KKWRC and from museum material. The results were surprising. The captive breeding-groups identified by earlier authors as "*saudiya*" proved to be quite different from the genuine (i.e., the museum) *saudiya*—the former had sequences typical of either *G. subgutturosa* (three out of the four Al Areen specimens) or of "*G. bennetti*" (actually, *G. fuscifrons* in our present taxonomic scheme: the fourth Al Areen animal, plus all three from Al Ain). A specimen in the Harrison Museum, identified as *G. saudiya*, proved to be *G. marica*. Meanwhile, the three Al Wabra specimens—which had been informally ascribed to *G. saudiya*, simply because they were rumored to come from Saudi Arabia, not because they resembled *G. saudiya* phenotypically (see above)—proved to be *G. dorcas*, exactly as they always appeared to be.

The real *G. saudiya* turned out, after all, to be related to *G. dorcas*. And, alas, they are totally extinct, unless some of the captive stock contain their genes (recall that only mtDNA was sequenced).

Gazella pelzelni Kohl, 1886

Pelzeln's gazelle

Very large in size for the *G. dorcas* group; horns long, with the tips hardly (or not at all) turned in; horns not widely flared, and sparsely ringed; color bright sandy ochre, with a well-marked, dark lateral flank band and dark face-stripes; dark knee-tufts; always at least a trace of a dark nose-spot.

N Somalia and the Danakil country of Ethiopia and S Eritrea (see below).

Groves (1981c) included this species in *G. dorcas*, on the grounds that specimens from the Danakil country are intermediate between it and "true" *G. dorcas*, but *G. pelzelni* is, in fact, absolutely different from any of them, and the Danakil specimens may actually be a hybrid population.

Living animals, seen in Al Wabra, seem noticeably more slender and long-legged than *G. dorcas*.

Two museum specimens from Danakil, whose partial cytochrome *b* sequences were analyzed, fell into the *G. dorcas* clade (R. Hammond et al., 2001).

Gazella gazella Group

Differs from the *G. dorcas* group in the more weakly developed horns of the females; the more prominent, less close-set horn rings of the males; the (usually) deeper pit for the supraorbital foramina; the more arch-shaped nasofrontal suture; the longer median nasal tips, usually longer than the lateral prongs; and the less high and less rounded cranial profile.

Even under the Biological Species Concept, it came to be realized that this species-group cannot be united in a single species. Groves (1985) wrote of "darker, straighter-horned individuals" of what he called *G. g. cora* predominating along the Red Sea coast, and later Groves (1996a) separated this population as a full species (*G. erlangeri*), without yet associating the small, dark Thumamah gazelle with it (see below).

In R. Hammond et al.'s (2001) cytochrome *b* tree, members of the *G. gazella* group (museum specimens of *G. muscatensis*, plus a living KKWRC specimen of uncertain affinities) formed a sister clade to *G. dorcas* / *G. pelzelni* plus *G. saudiya*.

Table 41 gives the univariate statistics for males of this species-group.

Gazella gazella (Pallas, 1766)

mountain gazelle

$2n=34$ (females), 35 (males) (Greth et al., 1993).

Horn cores of the males elliptical in cross section; broad groove running up the anterior surface of the core, frequently with a second groove, less marked, just medial to it, and with another groove running up the posterior surface (Ducos, 1968; Tchernov, 1987). Horns of the females short (84–153 mm), nearly smooth, and fragile, easily broken in life. Ascending branch of the premaxilla only just contacting the nasals, or not at all. Nasals usually broader posteriorly, and not penetrating deeply between the frontals. Coronal suture double-bowed. Light facial stripes off-white, edged below with dark or black; midface dark, usually with a blackish nose-spot. Body gray-brown, with a thick, dark stripe separating this color from the white underparts, and a paler stripe above the dark one. Premolar length 37.5%–41% of the whole toothrow; lingual styles and valleys on the cheekteeth angular; talonid of the third lower molar short.

A relatively stockily built gazelle; weight (in kg), expressed as a percentage of the head plus body length (in cm), averaging 23.6% in nine specimens. Limbs relatively short, with short distal segments: tibia averaging 126.1% of the femur; metatarsal, 101.1% of the femur; hindleg, 301.9% of the skull length ($N=21$ or 22). Ear relatively short, its length averaging 11.7% of the head plus body length ($N=15$), and characteristically held upright, more or less parallel with the

Table 41 Univariate statistics for desert members of the *Gazella gazella* group: males (measurements for *G. farasani* are from Thouless & al Bassri 1991)

	Base br	Nas ant	Nas post	Nas l	Gt br	Pal br	Gt l	Braincase	Horns	Tip–tip	Span
acaciae											
Mean	66.33	19.33	25.67	54.40	80.40	48.00	195.60	58.33	236.80	111.00	130.20
N	3	3	3	5	5	3	5	3	5	3	5
Std dev	3.055	1.528	3.512	2.966	1.673	4.359	3.507	2.082	30.343	10.583	5.718
Min	63	18	22	50	79	45	192	56	208	99	123
Max	69	21	29	58	83	53	200	60	277	119	135
cora											
Mean	64.70	20.13	22.39	46.39	81.04	45.46	184.57	59.75	217.93	89.15	133.41
N	27	23	23	23	27	26	21	24	27	27	27
Std dev	2.163	1.866	1.751	5.263	2.312	2.121	5.861	2.364	16.920	18.340	12.879
Min	61	16	19	38	77	39	173	55	189	56	115
Max	69	23	26	58	87	49	195	65	255	129	170
dareshurii											
Mean	63.20	20.36	22.64	39.27	78.60	43.90	175.10	54.00	202.00	110.71	149.71
N	10	11	11	11	10	10	10	10	7	7	7
Std dev	3.615	.924	1.027	4.429	2.171	1.524	9.480	1.247	12.028	20.702	24.088
Min	55	19	21	33	75	42	152	52	188	89	117
Max	67	22	24	47	81	47	185	56	225	143	175
erlangeri											
Mean	60.00	21.80	23.60	42.60	81.20	48.60	176.60	55.80	212.60	109.80	120.60
N	5	5	5	5	5	5	5	5	5	5	5
Std dev	3.536	1.483	2.608	7.701	4.266	3.362	5.320	4.147	22.300	22.797	16.041
Min	55	20	22	35	75	45	169	52	177	85	96
Max	64	24	28	54	87	53	184	62	232	137	135
farasani											
Mean	57.8	—	—	—	—	46.5	165.6	51.0	193.2	81.2	86.0
N	12	—	—	—	—	4	3	7	6	6	6
Std dev	2.8	—	—	—	—	2.1	6.7	1.3	22.5	9.1	4.1
karamii											
Mean	58.00	21.00	22.00	41.00	84.00	49.00	173.00	58.00	228.00	190.00	210.00
N	1	1	1	1	1	1	1	1	1	1	1
muscatensis											
Mean	62.00	20.25	23.25	36.25	79.33	44.67	172.00	57.67	188.40	86.40	132.40
N	6	4	4	4	6	6	1	6	5	5	5

horns (all of these figures from Mendelssohn et al., 1997).

Details of the biology of this species have been given by Mendelssohn et al. (1995).

G. gazella has flourished under protection in N Israel and the Golan Heights. Its survival in Lebanon and Syria is unknown, but Dr. Tolga Kankiliç has kindly forwarded photos (pers. comm. to CPG) of a thriving population of *G. gazella* at Hatay, in Turkey near the Syrian border, just inland from the Mediterranean coast.

Other Species of the *Gazella gazella* Group

Notes on the distribution and present status of the Arabian peninsular species are given by Greth et al. (1993).

The three desert taxa are distinct from each other on multivariate analysis, and univariate statistics are given in table 41.

G. erlangeri is smaller and relatively broader than *G. acaciae* and *G. cora*, its nasals narrower anteriorly; *G. acaciae* is larger and relatively narrower, its nasals wider posteriorly and narrower anteriorly. (Note, of course, that the sample size of *G. acaciae* is small).

G. erlangeri has a wider tip-to-tip interval and a narrower span (i.e., the horns are straighter and less divergent), and the horns are narrower across the bases than the other two; *G. acaciae* averages wider across both the tips and the bases.

The three Persian Gulf taxa can now be brought into the comparisons. Of the skulls in the sample of

G. muscatensis, most are incomplete (primarily lacking the premaxillae). *G. muscatensis* and *G. karamii* have wide skulls and short nasals, and are perhaps slightly narrower across the horn bases; *G. dareshurii* is the opposite, with *G. erlangeri* in between. *G. erlangeri* and *G. karamii* have nasals that are wide anteriorly, compared with posteriorly, and have narrow horn bases.

As far as horn variables are concerned, *G. muscatensis* and *G. dareshurii* have relatively short horns, with a wide span, but they are narrow across the tips (the horns are initially bowed outward, then converge toward the tips) and have relatively wide bases; *G. dareshurii* averages larger than *G. muscatensis* in all horn dimensions. *G. karamii* is also large in all dimensions, but it has a large tip-to-tip interval compared with the span, and long horns compared with their basal breadth.

In summary, the cranial and horn differences among these desert gazelles are as follows (mentioning mainly the extremes in each case):

—In size, *acaciae* > *cora* > others, but there are overlaps between neighboring taxa.
—*G. karamii* is exceptionally narrow across the horn bases, compared with a wide braincase.
—*G. acaciae* has nasals that are exceptionally narrow anteriorly and wide posteriorly.
—*G. dareshurii* and *G. muscatensis* are alike in having very short nasals.
—*G. karamii* and *G. erlangeri* both have a wide palate.
—*G. muscatensis* has comparatively short horns; *G. karamii* and *G. acaciae* have long horns.
—*G. muscatensis* and *G. cora* have a very narrow tip-to-tip distance, especially compared with the horn span; *G. acaciae*, and especially *G. erlangeri*, have a fairly narrow span, with less inturning of the tips.

Relationships among these species are not geoclinal. Probably *G. cora* and *G. acaciae* are more closely related to each other, while the other four seem to form a second morphological cluster, though all are diagnosably distinct, and we recognize them all as full species here.

G. karamii has perhaps a superficial similarity to the *G. subgutturosa* group in its skull, in that it has a rather wider skull, especially the palate, compared with the *G. gazella* group, though the widths are not as great as in the *G. subgutturosa* group. The horns *G. karamii*, however, are characteristically those of the *G. gazella* group; indeed, they are very like those of *G. dareshurii*—much wider between the tips, compared with span, and especially much wider across the horn cores—than in any member of the *G. subgutturosa* group.

Dollman (1927) described *Gazella arabica hanishi* from Great Hanish Island, in the S Red Sea, describing it as differing from "the typical *Gazella arabica arabica*" in its larger, more sharply defined nose-spot; at the base of horns, there "are two black spots, which extend down the face almost as far as the dark nasal marking." *Gazella arabica hanishi* is said to be less reddish, and the flank stripe is more developed, approaching *G. erlangeri*, but *hanishi* is less rufous and has a larger nose-spot. The type of *G. a. hanishi* was the mounted head of an adult male in the collection of the Duke of York: horns 9 in. long (= 229 mm), tip-to-tip distance 4-1/8 in. (= 124 mm). The following year, Dollman (1928) redescribed the specimen, and this time he illustrated it. A second specimen was mentioned, but not pictured; this was said to have horns 6-3/4 in. (= 173 mm) long.

It is somewhat difficult to judge this described taxon without seeing the type specimen, which is not, at present, in the Natural History Museum in London, although it may be in a private royal collection somewhere in Britain. It is clearly not *G. cora*, because the horns are nearly straight (just somewhat inturned at the tips) and the face is much too dark. *G. arabica* and *G. bilkis* are dark, but, at least in the latter, the light face-stripes are much more infused with ocher, and the midface is so dark that any nose-spot cannot be detected. In the type (and only known specimen) of *G. arabica*, the horns are 272 mm long; their span is 102 mm; the tip-to-tip distance is 92 mm; and they have 15 rings. In *G. bilkis*, the horns are 204–254 mm long in the males; the greatest horn span is 86–102 mm; the tip-to-tip distance is 52–86 mm; and there are 12–15 rings ($N=2$). The horns in both are shorter and more robust than those of the specimen in Dollman's photo; there are fewer and less prominent rings; and the tips do not turn in, as they do for the specimen in the photo. The closest resemblance of the Hanish gazelle is with *G. erlangeri*, which has a similarly dark face, but with a large, clearly defined nose-spot; long, slender horns (length 177–232 mm), tending to turn in at the tips (span 96–135 mm, versus tip-to-tip distance 85–137 mm); and, usually, fewer than 15 prominent rings.

Gazella acaciae Mendelssohn et al., 1997

acacia gazelle

A very slender, rangy gazelle; weight (in kg), expressed as a percentage of the head plus body length (in cm), only 16%–18% (N=3). Distal limb segments extremely elongated: tibia length 135.7% of the femur length, metatarsal length 114.9% of the femur length; hindleg length 306.8% of the skull length (N=1). Neck long. Ear long, 11.4%–13.5% of the head plus body length (N=3), reaching to the nostrils when folded forward on the flat skin, and broad, in life held slanting sideways at a 45° angle. Tail long, bushy; body color dark earth-brown, lighter in summer than in winter; pygal stripe black; white of the rump reaching relatively high up on either side of the tail; facial markings strongly expressed, the dark stripes broad, the midface less dark than in *G. gazella*, the light face-stripes nearly white, with a very broad, conspicuous, black nose-spot; white stripe on the inner surface of the hindleg reaching down to the hoof. (Following Mendelssohn et al., 1997).

Nasopremaxillary contact longer than usual in this species-group. Premolars relatively long. Tympanic bullae large.

Tchernov et al. (1987) could not examine horn cores of this species, so these authors were unable to confirm that this was the species that inhabited the Sinai until the increasing aridification after the Pre-Pottery Neolithic B period. At any rate, after that time, a member of the *G. gazella* species-group was replaced by *G. dorcas* in the Sinai and Israel.

Gazella cora Hamilton Smith, 1827

Arabian desert gazelle (sometimes known, inappropriately, as "mountain gazelle")

Groves (1983b) showed that the name *arabica*, "traditionally" used for this species, is actually not applicable. The earliest available name seems to be *Antilope cora* Hamilton Smith, 1827, although, given the uncertainty, it is probably appropriate that in the near future a neotype should be selected.

$2n = 34$ (females), 35 (males) (Greth et al., 1993).

Limbs elongated—especially the hindlimb, which is 317.6% the length of the skull—but less elongated distally than in *G. acaciae*: tibia length 125% of the femur length; metatarsal length 105% of the femur length (N=1). These figures are not very different from *G. gazella*. Ear 119% of the head plus body length (N=3), more or less like *G. gazella*, but held slanting outward like *G. acaciae*. (Following Mendelssohn et al. 1997).

Color medium to very light fawn; flank-stripe varying from a thin black line to just a very slightly darker zone, with a lighter zone above it on the flanks, which is more marked in the darker skins but less so in the paler ones. Weakly marked transition on the haunch from the darker fawn above to the paler fawn below, continuing down the hindlegs. Little or no pygal band. Forehead and nose varying from dark chestnut to merely a slightly darker fawn; nose-spot, if any, vague; forehead often sprinkled with whitish hairs; face-stripes clear, broad, white, bordered below with a dark brown line. Tail relatively long and thin. Ear, when folded forward, reaches halfway or completely to the nostrils.

Gazella erlangeri Neumann, 1906

Arabian coastal gazelle

$2n = 34$ (females), 35 (males), as in *G. cora* and *G. gazella* (Greth et al., 1993).

Overall color very dark gray-brown; legs a lighter, more ochery brown, with a sharp transition between this and the darker upper-body color on the haunch; face dark, with a large, clearly defined nose-spot, these being the only real differences in color from *G. muscatensis*. Tail with a thick, black tuft. Ear long, reaching halfway between the eye and the nostril when folded forward. Long, slender horns (length 177–232 mm), tending to turn in at the tips (span 96–135 mm, versus tip-to-tip distance of 85–137 mm), and usually fewer than 15 prominent rings. Females with relatively long horns. Very wide palate.

Found sporadically along the W Arabian coast, from Aden an unknown distance N along the Red Sea coast.

The description of *Gazella gazella farasani* Thouless & al Bassri (1991), from Great Farasan Island and the nearby island of Zifaf, is extraordinarily reminiscent of *G. erlangeri*, and photos seen by CPG reinforce this impression. The means and the standard deviations given by Thouless & al Bassri (1991) have been added to table 41; they denote an animal clearly smaller, on average, than *G. erlangeri*, and—provided always that this difference is not purely a result of phenotypic plasticity—indicate a valid subspecies. Vassart, Granjon, et al. (1994), however, found that *G. g. farasani* clustered more closely with *G. cora* than with *G. erlangeri* on the basis of allele frequencies. In this study and a subsequent one (Vassart, Granjon, et al., 1995), *G. erlangeri* was found to be fully monomorphic (17 individuals); the same study found its karyotype identical to other species of the *G. gazella* group.

Gazella muscatensis Brooke, 1874

muscat gazelle

Very small, like *G. erlangeri*; but the horns very slender, bowed outward and then inward at the tips. Females with comparatively long horns, like *G. erlangeri*. Very dark gray-brown, often with an almost silvery sheen, resembling *G. erlangeri* very closely in color. Flank-stripe in the form of a conspicuous darkening along the flank, with a lighter, pinky gray zone above it. Pygal stripe a clear, dark rim to the white buttocks. Very sharp, oblique transition on the haunch between the dark body color above it and the fawn-colored haunch and hindleg. Forehead dark chestnut; nose-stripes white, becoming pure white above and behind each eye, bordered below by a smudgy dark stripe. Nose-spot smudgy, small, black. Ear-backs dull fawn; ear, when folded forward, reaching halfway between the eye and the nostril. Tail with a long, thick, black tuft. Nasal bones very short.

Along the Batinah coast of Oman.

Gazella dareshurii Karami & Groves, 1993

Farrur gazelle

Most closely related to *G. muscatensis*, but the horns of the males longer; horns of both sexes broad across the base; skull narrower; nasal bones posteriorly narrow. Ascending branch of the premaxilla apparently never reaching the nasals. Photographs of these animals in the wild show a pale, sandy brown animal.

As far as is known, *G. dareshurii* is restricted to the island of Farrur (Jazireh-e-Forur), 26° 30' N, 54° 30' E, in Iran. Karami & Groves (1993) discussed its possible existence on the mainland.

Gazella karamii Groves, 1993

Bushehr gazelle

Distinguished by its small size; very narrow across the bases of the horns; palate very broad. The skin of the type, found during a visit by CPG to Berlin in October 2009, is as follows: color very dark, resembling *G. muscatensis*, and with the same well-tufted tail; face-stripes off-white (more strikingly white above the eye), bordered below by a thin black band; nose-spot small, dark, smudgy; forehead sprinkled with white hairs.

The type, a zoo specimen that died in 1928, was caught E of Borazjan, near Bushehr, on the Persian Gulf. CPG, during his earlier visit to Berlin in the 1980s, when he first studied the skull, must have been having an off day (or a hangover), because he later described it as a subspecies of *G. bennetti*!

Photos taken in the wild by Dr. M. R. Hemami (copies kindly sent to CPG) of gazelles in the Nayband Protected Area, 27° 22' N, 52° 38' E, about 250 km S by SE of Bushehr, show a brightly colored, reddish sandy gazelle, with poorly expressed transitions between the dark and light zones on the flanks and haunch; poorly marked on the flanks, with only a slightly darkened pygal band; but a clear dark nose-spot and dark forehead. The horns of the males are thick, straight, and well ringed, only slightly divergent; those of the females are straight, slender, and well formed. Evidently, the gazelles here are of the *G. bennetti* group, perhaps closest to *G. shikarii*, but appearing to be more brightly colored.

Gazella bilkis Groves & Lay, 1985

Bilkis gazelle

Size very large; horns relatively short, straight, and relatively poorly ringed in the males, but very large and ringed in the females. Color very dark, sharply paler on the flanks, the haunches, and the legs; flank-stripe very thick, black, with a thin red line below; white on the inside of the thigh, continuing down the shank. Horns 204–254 mm long in the males; greatest horn span 86–102 mm, tip-to-tip distance 52–86 mm; 12–15 rings ($N = 2$).

Greth (1992) recorded seeing a pair of horns in the National Museum in Taez, Yemen, which were nearly straight, "about 25–28 cm" long, with a very narrow span and 14–15 rings. This seems most consistent with *G. bilkis*, but there was no evidence of the species' survival in Yemen.

Gazella arabica Lichtenstein, 1827

large Farasan gazelle

Groves (1983b) showed that this name does not belong to the "common" Arabian gazelle, but to a gazelle formerly(?) existing on the Farasan Islands.

Large, very dark, with a red line between the dark flank band and the white of the underside, and long, very straight horns. In the type (and only known specimen), the horns 272 mm long; the span 102 mm; the tip-to-tip distance 92 mm; 15 rings.

Farasan Islands; possibly extinct.

Gazella spekei Blyth, 1863

Speke's gazelle

In the phylogeny of Effron et al. (1976), *G. spekei* is placed as sister to the *G. dorcas* plus *G. gazella* clade; $2n = 33$ (males), 32 (females). We provisionally (and for convenience) place it here in the *G. gazella* group.

Horn pedicels much more compressed than in the *G. dorcas* group; ratio of the transverse to the anteroposterior diameter averaging 61.7 mm, with a range of 58–66 mm (A. W. Gentry, 1964). Anterior basioccipital tuberosities tending to resemble those of the *G. bennettii* group in size and position.

Gestation is 169–190 days (mean 178.6 days); the age of the male at first fertile mating, 480 days; of the female, 504 days, but as early as 240 days (Furley, 1986).

Gazella bennetti Group

chinkara

$2n$ =50 (females), 51 (males). Furley et al. (1988) show that the chromosomes of the *G. bennetti* group (based on specimens of unknown origin, but very likely from Pakistan) are quite different from those of the *G. dorcas*, the *G. gazella*, and the *G. subgutturosa* groups.

In a study by Kumamoto et al. (1995), three females from Pakistan had $2n = 51$, and a further female (of unknown origin) had $2n = 50$, while a male had $2n = 52$; a male from Iran had $2n = 49$, and a female had $2n = 52$. The Pakistani series was polymorphic for a Robertsonian translocation between chromosomes 8 and 14; the Iranian sample was polymorphic for three independent Robertsonian translocations. Chromosomes 22 and 25 were metacentric in the Iranian specimens and acrocentric in the Pakistani ones.

The *G. bennetti* group differs from the *G. dorcas* and the *G. gazella* groups in their horn rings: more widely spaced, not angulated, much more prominent posteriorly as well as anteriorly, and not connected with the longitudinal ridges. Horn tips pointing somewhat outward. Supraorbital foramina 2–3. Nasofrontal suture, as in the *G. dorcas* group, tending to be triangular in shape, penetrating well into the frontals. Glandular knee tufts hypertrophied.

Anterior basioccipital tuberosities fairly small, as in the *G. dorcas* group; longitudinal ridges behind them tending to be as bulky as the tuberosities themselves; tuberosities lying behind the level of the foramen ovale (A. W. Gentry, 1964).

Our discriminant analysis, using a combination of nine cranial variables (greatest skull length, braincase length, preorbital length, braincase breadth, palate breadth, biorbital breadth, nasal length, posterior nasal breadth, and anterior nasal breadth) distinguished all taxa, except for an overlap between *G. shikarii* and *G. salinarum*.

The largest species, *G. salinarum*, may have a slightly greater sexual dimorphism than the other species. Differences in horn length among the females are much more interesting. The horns in the females of the two Indian species are very short; in *G. shikarii* the females have much longer horns. When we look at horn length versus skull length for the males, it is clear that *G. fuscifrons* has relatively short horns compared with its general skull size, whereas the opposite is true for others, especially *G. salinarum*.

Gazella shikarii Groves, 1993

western jebeer

Color light, resembling *G. subgutturosa*, with which it is sympatric in W and N Iran. In fact, presumably for this reason, this species had long been misidentified as *G. subgutturosa* (Groves, 1993a). Horns strikingly long in the females; more divergent in both sexes than in other species of the *G. bennetti* group.

The Iranian plateau, from the Tehran region E to the Touran Biosphere Reserve (35° 39' E, 56° 42' E) and S to Deh Kuh (27° 53' N, 54° 22' E). Animals from Semnan and Kerman provinces, where the range approaches that of *G. fuscifrons*, are entirely typical, and do not approach the latter in appearance (Karami et al., 2002).

Gazella fuscifrons Blanford, 1873

eastern jebeer

Dark in color; prominent flank band and face-stripes; conspicuous blackish nose-spot; very dark color on the forehead. Skull and horns slightly smaller in the males, but much smaller in the females, than in *G. shikarii*; horns, however, very much less divergent.

SE Iran, into the Makran coastal region of Pakistan. A specimen from 100 km S of Lar (27° N, 54° 20' E), Hormozgan Province, where its range closely approaches that of *G. shikarii*, is entirely typical of the species, and does not approach the latter in appearance (Karami et al., 2002).

Gazella bennettii (Sykes, 1831)

Deccan chinkara (following Groves, 2003)

Dull reddish brown; median dorsal region and lower flanks abruptly darker, tawny; coat longer in winter than in summer. Horn tips of the males slightly less divergent, compared with the span, than in *G. fuscifrons*; horns of the females much shorter; nasals longer.

Deccan Plateau and the Ganges Valley.

Gazella christyi Blyth, 1842

Gujarat chinkara (following Groves, 2003)

Very pale, almost silvery drab brown, with only the very restricted median dorsal and lower flank zones slightly darker. Little seasonal difference in coat length. Males larger, with longer and, on average,

Table 42 Univariate statistics for the *Gazella bennetti* group

	Horn l	Tip–tip	Span	Nas l	Gt l	Gt br
Males						
bennettii						
Mean	256.75	120.68	127.84	54.31	181.75	84.00
N	20	19	19	13	8	19
Std dev	31.817	33.657	31.360	5.345	7.996	4.096
Min	177	57	70	47	173	77
Max	338	181	194	67	198	93
christyi						
Mean	272.94	129.06	136.15	54.09	191.14	85.27
N	33	33	33	11	7	33
Std dev	25.747	29.830	26.680	4.847	2.610	2.491
Min	217	45	81	47	187	81
Max	350	181	190	63	195	91
fuscifrons						
Mean	254.45	116.36	126.09	51.76	185.27	84.48
N	22	22	22	17	15	21
Std dev	28.892	30.335	30.638	4.409	4.008	2.400
Min	194	55	41	44	178	81
Max	304	169	173	59	192	89
salinarum						
Mean	283.47	140.36	145.36	57.22	198.00	87.00
N	15	14	14	9	7	15
Std dev	16.115	35.454	36.097	3.383	5.292	1.813
Min	254	96	107	52	190	84
Max	305	235	242	61	205	91
shikarii						
Mean	263.50	134.25	150.75	53.67	192.50	85.75
N	4	4	4	3	4	4
Std dev	25.878	31.117	27.921	3.215	3.317	2.217
Min	247	99	124	50	188	83
Max	302	174	188	56	196	88
Females						
bennettii						
Mean	116.86	53.33	63.67	50.62	179.50	80.20
N	7	3	3	8	6	10
Std dev	18.032	9.292	6.506	5.397	7.817	2.098
Min	101	47	57	44	172	78
Max	149	64	70	61	190	85
christyi						
Mean	118.00	87.00	91.00	46.00	184.00	81.00
N	3	1	1	3	2	4
Std dev	23.516	—	—	7.810	—	2.000
Min	102	—	—	41	184	78
Max	145	—	—	55	184	82
fuscifrons						
Mean	139.90	58.33	72.67	49.20	181.10	81.45
N	10	6	6	10	10	11
Std dev	60.154	15.201	15.642	5.160	4.630	2.544
Min	0	38	52	43	174	77
Max	230	77	94	58	188	86
salinarum						
Mean	136.00	53.00	58.00	55.50	191.50	82.67
N	3	1	1	2	2	3
Std dev	29.462	—	—	—	—	—
Min	118	—	—	55	186	79
Max	170	—	—	56	197	85

(continued)

Table 42 (continued)

	Horn l	Tip–tip	Span	Nas l	Gt l	Gt br
shikarii						
Mean	188.00	85.50	67.00	60.00	191.00	80.50
N	2	2	2	1	2	2
Min	182	78	57	—	191	79
Max	194	93	77	—	191	82

more divergent horns than *G. bennetti*; females hardly different in size, their horns short, as in *G. bennetti*, but much more divergent (at least in the single available specimen).

Thar Desert, Gujarat, and probably Rajasthan.

Gazella salinarum Groves, 2003

Salt Range gazelle

Color rich tobacco-brown, with no contrasting zone on the back, but a contrasting flank band. Largest species of the *G. bennetti* group in the males (the females are the same size as the females of *G. shikarii*), with the longest horns, the tips not inturned, with the tip-to-tip distance thus about the same as the span; horns of the females much shorter than in *G. shikarii* (about the same size as those of the considerably smaller *G. fuscifrons*), and straight, like those of the males. Especially long nasal bones.

Salt Range, E to Delhi.

Eudorcas Fitzinger, 1869

$2n=58$. Like *Nanger*, there is not only the usual gazelle X-autosome translocation, but also a Y-autosome translocation (Vassart, Séguéla, et al., 1995). There is little, if any, difference in chromosomes between *E. rufifrons* and *E. thomsoni*.

Differs from other genera of the *Antilope/Nanger/Gazella* clade as follows: horn rings extremely prominent, widely spaced; horn cores more compressed than the others, except for *Nanger*; molars not elongated, the styles more developed; lingual wall of the mandibular P^4 open mesially, but fully closed distally; lacrimal bone more extensive on the face than in *Gazella*; nasofrontal suture with an anterior V in the midline; median nasal tips unusually long, much longer than the lateral prongs; pterygoid plates with large hamuli, as in *Nanger*; horns slope back at a 45° angle in the males, as in *Antilope*; rhinarium restricted to the median portion of muzzle. (Following Groves, 2000).

Anterior basioccipital tuberosities spaced much more widely apart than in *Gazella*, without longitudinal ridges behind them. Auditory bullae comparatively small. Preorbital fossa very large, its inferior edge being directed obliquely downward, not anteriorly, as in *Gazella*. Supraorbital pits larger than in *Gazella*. (These skull characters are as first described by A. W. Gentry, 1964).

Groves (1979a) included the species of *Eudorcas* in *Gazella*, and even made them conspecific with *G. cuvieri* (!). He also included the E African gazelles (here, *E. thomsoni* and relatives) in this species, as well as NE African *E. tilonura*, but not the enigmatic *E. rufina*.

Lange (1972) did not agree that *E. thomsoni* and *E. nasalis* belong to *E. rufifrons*, drawing attention, in particular, to the very long premaxillary contact with the nasal, which sometimes excludes the maxilla from any contact, and the idiosyncratic form of the nasofrontal suture. After some discussion, however, Lange concurred with Groves (1979a) in making *tilonura* a subspecies of *E. rufifrons*. He also argued that *E. rufina* is related to the *E. rufifrons* group.

Eudorcas rufifrons (Gray, 1846)

red-fronted gazelle

Reddish to ochery in color, with a thin, black flank-band, bordered below by a band of the body color (or slightly redder), separating it from the white of the underside. White of the face most clearly expressed as a broad, white eye-ring, with a less well-marked stripe continuing down to the nose, and little expression of the usual gazelline dark stripe below that. Ratio of the transverse to the anteroposterior diameter of the horn cores averaging 69.1 mm, with a range of 61–75 mm (A. W. Gentry, 1964).

The growth and physical characteristics of *E. r. kanuri* in Waza National Park, Cameroon, have been recorded by Nchanji & Amubode (2002). Color reddish, deeper in the midfacial region. Mean shoulder height 69 cm in both sexes ($N=95$); body weight 29.7 kg. Horn length averaging 39.9 cm; interestingly, virtually the same in the two sexes. One pair of inguinal glands; interdigital glands in both feet.

Groves (1975a) found that *E. r. kanuri* is different from both the smaller, more reddish, more sexually dimorphic W nominotypical *E. r. rufifrons* and the larger, more ochery, less sexually dimorphic E *E. r. laevipes*. The three subspecies are marginally distinct, and restudy is needed to confirm their status.

Eudorcas tilonura (Heuglin, 1869)

Eritrean gazelle

Distinguished by its very small size, and much reduced sexual dimorphism; horns more slender, much shorter in the males, and suddenly hooked in toward the tips. More rufous in color than *E. rufifrons*; no nose-spot; pygal band absent; knee-brushes larger; light lateral face-stripes nearly obsolete, except around the eye.

Sudan, E of the Nile and N of the Setit, as far NE as the Bogos River in Eritrea.

Eudorcas albonotata (Rothschild, 1903)

Mongalla gazelle

Size similar to *E. rufifrons*, but the skull narrower, and the horns of the females much shorter, averaging only 50% the length of those of the males (compared with 60%–70% in *E. rufifrons* and *E. tilonura*). Nose-spot usually present; knee-brushes large; forehead often entirely white; flank band very wide, with only a slight rufous stripe below it, if that; pygal band present. Compared with *E. thomsoni* and *E. nasalis*, horns of the males shorter and curved forward more at the tips.

Upper Nile, S of the Sudd, and N of the Uganda border.

Eudorcas thomsoni (Günther, 1884)

eastern Thomson's gazelle

Premaxilla making an extremely long contact with the nasal, generally longer than that of the maxilla (generally shorter in *E. rufifrons*). Horns of the females only 39% as long as those of the males. Horn pedicels much more compressed than in *E. rufifrons*; ratio of the transverse to the anteroposterior diameter averaging 62.8 mm, with a range of 57–69 mm. Posterior expansion of the nasals greater than in *E. rufifrons*, reaching across the area of the ethmoidal fissure. Nasofrontal sutures strongly concave, suddenly extending backward in the midline, together making a backward-pointing "relief nib." Nasals longer in the females than in the males (not so in *E. rufifrons*). Length of the premolar row noticeably shorter than in *E. rufifrons* or, indeed, in the genera *Gazella* and *Nanger*; premolar to molar row length 51–64 mm (mean 57.3 mm), while being generally above 60 mm in *E. rufifrons* and the other gazelle genera. (These differences were first demonstrated by A. W. Gentry [1964]).

Compared with *E. nasalis*, *E. thomsoni* is considerably larger, with the horns spreading more in the males, and almost no white on the face, whereas the nose-spot is less conspicuous and the eye stripes are less marked (Brooks, 1961).

Furley (1986) gives gestation as about 180 days in the wild, but there are records of up to 230 days in captivity; the earliest age of the male at first fertile mating is 480 days; of the female, 204–364 days; twinning is extremely rare.

E of the Rift Valley in Kenya and Tanzania, S to the Arusha district, and turning SW to Lake Eyasi, the Wembere Plains, and Shinyanga (Brooks, 1961).

Eudorcas nasalis (Lönnberg, 1908)

Serengeti Thompson's gazelle

Horns of the females very short, only 32% as long as those of the males. Size much smaller in both sexes than in *E. thomsoni*, but the horns not much less; horn tips less widely apart, on average. Brooks (1961) recorded that in the field, *E. nasalis* appears redder than *E. thomsoni*, but this difference could not be recognized in museum skins. Face-stripes darker; nose-spot much more prominent. White crown, extending as a V down the center of the forehead, often making the face virtually white, except for the dark eye stripes and nose-spot (Brooks, 1961); in this, *E. nasalis* resembles *E. albonotata*.

The Serengeti ecosystem, extending into the Kenya Rift Valley.

Eudorcas rufina (Thomas, 1894)

red gazelle

1894 *Gazella rufina* Thomas. Algiers (purchased).

1895 *Antilope (Dorcas) pallaryi* Pomel. Oran (purchased).

E. rufina is still very difficult to classify, but it does possess many of the features of the genus: compressed horns with very strong, well-spaced ridges, sloping back at a 45° angle; long median nasal tips; well-developed styles on the molars; lower margin of the preorbital fossa slanting inferiorly. On the other hand, the anterior basioccipital tuberosities larger than in other species, with developed ridges behind them.

As far as is known, *E. rufina*, which presumably formerly lived in N Algeria (although even this is not known for certain), is extinct.

Dorcatragus Noack, 1894

Apparently more closely related to dik-diks than to gazelles, but slightly larger, without an inflated muzzle and no crown tuft; no face-glands; hoofs with small posterior pads; sexes the same size; horns extremely long for the small size of the skull; skull with very large auditory bullae.

Dorcatragus megalotis (Menges, 1894)

beira

Color a delicate reddish gray, with contrastingly ochery legs and underside; noticeable dark, gazelline flank-band; head ochery, with white eye-rings.

Along the Somali coast from the Horn of Africa W to just over the Ethiopian border, and N into SE Djibouti, where there is an isolated population in the Randa region and somewhat farther E, just N of the Gulf of Tajoura (Künzel & Künzel, 1998).

Madoqua Ogilby, 1837

dik-dik

Tiny antelope with very swollen nasal cavities, short nasals, and somewhat shortened premaxillae; males with small horns, females hornless; tuft of hair on the crown; females in most species larger than the males.

This genus is sometimes divided into two different genera, or at least subgenera: a brightly colored group with a less swollen nasal region, and a more dull-colored group with an extreme nasal enlargement (the putative genus, or subgenus, *Rhynchotragus*). Here, they are regarded simply as two species-groups, but more detailed study is needed.

Madoqua saltiana Group

Small dik-diks (often separated as the subgenus *Madoqua*), all with a gray-agouti back and upper flanks, sharply distinct from the reddish-colored lower flanks, underside, and legs, this reddish area generally extending to the shoulders (the "shoulder flares" of Yalden, 1978).

Yalden recognized *M. piacentinii* as a distinct species, and all other taxa as subspecies of *M. saltiana*. A close reading of Yalden's paper indicates that the taxa relegated to subspecies status are, in fact, extremely sharply differentiated, though they form an interesting geocline (see his figure 5). Therefore, we recognize a number of different species. Our descriptions (below) are closely based on those in Yalden (1978).

Madoqua saltiana (Desmarest, 1816)

Salt's dik-dik

1860 *Antilope saltiana* Desmarest. 1816. Abyssinia.

1909 *Madoqua cordeauxi* Drake-Brockman. Dire Dawa, 9° 35' N, 41° 52' E.

Largest of the species in the *M. saltiana* group. Color rather drab; upper parts agouti–reddish gray; legs sandy, not agouti.

From the Atbara River to the southernmost Red Sea hills of Sudan, through Eritrea and Ethiopia N of the highlands, as far SE as the Chercher Mountains.

Madoqua phillipsi Thomas, 1894

Phillips's dik-dik

1894 *Madoqua phillipsi* Thomas. Dobwain, 40 mi. S of Berbera, N Somalia, ca. 10° N, 45° E.

1909 *Madoqua phillipsi gubanensis* Drake-Brockman. Hul Kaboba, Golis foothills, 35 mi. S of Berbera, 10° 03' N, 45° 07' E.

Averaging considerably smaller than *M. saltiana*; somewhat paler and more contrastingly colored. Upper parts pale gray–agouti, with pale orange shoulder flares; shoulder flares seeming to be less marked toward the W of the range and much more marked toward the E.

N Somalia, from the border with Ethiopia and Djibouti, S at least to the Golis Mountains.

Madoqua hararensis Neumann, 1905

Harrar dik-dik

1905 *Madoqua hararensis* Neumann. Kumbi, Ennia-Gallaland, 8° 09' N, 41° 39' E.

Size as in *M. phillipsi*; brighter and redder than *M. phillipsi* or *M. saltiana*. Much darker than most specimens of *M. phillipsi*; flanks and shoulder-flares bright reddish brown, the reddish tint spreading to the back to give a ginger-agouti tone.

Apparently with an inland range, S of the Golis Mountains and E of the Chercher Mountains, as far S as the Webi Shebeyli River. Specimens of *M. hararensis* were obtained with *M. phillipsi* at Daberhick, at the N end of the Golis Mountains, and with *M. swaynei* at Milmil, 60 km E of Harar; these records may indicate sympatry or parapatry, although Yalden (1978:262) suggests that, alternatively, this might indicate "a single species, somewhat variable in color."

Madoqua swaynei Thomas, 1894
Swayne's dik-dik

1894 *Madoqua swaynei* Thomas. Type purchased in Berbera, but almost certainly from somewhere S of the River Web (Yalden, 1978).

1905 *Madoqua erlangeri* Neumann. Near Sheik Hussein, 7° 36' N, 40° 40' E.

1922 *Madoqua citernii* de Beaux. Dolo, 4° 11' N, 42° 05' E.

Size apparently as in *M. phillipsi*. Upper parts "grey brown agouti, the plain sandy color being confined to the legs, virtually no shoulder flares" (Yalden, 1978:252); thus somewhat similar to *M. saltiana* in its dull color tones.

Yalden (1978) discussed the problem of identifying the "true" *M. swaynei*, not least because its actual type locality is unknown. He mapped this species from the Webi Shebeyli S to beyond the Juba River; the range is apparently limited to the W by the aridity of the Ogaden.

Madoqua lawrancei Drake-Brockman, 1926
Lawrance's dik-dik

1926 *Madoqua phillipsi lawrancei* Drake-Brockman. Eil Hur, near Obbia, Somalia, 5° 00' N, 48° 17' E.

Size as in *M. phillipsi*. Body color pale silvery agouti; legs and shoulder flares deep reddish orange, sharply distinct from the back color.

Mapped by Yalden (1978) from the N and C parts of the E coastal region of Somalia, its distribution to the W apparently limited by arid country.

Madoqua piacentinii Drake-Brockman, 1911
silver dik-dik

1911 *Madoqua piacentinii* Drake-Brockman. Gharabwein, near Obbia, Somalia, 5° 25' N, 48° 25' E.

The smallest species, averaging somewhat smaller than *M. phillipsi*. Body silver, more finely agouti-banded than other species; legs pale sandy; no shoulder flares. Distinctive black rim around the back surface of the ears.

A few localities on the C Somali coast, nearly as far S as the Webi Shebeyli.

Madoqua kirkii Group
This is the group sometimes separated as the genus (or subgenus) *Rhynchotragus*, distinguished by its very enlarged proboscis, with an enlarged nasal cavity, underlain by extremely short nasals and short, curved (S-shaped) premaxillae.

It was pointed out by Ryder et al. (1989) that groups of dik-dik (referring to breeding-groups in the San Diego Zoo), which may be almost indistinguishable externally, may nonetheless have quite different karyotypes and so would not be interfertile.

Of three *M. guentheri* specimens, two had $2n = 50$, and the third had $2n = 48$, the difference being a simple Robertsonian fission/fusion; offspring had, as expected, $2n = 49$. But it was dik-diks ascribed to *M. kirkii* that appeared most bewildering. The authors divided them into two cytotypes, A and B. Cytotype A had $2n = 46$; B had $2n = 47$, differing by a translocation between the X chromosome and chromosome 10 and, in addition, by several inversions and tandem fusions. The first group had been imported from an unknown location in Kenya; the second, from Garissa (and a further specimen was later imported from Mbalambala, along with a group of *M. guentheri*). Attempts to hybridize them resulted in fertile females, but sterile males.

The karyotype of *M. damarensis* (at that time considered a subspecies of *M. kirkii*) was studied by Kumamoto et al. (1994), who assigned it to a further type (cytotype D), differing from cytotype A by a tandem fusion and a pericentric inversion, and from cytotype B by a pericentric inversion, two additions/deletions of heterochromatin, and the lack of an X-autosome translocation.

The *M. kirkii* and the *M. guentheri* groups separate completely when we compare male skulls. They differ in orbitonasal length compared with greatest skull length, biorbital breadth, and nasal length; the major difference between the *M. guentheri* and the *M. kirkii* groups is the much longer premaxilla, and the shorter orbitonasal distance, of the former (in this relationship, *M. damarensis* is not different from the *M. kirkii* group). The taxa within each group differ in the degrees of contrast between greatest length, biorbital breadth, nasal length, and premaxillary length against orbitonasal length and (to a slight degree) nasal length.

In the *M. guentheri* group, there is an absolute difference between *M. guentheri* and *M. smithii*. Within the former, the Kenyan and Torit samples differ from the Ethiopian sample in their large size, but essentially do not differ from each other. Hence we can recognize just two species, with their distributions as follows:

1. *Madoqua guentheri*—most of the Ethiopian range, including *M. guentheri* Thomas, 1894;

M. *wroughtoni* Drake-Brockman, 1909; and *M. hodsoni* Pocock, 1926.
2. *Madoqua smithii*—Lake Stephanie S to Lake Baringo, W to the Nile.

The female samples differ in much the same way as the males.

In the *M. kirkii* group, three groupings clearly emerge: Somalia; Kenya Rift Valley and highlands plus Tanzania; and E Kenya plus Tsavo, although there is also some difference between the NE Kenya and Tsavo samples. We can recognize the following four species, one of them distinct on pelage, the other three on pelage plus craniometrics:

1. Somalia—specimens from N and E of the Tana River are poorly distinguished from specimens from here. This is the "true" *Madoqua kirkii*.
2. Tsavo and elsewhere in E Kenya—the prior available name is *Madoqua hindei*.
3. Kenya and the N Tanzania Rift Valley, including the Serengeti ecosystem—*Madoqua cavendishi*.
4. W Tanzania (mainly S of item #3 above)—this species, distinguished by pelage characters only (the skull mostly resembles item #1 above), would bear the name *Madoqua thomasi*.
5. Skulls from the San Diego Zoo of the known cytotype were kindly forwarded to CPG by Steve Kingswood and the late Arlene Kumamoto. Discriminant analysis showed clearly that cytotype A is the Rift Valley species (i.e., *M. cavendishi*); cytotype B is (almost certainly) the Tsavo species (i.e., *M. hindei*).

Skull lengths for taxa of the two groups confirm that *M. guentheri* is much smaller than *M. smithii*, and that *M. cavendishi* is the largest of the *M. kirkii* group, followed by *M. hindei*, with *M. thomasi* and *M. kirkii* being equal as the smallest species.

Madoqua thomasi (Neumann, 1905)
Thomas's dik-dik

1918 *Madoqua cavendishi* sensu Hollister.

Females slightly larger than the males. Rabbit-gray, with red tones. Flank transition zone usually narrow. Hair bases pinkish white; dark band long, sharp, black or brownish ; light band yellow or red-yellow. Head chestnut. Forehead and tuft chestnut, often mixed with black and off-white. Eye-rings usually going right around the eyes. Rump and thighs more gray. Legs chestnut, with the chestnut zone extending to the shoulders. White on the limbs extending to the knee, and (usually) beyond the hock.

NW Tanzania. We have seen specimens from Mwanza, Olduvai, Engaruka, 25 mi. E of Ngorongoro, Mto wa Mbu, Kondoa, Dodoma, Bugogo, Irangi, Tabora.

Madoqua hindei Thomas, 1902
Hinde's dik-dik

1913 *Rhynchotragus kirki nyikae* Heller.

Unlike any of the other E African species, the males larger than the females. Red-olive; dorsum redder, more finely speckled. Transition zone on the flank very narrow. Hair bases darker; black band incomplete; light band red-yellow. Less white on the underside, the limbs, and the face. Becoming more red, less olive, toward the midback. Chestnut not extending to the shoulder; legs paler chestnut (restricted in its extent). White on the legs reaching less far down. Muzzle paler, whitish. Forehead and tuft marbled with black. Eye-ring reaching farther in front of the eye.

SE Kenya, including the Ngong Hills, Tsavo, Ukamba, and the foot of Mt. Kilimanjaro. The range is probably bounded by either the Tana River or the Athi-Galana River, the coast, the Pare-Usambaras, and the Rift highlands.

Madoqua kirkii (Günther, 1880)
Kirk's dik-dik

1912 *Rhynchotragus cavendishi minor* Lönnberg.

Females somewhat larger than the males. Much paler (yellow-olive), but the black speckling varying. Transition zone on the flanks very narrow. Hair bases very pale; black band very pale; light band very pale. Nose white. Forehead and tuft marbled with black. Eye-ring reaching farther in front of the eye. Legs paler chestnut (restricted in its extent); white on the limbs reaching less far.

N Guaso Nyiro to Lamu, into Somalia as far as Mogadishu; the range is probably E and N of either the Tana River or the Athi-Galana River.

Madoqua cavendishi Thomas, 1898
Cavendish's dik-dik

1909 *Madoqua langi* J. A. Allen.

Largest of the E African species (greatest skull length >113 mm, versus <113 mm in the other species) and, uniquely, lacking sexual dimorphism in size. General color gray; hair bases more whitish; black band short, inconspicuous; light band yellow. Transition zone on the flank wide. On the head, only the nose chestnut. Eye-ring reaching farther in front of the eye. Forehead and tuft marbled with black. Legs pale chestnut (restricted in its extent); white on the limbs reaching farther down.

Kenya's Rift Valley to the Serengeti ecosystem: Lake Elmenteita; Enterit River, S end of Lake Nakuru; Lake Naivasha; Kijabe; Kedong River; Olorgesailie; Amala River; Loita Plains; 13 mi. NW of Narok; Sotik, S Guaso Nyiro; Grumeti River; Banagi.

***Madoqua damarensis* (Günther, 1880)**

Damara dik-dik

Females considerably larger than the males (apparently to a greater extent than any of the E African species); general size range about the same as in *M. cavendishi*. Uniquely, foot glands absent, but a rudimentary skinfold between the toes (the glands present in all other dik-diks). Anterior part of the body paler than in the E African members of the *M. kirkii* group; crest between the horns more blackish in color.

N and NW Namibia, and SW Angola

***Madoqua guentheri* Thomas, 1894**

Gunther's dik-dik

M. guentheri and *M. smithii* differ from all previous *Madoqua* species in their extremely reduced premaxillae and nasals, with a greatly enlarged proboscis. Virtually no sexual dimorphism in skull size, although in external measurements, the females noticeably larger than the males. Greatest skull length 98–109 mm. Finely speckled gray; not much chestnut on the shoulder; underparts pinkish buff. Forehead and crest strongly marbled with black.

Lowlands of S and E Ethiopia, NE from Lake Stephanie, and Somalia (except the coast).

***Madoqua smithii* Thomas, 1901**

Smith's dik-dik

Closely related to *M. guentheri*, with which it shares extreme cranial and nasal specializations and a lack of sexual dimorphism, but much larger; greatest skull length 109–120 mm. General color more reddish; underparts buffy white.

From Lake Stephanie S to Lake Baringo, and W to the Nile.

Ourebia Laurillard, 1842

oribi

This genus is unusual among the Antilopini in that the lingual walls on the lower molars are not flat, the nasal tips lack lateral prongs, and the premaxilla fails to contact the nasals. There are also, as noted by A. W. Gentry (1992), some resemblances to the Reduncini, which must be interpreted as convergences; namely, the bare glandular patch below the ear, and the deep labial valley on the posterior lower premolar.

We recognize four species, each with a wide range; although all are rather variable in pelage and other characters, we cannot seem to make any convincing divisions within any of them. Undoubtedly there is much to be learned about this genus, and our arrangement must be taken as provisional.

Table 43 gives the univariate statistics, and table 44 lists the characteristics of each species. Full synonymies are not given here.

***Ourebia quadriscopa* (Hamilton Smith, 1827)**

Skull quite distinct (see table 43); color varies, but quite distinct; apparent difference between the sexes in the presence or absence of the dark crown mark.

W Africa; exact E boundary is not known.

***Ourebia montana* (Cretzschmar, 1826)**

1907 *Ourebia goslingi* Thomas & Wroughton.
1912 *Ourebia montana aequatoria* Heller.
1913 *Ourebia gallarum* Blaine.
1921 *Ourebia montana ugandae* de Beaux.
1926 *Ourebia pitmani* Ruxton.

Color varies; dark crown mark variable; considerably larger than *O. quadriscopa*, but the horns not longer. Tail sometimes darker, but not usually black (unlike other species); only the Ankole population ("*O. pitmani*") sometimes with a black tail.

From N Nigeria E into Ethiopia, S into Uganda.

***Ourebia hastata* (Peters, 1852)**

1895 *Neotragus haggardi* Thomas.
1905 *Ourebia kenyae* Meinertzhagen.
1908 *Ourebia cottoni* Thomas & Wroughton.
1922 *Ourebia rutilus* Blaine.

Distinguished particularly by the much heavier banding of richer, more yellow hairs; tail black, even on the underside. Size the same as in *O. montana*, but the horns considerably longer.

Kenya S to Mozambique and E to Angola.

***Ourebia ourebi* (Zimmermann, 1783)**

Skull quite distinct. Remarkable as the only *Ourebia* species with the males larger than the females, instead of the females being larger. Color more rufous, without conspicuous banding.

S of the Zambezi.

Table 43 Univariate statistics for *Ourebia*

	Gt l	Biorb	Max teeth	Pre orb	Horn l
Males					
quadriscopa					
Mean	156.875	69.500	45.333	83.800	102.000
N	4	4	6	5	1
Std dev	2.8395	.5774	2.2509	1.3038	—
Min	154.0	69.0	43.0	82.0	—
Max	160.0	70.0	48.0	85.0	—
goslingi					
Mean	169.182	74.245	48.327	92.833	97.822
N	11	11	11	6	9
Std dev	4.4118	3.5472	2.4617	4.6224	14.6423
Min	166.0	69.0	42.5	89.0	81.5
Max	180.5	79.0	50.8	102.0	132.0
aequatoria					
Mean	169.325	74.289	49.050	91.188	108.550
N	20	19	20	8	20
Std dev	5.1126	3.3221	2.7140	5.1543	16.0361
Min	159.5	69.0	44.0	81.0	79.5
Max	179.0	81.5	53.0	99.0	143.0
montana					
Mean	167.414	74.773	257.657	90.440	106.306
N	35	33	35	25	36
Std dev	5.0534	2.8009	1234.0872	3.9773	12.2102
Min	155.0	70.5	44.0	82.0	84.0
Max	176.5	81.5	7350.0	99.5	134.0
gallarum					
Mean	167.889	77.306	50.184	88.100	105.222
N	18	18	19	10	18
Std dev	4.6827	3.3437	1.9380	4.0947	11.7301
Min	160.5	71.0	45.0	83.0	83.0
Max	175.5	82.5	52.0	95.0	126.0
cottoni					
Mean	170.886	76.913	49.205	86.800	120.667
N	22	23	22	10	21
Std dev	2.8323	1.9344	2.0334	13.1322	11.8894
Min	166.0	74.0	45.0	50.0	97.0
Max	178.5	83.0	53.0	96.0	145.5
pitmani					
Mean	167.750	74.711	48.275	91.500	113.906
N	16	19	20	17	16
Std dev	4.3321	2.1751	1.8812	5.5199	13.4959
Min	161.0	70.0	45.0	85.5	88.0
Max	176.0	77.5	52.0	109.0	130.0
kenyae					
Mean	168.083	77.571	48.643	88.167	123.571
N	6	7	7	6	7
Std dev	1.3571	2.2809	1.4920	1.5055	5.6600
Min	166.0	75.0	46.0	85.5	116.0
Max	169.5	81.0	50.5	90.0	134.0
haggardi					
Mean	162.600	73.600	49.800	87.900	122.900
N	5	5	5	5	5
Std dev	6.1887	1.4318	2.0187	3.8471	11.9185
Min	155.0	71.5	47.0	84.0	108.0
Max	169.0	75.5	52.0	94.0	137.0

Table 43 (continued)

	Gt l	Biorb	Max teeth	Pre orb	Horn l
hastata					
Mean	173.250	75.583	50.250	95.667	105.800
N	6	6	6	6	5
Std dev	2.8240	2.3752	2.4444	2.4014	16.8434
Min	168.0	72.0	45.5	93.0	83.0
Max	175.0	79.0	52.0	99.0	126.0
W Zambia					
Mean	168.350	73.156	50.563	91.375	123.206
N	10	16	16	8	17
Std dev	3.1096	1.9470	2.1282	2.1835	14.3745
Min	163.0	70.5	45.0	87.0	100.0
Max	174.0	77.0	53.0	94.0	157.0
rutila					
Mean	169.000	74.750	50.417	92.667	139.643
N	4	6	6	3	7
Std dev	2.0412	1.2550	2.8708	1.5275	17.7570
Min	167.5	73.5	46.0	91.0	125.5
Max	172.0	77.0	55.0	94.0	177.0
ourebi					
Mean	166.667	75.500	49.333	89.750	117.000
N	3	3	3	2	3
Std dev	.5774	1.8028	.5774	—	17.5214
Min	166.0	74.0	49.0	86.5	99.0
Max	167.0	77.5	50.0	93.0	134.0
Females					
quadriscopa					
Mean	162.611	71.125	46.500	88.222	—
N	9	8	9	9	—
Std dev	4.0603	3.7201	2.4495	2.5386	—
Min	156.0	64.0	43.0	85.0	—
Max	167.0	76.0	50.5	92.0	—
goslingi					
Mean	173.500	72.542	49.731	95.200	—
N	12	12	13	5	—
Std dev	5.4897	1.9005	3.8058	2.7749	—
Min	166.0	69.0	42.0	92.0	—
Max	180.0	76.0	54.0	99.0	—
aequatoria					
Mean	172.818	71.091	50.292	94.071	—
N	11	11	12	7	—
Std dev	4.4737	2.3856	3.5958	2.7903	—
Min	164.0	66.5	45.0	90.0	—
Max	179.0	74.5	56.0	98.5	—
montana					
Mean	171.214	71.133	49.200	91.611	—
N	14	15	15	9	—
Std dev	6.1354	2.0131	2.2975	9.6559	—
Min	157.5	67.0	45.5	67.0	—
Max	179.0	74.0	53.0	101.0	—
gallarum					
Mean	170.500	72.786	48.214	92.667	—
N	7	7	7	6	—
Std dev	7.3144	1.8225	2.0587	5.1737	—
Min	161.5	71.0	46.5	86.5	—
Max	185.0	75.5	52.5	102.0	—

(continued)

Table 43 (continued)

	Gt l	Biorb	Max teeth	Pre orb	Horn l
cottoni					
Mean	173.750	73.000	50.000	89.250	—
N	6	6	6	2	—
Std dev	4.6449	.9487	2.0736	1.0607	—
Min	167.0	72.0	48.0	88.5	—
Max	179.0	74.0	53.0	90.0	—
pitmani					
Mean	172.080	82.056	48.800	94.556	—
N	10	9	10	9	—
Std dev	4.2931	34.1462	2.1499	3.0867	—
Min	165.0	68.5	46.0	90.0	—
Max	180.0	173.0	51.0	100.0	—
kenyae					
Mean	174.250	74.500	48.000	92.250	—
N	2	2	2	2	—
Std dev	6.7175	3.5355	2.1213	1.7678	—
Min	169.5	72.0	46.5	91.0	—
Max	179.0	77.0	49.5	93.5	—
haggardi					
Mean	165.000	75.000	46.500	89.000	—
N	1	1	1	1	—
Std dev	—	—	—	—	—
Min	165.0	75.0	46.5	89.0	—
Max	165.0	75.0	46.5	89.0	—
hastata					
Mean	174.500	70.000	49.750	97.750	—
N	2	2	2	2	—
Std dev	2.1213	.0000	.3536	1.0607	—
Min	173.0	70.0	49.5	97.0	—
Max	176.0	70.0	50.0	98.5	—
W Zambia					
Mean	173.625	73.125	49.400	95.333	—
N	4	4	5	3	—
Std dev	4.8713	1.4361	3.4533	4.3108	—
Min	168.5	71.5	45.0	91.5	—
Max	179.5	75.0	53.0	100.0	—
rutila					
Mean	173.000	71.833	50.500	95.667	—
N	3	3	3	3	—
Std dev	5.2915	1.2583	.5000	2.3094	—
Min	167.0	70.5	50.0	93.0	—
Max	177.0	73.0	51.0	97.0	—
ourebi					
Mean	163.500	75.833	50.000	94.250	—
N	3	3	3	2	—
Std dev	15.1740	1.6073	2.6458	1.0607	—
Min	146.0	74.0	47.0	93.5	—
Max	173.0	77.0	52.0	95.0	—

Procapra Hodgson, 1846

Females without horns. Horn rings of the males prominent, closely spaced, equally prominent on the back and the front, connected by only very poor longitudinal ridges. Horn cores not compressed. Anterior premolar (P^2) large in both jaws; styles on the other premolars and on the molars small; molars not elongated. Mandibular P^4 elongate, with the lingual wall closed both mesially and distally; mandibular

Table 44 Characteristics of each species of oribi

	quadriscopa	*montana*	*hastata*	*ourebi*
Color	Gray-fawn to deep orange	Dark clay to orange-buff	Rich yellow-ochre	Bright sandy rufous
Speckling?	Speckled	Speckled	Heavy banding	Uniform
Crown mark	Small in the females, absent in the males	Varies from conspicuous to absent	—	Small, conspicuous
Tail	Black	Darker, at most	Black, including the underside	Black
Horns	—	—	Compressed, ridged	—
Horn length	102	106	124	117
Skull length, males	159	169	168	167
Skull length, females	163	174	174	164
Tail length	—	83–95	—	75

molars not elongate, their buccal midwall convexities angular, but not developed into stylids, their lingual valleys hardly developed; distal lobe on the posterior lower molar simple. Preorbital fossa totally absent. Supraorbital foramina flush with the forehead; lacrimal bone extending onto the face; ethmoid fissure very long, extending along the sides of the nasals; lacrimal bone with no fenestrae; median nasal tips very long, the lateral prongs absent. Pterygoid plates with large hamuli. Cranial profile fairly short and high; facial profile convex or straight; horns placed well behind the posterior margin of the orbit, sloping at a 60° angle (in *P. gutturosa*, *P. przewalskii*), or at >60° (in *P. picticaudata*); tips of the premaxillae well above the occlusal level; preorbital foramen rounded, above the middle premolars; auditory meatus oval. Glandular knee-tufts absent in *P. picticaudata*, but present in *P. gutturosa*; preorbital glands present only in *P. gutturosa*; foot and inguinal glands present; rhinarium absent; two pairs of mammae; tail extremely short. (Following Groves, 2000).

On this genus, see Groves (1967c).

In apparent conflict with morphological data, molecular data (mitochondrial 12S and 16S rRNA sequences) found that *P. przewalskii* and *P. gutturosa* form a clade with respect to *P. picticaudata*, with 100% bootstrap support (Lei, Hu, et al., 2003).

Procapra picticaudata Hodgson, 1846

Tibetan gazelle

Smallest species; relatively long rostrum; broad zygomata; short braincase; narrow nasal bones. Horns very long, slender, curved in only one sagittal plane. No preorbital glands or carpal tufts; no inguinal glands; glands behind the horns.

P. picticaudata is evidently found all over the Tibetan plateau, and overlaps with *P. przewalskii* in the valley of the Buha River, which flows into Qinghai Lake. No geographic differentiation could be detected in the skulls or the horns by Groves (1967c), but mitochondrial control region and cytochrome *b* sequences strongly differentiated five specimens from the Ruoergai Nature Reserve, Sichuan, from 41 specimens from Tibet and Qinghai Lake (Zhang & Jiang, 2006). The Sichuan clade had a bootstrap value of 91% in the control region tree and 96% in the cytochrome *b* tree; the non-Sichuan clade had bootstraps of 100% in both cases.

Procapra przewalskii (Büchner, 1891)

Przewalski's gazelle

Much larger than *P. picticaudata*; rostrum relatively short, zygomata less wide, braincase less steeply descending posteriorly. Horns relatively short, curved backward and inward at the tips. As far as is known, complement of glands as in *P. picticaudata*.

In the 1990s, this species was known only from the N and E shores of Qinghai Lake (Kukunor), and a small area (Bird Island) at the W end (Jiang et al., 1995); in the mid-2000s, two small populations were identified S of the lake, and along the Buha River ("Bukhain Gol" of early Russian authors), which supplies the lake from the W.

Lei, Hu, et al. (2003) studied the hypervariable region of the mtDNA control region from all four of the surviving population isolates that were known at that time: Bird Island (now a peninsula as a result of the recent progressive shrinkage of Qinghai Lake), Hudong-Ketu, Yuanzhe, and Shadao-Gahai. The Bird Island population is at the W end of the lake; the second and third populations are at its E end; and the fourth population is on its N shore. Present distances between them are not great, but they are strongly isolated, and the degree to which the isolation is entirely due to the intrusion of residential areas and roads, as opposed to being partially natural, is unclear. In the minimum-spanning network, the Shadao-Gahai population proved to be central, with the other three radiating out from it: Yuanzhe separated from it by 4 substitutions, Hudong-Ketu by 6, and Bird Island by 14. The authors discussed reasons for the high degree of differentiation, concluding that an interruption of gene flow between them by human activity, and a founder effect, due to the reduction in number, were most likely. It is, nonetheless, most striking that the Hudong-Ketu and Yuanzhe populations, just 9 km apart, not only shared no haplotypes, and had several fixed differences between them, but were related only through one of the Shadao-Gahai haplotypes.

Procapra przewalskii przewalskii (Büchner, 1891)

1891 *Procapra przewalskii przewalskii* (Büchner). Bukhain Gol (= Buha River), which runs into Qinghai Lake.

Populations still persist in this general region, but they are critically endangered.

Procapra przewalskii diversicornis (Stroganov, 1959)

1959 *Gazella przewalskii diversicornis* Stroganov. Oasis of Sin-Zhin-Pu, Gansu.

Size larger; horns more slender, strongly divergent, with the tips only a little incurved; color darker.

This is known only from two collections: that from the type locality, obtained by Pyotr Kozlov in 1909, and another, evidently from nearby, by Fenwick Owen in 1913. The differences from nominotypical *P. p. przewalskii* are very great, and do not overlap. One would like to classify them as separate species (and this might be justified), but recent evidence (Jiang Zhigang, pers. comm. to CPG) indicates that matters are not quite so simple.

Procapra gutturosa (Pallas, 1777)

Mongolian gazelle

$2n = 60$, all being acrocentrics; this is almost identical to the saiga, except that in the latter, the Y chromosome has very short arms (Soma et al., 1979).

Much larger than either *P. picticaudata* or *P. przewalskii*; small preorbital glands; carpal tufts, which may be glandular; inguinal glands large. No glands behind the horns. Large throat pouch in the males.

A detailed description of this species is given by Sokolov & Lushchekina (1997).

Tribe Reduncini Knottnerus-Meyer, 1907

Reduncini Knottnerus-Meyer, 1907, is antedated by several other names, some of them ranking as *nomina oblita*, but one other, Peleini Gray, 1872 (original spelling "Peleadae"), may be a threat to priority. Grubb (2004a) suggested retaining Reduncini on the grounds of familiarity, pending a submission to the ICZN.

The horn cores are transversely ridged; the temporal lines are often closely approximated on the skull roof; there are prominent maxillary tuberosities; the cheekteeth are relatively small, but they have basal pillars and (on the upper molars) ribs between the styles.

It is evident that the reedbuck/kob/lechwe/waterbuck group belongs in a clade with the Antilopini, the Alcelaphini, and the Hippotragini, but constitutes the sister group to them (Matthee & Robinson, 1999a; Ropiquet, 2006). The only demurral is Kuznetsova et al. (2002), who found them in a clade with the Cephalophini, supported by bootstrap values of >90% in the combined 12S and 16S sequences.

Within the Reduncini, cytochrome *b* sequences gave two clear clades, corresponding to the genera *Kobus* and *Redunca*, with 98% and 97% bootstrap support, respectively (Birungi & Arctander, 2001). When *Pelea* was included, it formed a sister clade to *Kobus* plus *Redunca*, but the latter clade had only 53% bootstrap support.

Redunca Hamilton Smith, 1827

reedbucks

Distinguished from *Kobus* by the long diastema; the very short midsection of the horns; the presence of a subauricular gland (marked by a patch of bare skin); and a wholly bushy tail, lacking a smooth, elongate section (Vrba et al., 1994). The face–braincase angle is low, the basicranial tuberosities are wide, preorbital glands are absent, foot glands are present, and inguinal glands are present (very strongly developed).

The species / species-groups of this genus form a morphocline from *R. fulvorufula* through *R. redunca* to *R. arundinum*, with *R. cottoni* forming a branch diverging from *R. redunca* (Grubb, 2000a).

On the basis of cytochrome *b* sequences, *R. fulvorufula* was found to be sister to *R. redunca* plus *R. arundinum* (Birungi & Arctander, 2001); note that the old three-species arrangement was used by these authors.

Redunca redunca Group

common reedbucks

The reedbucks in the *R. redunca* group are relatively large in size, with small bullae and relatively small orbits; the horns are evenly concave forward. This group is commonly divided into two species, *R. arundinum* (corresponding to the first two species in the present arrangement, *R. occidentalis* and *R. arundinum*) and *R. redunca* (the remaining species). There is, in fact, less overall difference between *R. arundinum* and *R. redunca* than this traditional arrangement might suggest, despite the fact that there is overlap between them in S Tanzania. What has generally been called *R. r. cottoni* (here ranked as a full species) is, in its main features, very like *R. arundinum*, to the extent that Lydekker & Blaine (1914a:209) even referred a specimen of *R. cottoni*, from Wau, Bahr-el-Ghazal, to *R. arundinum*.

Redunca arundinum (Boddaert, 1785)

southern reedbuck

1785 *Redunca arundinum* (Boddaert). "Cape."

Light grayish fawn, grizzled with brown, with a fulvous vtinge, especially on the head and the neck.; base of the ears white; round forepart of the eyes whitish; chin, upper throat, and underparts white; forelegs generally black in front, from the knee to the hoof; hindlimbs frequently with black markings on the lower part of the shanks. Tail thick, bushy.

E Cape and the E and N provinces of South Africa, Zimbabwe, N Namibia, and W Zambia (W of the Muchinga Escarpment), extending into SW Tanzania.

Redunca occidentalis (Rothschild, 1907)

Malawi reedbuck

Paler and grayer; pale rusty gray on the limbs, the tail, and the body.

R. occidentalis has a longer greatest skull length compared with condylobasal length than does *R. arundinum*, indicating a larger occipital crest; skull relatively narrow; nasals longer and broader.

Malawi; Zambia E of the Muchinga Escarpment.

Redunca redunca Pallas, 1767

western bohor reedbuck

Very small; relatively broad skull; horns short, not spreading widely; short-horn tips hooked strongly forward and inward. Hair relatively long. Dark yellowish fawn. No dark markings on the limbs.

W Africa (Senegal, Gambia, Guinea Bissau).

Redunca nigeriensis Blaine, 1913

Nigerian reedbuck

Light fulvous-fawn, slightly darker along the midline of the back. Paler flanks merge into the white of the underparts. Pale dusky stripe down the front of the lower portion of the forelegs. Hair short, close.

Nigeria (Sokotu, Wase, Ibi), Chad.

Redunca cottoni Rothschild, 1902

Sudan reedbuck

Very long, spreading, but thin horns the most distinctive feature of *R. cottoni*, the only overlaps with other reedbucks being provided by old individuals, with worn horn tips. Color light. Stripe extending down the front of the leg pale, usually mouse-gray.

Sudan, on both sides of Nile, from Bahr-el-Ghazal to the Dinder and Setit rivers, S to Mongalla.

Redunca bohor Rüppell, 1842

eastern bohor reedbuck

1900 *Cervicapra redunca wardi* Thomas.

1913 *Cervicapra bohor ugandae* Blaine.

1913 *Redunca redunca tohi* Heller.

Grizzled yellowish fawn; dark limb markings sometimes present. Underparts white, sharply defined. Fairly close in general characters to *R. nigeriensis*, but separated geographically by the very different *R. cottoni*.

Ethiopia, Kenya, Uganda.

Another Species in the *Redunca redunca* Group

There is evidently an undescribed further species from Tanzania (Kigoma S to the Ruaha Plains, Lake Rukwa), distinguished by its strong sexual size difference. *R. arundinum* also occurs in this region.

Redunca fulvorufula Group

mountain reedbucks

Size small; auditory bullae very large; orbits large, tubular; horns slender, short, with only weak rings.

Redunca fulvorufula (Afzelius, 1815)

1815 *Redunca fulvorufula* (Afzelius). E Cape.

Grizzled grayish fawn, tinged with rufous, especially on the head and the neck; chin, upper throat, underparts (extensively), and inner limbs white.

S region of the E Cape, and, farther N, in KwaZulu-Natal, Free State, and the former Transvaal, just over the borders into SE Botswana, Lesotho, and S Mozambique.

Table 45 Univariate statistics for skulls in the *Redunca redunca* group

	Bas l	Gt l	Biorb	Zyg br	Nas l	Nas br	Max teeth
Males							
W Africa							
Mean	206.00	226.60	97.88	87.52	86.80	19.66	54.86
N	5	5	5	5	5	5	5
Std dev	4.301	2.702	1.171	2.890	6.181	1.711	1.816
Min	201	224	96	85	77	17	52
Max	211	231	99	91	93	22	57
Nigeria							
Mean	221.86	245.78	105.00	93.70	95.80	20.50	59.55
N	7	9	11	10	10	10	11
Std dev	4.914	6.591	2.530	4.990	8.135	.707	3.267
Min	216	238	101	84	82	19	55
Max	229	256	109	103	110	21	65
Chad							
Mean	216.40	235.40	99.40	89.33	99.00	20.60	57.00
N	5	5	5	3	5	5	5
Std dev	5.857	7.987	4.615	4.041	6.595	2.510	2.739
Min	208	226	93	87	89	19	53
Max	222	245	106	94	107	25	60
Sudan W of Nile							
Mean	227.31	251.23	105.15	96.15	100.15	22.46	59.92
N	13	13	13	13	13	13	13
Std dev	7.994	8.457	5.320	3.671	8.019	2.570	1.801
Min	213	237	91	91	91	18	57
Max	240	262	111	101	112	26	63
Sudan E of Nile							
Mean	231.00	254.00	103.54	95.92	106.80	23.33	60.92
N	11	12	13	13	10	12	12
Std dev	8.922	10.180	3.332	3.883	8.677	2.146	4.719
Min	215	235	98	91	97	20	51
Max	241	267	108	104	120	28	70
Ethiopia							
Mean	226.55	249.09	108.25	96.73	97.45	21.27	61.36
N	11	11	12	11	11	11	11
Std dev	5.520	9.534	3.166	4.474	5.184	2.195	2.942
Min	218	230	104	87	91	19	59
Max	233	260	115	104	110	27	68
SW Uganda							
Mean	221.46	244.56	104.70	96.46	97.54	21.59	59.00
N	24	26	27	28	28	27	28
Std dev	5.672	6.800	3.383	2.782	5.935	1.760	2.293
Min	210	234	96	92	84	18	54
Max	233	261	113	102	107	25	63
E Uganda, W Kenya							
Mean	231.30	249.73	106.92	98.58	95.92	23.25	60.67
N	10	11	12	12	12	12	12
Std dev	19.189	7.268	3.476	2.843	6.640	1.288	3.312
Min	211	234	101	94	82	21	55
Max	282	262	114	105	104	25	65
E Kenya							
Mean	227.00	249.50	107.00	97.50	96.50	20.50	56.50
N	2	2	2	2	2	2	2
Min	225	246	104	95	96	20	54
Max	229	253	110	100	97	21	59

Table 45 (continued)

	Bas l	Gt l	Biorb	Zyg br	Nas l	Nas br	Max teeth
Tanzania							
Mean	234.63	257.00	109.44	99.22	104.56	21.89	60.44
N	8	8	9	9	9	9	9
Std dev	9.334	12.444	5.570	4.024	7.892	2.147	4.447
Min	220	240	98	91	97	18	53
Max	251	280	119	104	123	25	68
Females							
Nigeria							
Mean	—	—	108.00	96.00	—	—	—
N	—	—	1	1	—	—	—
Chad							
Mean	210.00	233.00	84.00	86.00	84.00	18.00	58.00
N	1	1	1	1	1	1	1
Ethiopia							
Mean	216.00	244.50	102.00	94.50	98.00	21.50	61.50
N	2	2	2	2	2	2	2
Min	216	237	97	91	96	21	58
Max	216	252	107	98	100	22	65
SW Uganda							
Mean	220.00	234.67	92.33	92.00	84.33	21.00	58.33
N	3	3	3	3	3	3	3
Std dev	16.371	6.028	3.512	6.557	3.786	3.464	.577
Min	206	229	89	86	80	19	58
Max	238	241	96	99	87	25	59
E Uganda, W Kenya							
Mean	221.25	246.00	101.38	94.25	97.25	21.33	60.00
N	8	8	8	8	8	9	9
Std dev	14.878	15.043	5.999	4.464	9.377	2.179	3.122
Min	195	223	91	89	86	18	56
Max	243	269	110	100	113	25	64
E Kenya							
Mean	206.00	227.00	92.00	92.00	93.00	17.00	62.00
N	1	1	1	1	1	1	1
Tanzania							
Mean	211.50	235.50	95.00	94.00	86.50	19.00	61.00
N	2	2	2	2	2	2	2
Min	207	227	93	91	82	18	60
Max	216	244	97	97	91	20	62

It seems likely that two species are actually represented under the *Redunca fulvorufula* heading: two specimens (one incomplete) from the Cape are smaller than the others; thus somewhat resembling *R. chanleri*.

Redunca chanleri (Rothschild, 1895)

1895 *Redunca chanleri* (Rothschild). Kenya.

Color lighter and grayer than in *R. fulvorufula*, with very little reddish tint; size smaller, with relatively longer horns. "The break in distribution between the southern and northern mountain reedbucks is so wide and the proportions of their skulls so different that I cannot see why they should be regarded as of the same species: they could well have evolved separately from the species occurring in the plains in their respective areas" (A. Roberts, 1951:293). The difference in craniometrics is, in fact, mostly one of smaller size, but with relatively longer horns. Similar suggestions have been made in the past about other stenotopic species with consequently sporadic, isolated distributions, such as sitatunga. This may or may not be justified, but it is necessary to be cautious, because of what we now know about the shifting distributions of habitats within Africa in the past.

Table 46 Univariate statistics for horns in the *Redunca redunca* group: males

	Horn l curve	Span	Tip–tip	Bas diam	Horn l str
W Africa					
Mean	195.00	135.20	82.00	39.60	140.50
N	5	5	5	5	4
Std dev	25.000	31.396	31.757	2.074	25.489
Min	160	102	57	38	114
Max	230	187	127	43	175
Nigeria					
Mean	226.00	186.50	130.10	42.70	177.63
N	10	10	10	10	8
Std dev	19.692	27.355	36.214	3.302	20.199
Min	185	132	90	39	133
Max	250	226	191	48	193
Chad					
Mean	216.40	163.20	155.40	38.67	173.75
N	5	5	5	3	4
Std dev	17.743	82.087	24.027	2.082	8.057
Min	200	19	118	37	167
Max	240	217	184	41	183
Sudan W of Nile					
Mean	270.00	254.00	176.58	42.17	204.08
N	12	12	12	12	12
Std dev	27.303	40.779	58.405	2.480	19.148
Min	215	194	77	38	176
Max	310	330	282	46	236
Sudan E of Nile					
Mean	267.67	261.58	209.08	41.92	203.09
N	12	12	12	12	11
Std dev	39.032	55.590	57.270	2.109	27.566
Min	195	170	100	39	147
Max	330	357	287	46	247
Ethiopia					
Mean	214.50	196.78	165.22	40.50	172.40
N	10	9	9	10	10
Std dev	18.326	24.278	32.026	1.958	12.465
Min	185*	151	126	37	153
Max	240	228	210	43	194
SW Uganda					
Mean	196.05	153.90	109.38	40.46	154.26
N	19	21	21	24	23
Std dev	19.118	22.246	32.549	3.551	17.353
Min	160	104	41	36	115
Max	220	189	162	47	184
E Uganda, W Kenya					
Mean	215.00	163.60	111.70	45.10	168.09
N	11	10	10	10	11
Std dev	24.393	22.653	27.841	4.149	23.437
Min	185	130	68	39	132
Max	265	198	152	51	214
E Kenya					
Mean	195.00	206.00	171.00	39.00	150.00
N	1	1	1	1	1
Tanzania					
Mean	224.38	177.14	128.86	41.88	179.50
N	8	7	7	8	8
Std dev	21.287	26.448	27.631	2.031	23.028
Min	200	154	84	39	145
Max	260	226	164	45	208

*This includes some aged specimens with very worn tips.

Table 47 Univariate statistics for the *Redunca fulvorufula* group

	Gt l	Bas l	Biorb	Nas l	Nas br	Max teeth	Horn l str	Span	Tip–tip	Horn l curve
Males										
Cape										
Mean	223.00	198.00	101.50	90.00	23.00	58.50	127.00	108.00	86.50	152.50
N	1	1	2	2	2	2	2	2	2	2
Min	—	—	100	88	22	58	125	102	84	150
Max	—	—	103	92	24	59	129	114	89	155
fulvorufula										
Mean	231.60	205.93	105.77	88.86	23.07	54.70	156.57	128.33	—	—
N	15	14	15	7	7	15	14	6	—	—
Std dev	3.180	2.786	4.292	5.336	1.644	2.313	20.879	22.250	—	—
Min	226	201	99	81	21	50	134	102	—	—
Max	238	211	113	98	26	59	213	156	—	—
chanleri										
Mean	218.35	192.94	98.65	92.50	22.00	55.50	136.88	127.50	83.50	185.00
N	10	9	10	2	2	10	8	2	2	2
Std dev	5.869	4.333	2.109	—	—	3.171	27.368	—	—	—
Min	209	185	96	89	19	50	113	95	53	130
Max	229	199	103	96	25	60	199	160	114	240
adamauae										
Mean	—	180.25	90.50	—	—	—	127.50	—	65.00	—
N	—	2	2	—	—	—	2	—	1	—
Min	—	180	90	—	—	—	125	—	65	—
Max	—	181	92	—	—	—	130	—	65	—
Mongalla										
Mean	210.00	184.00	93.00	—	—	54.00	119.00	—	—	—
N	1	1	1	—	—	1	1	—	—	—
Females	—	—	—	—	—	—	—	—	—	—
fulvorufula										
Mean	228.83	200.83	98.50	90.50	20.25	56.33	—	—	—	—
N	6	6	6	2	2	6	—	—	—	—
Std dev	2.317	1.722	2.881	2.121	2.475	3.327	—	—	—	—
Min	226	199	94	89	19	50	—	—	—	—
Max	233	204	101	92	22	59	—	—	—	—
chanleri										
Mean	230.00	—	—	—	—	—	—	—	—	—
N	1	—	—	—	—	—	—	—	—	—

Through the mountainous areas of Kenya and Ethiopia, and into N Tanzania.

***Redunca adamauae* Pfeffer, 1962**

Size still smaller. A taxonomically undescribed specimen from Mongalla is cranially intermediate between *R. adamauae* and *R. chanleri*.

Pfeffer (1962) described *R. adamauae* as a new subspecies (*Redunca fulvorufula adamauae*) of smaller size than the others, having a bright reddish coloration, with multibanded, red-tipped hairs; red color becoming more yellowish on the flanks, and white on the belly and the internal aspects of the upper limb segments. Horns more slender than other reedbuck, practically parallel or only slightly divergent, and very little recurved forward. Compared with nominotypical *R. f. fulvorufula*, the skull much broader, with somewhat longer nasals and frontals. In the only two specimens available at the time of the type description, there are only two premolars in the upper jaw, whereas all other reedbuck, of this group or any other, have three (the lower jaw of *R. adamauae* has three).

Pfeffer opined that the distribution of *R. adamauae* was confined to the Massif of Adamawa (he gave an altitude of 1923 m), where it is sympatric with a member of the *R. redunca* group, which he identified as *R. nigeriensis*.

Kobus A. Smith, 1840

Distinguished from *Redunca* by the relatively much shorter diastema; the long, sweeping midsection of the horns; the absence of the subauricular gland; and a smooth, elongate section on the tail (Vrba et al., 1994). The face–braincase angle is high, the basicranial tuberosities are narrow, and foot glands are present, while preorbital glands and inguinal glands vary in different groups of species.

On the basis of complete mitochondrial cytochrome *b* sequences, Birungi & Arctander (2001) found that this genus falls into two clades: one for the lechwe, and one for waterbuck plus kob. The names *Onotragus*, *Kobus*, and *Adenota* are available for these three groups, respectively, if they should prove generically or subgenerically distinct.

Kobus leche Group

lechwe

Members of this group are distinguished from the *Kobus kob* group by the much longer midsection of the horns; the longer, rougher pelage; the lack of a preorbital gland; and the presence of inguinal glands (Vrba et al., 1994). Within the group, *K. megaceros* is sister to the S clade, within which *K. smithemani* is sister to *K. leche* plus *K. kafuensis*, with 92% support for the latter clade (Birungi & Arctander, 2001).

For convenience, a few measurements from Cotterill (2005) are tabulated below:

There are strong average differences (note particularly the enormous horns of *K. kafuensis*, and the narrowness of the skull in *K. anselli*), but the ranges overlap at their extremes. Nonetheless, the four well-represented species (i.e., excluding *K. robertsi*) separated very cleanly on discriminant analysis (Cotterill, 2005:figure 6).

Kobus leche Gray, 1850

red lechwe

Color light to bright red-brown, paler on the sides of the neck and the flanks. Ear-backs light red-brown. Foreleg-stripe broad, very dark from the hoof to the elbow. White throat stripe continuous with the white hair of the chin and the underside.

A very wide distribution in the wetlands of Zimbabwe, Botswana, and much of Zambia.

Kobus anselli Cotterill, 2005

Upemba lechwe

Color light to bright red-brown, paler red on the sides of the neck and the flanks. Underside, lips, chin, and around the eyes and the ears white. White of the inner and anterior aspects of the ears extending onto the ear-backs. Dark striping along the anterior foreleg reduced to a thin stripe on the lower leg, or even just a knee-patch. White throat stripe reduced or absent. Relatively small in size; gracile. Skull small and narrow; horns more slender than in other taxa, but relatively long. (Following Cotterill, 2005).

Wetlands of the Upemba region, Katanga.

Kobus kafuensis Haltenorth, 1963

Kafue Flats lechwe

Adult males developing large, dark shoulder-patches at the top of the thick black leg-stripes. White throat stripe continuous with the white chin and the white underside. Size relatively large; horns extremely large.

Restricted to the Kafue Flats, Zambia.

Kobus smithemani Lydekker, 1900

black lechwe

Horns relatively short; tips more divergent. Body in the adult males (in maximum development over 4 years of age) suffused with dark hairs on the face, the

Table 48 Measurements for lechwe, from Cotterill (2005)

	Horn l	Tip–tip	Gt skull l	Zyg br
anselli	546	315	280.6	118.8
	410–667 (15)	194–474 (15)	269–292 (10)	114–125 (10)
kafuensis	712	443	300.3	126.4
	588–812 (75)	181–756 (74)	282–317 (54)	116–134 (73)
leche	575	292	298.1	124.6
	500–694 (39)	125–539 (39)	287–320 (28)	116–131 (34)
smithemani	529	359	286.0	119.1
	412–600 (37)	208–517 (35)	271–303 (28)	108–127 (32)
type of *robertsi*	449	254	281.3	113.5

neck, the shoulders, and the flanks. White throat stripe continuous with the white chin and the white underside. Size smaller than *K. leche* and *K. kafuensis*, and somewhat larger than *K. anselli*.

Lake Bangweulu and the Bangweulu district, NE Zambia.

Kobus robertsi Rothschild, 1907

Roberts's lechwe

Dark shoulder-patches in the adult males, extending onto the throat. Continuous white throat-stripe, extending forward to the chin and backward to the underside. On the evidence of a single skull, size small (about the same as *K. anselli*); horns very short.

Formerly from a small area in NE Zambia; now apparently extinct.

Kobus megaceros (Fitzinger, 1855)

Mrs. Gray's waterbuck

The females and the juveniles (and the subordinate adult males) fawn, but the (dominant) adult males blackish. Large white patch on the shoulders; white on the lips, the underside of the jaw, the midline of the abdomen, and the inner aspect of the limbs. Hair reversed along the midline of the neck. Slightly smaller than the S species of lechwe, but with extremely long horns.

As shown by Vrba et al. (1994), *K. megaceros* has a higher face–braincase angle than other lechwe or kob, more similar to that of waterbuck.

Falchetti et al. (1995) studied the relationship between color pattern and dominance in the males in the Rome Zoo. The males are sandy fawn, like the females, until 2 years of age, after which they start to turn black, and the nape, the neck, and the area between the horns all begin to become lighter; the white areas contrast strongly with the remaining blackish body areas. The authors pointed out that such a color pattern not only acts as a visual symbol in itself, and emphasizes a male's threat display, but also makes for an easy assessment of horn size. Three bachelor males, housed together, changed their color between 2.6 and 3 years of age, but two of them, after losing fights, returned to the subadult coloration; one of them later regained the adult color after being separated from the others. In a second enclosure, two young males, housed with an adult male as well as females and other young males, "acquired a dull adult coloring after three years," but lost it again after a short while; when the adult male died, one of them reacquired the adult coloration.

Swamps of the White Nile, Sudan.

Kobus kob Group

kob

The horn core midsection is shorter than in the *K. leche* group; preorbital glands are present; the inguinal glands are more strongly developed; and the adult pelage is short and smooth (Vrba et al., 1994). Because our arrangement is new, we here give full synonymies.

Kobus kob (Erxleben, 1777)

Buffon's kob

1777 *Antilope kob* Erxleben. Upper Guinea, "ad Senegal."
1827 *Antilope adenota* Hamilton Smith. Senegal.
1827 *Antilope forfex* Hamilton Smith. Gambia.
1840 *Kobus adansoni* A. Smith. W Africa.
1842 *Antilope annulipes* Gray. Gambia.
1869 *Adenota buffonii* Fitzinger. Senegambia.
1869 [*Pseudokobus forfex*] *fraseri* Fitzinger. W Africa.
1899 *Cobus cob* Lydekker.
1899 *Cobus nigricans* Lydekker. Kafari, 80 mi. in [i.e., to the] W of Freetown, Sierra Leone.
1905 *Adenota koba* Neumann.
1914 *Adenota kob riparia* Schwarz. Kpandu district, W Togo.

Foot glands absent. Color orange-fulvous or tawny. Whitish ring around each eye, and another around the base of each ear. Ears fulvous on the back, with indistinct black tips. Indistinct blackish stripe down the front of the legs, usually interrupted by a white band just above the hoofs. Leg lines thin. Smaller than other taxa of the *K. kob* group, though with slight overlaps in individual measurements.

W Africa, more or less W from Lake Chad. Samples from far W Africa, Togo, and Nigeria cannot be distinguished from each other either externally or craniometrically.

The type of *Cobus nigricans* (from Sierra Leone) has a more dusky color than usual, with the middle and posterior part of the back being chocolate-brown. This coloration is an approach toward the typical black color of the dominant male in *K. thomasi* populations, and thus is perplexing. Intuitively, the strong sexual dimorphism seen in *K. thomasi*, as well as in its sympatric congener *K. megaceros* (perhaps even mimicking it?), would be a derived condition, responding to sexual selection; if so, its occurrence in the far W of Africa might be interpreted as a homologous mutation. Alternatively, the blackened tone might once have characterized all of the *K. kob* group, having been lost in all but *K. thomasi*; Kingdon (1997) implied that the black coloration might be selected against, because it is more conspicuous to predators.

Kobus loderi Lydekker, 1900
Loder's kob

1900 *Cobus vardoni loderi* Lydekker. Type locality: unknown.
1905 *Adenota pousarguesi* Neumann. Sanaga Valley.
1913 *Adenota kob adolfi-friderici* Schwarz. W side of the mouth of the Shari River.
1913 *Kobus kob alurae* Heller. Radio Camp, Lado.
1913 *Kobus kob bahrkeetae* Schwarz. Bahr Keeta, Upper Shari River district.
1913 *Kobus kob neumanni* Rothschild. Lake Albert.
1913 *Kobus kob ubangiensis* Schwarz. Duma, near Libenge, Oubangui.
1914 *Kobus kob adolfi* Lydekker & Blaine.

Tawny; thick, deep, black line down the front of the foreleg below the elbow; thinner black line down the hind shank; black line crossing the anterior pasterns, separating the white of the pasterns into two spots. Leg lines varying from thick to thin, ending fairly abruptly on the foreleg, just below the level of the stifle. Ears with the inside tending to be white.

From the Benue River, via the Shari River, to the upper Bahr-el-Ghazal region.

Samples in Frankfurt that are referable to Schwarz's various described taxa plus *Adenota pousarguesi*, to which are added a few from corresponding localities in the Berlin and Powell-Cotton museums, cannot be distinguished from one another; in craniometrics, they overlap widely, and, in the absence of pelage differences, they are here synonymized.

Some of the E specimens, such as the type of *K. k. ubangiensis*, may, however, be darker, with a distinct black suffusion; whitish on the outer half of the ears; and have a more conspicuous white eye-ring. Horn span is slightly less in C African than in Sudanese specimens. There is, thus, some slight clinal variation.

Kobus thomasi Sclater, 1896
Uganda kob

1896 *Cobus thomasi* Sclater. Berkeley Bay, Lake Victoria, on the Uganda–Kenya border.

Large in size. Markings very distinct, deep black, but not very extensive (ending in a point, at the level of the stifle, on the legs). White area around the eyes very much larger than in most specimens of *K. loderi*, tending to become buff below the eye. Backs of the ears showing a tendency to whiteness.

From the Guasin-Gishu Plateau W through Uganda as far as the Semliki Plains. Semliki kob have perhaps less white on the ear than those elsewhere.

Craniometrically, Uganda kob are distinct from those from W Sudan and C Africa, mainly in their small, less spreading horns in comparison with their slightly larger skull size; the horns, however, are stouter.

Kobus leucotis (Lichtenstein & Peters, 1854)
white-eared kob

1854 *Antilope leucotis* Lichtenstein & Peters. Sobat River, Sudan.
1854 *Kobus leucotis* (Lichtenstein & Peters). White Nile, 100 km S of Khartoum.
1863 *Adenota kul* Heuglin. Plains of the Sobat River.
1863 *Adenota wuil* Heuglin. Sobat River plains.
1899 *Adenota nigroscapulata* Matschie. Bahr-el-Gebel, W of Mongalla.
1906 *Cobus vaughani* Lydekker. Wau, Bahr-el-Ghazal.
1913 *Adenota kob notata* Rothschild. Gebel Achmed Agha, upper White Nile.

Adult males deep black; females and juveniles, and some adult males, fawn, like other species of *Kobus*. Large white patch around the eyes and the ears, and, as a continuum, through the muzzle, the chin, the upper throat, the chest, and the inner sides of the upper portion of the limbs; white eye–ear zone separated from the white chin–muzzle–throat zone by just a relatively thin, but well-marked, black stripe, extending from the black midface to the side of the neck. Females and other non-black individuals with more white around the eye than in other species of *Kobus*, with this continuing forward on the snout; comparatively light brown leg-stripe reaching the shoulder and going forward to the neck to a variable extent; however, a similar black patch on the stifle not connected to the hindleg line.

On the evidence of very few individuals, the horns extremely long and widely spreading, in the upper range of other kob.

Along the White Nile, N from the Uganda border and E to about the Ethiopian border.

Selous (1908) thought the black coloration in the adult males was seasonal; Roosevelt and Heller (1914) thought that it was age related, because black and non-black were equally common during the same season. Analogous to *K. megaceros*, one presumes that this coloration is more likely to be an aspect of dominance. The hypothesis irresistibly arises that *K. leucotis* is a mimic of the somewhat larger, much longer-horned *K. megaceros*, which lives in much the same region.

There is a magnificent photo (http://i.livescience.com/images/070612_kob_herd_02.jpg), taken in Boma

National Park, about 6° N, in Sudan near Ethiopian border, of about 30 animals, mostly or entirely males, including four strikingly black males with white ear–eye and muzzle–throat–inside limb zones; three or four that are less black, mainly with a deep red-brown tinge; and others that are tawny, but with the same very white markings. At least one young male can be seen in which the white markings are muted and less distinct.

Revision of the *Kobus kob* Group

Comments on the table leading to our new revision of this group are as follows:

- —When differentiating the W Sudan versus the C Africa (both *K. loderi*) versus the Uganda (*K. thomasi*) samples with horn as well as cranial measurements, Uganda is distinct (just), while Sudan and C Africa are intermixed. The types of *K. k. pousarguesi* and *K. k. ubangiensis*, as well as a specimen from Bahr-el-Arab, are all within the C African range of variation. Uganda is distinguished from the other samples by skull size, including both zygomatic breadth and greatest length. The W Sudan sample of *K. loderi* is distinguished from the C African one by its horn span and (slightly) lower palate length and greatest length.
- —Horn measurements (length [straight], span, tip-to-tip, and basal diameter) distinguish the two E Sudan specimens (*K. leucotis*) strongly from the 10 from W Sudan.

Table 49 gives the univariate statistics for the group. The overall small size of *K. kob* can be seen.

Using the mitochondrial control region, Birungi & Arctander (2000) found that *K. thomasi* is paraphyletic with respect to *K. kob*. Lorenzen et al. (2007) compared mtDNA control region sequences and seven microsatellites in three of the four taxa recognized here: *K. kob* (Ghana and Ivory Coast), *K. thomasi* (Queen Elizabeth National Park, Semliki, and Murchison Falls National Park) and *K. leucotis* (Sudan, both E and W of the Nile). The microsatellites gave some differentiation between the three species, except that *K. thomasi* from Murchison Falls fell into the *K. leucotis* grouping (note that no microsatellites were available from Semliki). The control region haplotypes fell into two strongly supported clades: clade 1 contained both *K. thomasi* from Queen Elizabeth National Park and *K. kob*, and clade 2 contained *K. leucotis*. There were, however, some "wrong" associations: 5 out of the 40 *K. leucotis* specimens were part of clade 1, and 13 out of the 17 *K. thomasi* ones from Murchison Falls were part of clade 2. Out of the 15 samples of *K. thomasi* from Semliki, geographically between Murchison Falls and Queen Elizabeth National Park, 9 fell into clade 1, and 6 into clade 2. The authors explained this by noting that the *Kobus kob* group was, at times during the Pleistocene, divided into W and E African refuges (proto-*kob* and proto-*leucotis*). Later, the W population dispersed into E Africa as *K. thomasi*. Finally, the migratory *K. leucotis* spread S, hybridizing with N populations (i.e., Murchison and, to a lesser extent, Semliki) of *K. thomasi*.

We might propose further that continued backcrossing of the hybrids with indigenous *K. thomasi* would serve to genetically swamp any trace of the *K. leucotis* phenotype, though the continued predominance of *leucotis* microsatellites in Murchison Falls suggests selection one way or the other—either to eliminate the *leucotis* external characters or to maintain the latter's microsatellites. If Kingdon's (1997) inference is correct, that the black coloration is too conspicuous to predators, then there would additionally be selection against it outside the White Nile region.

***Kobus vardoni* (Livingstone, 1857)**

puku

Foot glands present. Ingles (1965) recorded that the young males have two pairs of mammae in front of the scrotum and (laterally placed) inguinal glands, whereas in the females, one pair was anterior and one posterior to the inguinal glands. Color golden–buffy yellow, brownish only at the hoofs, and whitish inside the ears, around the eyes, the upper lips, the chin, and the upper throat, and again on the underside of body and inside the upper part of the legs. Females with a brownish crown. Black ear-tips small.

Birungi & Arctander (2001) found very little difference from the *K. kob* group in either cytochrome *b* or two nuclear pseudogene sequences, and they even suggested that members of this group could be only subspecifically distinct. Earlier, the same authors (Birungi & Arctander, 2000) had even found puku paraphyletic with respect to kob.

Ngamiland, Zambia, to Malawi.

Table 50 gives the univariate statistics for *K. vardoni*. Data for females are more abundant than those for males. No geographic variation can be detected, with the possible exception of the very broad skull and short nasal bones of the single specimen of each sex from Ifakara.

Table 49 Univariate statistics for the *Kobus kob* group

	Bas l	Gt l	Pal l	Zyg br	Nas l	Teeth	Span	Tip–tip	Bas diam	Horn l str
Males										
W Africa										
Mean	243.42	268.94	143.12	99.26	99.06	67.83	249.10	184.20	43.48	299.31
N	12	17	17	19	16	18	21	20	21	16
Std dev	6.748	9.236	4.428	3.445	8.752	2.550	33.602	44.915	4.589	31.419
Min	234	253	134	95	87	64	186	100	35	254
Max	254	283	148	105	119	73	312	268	55	346
C Africa										
Mean	262.55	290.70	154.17	104.88	109.35	73.25	305.39	237.74	48.50	385.00
N	29	30	30	32	31	32	33	31	32	30
Std dev	6.495	8.078	5.120	3.536	7.387	3.894	31.282	63.493	4.565	44.830
Min	252	272	142	98	92	65	234	147	41	298
Max	281	312	167	111	122	81	377	377	66	457
W Sudan										
Mean	263.36	292.23	152.27	107.31	108.00	72.69	353.77	262.46	49.85	417.62
N	11	13	11	13	12	13	13	13	13	13
Std dev	9.678	8.584	6.051	3.326	5.815	3.401	46.332	57.851	3.805	27.807
Min	250	273	141	99	98	67	280	193	44	373
Max	283	305	162	111	120	78	433	400	57	470
Uganda										
Mean	269.29	298.82	155.33	111.45	112.10	73.64	306.67	223.78	53.11	384.00
N	7	11	9	11	10	11	9	9	9	9
Std dev	12.175	10.088	6.538	4.344	9.351	3.613	42.796	87.692	3.444	21.131
Min	257	285	146	105	96	68	256	90	47	347
Max	287	312	167	120	122	79	376	376	58	410
E Sudan										
Mean	260.00	285.00	150.50	107.00	109.33	73.67	381.33	291.00	50.00	462.00
N	2	3	2	3	3	3	3	3	3	3
Std dev	—	12.000	—	2.000	9.238	2.082	17.039	85.539	2.646	11.136
Min	248	273	149	105	104	72	365	207	47	450
Max	272	297	152	109	120	76	399	378	52	472
Females								—	—	—
W Africa										
Mean	227.25	257.25	142.25	95.25	88.75	63.25	—	—	—	—
N	4	4	4	4	4	4	—	—	—	—
Std dev	7.089	8.808	2.872	.957	11.295	2.363	—	—	—	—
Min	217	245	138	94	75	60	—	—	—	—
Max	232	266	144	96	98	65	—	—	—	—
C Africa										
Mean	242.00	269.00	149.00	102.00	96.00	69.00	—	—	—	—
N	1	1	1	1	1	1	—	—	—	—
W Sudan										
Mean	243.00	277.00	153.00	99.00	110.00	72.50	—	—	—	—
N	2	2	2	2	2	2	—	—	—	—
Min	240	275	150	98	107	68	—	—	—	—
Max	246	279	156	100	113	77	—	—	—	—
Uganda										
Mean	248.67	277.33	149.67	105.67	103.50	71.00	—	—	—	—
N	3	3	3	3	2	3	—	—	—	—
Std dev	9.504	7.234	3.055	3.055	13.435	2.000	—	—	—	—
Min	239	269	147	103	94	69	—	—	—	—
Max	258	282	153	109	113	73	—	—	—	—

Table 50 Univariate statistics for *Kobus vardoni*

	Gt l	Cb l	Gt br	Nas l	Nas br	Max teeth	Horn l curve	Horn l str
Males								
Ngamiland								
Mean	303.00	282.00	112.00	116.00	23.50	75.00	362.50	362.50
N	1	1	1	1	1	1	2	2
Min	—	—	—	—	—	—	347	347
Max	—	—	—	—	—	—	378	378
W+C Zambia								
Mean	—	—	113.00	111.00	28.00	73.00	480.00	400.00
N	—	—	1	1	1	1	1	1
Luangwa+Malawi								
Mean	—	—	—	129.00	33.00	74.00	360.00	302.00
N	—	—	—	1	1	1	1	1
Ifakara								
Mean	303.00	286.00	119.00	97.00	31.00	73.00	430.00	373.00
N	1	1	1	1	1	1	1	1
Females								
Ngamiland								
Mean	288.33	273.00	102.83	107.67	22.00	72.67	—	—
N	3	3	3	3	3	3	—	—
Std dev	1.528	5.000	5.795	7.371	2.000	3.786	—	—
Min	287	268	98	102	20	70	—	—
Max	290	278	109	116	24	77	—	—
W+C Zambia								
Mean	280.17	266.83	106.33	105.20	24.00	73.33	—	—
N	6	6	6	5	5	6	—	—
Std dev	5.419	4.535	3.559	9.680	3.674	6.121	—	—
Min	276	261	102	92	21	65	—	—
Max	291	274	111	117	30	78	—	—
Luangwa+Malawi								
Mean	278.00	266.50	104.00	109.67	22.67	74.67	—	—
N	3	2	3	3	3	3	—	—
Std dev	6.245	3.536	6.245	9.609	2.082	2.517	—	—
Min	271	264	99	101	21	72	—	—
Max	283	269	111	120	25	77	—	—
Ifakara								
Mean	283.00	267.00	112.00	104.00	23.00	70.00	—	—
N	1	1	1	1	1	1	—	—

Kobus ellipsiprymnus Group

waterbuck

Members of the *K. ellipsiprymnus* group have more slender horn cores, with a lower angle between the bases, than in other species-groups; the braincase is shorter; the anterior basicranial tuberosities are narrower; the horn cores do not form a distinct angle from the base to the midsection; the midsection of the horns is long, as in the *K. leche* group; the face–braincase angle is high, as in *K. megaceros*. Preorbital glands, foot glands, and inguinal glands are absent. Vrba et al. (1994), who listed these characters, maintained that waterbuck are paedomorphic in the characters of the horns, the short braincase, the small basicranial tuberosities, the absence of inguinal glands, their small group size, and their lack of migration.

Kobus ellipsiprymnus (Ogilby, 1833)

ellipsen waterbuck

$2n = 50–52$ (Kingswood, Kumamoto, Charter, Aman, et al., 1998). The sample of 26 individuals was polymorphic for a centric fusion between chromosomes 7 and 11; a fusion between 6 and 18, for which *K.*

defassa was polymorphic, was fixed in the sample of *K. ellipsiprymnus*.

White ring around the rump.

Craniometrically, specimens from South Africa, Malawi, and Somalia can be absolutely distinguished on very small samples, but larger samples may alter this. Specimens from Gorongoza and from two localities in Kenya (Tsavo and Thika) fall broadly within the range of those from Malawi.

Those from South Africa are the largest, followed by those from S Tanzania; the others (Gorongoza, Malawi, E Kenya) are small. South African skulls are exceptionally broad—all others (except Gorongoza) are below the South African range.

Those from Kenya have very small teeth.

Kobus defassa (Rüppell, 1835)

defassa waterbuck

$2n=53–54$ (Kingswood, Kumamoto, Charter, Aman, et al., 1998). The sample of 26 individuals—most from zoos but including three wild individuals from Lake Nakuru—was polymorphic for a centric fusion between chromosomes 6 and 18; a fusion between 7 and 11, for which *K. ellipsiprymnus* was polymorphic, was fixed in the sample of *K. defassa*.

Very similar to *K. ellipsiprymnus*, but the entire back of the rump white, instead of having a white ring.

SW Tanzanian skulls are large and short-faced, with short nasals. Zambian skulls are small and long-faced, with long nasals. Those from Serengeti are small, with short horns, a narrow palate, and a long snout. Those from Laikipia in Kenya are small, but with a broad palate and a wide horn span. Sudan, Ankole, and Rutshuru skulls are large but relatively short-faced, differentiating them somewhat from others, including those from W Africa.

Table 51 gives univariate statistics for all waterbuck. Despite its larger range, *K. defassa* seems less geographically variable in skull measurements than *K. ellipsiprymnus*.

Differences in the horns are more noticeable than in the skulls, though horn measurements are rarely different more than on average. Those from Rutshuru and Ankole are very long (627–796 mm), and those from both Mongalla and from Zambia/Angola are very short (504–656 mm). Horn span is widest in Ethiopia/Sudan and Ankole, averaging 559 and 528 mm. respectively; all others have an average horn span of under 500 mm. Zambia/Angola animals have tips that do not turn in much: the mean tip-to-tip distance is only 43 mm, less than the mean span. In Ethiopia/Sudan, Rutshuru. and Mongalla, tip-to-tip distance is 49 mm less; in W Africa, 76 mm less; and in Ankole, Laikipia, and Serengeti, 95–100 mm less.

A study of waterbuck by Lorenzen, Simonsen, et al. (2006) found that mitochondrial control region haplotypes sort into four clusters: two consist entirely of *K. defassa* (W and E African samples were not distinguished), one contains both *K. ellipsiprymnus* and hybrids from Samburu and Nairobi National Park, and the fourth contains samples of all three groups (*K. defassa*, *K. ellipsiprymnus*, and the hybrids). There was very little geographic structuring within either of the two species.

Pelea Gray, 1851

This genus is certainly a sister group to *Kobus* and *Redunca*. Horns very slender, upright, and straight, with no division into basal, midsection, and terminal portions; braincase short; face–braincase angle low, basicranial tuberosities small, narrow; foot glands strongly developed; other skin glands absent; tail like *Redunca*.

Pelea capreolus (Bechstein, 1800)

rhebok

Small in size, with a pale gray, woolly coat; rhinarium somewhat swollen, extending back behind the nostrils. Horns short, spikelike, nearly straight or slightly bent back.

Found spottily from the Cape, NE through the Free State, through the high veldt of the NE parts of South Africa.

Tribe Hippotragini Sundevall, 1845

The author and date of this name are corrected after Grubb (2004a).

The females are almost or quite as well-horned as the males; the horns, which are ringed, are generally more or less parallel; the frontal sinuses extend into the horn pedicels; the cheekteeth are hypsodont; the premolars are not very reduced; and the molars have basal pillars.

Hippotragus Sundevall, 1845

Grubb (2004a) gave something of the complex history of this name.

Hippotragus contains the largest members of the tribe with upright horns curving back in a close arc, so that the tips point downward in large males. There is always a mane along the neck, reaching to a variable distance along the back.

Table 51 Univariate statistics for waterbuck: males

	Gt l	Gt br	Preorb	Nas l	Nas br	Max teeth	Pal br	Horn l str	Horn span	Tip–tip
S Africa *ellips*										
Mean	413.40	172.67	—	170.67	43.67	106.83	—	—	—	—
N	5	6	—	6	6	6	—	—	—	—
Std dev	16.365	3.559	—	9.416	3.488	4.956	—	—	—	—
Min	398	168	—	157	39	100	—	—	—	—
Max	438	177	—	186	49	114	—	—	—	—
Gorongoza *ellips*										
Mean	384.00	168.00	223.00	136.00	44.00	100.00	94.00	522.00	383.00	247.00
N	1	1	1	1	1	1	1	1	1	1
Malawi *ellips*										
Mean	394.00	158.00	250.00	174.00	40.00	100.00	94.00	475.00	397.00	293.00
N	1	1	1	1	1	1	1	1	1	1
S Tanzania *ellips*										
Mean	400.00	158.50	230.50	151.00	41.50	108.00	93.00	546.00	357.00	195.00
N	1	2	2	2	2	2	2	1	1	1
Min	—	156	220	143	41	107	85	—	—	—
Max	—	161	241	159	42	109	101	—	—	—
Tsavo *ellips*										
Mean	382.00	162.00	230.00	162.00	40.00	99.00	101.00	—	—	—
N	1	1	1	1	1	1	1	—	—	—
Thika *ellips*										
Mean	376.67	159.67	227.33	145.33	34.00	93.00	92.67	496.00	436.00	403.00
N	3	3	3	3	3	3	3	2	2	2
Std dev	18.583	7.024	12.014	6.028	1.000	10.149	1.155	—	—	—
Min	356	153	215	139	33	82	92	493	435	401
Max	392	167	239	151	35	102	94	499	437	405
W Africa, W Sudan *defassa*										
Mean	393.56	155.95	242.17	153.21	41.74	99.26	98.21	614.94	475.88	400.19
N	18	19	18	19	19	19	19	17	16	16
Std dev	20.999	6.753	16.336	9.549	3.694	4.241	6.294	67.389	66.369	107.767
Min	375	140	226	138	36	93	86	520	363	210
Max	471	166	290	181	48	111	113	750	592	546
Ethiopia, NE Sudan *defassa*										
Mean	393.50	160.00	234.33	156.86	40.43	98.57	96.86	643.00	558.60	510.20
N	6	7	6	7	7	7	7	6	5	5
Std dev	5.206	3.266	3.204	5.460	6.655	5.740	2.673	53.062	86.382	131.551
Min	384	155	229	151	32	91	93	554	453	343
Max	398	164	237	167	51	106	101	687	682	682
Rutshuru *defassa*										
Mean	403.40	165.60	239.80	160.30	47.30	100.60	100.10	692.83	528.13	478.64
N	10	10	10	10	10	10	10	12	8	11
Std dev	11.433	4.326	8.664	8.858	5.012	7.090	5.820	49.883	50.654	73.320
Min	389	157	228	146	40	88	88	627	467	374
Max	422	170	251	173	53	107	111	770	635	635
Ankole *defassa*										
Mean	402.71	162.86	242.57	159.00	45.57	105.00	101.29	702.00	510.50	414.25
N	7	7	7	7	7	7	7	5	4	4
Std dev	6.800	2.478	4.894	6.377	4.353	3.830	3.684	59.317	23.700	56.287
Min	396	160	238	150	40	100	96	636	482	356
Max	415	166	252	168	54	110	106	796	535	490
Mongalla, Karamoja *defassa*										
Mean	387.00	155.00	232.67	145.50	39.25	101.50	91.00	564.50	426.00	367.67
N	3	4	3	4	4	4	4	4	3	3
Std dev	13.892	6.880	2.517	8.544	3.202	6.608	6.164	73.903	90.714	138.609
Min	378	150	230	135	37	92	87	504	326	226
Max	403	165	235	155	44	106	100	656	503	503

(*continued*)

Table 51 (continued)

	Gt l	Gt br	Preorb	Nas l	Nas br	Max teeth	Pal br	Horn l str	Horn span	Tip–tip
Serengeti *defassa*										
Mean	419.44	164.18	253.00	158.73	43.73	107.91	102.91	630.17	474.57	375.50
N	9	11	10	11	11	11	11	6	7	6
Std dev	16.576	5.724	11.284	13.229	4.077	4.929	4.614	43.088	51.626	83.131
Min	398	152	239	144	39	98	97	580	407	260
Max	446	172	272	183	50	116	114	680	546	469
Laikipia, Baringo *defassa*										
Mean	409.20	161.40	253.40	166.40	45.00	98.20	103.00	612.67	477.00	381.67
N	5	5	5	5	5	5	5	3	3	3
Std dev	14.307	2.302	10.761	9.762	3.674	5.848	5.050	26.388	40.632	115.578
Min	398	158	242	158	41	90	97	595	432	260
Max	433	164	271	183	50	106	107	643	511	490
Zambia, Angola *defassa*										
Mean	389.78	159.70	234.56	157.30	42.10	102.50	98.00	569.29	467.00	424.14
N	9	10	9	10	10	10	9	7	7	7
Std dev	33.815	10.904	25.100	19.015	6.027	9.443	10.025	39.225	79.484	122.380
Min	304	135	170	111	33	79	79	515	393	291
Max	425	171	256	178	52	110	113	611	594	594

Hippotragus leucophaeus (Pallas, 1766)

blaubok

Smallest species of the genus. As we have only very old (late 18th century), presumably faded, skins, the color, beyond vaguely "blue-gray," is difficult to reconstruct, including its degree of disruptive patterning. Only two skulls are known (Groves & Westwood, 1995).

H. leucophaeus seems to have had a wide distribution at one time; Loubser et al. (1990) argue that paintings of it occur in a rock shelter near Ficksburg in the Caledon River Valley in the Free State. These authors also mapped its former distribution, as known by archaeological deposits, along the coast of the W Cape, as far N as Elands Bay, and they drew attention to an eyewitness description of what must have been this species as late as 1853, in the Bethlehem region, not far N of Ficksburg.

A short cytochrome *b* sequence was extracted from the skin of *H. leucophaeus* by Robinson et al. (1996). Whether by parsimony, neighbor-joining, or maximum-likelihood, phylogenetic analysis (using *Damaliscus* as the outgroup) made this species sister to *H. equinus* plus *H. niger.*

Hippotragus equinus (Desmarest, 1804)

roan antelope

Largest species of the genus, but with relatively short, strongly curved horns. Overall body color fawn, varying from nearly white to strongly reddish; facial region contrastingly black, with white preorbital streaks and a white muzzle. Ears very long, the tips downturned and strongly tufted.

H. equinus is traditionally divided into a number of subspecies, but the exact number is disputed. Available names, and their approximate type localities, are as follows:

Aegocerus koba (Gray, 1872). Gambia.
Antilope equina (Desmarest, 1804). South Africa.
Egocerus equinus scharicus (Schwarz, 1913). Lower Shari River.
Hippotragus bakeri Heuglin, 1863. Sennar, Sudan.
Hippotragus equinus cottoni Dollman & Burlace, 1928. Cuanza River, Angola.
Hippotragus equinus dogetti de Beaux, 1921. Gondokoro, White Nile.
Hippotragus equinus gambianus Sclater & Thomas, 1899. Gambia.
Hippotragus langheldi Matschie, 1898. Tabora, Tanzania.
Hippotragus rufopallidus Neumann, 1899. Upper Bubu River, Tanzania.

The applicability of Gray's name is in some question; it is based on "le koba" of Buffon, which in fact referred to both roan antelope and korrigum, in both cases from Senegal. In recent times, however, most authorities have felt at ease with its use for the W African roan antelope.

Matthee & Robinson (1999b), in their DNA phylogeny, found that roan antelope divide into two clades: *H. e. koba* and the rest. The non-*koba* clade had 100% bootstrap support in the maximum-likelihood and neighbor-joining trees, but only 82% in the parsimony tree. Likewise, Alpers et al. (2004), using mitochondrial control region sequencing and microsatellite genotyping, found the major division to be between W Africa (Benin, Ghana, and Senegal) and the rest; samples from Kruger National Park, Namibia, Zambia (both E and W of the Muchinga Escarpment) and various parts of Tanzania were scattered through the E/S clade; and, interestingly, the single specimen from Cameroon was linked to this portion, not to the W African one, although it was slightly outside the E/S group (their figure 3).

The results of our craniometric analysis are as follows:

Skulls from South Africa, Mozambique, Zimbabwe, Angola, Katanga, and (W and C) Zambia do not separate. Skulls from S Tanzania and Uganda are slightly more separated, but there is still no absolute difference. Maasailand skulls (*H. e. langheldi*) are larger in both length and breadth, but greatest length is large compared with condylobasal length (i.e., there is more occipital crest) and with breadth across the horn bases. Skulls from the White Nile (*H. e. doggetti*), and from Eritrea and W Ethiopia (*H. e. bakeri*), are somewhat different, on average.

There is little difference among populations in condylobasal length, but only Ankole and W Africa fail to show a marked median occipital protuberance (giving rise to a reduced greatest skull length); might this be related to their smaller horns? The average width across the horn bases is smaller in *H. e. doggetti* than in other populations.

Sexual dimorphism in horn size is enormous in *H. e. doggetti* (though sample sizes are small, as $N=4$ [males], 2 [females]), very large in *H. e. bakeri* and *H. e. scharicus*, much less in others, and apparently nonexistent in *H. e. koba* (again, note the small sample sizes, as $N=6$, 4).

Sexual dimorphism in skull length is greatest in *H. e. doggetti* ($N=6$, 4), though this is not as marked as it is for horn length; it is unexceptional in others.

As far as external characters are concerned, we find the following to vary significantly individually, but not geographically: length of the mane, ranging from just behind the shoulders to well along the back; foreleg-stripe from poorly marked to very thick and black, either extending to the knee, or not.

The following external features do vary geographically, but they are not highly consistent:

S forms tend to be paler; the two W African forms are much redder in color, on average; *H. e. bakeri* is more distinctively colored. Forehead (and to some extent face) color is sexually dimorphic in *H. e. langheldi*.

At the moment, we are not inclined to divide up *H. equinus*. The geographic variations in color and skulls are not well-enough defined, and, with the exception of the W African form, the DNA haplotypes are all intermixed. It is possible that a W African taxon could be recognized, distinguished solely by mtDNA, but more research needs to go into this.

Hippotragus niger Group

sable antelope

Sable antelope are rather smaller than roan antelope, but they have much larger horns, which are also more sexually dimorphic: those of the males are very long, and form what is more or less a semicircle, the tips pointing downward; while those of the females are much shorter, and make only a relatively short arc. The juveniles are bright sandy fawn in color, with a white underside, a white muzzle, and white stripes from near the bases of the horns down toward the muzzle. Both sexes darken with age, becoming nearly black in all of the males, and in the females of some populations; in other populations, the degree of darkening is much less (see below).

J. H. Grobler (1980) has studied growth in this species in Zimbabwe. Females in good condition weighed 160–180 kg, and more if in "very good condition" or pregnant; males, 180–200 kg, but with their weight seeming to respond more to their general condition. Both sexes became sexually mature before physical maturity; males reached physical maturity at 7–8 years of age, and females at 5–6 years. Full dental eruption was complete at 44 months. The horns of the females appeared to reach their asymptotic length, some 670–690 mm, at about 8 years of age; and those of the males, circa 940–970 mm, at about the same time.

All sable antelope are usually placed in a single species, but we find good reason to divide them into two, *H. niger* and *H. roosevelti*; not, however, to divide them into more than two, the described taxa *H. n. kirkii*, *H. n. variani*, and *H. n. anselli* not being fully diagnosably different from each other and from the nominotypical *H. n. niger* (though there is only a little overlap in the case of *H. n. variani*).

Matthee & Robinson (1999b), using the mtDNA control region, found that sable antelope form two 100% supported clades: *H. roosevelti*, and the rest. Within the non-*roosevelti* clade, samples of *H. n. niger*,

Table 52 Univariate statistics for roan antelope

	Horn	Gt l	Cb l	Orb	Bases	One base
Males						
equinus						
Mean	518.70	450.52	435.59	180.20	135.19	51.04
N	63	84	78	92	90	74
Std dev	46.543	15.635	14.009	7.711	8.333	5.313
Min	489	401	397	158	118	37
Max	606	486	467	197	155	67
langheldi						
Mean	539.00	436.71	415.50	183.00	131.43	47.60
N	6	7	6	7	7	5
Std dev	57.124	11.011	4.637	4.546	6.949	2.608
Min	488	428	408	177	120	46
Max	650	460	422	191	143	52
Ankole						
Mean	497.33	439.20	429.38	178.30	133.10	48.20
N	6	10	8	10	10	5
Std dev	42.387	14.987	12.850	4.900	7.978	5.020
Min	428	411	414	171	117	42
Max	553	460	453	188	144	56
bakeri						
Mean	555.33	451.36	435.92	179.31	133.25	52.44
N	9	14	13	16	16	16
Std dev	63.250	20.075	19.619	7.002	5.905	4.427
Min	460	425	411	165	124	47
Max	645	493	474	187	145	62
doggetti						
Mean	509.50	437.75	431.67	174.80	124.20	48.40
N	4	4	3	5	5	5
Std dev	76.405	17.914	21.197	6.140	10.474	7.537
Min	418	415	409	166	111	37
Max	605	458	451	182	140	57
scharicus						
Mean	530.72	456.10	444.29	182.06	132.89	50.91
N	18	29	28	35	35	34
Std dev	31.470	15.488	13.410	6.301	6.597	4.323
Min	476	426	412	159	110	41
Max	584	481	469	192	145	59
koba						
Mean	481.75	440.86	426.50	172.70	128.70	49.11
N	4	7	6	10	10	9
Std dev	25.025	8.275	7.450	5.982	7.212	5.904
Min	449	431	416	164	119	40
Max	505	455	435	183	140	55
Females						
equinus						
Mean	483.68	447.70	432.96	175.38	122.85	41.59
N	37	47	45	56	55	46
Std dev	44.412	11.779	11.076	5.839	5.596	3.429
Min	391	422	408	162	110	35
Max	608	469	455	187	137	49
langheldi						
Mean	504.40	433.80	423.40	172.00	121.83	41.50
N	5	5	5	6	6	4
Std dev	69.230	15.659	17.009	12.586	6.113	4.041
Min	443	415	401	149	116	38
Max	593	458	448	185	130	47

Table 52 (continued)

	Horn	Gt l	Cb l	Orb	Bases	One base
Ankole						
Mean	494.00	433.71	422.29	168.29	121.71	39.67
N	1	7	7	7	7	6
Std dev	.	17.270	15.798	7.825	7.181	4.082
Min	494	405	400	155	113	36
Max	494	463	449	180	134	47
bakeri						
Mean	490.00	447.67	432.67	166.00	114.00	36.00
N	2	3	3	3	2	2
Std dev	52.326	7.638	7.767	10.149	8.485	4.243
Min	453	441	424	155	108	33
Max	527	456	439	175	120	39
doggetti						
Mean	362.00	425.33	412.83	167.00	118.33	38.50
N	2	6	6	2	6	6
Std dev	35.355	21.323	18.335	11.314	6.623	2.811
Min	337	395	383	159	108	36
Max	387	457	434	175	125	42
scharicus						
Mean	487.82	447.14	434.36	173.64	121.50	42.30
N	11	14	14	14	14	10
Std dev	53.671	14.469	15.599	5.786	6.992	4.877
Min	396	420	405	161	109	33
Max	572	470	460	184	131	49
koba						
Mean	493.83	438.25	425.33	170.67	118.89	41.00
N	6	8	6	9	9	5
Std dev	38.851	14.840	15.371	7.778	3.790	2.550
Min	435	415	405	153	114	37
Max	533	466	449	178	126	43

Table 53 External features of roan antelope, describing inconsistent geographic variation

	N	Color	Face	Forehead
S Africa *equinus, cottoni*	42	Whitish with red tones to medium brown-gray or red-brown	Chocolate to black	Light brown to black, especially in the females
C+N Tanzania, Kenya *langheldi*	16	Very pale red-fawn to grayish or olive-brown	Males: black Females: usually more dark brown	Males: black or chestnut (sometimes a few brown hairs) Females: always chestnut
Ankole	4	Pale roan or gray-fawn	Black-brown to very black	Chestnut (both sexes)
E of White Nile *doggetti*	10	Light fawn to gray / rusty gray or red-brown	Dark brown to black	Male (*N*=1): redder, browner Females and sex uncertain: tan to nearly black
Dinder, Setit, Eritrea *bakeri*	8	Pale to darker red to honey-colored	Dark chocolate to blackish	Brown to nearly black
Shari *scharicus*	11	Pale red to gray-red-tawny	Deep brown to very black	Chestnut to dark brown
W Africa *koba*	7	Pale red-brown to sandy red-gray	Dark brown to black	Brown to chestnut

H. n. kirkii, and *H. n. variani* were intermixed (no samples of *H. n. anselli* were studied). Interestingly, in view of subsequent findings (those of Pitra et al., 2002, 2006), all their samples of *H. roosevelti* were from W Tanzania.

Pitra et al. (2002) sequenced the control region of 95 individuals. Like Matthee & Robinson (1999b), they found a wide separation between *H. roosevelti* (from the Shimba Hills and three localities in E Tanzania: Sadani, Songea, and Selous), on the one hand, and the S African samples (Zambia, Malawi, Katanga, and farther S), on the other. They also found an additional, even more divergent clade, restricted to W Tanzania, from Wembere S to Rungwe; interestingly, 4 out of the 40 samples from W Tanzania assorted not with this clade, but with the S African clade. The authors found the same divisions in their more limited sample of cytochrome *b*. To explain this, they suggested that there was an initial division between a W Tanzanian population and the rest, followed by the subsequent division of "the rest" into an E coastal population and a S population; after that, individuals from the S population spread N and introgressed into the W Tanzanian population.

The Pitra et al. (2002) study did sample a few specimens from Luangwa (presumably representing *H. n. anselli*), which nested well within the S clade. The authors did not sample any specimens of *H. n. variani*.

Later, part of the same team (Pitra et al., 2006) managed to obtain samples of *H. n. variani*, which formed a distinct subclade within the S clade. (The other two clades were as before). The S clade, more clearly than in the earlier study, proved to have distinct subclades, corresponding approximately to Zambia, Zimbabwe–Botswana–Namibia, and W Tanzania *H. n. variani*, although only the last of these was strongly supported (98% bootstrap). The finding that the *H. n. variani* sequences form a sister group to the minor component of the W Tanzanian taxon may possibly suggest a source for the latter.

Unlike roan antelope, sable antelope have very definite, geographically varying morphological features, which may be conveniently summarized in the following table:

In this table, premaxilla type b means that the ascending branch of the premaxilla barely makes contact with the nasal bone; type c means that it forms a substantial suture with the nasal.

Evidently, according to this summary table, there are at least four distinct taxa here: S of Zambezi, Zambia/Katanga/Malawi / SE Tanzania, Angola, and Kenya. The first two seem different on average only, and perhaps the Angolan one as well; the Kenya one is very different. The W Tanzania population may be a hybrid swarm, just as was indicated by the DNA data.

Table 55 presents the results of our craniometric analysis.

In the skull, as in the external features, *H. n. niger*, *H. n. kirkii*, and *H. n. anselli* differ on average only. *H. n. kirkii* averages larger than *H. n. niger* or *H. n. anselli*; *H. roosevelti* is very small, as is the W Tanzania sample; *H. n. variani* averages larger than other taxa, but it is not outside the range. Horns of *H. n. kirkii* average considerably smaller than those of *H. n. niger*, but the only horn of *H. n. variani* available to be measured by us is not quite as large as the largest horn of *H. n. niger*. The horns of *H. roosevelti* are very small (on average), and an especially notable feature of *H. roosevelti* is the narrow distance across the horn bases. *H. n. kirkii* and *H. n. anselli* differ from *H. n. niger* in their high condylobasal length compared with greatest length (presumably this relates to the development of the occipital ridge), and the wide distance across their horn bases compared with biorbital breadth. *H. n. niger* and *H. n. kirkii* differ from *H. n. anselli* in their wide biorbital breadth compared with small greatest length, and the small distance across their horn bases. The two available W Tanzanian skulls fall (more or less) within the range of *H. n. roosevelti* from Kenya.

As far as skull length is concerned, the size in S African taxa increases in the series *niger-kirkii-variani*; *H. n. anselli* (including the Tendaguru sample) is smaller again, and *H. roosevelti* (also the enigmatic W Tanzanian sample) is extremely small, though in size it does overlap with other samples. The limited data also suggest (for future investigation) that *H. roosevelti* (if it includes the W Tanzanian sample) is less sexually dimorphic than other taxa.

Consequently, we classify sable antelope as given below.

Hippotragus niger (Harris, 1838)
southern sable

Size large, horns long and wide across the bases; females becoming dark at maturity, either black or dark red-brown, but never golden-red.

Hippotragus niger niger (Harris, 1838)

Females, and both adult and young males, very dark red-brown, nearly but not quite black. Estes (2000) engagingly calls this the "black black sable."

S of the Zambezi.

Table 54 Summary of the geographically varying features of sable antelope

		Premax					
	N	b	c	Color of males	Color of females	Face stripes	Literature on female color
South Africa	7/6	5	2	Rich deep red-brown to very dark black-brown	Deep red-brown to black-brown	Normal; occasionally vague (*N*=1)	Deep chestnut-brown, verging on black
Zimbabwe	3/1		4	Deep reddish brown-black to purplish-black	Deep reddish black-brown	Normal (*N*=3); narrow, invaded (*N*=1)	Many females nearly as black as the males (Victoria Falls NP, Zambia: Ansell 1971)
W Zambia	6/5		6	Deep brown-black to nearly black	Fairly light red-brown to deep red-brown	Normal (*N*=4); fairly obfuscated (*N*=3); almost gone (*N*=1)	Deep chestnut to dark brown (Kafue NP: Ansell 1971); tawny (Batoka: Ansell 1971)
Katanga	3/2	2	11	Deep blackish red-brown to black	Dark red-brown, often with blackish tones	Only eye-spots (*N*=1); just traces on the nose (*N*=2); complete, but yellowed (*N*=2)	—
Malawi	5/—	9	2	Deep brown	—	Normal, even very broad	Brown (E Province, Zambia: Ansell 1971)
Iringa, Tunduru (SE Tanzania)	1/2	9	13	Nearly jet-black	Deep brown, often with blackish tones	Normal	—
Rukwa, W Tanzania	6/5	2		Blackish or dark chestnut-brown	Darkish golden-brown to light reddish-brown	Thin, disappearing (just) before the muzzle or halfway down; or to the muzzle but much yellowed	—
Kenya	1/6*	1	1	Blackish	Light golden-brown	—	Chestnut; face stripes buffy yellowish (Shimba Hills)
Angola	9/4	1	13	Black	Deep chestnut- to blackish-brown	Only eye-spots (*N*=4); or vague lines (*N*=9)	—

*Based on one museum skin (the type specimen, female) and a wild group seen in the Shimba Hills.

Hippotragus niger kirkii Gray, 1872

Color of the females usually rich chocolate-brown, darker than in nominotypical *H. n. niger*, with a few darker females present in the W part of the distribution.

N of the Zambezi, in Katanga and Zambia W of the Luangwa.

Hippotragus niger anselli Groves, 1983

As described by Groves (1983d), *H. n. anselli* differs strongly on average, but not absolutely, from other S sable antelope. Color of the females resembling that of *H. n. kirkii*; white face-stripes averaging broader than other taxa; skull averaging narrower than other taxa; premaxilla making a much shorter suture with the nasal.

Zambia E of the Luangwa, Malawi, N Mozambique, SE Tanzania (Tendaguru region).

Hippotragus niger variani Thomas, 1916

giant sable

Larger and longer-horned on average than other S sable; white face-stripe restricted to a white oblong in front of the eye, or very vaguely continuing to the snout. Females usually more chestnut-colored than in other *H. niger* subspecies.

Now restricted to two reserves in Angola. They have had to be reintroduced to one of the reserves after the remnant population of females, male-deprived, apparently all hybridized with a male roan antelope.

Hippotragus roosevelti (Heller, 1910)

Roosevelt's sable

Noticeably smaller than other sable, with shorter horns, and less wide across the horn bases; females almost

Table 55 Univariate statistics for the *Hippotragus niger* group

	Horn l	Gt l	Cb l	Orb	Max teeth	Max premolar	Mand teeth	Mand premolar	Bases	One base
Males										
niger										
Mean	765.96	422.58	408.39	155.69	111.07	45.75	117.33	46.83	128.67	53.51
N	24	33	18	45	28	20	6	6	30	23
Std dev	99.405	16.072	14.809	5.608	4.422	2.337	4.033	2.483	7.136	5.018
Min	555	394	387	145	104	40	114	44	115	42
Max	1060	457	438	168	120	50	125	51	141	61
kirkii										
Mean	708.27	442.27	419.36	159.09	114.94	49.06	123.00	49.33	132.89	52.89
N	15	15	11	23	18	17	3	3	19	18
Std dev	60.879	15.632	11.952	7.982	6.530	2.221	2.646	2.082	6.871	4.639
Min	555	413	397	137	102	45	121	47	122	43
Max	790	465	432	183	124	53	126	51	145	60
anselli										
Mean	—	425.26	406.00	153.58	113.71	48.00	121.33	47.50	125.86	49.57
N	—	23	5	24	7	7	6	6	7	7
Std dev	—	12.632	6.364	5.875	4.112	3.162	3.445	2.588	5.429	3.409
Min	—	396	397	140	108	44	118	45	116	46
Max	—	448	412	164	120	53	127	52	132	55
Tendaguru										
Mean	732.00	424.23	402.54	156.93	108.64	43.14	114.00	43.50	131.71	52.17
N	10	13	13	14	14	7	4	4	14	6
Std dev	48.605	13.492	9.527	4.969	5.583	3.132	4.690	2.887	4.665	1.722
Min	667	394	380	149	97	39	109	40	125	50
Max	820	447	417	165	115	47	120	47	141	55
W Tanzania										
Mean	720.00	410.67	396.00	160.33	107.00	46.50	117.50	46.00	124.75	48.67
N	3	3	3	3	4	2	2	2	4	3
Std dev	24.576	5.859	3.606	6.429	1.414	—	—	—	4.425	2.309
Min	692	404	392	153	106	46	115	44	120	46
Max	738	415	399	165	109	47	120	48	129	50
roosevelti										
Mean	678.00	407.60	373.00	154.17	107.33	41.33	115.00	43.00	120.33	47.20
N	4	5	1	6	6	3	1	1	6	5
Std dev	33.596	14.170	—	5.193	3.327	2.517	—	—	8.959	5.215
Min	641	388	—	149	103	39	—	—	109	43
Max	715	422	—	164	112	44	—	—	134	55
variani										
Mean	980.00	453.50	438.14	160.50	116.58	47.27	126.00	47.50	132.93	55.23
N	1	10	7	14	12	11	2	2	14	13
Std dev	—	11.674	11.067	4.398	5.728	2.936	4.243	.707	5.717	6.796
Min	—	430	422	153	104	41	123	47	123	45
Max	—	467	456	168	123	51	129	48	143	69
Females										
niger										
Mean	623.80	401.17	395.50	146.41	109.54	44.46	115.50	44.50	113.27	42.42
N	10	23	8	27	13	13	4	4	11	12
Std dev	48.540	14.825	17.889	5.679	5.043	3.017	4.123	2.646	5.815	2.392
Min	558	376	361	136	102	40	110	41	106	38
Max	732	434	414	158	118	50	120	47	125	45
kirkii										
Mean	611.57	422.60	409.00	153.30	110.14	49.17	111.33	48.67	117.50	41.00
N	7	5	2	10	7	6	3	3	8	8
Std dev	71.575	17.111		4.498	5.460	4.355	14.572	2.517	2.563	2.268
Min	462	407	400	148	103	44	95	46	114	37
Max	665	444	418	160	117	55	123	51	121	44

Table 55 (continued)

	Horn l	Gt l	Cb l	Orb	Max teeth	Max premolar	Mand teeth	Mand premolar	Bases	One base
anselli										
Mean	—	415.83	401.00	147.58	113.00	45.80	119.75	45.75	113.20	39.60
N	—	12	5	12	5	5	4	4	5	5
Std dev	—	12.918	8.276	5.760	6.892	2.387	7.500	3.862	6.535	2.793
Min	—	393	389	140	103	42	110	40	103	37
Max	—	438	412	162	120	48	128	48	120	43
Tendaguru										
Mean	634.09	409.38	389.69	150.23	108.15	44.00	113.67	44.67	115.08	44.00
N	11	13	13	13	13	3	3	3	13	3
Std dev	35.498	6.475	6.473	5.876	3.760	3.000	4.041	1.155	4.804	4.000
Min	570	401	377	138	101	41	109	44	108	40
Max	682	419	402	160	115	47	116	46	123	48
W Tanzania										
Mean	581.50	397.50	379.50	148.00	106.67	44.33	112.50	43.50	113.33	41.67
N	2	2	2	2	3	3	2	2	3	3
Std dev	16.263	—	—	—	6.110	5.686	—	—	3.055	4.726
Min	570	391	376	147	100	38	107	39	110	38
Max	593	404	383	149	112	49	118	48	116	47
roosevelti										
Mean	516.00	—	—	152.00	108.00	44.00	—	—	112.00	43.00
N	1	—	—	1	1	1	—	—	1	1
variani										
Mean	—	—	—	150.00	114.33	46.33	127.00	48.00	112.67	38.33
N	—	—	—	2	3	3	1	1	3	3
Std dev	—	—	—	—	3.215	1.528	—	—	4.041	.577
Min	—	—	—	148	112	45	—	—	108	38
Max	—	—	—	152	118	48	—	—	115	39

invariably a relatively light golden-red color, although Estes (2000) illustrates a very dark female.

There are also appears to be an enigma with regard to the W Tanzanian sable, which may be a mixture between *H. roosevelti* and a now-vanished population related to *H. n. variani*.

Addax Rafinesque, 1815

The spiral-horned representative of the tribe; stockily built, with extremely large hoofs.

Addax nasomaculatus (de Blainville, 1816)

addax

A. nasomaculatus was described in some detail by Dolan (1966), in the context of the establishment of successful captive breeding groups and of a studbook. He also briefly discussed the question of whether there could be geographic variation across the Sahara, concluding, on the basis of an examination of material in the American Museum of Natural History, as well as of living animals, that there is no evidence for it.

Oryx de Blainville, 1816

oryx

Long-bodied and relatively short-legged, compared with *Hippotragus*; horns long, straight or only slightly curved, going back and continuing the line of the face instead of being upright. Color always light, usually with a black facial mask and a black line separating the darker flanks from the white underside.

Table 56 gives the univariate statistics for the genus.

The species of *Oryx* can be arranged in a morphocline, from primitive to derived, in horn length and in skull breadth (associated with body size), running from Arabia through E Africa into S Africa, branching to give off the Saharan *O. dammah* (Grubb, 2000a).

Iyengar et al. (2006), using the entire control region sequence with addax as the outgroup, found that the gemsbok is sister to the Arabian plus scimitar-horned oryx, with 95% bootstrap support for the non-gemsbok clade; using a larger dataset, but with only 750 control region nucleotides, the gemsbok was

Table 56 Univariate statistics for the genus *Oryx*

	Gt l	Biorb	Teeth	Horn l	Horn br	Tip–tip	Rings
type of *beisa*							
Mean	376.00	147.000	99.000	—	—	—	—
N	1	1	1	—	—	—	—
Awash							
Mean	375.17	147.857	105.063	820.00	114.00	188.00	20.00
N	6	7	8	8	1	1	1
Std dev	8.424	4.4508	4.5233	53.809	—	—	—
Min	362	140.0	97.0	738	—	—	—
Max	387	154.0	110.0	901	—	—	—
gallarum							
Mean	374.07	149.102	103.491	745.05	108.54	170.61	21.17
N	46	54	54	40	14	14	12
Std dev	11.328	6.4832	6.6115	47.866	7.672	48.939	—
Min	347	134.0	90.0	654	95	72	17
Max	398	165.0	139.0	857	121	244	24
callotis							
Mean	377.40	157.361	105.944	743.42	127.50	293.00	16.25
N	15	18	18	19	5	4	4
Std dev	9.156	8.0308	4.1759	70.206	10.308	81.273	—
Min	358	144.0	98.0	643	118	231	13
Max	393	175.0	113.0	895	143	409	23
Angola *gazella*							
Mean	409.33	158.000	121.250	892.75	127.63	362.50	19.75
N	3	4	4	4	4	4	4
Std dev	11.504	2.5820	3.2016	58.864	3.351	40.927	—
Min	398	155.0	119.0	818	124	306	17
Max	421	161.0	126.0	946	132	402	24
Namibia *gazella*							
Mean	431.24	167.087	115.833	895.74	133.31	398.33	20.13
N	21	23	21	19	8	6	8
Std dev	11.924	5.6722	6.7088	86.700	6.447	71.676	—
Min	403	158.0	104.5	705	125	319	11
Max	453	175.0	126.0	1060	145	514	25
Kalahari *gazella*							
Mean	425.50	164.875	116.833	933.00	131.50	477.50	20.67
N	8	12	9	13	6	6	6
Std dev	11.339	6.2126	5.2915	65.986	6.340	89.063	—
Min	401	157.0	108.0	794	123	319	17
Max	438	177.0	123.0	1010	142	545	27
Nata River *gazella*							
Mean	363.00	156.333	99.000	857.00	122.33	318.00	22.00
N	2	3	2	3	3	3	3
Std dev	—	5.1316	—	184.854	9.251	154.234	—
Min	349	152.0	93.0	703	117	224	19
Max	377	162.0	105.0	1062	133	496	27
leucoryx							
Mean	310.67	124.250	96.250	610.71	85.00	—	30.33
N	6	8	8	7	4	—	3
Std dev	35.814	10.4745	7.4785	260.401	5.050	—	—
Min	270	108.0	84.0	89	80	—	27
Max	375	138.5	105.0	890	91	—	34
dammah							
Mean	367.86	142.357	101.167	916.00	105.63	—	37.00
N	7	7	6	7	4	—	4
Std dev	10.254	2.4785	9.5376	65.087	4.423	—	—
Min	354	138.0	92.0	820	99	—	34
Max	380	145.5	113.0	1000	108	—	41

sister to the Arabian oryx, but in this case the bootstrap support was low.

Masembe et al. (2006) found that mtDNA (control region and cytochrome *b*) sequences distinguish *O. callotis* strongly from *O. gallarum* (which they referred to as *Oryx beisa beisa*), except for a single haplotype from Samburu, which assorted with *O. callotis*. Two widely divergent lineages of the control region occurred in the Samburu sample of *O. gallarum*; the authors interpreted this as a relic of secondary contact between two formerly isolated populations.

Oryx beisa (Rüppell, 1835)

Beisa oryx

Skins seen: 8.

More ochraceous (a pinkish wash); this color not extending below the black flank band, which forms the boundary between the gray flanks and the white belly. Face colored like the body; facial bands generally joined up with each other and with the gorge band. No ear tufts. Flank band 39–58 mm thick in four skins. Dorsal stripe 56 and 71mm thick in two males, and 31 and 46 mm in two females; this stripe sometimes extending fully up the neck, or for three-quarters of the neck, or just to the withers. Toothrow short, compared with other taxa (on the evidence of a single skull).

Somalia and extending into Ethiopia. The localities for specimens seen by us are Hullieh, Mersi, Haud, Bale, Berbera.

There is a fairly distinctive form, closest to this species, in the Awash Valley of Ethiopia. It differs in that the face is deeper ochraceous in color; the legs are paler; the flank band in 8 skins is only 23–45 mm thick; the dorsal stripe is 24–40 mm thick; and the toothrow is not shortened.

The horns, compared with *O. gallarum* and with most (but not all) *O. callotis*, are relatively very long for the skull length. Very unfortunately, the only available skull of *O. beisa* (the type) lacks the horn sheaths, so it cannot be compared with others.

Oryx gallarum Neumann, 1902

1910 *Oryx annectens* Hollister.

Probably *Oryx gazella subcallotis* Rothschild, 1921, is another synonym.

Skins seen: 16.

$2n = 58$; differing from *O. gazella* in lacking the centric fusion between chromosomes 2 and 17 (Kumamoto et al., 1999). The animals these authors sampled were identified as *Oryx gazella beisa*, but as they were reported as being from "eastern Africa, possibly Kenya," it seems more likely that they belonged to the present species.

Paler, a purer gray; this color extending below the black flank band, and therefore not a boundary between the body color and the white underside. Face a little paler than the neck, or even whitish. Facial and gorge bands separated (sometimes the two nasal bands joined); median bands black; nose streak diffuse. Legs white, with only a trace of stripes. Flank band 20–44 mm thick (12 skins), much as in the Awash taxon. Dorsal stripe 30–43 mm thick (12 skins), but generally very vaguely expressed, and totally lacking in one specimen.

The localities of specimens seen by us are N Guaso Nyiro, Guasinarok, Kahlbin, Ngare Ndare, Guida, Laikipia, Nyeri, and Archer's Post.

Oryx callotis Thomas, 1892

fringe-eared oryx

$2n = 58$; the karyotype is identical to that of *O. gallarum* (Kumamoto et al., 1999).

Darker, duller, and browner than in *O. gallarum*. Face deep ochery, except for the white muzzle stripes. Ears with long tufts or tassels. Usually no connection between the nasal and the median face bands. Legs not paler below the carpus/tarsus. Flank band 30 and 44 mm thick in two skins. Dorsal stripe very reduced, only 25 and 30 mm thick in two skins, confined to the rump only, or very faint. Skull, though the same overall size as in other NE African species, comparatively broad. Horns relatively short, but very thick at the base; tips comparatively wide apart.

We have seen specimens from S of Mt. Longido, and from 100 mi. S of Kilimanjaro.

Oryx gazella (Linnaeus, 1758)

gemsbok

$2n = 56$; differing from *O. beisa* and *O. callotis* in having a centric fusion between chromosomes 2 and 17 (Kumamoto et al., 1999).

Flank band 119–229 mm thick. Dorsal stripe 90–116 mm thick; in two juveniles, this thickness only 54 and 74 mm. Skull larger than in the NE African species. Horns thick, like those of *O. callotis*, and longer than in the NE African species, with the tips farther apart.

Samples from Angola, Namibia, and the Kalahari are, in general, very much alike, but three specimens from the Nata River (far NE Botswana, near the Zimbabwe border) are unexpectedly small (see table 56), with small teeth and less spreading horns.

Oryx dammah Cretzschmar, 1826

scimitar-horned oryx

$2n = 56–58$; polymorphic for a centric fusion between chromosomes 2 and 15 (Kumamoto et al., 1999).

Very pale species, with "washed-out" markings; white, with light reddish face markings and a reddish neck, this latter color continuing as a vague stripe along the lower flanks and onto the haunches. Skull comparable in size to that of most of the NE African species, but with rather small teeth. Horns more curved than in any other species, and very slender, with more rings.

Sahara; extinct in the wild, but breeding well in captivity.

Oryx leucoryx (Pallas, 1777)

Arabian oryx

$2n = 57–58$; polymorphic for a rare (two heterozygotes in 77 specimens) centric fusion between chromosomes 18 and 19 (Vassart et al., 1992; Kumamoto et al., 1999). Vassart et al. (1992) noted that the translocation occurs only in herds originating from Qatar, and ultimately from the Rub Al Khali. No genetic difference was detected by Vassart et al. (1991) between those from N and S Arabia.

White, but the typical oryx markings black and well expressed. Skull and teeth much smaller than in any other *Oryx* species; horns shorter and more slender. Number of horns rings intermediate between *O. dammah* and the other *Oryx* species.

Formerly ranged throughout the Arabian peninsula, N to Jordan. Became extinct in the wild, but reintroduced from captive stocks.

Tribe Alcelaphini Brooke in Wallace, 1876

This name is not to be displaced by the slightly earlier Connochaetini Gray, 1872 (original spelling "Connochetidae"), as noted by Grubb (2004a).

The females are almost as well horned as the males. The horn cores tend to be transversely ridged; there is a postcornual fossa; extensive frontal sinuses reach into the horn pedicels; the brain case is short and strongly bent downward; the temporal ridges diverge posteriorly; the facial skeleton is extremely elongated; the zygomatic arches are deepened beneath the orbits; the palatal foramina are wide; the teeth are very hypsodont; and the premolar rows are shortened.

This tribe contains four genera, of which *Connochaetes* is sister to the other three (Matthee & Robertson, 1999a; Ropiquet, 2006). On the basis of a cladistic analysis, Vrba (1979) proposed that the genus *Beatragus* should not be recognized, but should instead be sunk into *Damaliscus*; and that Lichtenstein's hartebeest was not part of *Alcelaphus*, but was more closely related to *Connochaetes*, so she revived the genus *Sigmoceros* for it. Later, Vrba (1997) reconsidered this, in part on the basis of a much fuller fossil record; she restored the genus *Beatragus*, and sunk *Sigmoceros* back into *Alcelaphus*, proposing the following splitting times based on the fossil record:

Alcelaphus versus *Connochaetes*—slightly <3 Ma
Alcelaphus/*Connochaetes* versus *Beatragus*—somewhat >5 Ma
Alcelaphus/*Connochaetes*/*Beatragus* versus *Damaliscus*—6 Ma

This order of splitting does not agree with molecular evidence (see above).

Alcelaphus de Blainville, 1816

hartebeest

One of the key early 20th-century papers on bovid taxonomy, and one that has survived the test of time with very little necessary modification, is Ruxton & Schwarz's (1929) revision of hartebeest. They may have recognized somewhat too many taxa, and they may have been wrong in ascribing almost all hartebeest to a single species (Schwarz was exhibiting his basic tendency toward lumping), but their analysis of hybridization, in particular, was ahead of its time.

As noted above, Vrba (1997) reversed her former decision to recognize the genus *Sigmoceros* for Lichtenstein's hartebeest. Matthee & Robinson (1999a) confirmed, on molecular grounds, that Lichtenstein's hartebeest does not constitute a distinct genus, but is part of *Alcelaphus*.

Some of the species of this genus form a morphocline as far as the elongation of the horn pedicels is concerned: *A. cokii* to *A. lichtensteini* on one branch, and to *A. lelwel* on another (Grubb, 2000a).

Arctander et al. (1999) took control region samples from all geographic groups of hartebeest, except for *A. buselaphus* and *A. tora*. They found a clean division between three quite distinct clades, all with high bootstrap values: clade I, *A. major*; clade II, an E African group (*A. lelwel*, *A. cokii*, *A. swaynei*); and clade III, a S and SE group (*A. caama*, *A. lichtensteini*). Within clade II, samples of the three taxa were partly separate: two haplotypes of *A. swaynei* were nested within *A. cokii*, whereas the other five formed a separate subclade, and three of *A. cokii* (out of quite a large number) nested within *A. lelwel*. Within clade III, the two component species were entirely separate.

A further study (Flagstad et al., 2001), with larger, more diverse samples (including some from museums), expanded somewhat on this. These authors found the same division between S, E, and W lineages; the E and W lineages together formed a major clade, with 91% bootstrap support against the S clade. Within the E lineage, the three available specimens of *A. tora* formed their own clade, the others being the main *A. swaynei* clade and a mixed *swaynei/cokii/lelwel* clade, within which each of the three species tended to occupy a separate subclade, but with a considerable number of "wrong" allocations. In addition, there was some paraphyly between E and W lineages, but not between them and the S lineage. The authors suggested that this favored an E African origin for the crown-group of the genus, and that the modern E species derived from a refugium within the range of present-day *A. lelwel*.

The studies of Cappellini (2007; Cappellini & Gosling, 2006) have added a great deal to our understanding of taxonomic variation in *Alcelaphus*, particularly as related to sexual dimorphism and sexual selection. Hartebeest males fight by dropping to their knees and battering each other, and, finally, by interlocking horns and wrestling. The elongated horn pedicels apparently displace the impact of the blows away from the braincase. Cappellini (2007) divided the species, broadly speaking, into heavily armed (*A. lelwel*, *A. lichtensteini*, *A. caama*, and *A. major*) and lightly armed (*A. buselaphus*, *A. tora*, *A. swaynei*, and *A. cokii*) groups. All species are sexually dimorphic, but to varying degrees: in skull length, there is very little in any (of the extant) species; in pedicle height, there is little in *A. tora* or *A. cokii*, more in *A. major*, and still more in the other four extant species; in horn length, little in *A. tora* or *A. caama*, enormous in *A. lichtensteini*, and intermediate in other species; in horn circumference, much greater in *A. lichtensteini* and somewhat greater in *A. caama* and *A. lelwel* than in other species; and, finally, in skull weight—an unusual metric, not commonly taken, but very revealing in this instance—low in *A. cokii*, *A. tora*, and *A. major*, and very great in the other four species. The length of the available breeding season in a given species was negatively correlated with pedicle height, skull weight, and (though not strongly) horn circumference. In other words, the shorter the breeding season, the more intense the male competition, so those hartebeest living in more seasonal environments, with more constrained breeding seasons, would have heavier skulls (and, perhaps, thicker horns) to deliver stronger blows, and longer pedicels to deflect the blows. The only other high correlation was between horn length and mean annual rainfall; evidently, male hartebeest grow relatively longer horns in more productive environments.

We give full synonymies in this genus, except for the two southernmost species, for which there is no taxonomic dispute.

Alcelaphus buselaphus (Pallas, 1766)

Bubal hartebeest

1766 *Antilope buselaphus* Pallas. Probably Morocco.
1767 *Antilope bubalis* Pallas. A renaming.
1836 *Bubalis mauritanicus* Ogilby.
1891 *Alcelaphus bubalinus* Flower & Lydekker.
1914 *Bubalis bubastis* Blaine. Abadiyeh, Egypt.

Males of this species with the shortest horns of all, though not more slender than in *A. tora*, *A. swaynei*, and *A. cokii* (Cappellini & Gosling, 2006). Greatest skull length 421 mm in a single male; horn span 97% of the basal length in the males; least frontal width 71%–72% of the biorbital width in two males (Ruxton & Schwarz, 1929).

Formerly lived in the Maghreb (Morocco, Algeria, and Tunisia N of the Atlas Mountains), and during dynastic times in Egypt. Hartebeest of unknown affinities formerly occurred in the Palestine/Israel region (Uerpmann, 1987).

Alcelaphus major (Blyth, 1869)

western hartebeest

1869 *Boselaphus major* Blyth. Gambia (fixed by Schwarz, 1920).
1914 *Bubalis luzarchei* Grandidier. Nieri Ko River, near the junction with the Gambia River, Senegal.
1914 *Bubalis major invadens* Schwarz. Garoua, Benue River.
1914 *Bubalis major matschiei* Schwarz. Kpandu, Togo.

Horns strongly twisted, U-shaped from in front. Greatest skull length 470–528 mm in the males; horn span 64%–83% of the basal length in the males; least frontal width 76%–87% of the biorbital width (Ruxton & Schwarz, 1929). Color uniform tan-brown, with no dark markings but, occasionally, a thin white band between the eyes. In Cappellini & Gosling's (2006) analysis, this species comes out as "middling" in all respects (relative horn length, horn circumference, pedicle height, skull weight).

From Senegal in the W to the sources of the Logone River, somewhat to the S of the W end of the distribution of *A. lelwel*.

Alcelaphus lelwel (Heuglin, 1877)

lelwel hartebeest

1877 *Acronotus lelwel* Heuglin. Jur River (fixed by Schwarz, 1920).
1892 *Bubalus jacksoni* Thomas. Between Lake Victoria and Lake Naivasha.
1904 *Bubalus jacksoni insignis* Thomas. Maanja River, W of Kampala.
1905 *Bubalis niediecki* Neumann. Gelo River, Upper Sobat River, Ethiopia.
1912 *Bubalis lelwel roosevelti* Heller. Gondokoro, Sudan.
1913 *Bubalis lelwel tschadensis* Schwarz. Ketekma, E of Tchekna, Bagirmi.
1914 *Bubalis lelwel modestus* Schwarz. Bahr Keeta, Upper Shari.

Horns forming a narrow, upright V from in front. Greatest skull length 455–562 mm in the males; horn span 52%–76% of the basal length in the males; least frontal width 68%–86% of the biorbital width (Ruxton & Schwarz, 1929). Color more reddish than in neighboring species.

The males in this species have relatively longer pedicels than in any other *Alcelaphus* species, and a heavy skull relative to its length (Cappellini & Gosling, 2006).

From Lake Chad E to E Sudan and possibly just into Ethiopia, SE to the borders of the E African Rift Valley, S to the Guasin-Gishu Plateau in Kenya and to Ankole in Uganda.

Alcelaphus tora Gray, 1873

tora hartebeest

1873 *Alcelaphus tora* Gray. Dembelas, Bogos, Eritrea.

Horn circumference low, relative to the skull length, in the males (Cappellini & Gosling, 2006). Greatest skull length 480–487 mm in three males; horn span 122%–128% of the basal length in the males; least frontal width 65%–67% of the biorbital width (Ruxton & Schwarz, 1929).

Formerly more or less along the Sudan–Ethiopia border from Lake Turkana N to W Eritrea. May now be extinct.

Alcelaphus swaynei (Sclater, 1892)

Swayne's hartebeest

1892 *Bubalis swaynei* Sclater. Haud, 100 mi. from Berbera, Somalia.
1905 *Bubalis noacki* Neumann. Suksuki River, S of Lake Zwai, Ethiopia.

As in *A. tora*, horn circumference in low, relative to skull length, in the males; horn length relatively low; pedicle length low in the females (as low as in *A. lichtensteini*), but not especially so compared with the males in other *Alcelaphus* species (Cappellini & Gosling, 2006). Greatest skull length 409–472 mm in the males; horn span 114%–148% of the basal length in the males; least frontal width 64%–69% of the biorbital width (Ruxton & Schwarz, 1929). Much deeper in color than other E African species, with striking black markings on the face and the body.

Formerly from N Somalia through the Ethiopian Rift Valley, at least as far S as Lake Zwai.

Flagstad et al. (2000) found that the surviving populations of this species still retain significant amounts of genetic diversity (mtDNA and microsatellites), and that two of the now-isolated surviving populations have diverged considerably.

Alcelaphus cokii Günther, 1884

kongoni

1884 *Alcelaphus cokii* Günther. Mlali Plains, E of Mpapwa, Tanzania.
1914 *Bubalis cokei sabakiensis* Zukowsky. Athi Plains, Kenya.
1914 *Bubalis cokei schillingsi* Zukowsky. Lake Jipe, SE of Mt. Kilimanjaro.
1914 *Bubalis cokei schulzi* Zukowsky. Olossirva Plateau, NE of Ngorongoro.
1914 *Bubalis cokei tanae* Matschie & Zukowsky. N of Nairobi.
1914 *Bubalis cokei wembaerensis* Zukowsky. Mkalamo, Wembere Steppe, Tanzania (selected by Ruxton & Schwarz, 1929).
1916 *Bubalis deckeni* Matschie & Zukowsky. Middle Rufu River, Tanzania.
1916 *Bubalis oscari* Matschie & Zukowsky. Mt. Gurui, Tanzania (selected by Ruxton & Schwarz, 1929).

Size small; greatest skull length 419–460 mm in the males; horn span 97%–128% of the basal length in the males; least frontal width 57%–72% of the biorbital width (Ruxton & Schwarz, 1929). Color brownish. As in *A. tora* and *A. swaynei*, horn circumference low, relative to skull length, in the males; horns relatively short, as in *A. swaynei* (Cappellini & Gosling, 2006).

An E African species, found from Lake Naivasha and the S flanks of Mt. Kenya SE through the S half of Tsavo East National Park, and SW to the Mara; in Tanzania as far S as the coast opposite Zanzibar, the S border running W to the Wembere Plains, then NW and N to Speke Gulf and the Serengeti. N of the W part of this range, from Lake Nakuru N through the

Rift Valley and W to Kavirondo and to the foot of Mt. Elgon, there is (or was) a zone claimed to consist mainly of hybrids with *A. lelwel*.

Alcelaphus lichtensteini (Peters, 1852)

Lichtenstein's hartebeest

Horn circumference, relative to skull length, in the males greater than in any other *Alcelaphus* species. Greatest skull length 458–502 mm in the males; horn span 67%–88% of the basal length in the males; least frontal width 79%–86% of the biorbital width (Ruxton & Schwarz, 1929).

Characteristic species of the *Brachystegia* zone, from the border with *A. cokii* (and perhaps interdigitating with it; see Kingdon, 1982b) to the Zambezi River in the W and E, Zimbabwe in the E, and (formerly) to KwaZulu-Natal, Mpumalanga, and Limpopo provinces of South Africa.

Alcelaphus caama (G. Cuvier, 1816)

red hartebeest

Much the longest horns relative to skull length, and a very heavy skull relative to its length, more so even than in *A. lelwel* (Cappellini & Gosling, 2006). Greatest skull length 464–505 mm in the males; horn span 87% of the basal length in a single male, least frontal width 75%–79% of the biorbital width (Ruxton & Schwarz, 1929).

From the Angolan border S to the Cape and SE to the Free State. Eliminated from most of this range by the early 20th century, but now widely reintroduced through much of it.

Alcelaphus Hybrids

Ruxton & Schwarz (1929) showed that hartebeest in the NE African Rift Valley are a hybrid swarm between *A. cokii* and *A. lelwel* (*Alcelaphus buselaphus jacksoni* in their arrangement): in absolute skull length, as well as in relative horn span, they cover part of the range of each of the parent species, and are more variable than either. The authors maintained that the taxa described below actually came from the hybrid zone:

Bubalis cokii kongoni Heller, 1912. Loita Plains, S Guaso Nyiro River.
Bubalis lelwel keniae Heller, 1912. 20 mi. NE of Nyeri, near Mt. Kenya.
Bubalis nakurae Heller, 1912. Lake Nakuru.

The hybrids, Ruxton & Schwarz (1929) stated, are predominantly of the *A. lelwel* type to the N, from Lake Baringo to Mt. Kenya; whereas to the S, from Lake Victoria to the S part of the Rift Valley zone, they tend to be more like *A. cokii*.

The region from the E side of Lake Turkana N to the Dinder River (Ethiopia/Sudan border) was identified by Ruxton & Schwarz as a hybrid zone between *A. lelwel* and *A. tora*, and they ascribed the following named taxa to such hybrids:

Bubalis neumanni Rothschild, 1897. E shore of Lake Turkana.
Bubalis tora digglei Rothschild, 1913. Ofat River, near Keili, upper Blue Nile.
Bubalis tora rahatensis Neumann, 1906. Shunfar Ambu and Shimmerler Jowee, W Ethiopia.

One of us (PG) has been inclined to think that the name *Bubalis neumanni* may actually designate a valid taxon, rather than a hybrid population. This question needs further investigation.

Ruxton & Schwarz (1929) maintained that the type of *Bubalus rothschildi* Neumann, 1905 (Adoshebai Valley, N of Lake Stephanie, Ethiopia), is probably a hybrid between *A. lelwel* and *A. swaynei*, although they could not be certain, because the type skin had been mislaid; the description mentions the lack of black markings on the face and the shoulders.

Beatragus Heller, 1912

$2n = 44$ (Kumamoto et al., 1996).

Large preorbital glands.

Beatragus hunteri (Sclater, 1889)

herola

Butynski (2000) has given a thorough description of this species, and what follows is largely based on his account. Smaller and lighter than a hartebeest, with a more horizontal back. Horns not raised on a pedicle; horns slender and lyrate, with long, upright tips. Pelage tawny, often with rufous tones, tending to darken with age, especially in the males; forehead, withers, dorsum, and lower segments of the limbs darker; underparts whitish tawny; inside of the ears and most of the long tail white, the terminal tuft mixed white and black; narrow, white eye-rings, connected by a white chevron across the face. Shoulder height 95–100 cm; horn length of the males 450–600 mm, of the females 350–490 mm; horn span of the males 350–400 mm, of the females 240–330 mm; tip-to-tip distance about the same as the span.

Butynski (2000) gave the known distribution (in about the middle of the 20th century) as between the Tana and Juba rivers; on the Tana River, it extended as far N as Garissa, and in Somalia to the Lag Der region inland from Afmadu. By 1996, the range in Kenya had shrunk to about 42% of its former extent; its fate in Somalia is unknown. In 1963, and again in 1996, a number of herola were introduced into Tsavo East National Park.

Damaliscus Slater & Thomas, 1894

$2n=36$ in *D. lunatus* and *D. jimela*, and $2n=38$ in *D. pygargus* and *D. phillipsi*. The karyotype differs from that of *Beatragus* by the fusion of acrocentric chromosomes 13 and 15, and 20 and 22, and the presence of autosomal chain complexes, one having 5 chromosomes and the other having 12 or 13 (Kumamoto et al., 1996). Robinson et al. (1991) studied the chromosomes of hybrids between *Damaliscus* and *Alcelaphus*; these hybrids were sterile.

In the control region phylogeny of Arctander et al. (1999), the *D. lunatus* group and E African *D. jimela* formed quite separate clades, though within the former, *D. lunatus* and *D. superstes* were not entirely separate (Peter Arctander, pers. comm. to CPG). The separation of *D. pygargus* from the other two clades was very deep.

Damaliscus pygargus Group

$2n=38$; a fusion between chromosomes 18 and 24, and one of the chromosome chains having only 12 members instead of 13, differentiates the karyotype from that of *D. lunatus* and *D. jimela* (Kumamoto et al., 1996).

Very small species, with a broadly white midfacial region and white on the legs.

Damaliscus pygargus Pallas, 1767

bontebok

The reasons why *Antilope dorcas* Pallas, 1766, is to be rejected were given by Rookmaaker (1991) and supported by Grubb (2004a).

Face between the eyes white down to the snout; rump white, extending around the base of tail to the legs and the underside; a light saddle contrasting with the dark brown pelage. Skull length 293–335 mm; horn length 320–356 mm (males), 291–331 mm (females); horn span 222–246 mm (males), 197–227 mm (females).

Known only from the S Cape.

Damaliscus phillipsi Harper, 1939

blesbok

Face between the eyes white; snout white; buttocks white, but not extending to the back of the rump; no light saddle; less extensively white underparts. Skull length 304–328 mm; horn length 335–384 mm (both sexes); horn span 209–268 mm (both sexes).

From the Karroo N to Botswana.

Damaliscus lunatus Group

$2n=36$ (for specimens from Transvaal; hence *D. lunatus*); the X chromosome submetacentric (as analyzed in Kumamoto et al., 1996).

Much larger species; (usually) dark midfacial strip; no white on the legs.

Damaliscus lunatus (Burchell, 1824)

western tsessebe

Smaller of the two species. Horns forming a semilunate profile. Color generally midbrown or pale brown-tan; facial blaze usually dark brown; shoulder- and haunch-patches mostly light gray to warm gray.

Horn length of the males 256–391 mm, of the females 271–318 mm. Tip-to-tip distance 84–415 mm (males), 190–364 (females); horn span 277–448 mm (males), 260–430 mm (females). Greatest skull length 365–429 mm (both sexes); maxillary toothrow 80.7–102.7 mm (both sexes). (Following Cotterill, 2003.)

Whether it is worth dividing this species into two subspecies is moot. On discriminant analysis (Cotterill, 2003:figures 4c and 4d), Botswanan and Zimbabwean samples have distinctly different "centres of gravity."

Zimbabwe, Botswana, the Caprivi Strip of Namibia.

Damaliscus superstes Cotterill, 2003

Bangweulu tsessebe

2003 *Damaliscus superstes* Cotterill. Muku Muku Flats, 12° 21' S, 30° E, SW Bangweulu Flats.

Larger than *D. lunatus*; horns thicker and wider spread, with broader pedicles. Horns growing symmetrically outward, with the tips curving inward, forming a sphere. Color chocolate-brown, darker than *D. lunatus*; facial blaze black; shoulder- and haunch-patches mostly dark gray.

Horn length of the males 351–400 mm, of the females 304–395 mm. Tip-to-tip distance 220–234 mm (males), 200–335 mm (females); horn span 378–446 mm (males), 342–430 mm (females). Greatest skull length 390–435 mm (both sexes); maxillary toothrow 88.9–105.3 mm (both sexes). (Following Cotterill, 2003.)

The two species of tsessebe separate well on discriminant analysis (Cotterill, 2003:figures 4c and 4d).

Bangweulu district, NE Zambia.

Damaliscus korrigum Group

2*n* = 36 in *D. jimela*, differing from that of the *D. pygargus* group by having one of the chain complexes involving 13 chromosomes instead of merely 12; and differing from *D. lunatus* in the X chromosome being acrocentric (Kumamoto et al., 1996).

Since the two sexes seem very alike, we attempted a discrimination between them craniometrically; 86.6% of the original grouped cases (85.4% of the cross-validated grouped cases) were correctly classified (47 males, 35 females). Sexing depends strongly on the diameter of the horn bases, contrasted with the horn span and the face length.

Our taxonomic analyses gave the results described below (see table 57).

Broadly speaking, they divide into a N group (*D. korrigum*, *D. tiang*), characterized by extremely robust horns, and the rest.

The three C African taxa distinguished by Schwarz (1920), using primarily the Frankfurt series, but incorporating a few other specimens from other collections, are not well differentiated. While 90% of the original grouped cases were correctly classified, this is almost entirely dependent on the small sample sizes of *D. korrigum purpurescens* (*N* = 5) and *D. k. lyra* (*N* = 2), although the sample size for *D. k. korrigum* (*N* = 13) is somewhat more substantial; and only 45% of the cross-validated cases were correctly classified. We conclude that no more than a single taxon can be recognized in C Africa. This, in turn, is not distinct from samples from Nigeria and Senegal.

Sudan *D. tiang* is slightly more extreme than W-C African *D. korrigum*; *D. selousi*, from the Guasin-Gishu, averages larger, with somewhat longer and narrower nasals and slightly longer, less spreading horns. *D. jimela* is absolutely distinct from the N taxa in its shorter, less robust horns, with slightly more divergent tips, and it is narrower across the zygomatic arches.

The more S groups, from E Africa, are better distinguished from one another than are the N groups. *D. eurus* differs from *D. jimela* and *D. ugandae*, particularly in its very long, broad nasals. *D. ugandae* is wider across the zygomatic arches, and has longer, more spreading horns than *D. jimela*. The coastal species, *D. topi*, is much smaller and shorter horned, with a narrower span, than the others.

Sexual size difference is greatest in *D. selousi*, then *D. ugandae*, then *D. topi*; both sexes of *D. topi* are much smaller than the others, which are all fairly similar in size. The sexual size difference in the horns is greater in *D. tiang* than in other taxa, although (on the more limited data) there also appears to be a considerable sexual size difference in the horns of *D. topi*.

Among females, the curiously small size of female *D. selousi*, barely larger than *D. topi*, is unexpected. Like the male skulls, the single specimen of a female *D. eurus* skull has long, broad nasals. Table 58 shows that the most approximated horn tips are in *D. jimela*. The N group has very robust horns. Relationships in horns among the females are much like those in the males.

Our division of this species-group into a surprisingly large number of species is given below.

Damaliscus korrigum (Ogilby, 1837)

korrigum

Bright orange-bay. Legs from the knees and the hocks to the hoofs cinnamon; above, body color dark ashy brown, which fades on the shoulders and the haunches into narrow, reddish gray patches, suffused with an ashy sheen. Facial blaze blackish gray. Horns very long and robust, with the tips somewhat converging; skull relatively broad.

A wide distribution, from Senegal E to the Nile. As far as we can detect, there are no striking differences, and certainly nothing that is in any way consistent, among different populations, even in C Africa (see above).

Damaliscus tiang (Heuglin, 1863)

tiang

Color reddish bay, suffused with a reddish purple bloom. Legs bright cinnamon. Shoulder- and haunch-patches ash gray, with a reddish tinge. Facial blaze blackish gray, with a reddish tinge. In skull and horn characters, slightly more extreme than *D. korrigum*, with somewhat more sexual dimorphism.

Apparently E of the Nile, extending into W Ethiopia. *D. tiang* is undoubtedly closely related to *D. korrigum*; we have only a limited number of specimens from E Sudan, and it is possible that further material may show that *D. tiang* is just an extreme clinal variant of *D. korrigum*.

Damaliscus selousi Lydekker, 1907

Guasin-Gishu topi

Males averaging larger than *D. korrigum* and *D. tiang*; nasals longer, narrower; horns slightly shorter, less spreading; sexual dimorphism the greatest in the *D. korrigum* group, the females being unexpectedly small in size. A *pygargus*-like white facial blaze has been recorded in this species.

Table 57 Univariate statistics for the skulls of taxa in the *Damaliscus korrigum* group

	Cb l	Gt l	Preorb	Pal l	Zyg br	Nas l	Nas br	Teeth
Males								
korrigum								
Mean	391.42	402.42	266.00	221.00	132.09	166.15	36.04	93.80
N	33	36	35	29	37	37	37	37
Std dev	9.788	10.979	8.971	6.308	4.870	9.896	2.593	3.890
Min	374	383	248	210	120	144	31	84
Max	410	431	293	233	143	188	43	100
tiang								
Mean	391.25	403.00	267.75	221.50	132.25	169.25	33.75	100.25
N	4	4	4	4	4	4	4	4
Std dev	8.302	6.377	3.096	3.109	.957	6.752	2.500	6.131
Min	382	398	265	219	131	163	31	94
Max	402	412	272	226	133	178	37	108
selousi								
Mean	391.00	401.80	267.40	218.14	133.42	169.00	31.64	95.09
N	9	5	5	7	12	10	11	11
Std dev	9.539	12.458	10.213	7.426	4.188	11.116	3.529	5.029
Min	373	384	255	208	128	152	26	87
Max	400	414	283	230	140	184	38	102
jimela								
Mean	383.32	393.00	261.33	214.28	128.00	169.95	36.32	92.68
N	19	9	9	18	19	19	19	19
Std dev	13.051	10.283	8.337	7.315	6.307	8.141	2.810	4.498
Min	363	376	246	202	118	153	32	86
Max	403	410	274	227	143	182	42	104
ugandae								
Mean	395.63	409.94	273.44	226.13	136.67	177.88	36.13	96.67
N	16	17	16	15	18	16	16	18
Std dev	8.763	8.467	8.718	17.402	7.647	9.715	3.793	7.268
Min	374	392	258	207	128	161	30	83
Max	407	420	284	284	156	196	45	109
eurus								
Mean	397.00	415.33	277.33	228.50	133.67	191.67	38.00	97.67
N	3	3	3	2	3	3	3	3
Std dev	17.349	19.553	12.583	7.778	2.309	9.292	1.732	5.686
Min	377	395	264	223	131	181	37	93
Max	408	434	289	234	135	198	40	104
topi								
Mean	370.20	377.00	249.60	208.20	125.80	155.20	31.80	90.00
N	5	5	5	5	5	5	5	5
Std dev	12.029	12.708	9.529	7.563	1.304	8.228	1.304	4.950
Min	355	365	239	199	124	144	30	83
Max	386	393	261	216	127	167	33	96
Females								
korrigum								
Mean	381.71	394.48	262.95	217.47	130.80	158.90	34.65	94.14
N	17	21	21	19	20	20	20	21
Std dev	9.904	12.852	11.373	7.268	4.753	15.307	2.033	5.859
Min	370	378	249	207	124	127	32	78
Max	405	426	290	231	139	202	39	109
tiang								
Mean	382.67	391.67	259.33	212.67	130.00	156.67	34.00	91.00
N	3	3	3	3	3	3	3	3
Std dev	18.448	20.817	13.650	12.503	7.000	11.060	4.583	1.732
Min	367	375	247	200	125	145	29	90
Max	403	415	274	225	138	167	38	93

Table 57 (continued)

	Cb l	Gt l	Preorb	Pal l	Zyg br	Nas l	Nas br	Teeth
selousi								
Mean	362.50	—	—	205.00	123.00	147.00	29.50	91.50
N	2	—	—	2	2	2	2	2
Min	360	—	—	205	122	147	29	90
Max	365	—	—	205	124	147	30	93
jimela								
Mean	374.86	399.33	270.17	211.00	125.67	163.89	33.11	93.44
N	7	6	6	7	9	9	9	9
Std dev	21.836	12.044	12.703	8.406	4.272	15.439	2.934	6.579
Min	356	383	251	202	120	145	29	84
Max	420	417	288	225	134	198	38	104
ugandae								
Mean	378.60	394.82	263.50	216.13	130.09	164.82	33.40	91.82
N	10	11	10	8	11	11	10	11
Std dev	10.047	15.329	14.246	6.770	7.162	10.989	4.300	5.326
Min	361	367	242	209	119	147	28	80
Max	401	420	286	228	146	181	40	98
eurus								
Mean	383.00	402.00	272.00	215.00	127.00	175.00	35.00	94.00
N	1	1	1	1	1	1	1	1
topi								
Mean	355.50	365.50	240.50	201.00	122.50	150.25	30.00	93.50
N	4	4	4	4	4	4	4	4
Std dev	4.796	7.047	7.000	6.055	3.000	9.179	.000	5.260
Min	351	359	233	194	119	140	30	86
Max	361	373	249	207	125	161	30	98

Known only from the Guasin-Gishu Plateau. The species may be extinct.

Damaliscus ugandae Blaine, 1914

Uganda topi

Color much darker, maroon suffused with an ashy sheen. Legs deeper cinnamon below the knees and the hocks; upper halves of the legs blue-black. Shoulder- and haunch-patches larger in area, steel-gray. Facial blaze blue-black, but occasional *pygargus*-like white facial blazes have been recorded.

Relatively smaller than *D. korrigum* and *D. tiang* but, like them, comparatively broad-skulled, with relatively long, spreading horns; horns of *D. ugandae*, however, considerably less robust. Sexual dimorphism relatively great, second only to *D. selousi*.

Ankole district of SW Uganda, extending to the Rutshuru Plains over the border in the DRC and to Akagera National Park in Uganda.

Damaliscus eurus Blaine, 1914

Ruaha topi

Size and color as in *D. ugandae*, but becoming lighter; bright reddish bay in the posterior dorsal region. In some respects intermediate between *D. jimela* and *D. ugandae* in skull characters, but with very long, broad nasals.

S Tanzania, in the Lake Rukwa / upper Ruaha region. It is unknown if there are any populations between this area and the Serengeti ecosystem, where *D. jimela* occurs.

Damaliscus jimela (Matschie, 1892)

Serengeti topi

Color not unlike *D. topi* (below), but shoulder-patches larger, and shoulder- and haunch-patches more clearly set off from the body color. Horns shorter, less robust than other *Damaliscus* species, with the tips not very divergent; skull very long, narrow.

Restricted to the Serengeti-Mara ecosystem.

Damaliscus topi Blaine, 1914

coastal topi

Color darker and richer than in *D. tiang*, heavily suffused with a mauve bloom, becoming lighter on the belly. Facial blaze blackish gray, with a reddish tinge, and sprinkled with white hairs. Much smaller and shorter-horned, with a narrower span, than other

Table 58 Univariate statistics for horn measurements for northern taxa of *Damaliscus*

	Span	Tip–tip	Diam	Horn l str
Males				
korrigum				
Mean	314.46	234.83	68.01	485.50
N	35	35	36	34
Std dev	29.955	50.275	5.279	46.300
Min	242	106	57	377
Max	375	329	84	553
tiang				
Mean	331.50	200.25	68.20	525.20
N	4	4	5	5
Std dev	51.733	49.742	12.398	56.962
Min	271	137	51	468
Max	389	257	78	600
selousi				
Mean	254.14	190.29	61.71	406.71
N	7	7	7	7
Std dev	16.906	30.707	6.047	34.067
Min	217	145	53	383
Max	266	228	69	478
jimela				
Mean	234.33	154.33	57.67	336.89
N	9	9	9	9
Std dev	10.296	25.199	3.640	19.310
Min	217	116	51	305
Max	254	185	63	377
ugandae				
Mean	275.25	190.83	62.92	377.23
N	12	12	13	13
Std dev	13.752	21.114	6.103	23.682
Min	245	149	49	345
Max	302	220	71	421
eurus				
Mean	258.67	194.00	62.33	329.67
N	3	3	3	3
Std dev	20.133	21.284	1.528	32.146
Min	240	175	61	293
Max	280	217	64	353
topi				
Mean	224.40	179.60	55.20	405.00
N	5	5	5	5
Std dev	16.742	24.846	5.718	15.379
Min	199	142	46	379
Max	239	203	61	419
Females				
korrigum				
Mean	278.00	181.19	56.00	453.67
N	21	21	21	21
Std dev	24.187	29.223	2.933	26.949
Min	207	125	51	402
Max	334	232	63	514
tiang				
Mean	257.33	177.00	59.00	436.00
N	3	3	3	3
Std dev	49.943	110.937	9.849	46.228
Min	228	94	48	383
Max	315	303	67	468
jimela				
Mean	230.00	148.40	53.17	348.17
N	5	5	6	6
Std dev	14.816	33.329	5.345	19.094
Min	218	111	50	322
Max	251	191	64	367
ugandae				
Mean	263.40	184.70	53.70	363.90
N	10	10	10	10
Std dev	13.159	20.870	4.945	21.037
Min	236	141	49	330
Max	284	210	67	389
eurus				
Mean	215.00	122.00	51.00	348.00
N	1	1	1	1
topi				
Mean	209.25	134.25	46.50	390.00
N	4	4	4	4
Std dev	7.136	18.464	1.915	11.916
Min	201	109	44	374
Max	218	153	48	402

species of the *D. korrigum* group, but sexual dimorphism greater than the other E African species, especially in horn size.

Coastal E Africa, from the Juba region of Somalia S at least to Malindi. The range is isolated from that of other taxa.

Connochaetes Lichtenstein, 1814

wildebeest, gnu

$2n = 58$; the X chromosome is a large acrocentric, the Y chromosome a small acrocentric. On the basis of mtDNA, a divergence date between *C. taurinus* and *C. gnou* of somewhat over 1 Ma was calculated, assuming a divergence rate of 2% per million years (S. W. Corbet & Robinson, 1991).

Quite different in appearance and posture from other genera of the tribe: the head is large, held low, with a convex nose on which there is a brush of short, stiff hair, and huge valvular nostrils. The horns are thickened and expanded at the base, and oriented either sideways or forward, curved downward and then up, not upright like other genera.

Usually, two species are recognized, *C. gnou* and *C. taurinus*, but within the latter there are actually four sharply distinct species: *C. taurinus*, *C. johnstoni*, *C. albojubatus*, and *C. mearnsi*. Interestingly, *C. mearnsi*, like *Damaliscus jimela*, is apparently endemic to the Serengeti-Mara ecosystem.

We thus recognize five species altogether.

Connochaetes gnou (Zimmermann, 1777)

white-tailed wildebeest, black wildebeest

Small; black; white tail; forward-swung horns.

Formerly from the Karroo N to the Orange and Vaal rivers; now preserved in only a few farms and reserves.

Connochaetes taurinus Group

Arctander et al. (1999) found almost complete separation between the four species in the *C. taurinus* group. The authors also had two samples from the Luangwa Valley (the taxon described as *C. cooksoni*), of which one clustered with *C. mearnsi*, the other with *C. johnstoni*. Two *C. mearnsi* (out of a large sample) clustered with *C. albojubatus*, and one *C. taurinus* clustered with *C. mearnsi*.

Table 59 gives univariate statistics for wildebeest of the whole *C. taurinus* group (we did not especially consider *C. gnou*).

The smallest species in skull length is *C. mearnsi*, but the skull is hardly narrower than in the other species of *Connochaetes*. The longest nasals are in *C. johnstoni*; the other species are all about the same. The two S species (*C. taurinus* and *C. johnstoni*) have much wider spans than the two N ones (*C. albojubatus* and *C. mearnsi*), and *C. mearnsi* has an extremely narrow span, although the distance between the tips is not much different (i.e., the tips do not turn inward as much).

In an effort to quantify the relative degree of downturn in the horns, we sat each skull on a table and measured, where possible—which was only in a relatively few individuals—the distance from the bottom of the curve of one horn to the table top. This averages only 45 mm in *C. taurinus* ($N=10$) and 50 mm in *C. johnstoni* ($N=3$), but 69 mm in *C. albojubatus* and *C. mearnsi* ($N=5$, in total). In *C. taurinus*, the horns can actually sweep down to touch the table top; in *C. johnstoni*, the minimum distance is 35 mm; and in the other two, the minimum is 58 mm. Size and shape relationships among females are very much as they are in the males, but *C. mearnsi* females have exceptionally short nasals. The span is least in the single female specimen of *C. johnstoni*.

In the two S species, the beard is black; in the two N ones, it tends to be creamy white. All have a pattern of thin, dark stripes on the neck, fading out along the flanks. The color is some shade of gray (generally darkish blue-gray in *C. taurinus*), with the face black (the forehead brownish white in the juveniles).

According to Roosevelt and Heller (1914), *C. mearnsi* "differs from *albojubatus* by the decidedly darker legs, which are olive brown or sepia in old males, and somewhat lighter in females. Body tends to become darker or quite blackish on chest, shoulders, and sides."

For Estes (1969), *C. mearnsi* (which he called "*C. hecki*") is the only geographic form that really stands out: smaller, darker, with a longer tail; horns with a more pronounced boss and a shorter spread; territorial males more vociferous, "with a call of a different, less metallic timbre."

In *C. albojubatus*, and *C. mearnsi*, the mane along the nape hangs limp, whereas it is erect in *C. taurinus*. Estes had not, at the time, studied any populations of *C. johnstoni*. From photos (and as later specified by Estes, 1991), the mane in *C. johnstoni* stands up like that of *C. taurinus*, and it is lighter gray in color, with a brownish tinge, the legs are light-colored, the tail is short. A white crescent across the face below the eyes often occurs in *C. johnstoni*; indeed, this was the chief distinguishing feature mentioned in the type description, but its presence is inconstant.

Connochaetes taurinus (Burchell, 1824)

brindled wildebeest, blue wildebeest

1893 *Connochaetes reichei* Noack.

1925 *Connochaetes mattosi* Blaine.

1933 *Connochaetes borlei* Monard.

N South Africa, presumably S Mozambique, Zimbabwe, Botswana, W Zambia.

Connochaetes johnstoni Sclater, 1896

Johnston's wildebeest

1911 *Connochaetes rufijianus* de Beaux.

1914 *Connochaetes cooksoni* Blaine.

E of the Luangwa River, Zambia; Malawi, N Mozambique, SE Tanzania.

Connochaetes albojubatus Thomas, 1892

eastern white-bearded wildebeest

1905 *Connochaetes hecki* Neumann.

1919 *Connochaetes babaulti* Kollman.

The supposed taxon *C. babaulti* was described on the basis of an unusual individual with a black beard.

Athi Plains in S Kenya, S to Tanzania, and W to the N shores of Lake Tanganyika.

Table 59 Univariate statistics for the *Connochaetes taurinus* group

	Gt l	Gt br	Nas l	Teeth	Span	Tip–tip
Males						
taurinus						
Mean	475.93	182.76	216.66	108.81	641.15	381.51
N	41	42	41	40	41	35
Std dev	22.517	7.556	18.920	6.706	53.860	56.936
Min	433	163	186	92	519	270
Max	524	199	291	123	740	478
johnstoni						
Mean	498.73	184.73	234.55	112.44	658.73	387.30
N	11	11	11	9	11	10
Std dev	16.402	4.077	14.201	5.981	26.127	60.988
Min	470	176	215	104	628	281
Max	520	191	252	124	718	459
albojubatus						
Mean	467.40	182.44	209.13	105.67	594.50	355.78
N	10	9	8	9	10	9
Std dev	13.786	10.806	12.722	4.123	61.562	64.591
Min	442	159	193	100	488	294
Max	493	194	234	112	670	476
mearnsi						
Mean	449.22	179.44	212.17	103.00	543.78	342.22
N	9	9	6	9	9	9
Std dev	10.779	8.457	8.134	3.464	28.146	43.746
Min	430	160	201	95	502	283
Max	458	189	222	106	600	443
Females						
taurinus						
Mean	467.77	171.77	211.62	102.58	493.38	275.25
N	13	13	13	13	13	8
Std dev	15.216	4.323	13.131	27.948	61.630	60.363
Min	445	164	190	11	333	215
Max	490	182	231	120	588	379
johnstoni						
Mean	480.00	176.00	218.00	103.00	418.00	175.00
N	1	1	1	1	1	1
albojubatus						
Mean	456.00	167.00	215.00	98.00	439.00	369.00
N	1	1	1	1	1	1
mearnsi						
Mean	421.20	162.00	185.00	99.40	432.20	267.40
N	5	5	4	5	5	5
Std dev	13.627	5.701	8.287	6.950	20.548	29.821
Min	406	154	179	90	414	225
Max	442	169	197	105	462	299

Connochaetes mearnsi (Heller, 1913)
Serengeti white-bearded wildebeest

1913 *Connochaetus albojubatus henrici* Zukowsky.
1913 *Connochaetus albojubatus lorenzi* Zukowsky.
1913 *Connochaetus albojubatus schulzi* Zukowsky.

The putative taxon *C. henrici* was from the Serengeti; *C. lorenzi* and *C. schulzi* were based on captive herds, both from Ngorongoro Crater, supposedly from the N and S parts of it, respectively(!), though there does not seem to be any real evidence that this was so.

Tribe Caprini Gray, 1821

Characterized by the short and anteroposteriorly compressed metapodials, the (almost always) horned females, the absence of lateral prongs on the nasal tips, the fusion on the posterior premolar of the metaconid and the entoconid, the strong mesostyles on the upper molars, and the presence of small additional cavities on the lingual side of the upper molars. (Following A. W. Gentry, 1992.)

The Caprini is now taken to include *Pantholops*, as well as those genera formerly included in a tribe of their own, the Rupicaprini, which, in fact, have little in common with each other except in lacking the cranial specializations of the "central" genera—*Ovis*, *Capra*, and their relatives.

In *Ovis*, *Capra*, and related genera, the morphology of the skull is correlated with that of the horns, which, in turn, is derived from the different fighting styles of the males, according to Schaffer & Reed (1972). The extension of the sinuses is likewise: extensive throughout the frontals, with complex septa, in *Ovis*, *Hemitragus*, *Ammotragus*, and *Pseudois*, but much less so in *Capra*, except that the sinuses extend throughout the horns, but lack complex septa, in *C. aegagrus*.

On the basis of chromosome morphology, Bunch & Nadler (1980) postulated the following phylogeny of those caprines studied by them:

> (*Capra* (*Pseudois* (*Hemitragus* (*Ammotragus* (*Ovis*))))).

Hartl et al. (1990) studied proteins of the Caprini and the Rupicaprini simultaneously, discovering that relationships among the genera of these two putative tribes were more complicated than had hitherto been supposed. The authors generated an interesting cladogram (their figure 4), which showed the following branching pattern:

> (*Ovis* ((*Rupicapra*, *Oreamnos*) (*Hemitragus* (*Ammotragus* (*Capra*))))).

P. Groves & Shields (1996) sequenced the mitochondrial cytochrome *b* gene of the Caprini, with particular reference to *Ovibos* and *Budorcas*, with a view to testing whether these two genera, in particular, are related. Though different methods of analysis gave somewhat different trees, a consensus of six trees, accepting as significant nodes only those with 67% or more support, was as follows:

> (*Saiga* (*Nemorhaedus*, *Capricornis*, *Ovibos*, *Oreamnos*) (((*Hemitragus*, *Capra*) (*Ovis aries* (*Budorcas* (*Ovis canadensis* group)))).

Much the same type of arrangement—with *Budorcas* as sister to *Ovis*, and *Hemitragus* to *Capra*—was found by Hassanin et al. (1998), but in this case *Oreamnos* also formed part of the *Ovis* clade. In addition, these authors studied *Ammotragus* and *Pseudois*, which formed part of the *Capra* clade.

A study of a few taxa in the Caprini by chromosome painting (Huang et al., 2005) confirmed that serow (*Capricornis*) and goral (*Nemorhaedus*) form a group when compared with goats and sheep, and these authors traced the chromosome changes that have occurred in the evolution of members of the serow/goral group.

Lalueza-Fox et al. (2005), in the course of a study on the extraction of mitochondrial nuclear sequences from the remains of *Myotragus*, generated an entire phylogeny of the Caprini, as follows (adopting their Bayesian tree for the concatenated mitochondrial segments):

> ((((*Capra*, *Hemitragus*), *Pseudois*) *Ammotragus* (*Ovis*, *Myotragus*) *Oreamnos* (*Rupicapra* (*Capricornis*, *Ovibos*) *Budorcas*) *Pantholops*)).

In other words, there is a virtual five-way split subsequent to the separation of *Pantholops*. But the authors went on from this to propose—using the 5.35 Ma separation of *Myotragus* from *Ovis* (based on geological data)—a general dating of the well-supported nodes on the tree: the *Ovis canadensis* group from other *Ovis* at 2.2 ±0.3 Ma; *Rupicapra pyrenaica* from *R. rupicapra* at 1.6 ±0.3 Ma; *Capra ibex* from *C. pyrenaica* at 0.6 ±0.1 Ma, and these from other *Capra* (except for *C. sibirica*, which separated much earlier) at 1.5 ±0.2 Ma; and, finally, *Pantholops* from all other Caprini at 6.2 ±0.4 Ma.

Ropiquet & Hassanin (2005) showed that the genus *Hemitragus* is polyphyletic; they restricted the genus to the type species, *Hemitragus jemlahicus*, and erected two new genera (*Nilgiritragus* and *Arabitragus*) to contain the other two species. They proposed that the Caprini had begun to diversify between 8.7 and 11.9 Ma, that is, much earlier than the date proposed by Lalueza-Fox et al. (2005); the next split was between *Rupicapra* and the rest, at between 7.5 and 10.4 Ma; followed by a division into two clades: *Ovis* plus *Nilgiritragus* (they did not study *Myotragus*), and *Capra* plus all remaining genera (*Budorcas*, *Oreamnos*, *Ovibos*, *Ammotragus*, *Arabitragus*, *Pseudois*, *Hemitragus*, and *Capra*). Later, the same authors (Ropiquet & Hassanin, 2006), in the context of demonstrating the hybrid origin of *Capra*, proposed some new dates, differing only slightly from the previous ones.

Finally, Ropiquet (2006) gave an updated version of this, recognizing the initial split (after the separation of *Pantholops*) as an undifferentiated comb with six branches: *Oreamnos*, *Budorcas*, (*Ammotragus*, *Arabitragus*, *Hemitragus*, *Pseudois*, *Capra*), (*Nilgiritragus*, *Ovis*), *Rupicapra*, and (*Nemorhaedus*, *Capricornis*, *Ovibos*). The six-way split is dated at about 8.5 Ma, and the split from *Pantholops* at about 9.5 Ma. Diversification within each of the three multigeneric branches of the six-way split began in the Pliocene, mostly at the beginning of this period.

Pantholops Hodgson, 1834

A. W. Gentry (1992) was the first to definitively separate this genus from *Saiga*, with which it had been traditionally placed, and include it in the Caprini. He pointed out that *Pantholops* shares various features with this tribe, in particular, the long, mediolaterally compressed, keeled horn cores; the frontal sinuses; the absence of lateral prongs on the nasal tips; the zygomatic arch expanded below the orbit; and the weak distal flanges of the metatarsals. His demonstration that *Pantholops* is a caprin has since been corroborated by Gatesy et al. (1997), followed by Lei, Jiang, et al. (2003), Ropiquet & Hassanin (2005), and all other molecular studies that included this genus. *Pantholops* is, nonetheless, in an isolated position; thus it is sister to the rest of the Caprini.

Pantholops hodgsonii (Abel, 1826)

chiru

The available information on this monotypic genus has been summarized by Leslie & Schaller (2008).

Oreamnos Rafinesque, 1817

$2n = 42$, with 18 metacentrics and 22 acrocentrics, quite different from the high numbers (with few or even no acrocentrics) of other "rupicaprins" (Soma & Kada, 1986).

The characters of this genus were described in detail by Dolan (1963).

Oreamnos americanus (de Blainville, 1816)

Rocky Mountain goat

A full description of this species—its general biology, distribution, behavior, ecology, and genetics—will be found in Rideout & Hoffmann (1975).

Dolan (1963) discussed the validity of the described subspecies, recognizing only *O. a. kennedyi* as different from *O. a. americanus*, distinguished mainly by its much larger size, constricted forehead in front of the horn bases, and perhaps more outwardly curved horns. We take no position on the existence (or otherwise) of subspecies, but note that further morphological study, as well as DNA study, is needed.

Budorcas Hodgson, 1850

takin

Though it has generally been linked with *Ovibos* into a common subfamily or tribe, *Budorcas* is more closely related to *Ovis*, while *Ovibos* is sister to *Nemorhaedus* (P. Groves & Shields, 1997).

Our descriptions and analysis below are based on observations not only of museum specimens, but of living, known-age animals, especially in the Shanghai Zoo breeding farm and the Tierpark in Berlin. There are striking age changes in both pelage and horns; these are far more marked in *B. taxicolor* and *B. tibetanus* than in the other two taxa.

Table 60 gives the univariate statistics for samples of *Budorcas*. Skull and horn characters appear to offer very little differentiation; perhaps the main varying character concerns the nasals, which are, on average, shorter and less arched ("Pal nas ht," i.e., the height of the most convex part of the nasals above the palate) in *B. bedfordi* than in others.

Li M. et al. (2003) studied the molecular phylogeny of the genus, using portions of the control region and cytochrome *b*. They had one sample of *B. taxicolor* from Gongshan; nine of *B. tibetanus* from Sichuan and four from Bailongjiang in Gansu; and 26 of *B. bedfordi* from Shaanxi. The three taxa separated well (with *B. taxicolor* as sister to the other two), except that the Gansu sample of *B. tibetanus*, curiously, fell into the *B. bedfordi* clade; but, as the authors pointed out, Bailongjiang is actually on the W edge of the Qinling Range, and takin from there have, at times in the past, been classified as *B. bedfordi* (although the authors gave no phenotype data for the specimens which yielded the samples).

Based on our morphological research, backed up by the DNA study of Li M. et al. (2003), we classify the genus into four species.

Budorcas taxicolor Hodgson, 1850

Mishmi takin

Juveniles, up to about 6 months of age, light brown, with an extensive black zone on the haunches, the belly, the legs, the lower neck, and the head, especially the face. Black extends with age to the whole of the face, under the neck, and to the lower flanks; mixed, brindled zone between these areas and the lighter (pale yellow) tone of the upper parts. Skulls of the males much larger than those of the females; horns also sexually dimorphic.

Table 60 Univariate statistics for *Budorcas*

	Gt l	Biorb	Span	Tip–tip	Teeth	Braincase	Nas l	Nas br	Pal nas ht
Males	—			—	—	—	—	—	—
bedfordi									
Mean	414.33	181.83	384.67	295.67	114.00	93.50	131.50	54.00	118.50
N	3	3	3	3	2	2	2	1	2
Std dev	25.027	11.751	71.501	35.642	—	—	—	—	—
Min	390	170	303	256	112	88	117	—	114
Max	440	194	436	325	116	99	146	—	123
Burma									
Mean	414.00	198.00	409.50	326.00	111.00	102.00	159.00	—	139.00
N	1	1	2	2	1	1	1	—	1
Min	414	198	393	302	111	102	159	—	139
Max	414	198	426	350	111	102	159	—	139
taxicolor									
Mean	394.50	190.17	371.80	247.40	118.00	100.20	155.17	75.50	137.50
N	2	6	10	10	5	5	6	6	6
Std dev	—	11.197	24.535	53.066	3.391	4.087	6.616	7.662	7.342
Min	384	168	341	160	114	93	145	62	126
Max	405	198	411	318	122	103	162	85	146
tibetanus									
Mean	428.17	192.63	406.09	299.45	119.20	95.80	145.25	68.33	140.80
N	12	16	11	11	5	5	4	3	5
Std dev	18.235	11.087	42.801	47.744	6.017	3.564	11.615	15.885	13.953
Min	392	170	298	226	109	92	135	50	121
Max	448	209	459	366	124	101	160	78	156
whitei									
Mean	420.00	176.19	319.56	247.17	120.16	103.80	145.90	67.00	123.00
N	2	7	8	6	5	5	5	2	3
Std dev	—	11.745	22.579	24.815	3.777	13.554	8.706	—	14.177
Min	413	161	297	202	114	97	132	52	112
Max	427	197	357	275	124	128	156	82	139
Females									
bedfordi									
Mean	394.40	166.80	301.75	204.00	119.67	92.00	130.00	57.00	115.00
N	5	5	4	4	3	3	3	3	2
Std dev	2.881	4.087	5.377	19.026	2.887	1.732	2.646	.000	—
Min	391	162	295	185	118	90	128	57	115
Max	398	172	307	230	123	93	133	57	115
Gongshan									
Mean	376.50	168.00	279.00	199.50	109.00	87.00	129.50	—	159.00
N	2	2	2	2	2	2	2	—	2
Min	375	166	276	192	105	86	128	—	122
Max	378	170	282	207	113	88	131	—	196
taxicolor									
Mean	—	176.00	281.33	215.33	117.00	87.00	137.00	58.00	121.00
N	—	1	3	3	1	1	1	1	1
Std dev	—	—	8.505	2.517	—	—	—	—	—
Min	—	—	275	213	—	—	—	—	—
Max	—	—	291	218	—	—	—	—	—
tibetanus									
Mean	394.00	162.40	287.00	201.75	119.00	90.50	125.00		107.00
N	3	5	4	4	1	2	1	—	1
Std dev	3.606	6.950	47.117	23.200	—	—	—	—	—
Min	391	153	218	175	—	89	—	—	—
Max	398	168	324	231	—	92	—	—	—
whitei									
Mean	408.00	172.40	280.75	149.00	120.80	—	140.75	—	—
N	1	2	2	1	1	—	2	—	—
Min	—	168	278	—	—	—	140	—	—
Max	—	177	284	—	—	—	142	—	—

Mishmi Hills, northernmost Burma, to Tengyueh in Yunnan. Specimens from Gongshan are considerably smaller, but otherwise fit into this taxon.

Budorcas whitei Lydekker, 1907

Bhutan takin

Juveniles and adults both similar to *B. taxicolor* juveniles in pelage; horns seemingly remain small, with a narrow span, through life; males small in size; teeth relatively large.

Bhutan and SE Tibet.

Budorcas tibetanus Milne Edwards, 1874

Sichuan takin

Golden-brown when young, the dark areas much less extensive than in *B. taxicolor*, and dark brown rather than black. Color continues to darken with age; dark golden-brown by the fourth year, with a conspicuous pale saddle, the dark brown zones extending onto the lower flanks, the lower part of the legs, and between the eyes; neck mane present in the adults. Black on the nose becoming extremely conspicuous with age. Aged females more contrasty than the males, the light areas becoming pale yellow-gray and the dark areas black. Colors somewhat darker in winter, browner, with the dark areas more blackish, than in summer. Sexual dimorphism obvious in the skull and the horns.

Sichuan.

Budorcas bedfordi Thomas, 1911

golden takin

Young animals light golden-brown, with dark zones only on the haunches; face wholly light-toned. With age, color a somewhat more red-gold tone, developing a paler golden saddle, but the face, in particular, never becoming dark. Nasal region less inflated than in other samples of *Budorcas*, and less sexually dimorphic.

Qinling Range.

Ammotragus Blyth, 1840

Barbary sheep

$2n = 58$.

Braincase descending steeply behind the horn. Tympanic tube long, pointing laterally; external auditory meatus relatively small, fairly circular. Frontals somewhat convex. Orbital margin merging with the zygoma. Nasofrontal suture broad, often concave backward. Anterior nasals long, with a trace of the lateral processes. Frontolacrimal suture angulated, as in *Arabitragus*. Ethmoid fissure present. Nasal branch of the premaxilla narrow, but approaching the nasal.

Ammotragus lervia Pallas, 1777

This species (if it is a single species) appears to be restricted to mountain ranges in the Sahara and the Maghreb, a distribution which would suggest that there should be strong geographic differentiation; but we have little knowledge of the described species and/or subspecies, and we know only that it requires investigation in the future.

Arabitragus Ropiquet & Hassanin, 2005

Arabian tahr

$2n = 58$.

Braincase behind the horn very short, descending at about a 45° angle. Tympanic tube very long, pointing laterally, as in *Ammotragus*; external auditory meatus a very large, vertical oval. Frontal region very convex. Orbital rim merging with the zygoma. Nasofrontal suture very broad, often slightly concave anteriorly. Nasal bones relatively short, with very small lateral processes. Frontolacrimal suture running medially, then angled, then running forward. Ethmoid fissure present, as in *Ammotragus*. Nasal branch of the premaxilla very narrow, widely separated from the nasal.

The Arabian tahr is sister to *Ammotragus*, not to other species of tahr, as was previously assumed (Ropiquet and Hassanin, 2005).

Arabitragus jayakari (Thomas, 1894)

1894 *Hemitragus jayakari* Thomas. Jebel Taw, Oman.

Diminutive species; skull length 201–219 mm; horns longer, more slender, and less ridged than in *Hemitragus*, but much shorter than in *Ammotragus*. Hair long on the flanks and the nape in the males; nape hair developing in the males into a black-tipped dorsal crest, this crest pale in the females. Hair much more shaggy in winter, forming ruffs on the legs. Light sandy brown; underside buffy whitish. Midline of the face broadly black; black stripe from the eye to the mouth; pale brown stripe in between.

Mountain ranges of E Oman, barely extending into the United Arab Emirates.

Hemitragus Hodgson, 1841

Himalayan tahr

$2n = 48$.

Braincase behind the horn descending steeply, almost vertically. Frontals slightly convex; zygoma separated from the orbit by a sharp orbital rim, with a trough below. Frontolacrimal suture forming a simple curve. No ethmoid fissure. Nasal branch of the premaxilla very broad, forming a broad suture with the nasal. Underside of the tail naked, like a goat.

Hemitragus jemlahicus **(Hamilton Smith, 1827)**
Large; heavily built; very sexually dimorphic. Horns compressed, nearly in contact at the base and then diverging, with the anterointernal edge raised into a sharp keel. Body hair long, shaggy, especially on the forequarters in the males, but short on the head. Color reddish or dark brown, darkening in old males; mane of old males becoming whiter.

Himalayas, from Kashmir E to Sikkim.

Pseudois Hodgson, 1846

blue sheep

$2n = 54$.

Taxonomy follows Groves (1978a).

Pseudois nayaur **(Hodgson, 1833)**

greater blue sheep

Large size; horns initially with backward-pointing tips, then, from about 10 years of age in adult males, curving around in a semicircle; long fur, with much underwool in winter; winter coat brown-gray, with a bluish cast; dark brown face-mask; black stripes down the foreleg and the hind shanks, sometimes down the whole hindleg, and along the flanks; underside white, occasionally golden-brown.

Skull length of the males in different populations averaging 252–266 mm; horn span in different populations averaging 596–734 mm; basal diameter of the horn core >80 mm, usually >90 mm.

Plateau country of the S and E rims of the Tibetan Plateau, from E Kashmir—via Nepal, Bhutan, the Himalayan regions of India, and S Tibet—to Yunnan, Sichuan, and Gansu.

Pseudois schaeferi **Haltenorth, 1963**

dwarf blue sheep

Males only half the weight of *P. nayaur*, the females less reduced in size; horns more upright, with simpler, upturned tips, not forming semicircles; shorter fur, with less underwool; drabby in winter, with a silvery sheen; dark markings poorly developed, except in old males; limbs dark overall; agouti bands on the hairs short, soon wearing off in the winter coat; underside less white.

Skull length of the males averaging 236.3 mm (maximum 247 mm); horn span averaging 543 mm (maximum 566 mm); basal diameter of the horn core <80 mm.

Arid, grassy slopes of the upper Yangtze Gorge, below the forest zone, from a little N of Batang and extending for an unknown distance to the S.

Habitats of plateau and dwarf blue sheep are separated by a belt of subtropical forest, at about 1000 m, and this almost (or completely) isolates the two populations. Sequences of the mitochondrial control region in four specimens of each showed an absolute difference of more than 12%; if a rate of divergence of 6%–12% per million years is followed, then they separated 1–2 Ma (Feng et al., 2001). The same authors found that, in contrast, ZFY intron sequences were not completely separate, which suggests that the occasional male greater blue sheep may make his way through the forest belt and then, dominating dwarf blue sheep rams, mate with the dwarf ewes.

Capra Linnaeus, 1758

goats

The species of the genus *Capra* divide easily into three phenetic groups: true goats, markhor, and ibex. They may be distinguished as follows:

1. True Goats—The horns are sickle-shaped, with a plane to heteronymous spiral, slender and bilaterally compressed, with a sharp, irregular keel along the front edge, tending to break off between the annual growth rings; at its insertion, the keel extends downward, below the plane of the horn base; the cores are smooth, with a strong anteromedian keel, and compressed, the transverse diameter of the core being about 70% of the anteroposterior diameter. The facial profile is strongly concave overall; the base of the nasals and the frontals rises up to form a sharp "forehead" anterior to the orbits, with a convexity at the base of the horns, so that the upper rims of the orbits do not interrupt the dorsal profile, and the upper margin of the rostrum is concave. The skull is relatively narrow; biorbital breadth is about 50% of the skull length. The nasals are short (about 31% of the skull length) and expanded posteriorly (their posterior breadth is about 44% of their length), forming a triangular suture with the frontals; their tips, in lateral view, are horizontal. The ethmoid fissure remains wide throughout life. The premaxilla forms only a short (about 10 mm) suture with the nasal; the premaxillary tips are not (or hardly) expanded. The braincase is very long and horizontal; the supramastoid crest is not

flared. The toothrows are gently outbowed. The median palatine notch tends to be narrow, extending forward to about halfway along M^3, about the same distance as the lateral notches. The basicranium is narrow; the pterygoid plates are straight and vertical (as also in *Capra nubiana*); the paroccopital processes are fairly long, reaching to the level of the lower margin of the condyles. Externally, there is (in wild forms) a strongly expressed body pattern, with a dark shoulder-stripe and dark face markings, as well as the usual black-and-white leg markings. True (wild) goats are among the smallest species of the genus, but their horns are among the longest.

2. Markhor—There is no precornual convexity, and the facial profile is not markedly concave in front of the orbits; the orbits usually interrupt the dorsal profile. The nasals are longer, sloping downward, and tending to turn down at the tips. The braincase is short, sloping downward somewhat. The palate is broad, so that the toothrows tend to be bowed outward more. The basicranium is broad, especially across the condyles, with small auditory bullae. In all these features, markhor resemble ibex; but there are, of course, many differences. The horns are spirally twisted and flattened, with a rounded anteromedial surface and a sharper posterior keel; insertion of the horns is not in one plane, but folded over the vertex of the skull; the cores are compressed, their transverse basal diameter only 60% of the anteroposterior diameter, and their median surfaces are longitudinally grooved (as also in *Capra pyrenaica*), bounded anteromedially by a weak keel and posteromedially by a stronger keel; the frontal-core angle is more obtuse than in any other species. The facial profile is nearly straight (or, at most, gently concave) from the horn bases to the interorbital region. The vertex of the skull is relatively far back, at the posterior margin of the horn core bases. The biorbital breadth is about 52% of the skull length (as in *Capra ibex*). The nasals are relatively short (about 37% of the skull length), but broad (the breadth 36% of the length). The ethmoid fissure is narrow, because the frontals intrude into it, often becoming obliterated with age by the intrusion of the frontal between the nasal and the lacrimal bones. The premaxilla forms a very long (about 50 mm) suture with the nasals; the premaxillary tips are not expanded. The median palatal notch is fairly narrow and reaches forward to halfway along M^3, much anterior to the level reached by the lateral notches. The paroccipital processes are rather shorter; the supramastoid crest is not flared. Externally, markhor are characterized by their long coats, with a throat fringe as well as a beard, and poorly expressed dark markings. They are among the largest species of the genus, and have among the longest horns.
3. Ibex—Ibex skulls have some similarities to markhor, as listed above, but they differ in many respects. The horns are not spiraled, or not tightly so, and typically are broad-fronted, with regularly spaced knots on the frontal surface; the cores lack keels or grooves (though there is comparatively little difference, in some cases), and they are hardly or not at all compressed, the transverse breadth of a core always being more than 74% of its anteroposterior diameter. There is no convexity at the horn base, but the facial profile is concave in the interorbital region (the orbits may or may not interrupt the line of the profile). The skull vertex is between the horn cores. The form of the nasals and the premaxillae varies; the nasopremaxillary suture is intermediate in length between true goats and markhor. The ethmoid fissure is narrow, but usually remains open. The supramastoid crest is flared.

How Many Species?

There is presumably only one (wild) species in the true goat group: the earliest name for any member of this group is *Capra hircus* Linnaeus, 1758, and that for a wild goat is *Capra aegagrus* Erxleben, 1777; wild and domestic goats are here kept apart.

At the moment only a single species of markhor is recognized: *Capra falconeri* (Wagner, 1839), but this may have to be modified.

The number of species of ibex has long been disputed. Generally, Western authors followed the lead of Ellerman & Morrison-Scott (1951) in assigning all except the Spanish ibex, *Capra pyrenaica* Schinz, 1838, and the Daghestan tur (which Ellerman & Morrison Scott called *Capra caucasica* Gueldenstaedt & Pallas, 1783) to a single species, *Capra ibex* Linnaeus, 1758. Heptner et al. (1961) split the ibex group into seven species: *C. pyrenaica*, *C. ibex*, *C. cylindricornis* (the Daghestan tur, the species incorrectly called *C. caucasica* by Ellerman & Morrison-Scott), the true *C. caucasica* (the Kuban tur), *C. sibirica*, *C. nubiana*, and *C. walie*. G. B. Corbet (1978) and Valdez (1985) proposed compromise schemes, recognizing both *C. cylindricornis* and *C. caucasica* as distinct species, but retaining *C. sibirica*, *C. nubiana*, and *C. walie* in *C. ibex*.

C. ibex of the Alps and *C. sibirica* of the C Asian mountains are undeniably more similar to each other than they are to the intervening Caucasian forms (the two species of tur); what unite alpine and Siberian ibex, in fact, are primitive (for ibex) features of the horns, while *C. caucasica* and *C. cylindricornis* are successively more derived. Moreover, while the two species of tur undoubtedly do share derived features of the skull and the horns (*C. caucasica* is an ideal intermediate between the alpine/Siberian type of ibex and the highly derived *C. cylindricornis*), they would appear to be reproductively isolated (see below). On these grounds alone, there is little alternative to recognizing all seven as distinct species: they are all diagnosably distinct, and the molecular data (see below) make the multispecies solution mandatory.

Ibex separate fairly well phenetically into a S and a N group. The distinctive Spanish ibex, *C. pyrenaica*, fits, if somewhat uneasily, into the latter.

S ibex have a longer, less-sloping braincase; the horn cores are less expanded (the transverse diameter is <80% of the anteroposterior diameter); the horns are rounded-rectangular in cross section, and somewhat compressed laterally, with prominent and discrete transverse knobs across their anterior surface; the nasals are short (only 34% of the skull length), somewhat oval, with parallel borders, and usually run horizontally before turning down at the tips; the toothrows are less convex; in most cases the median palatal notch does not reach farther forward than the laterals; the biorbital breadth is about 50% of the skull length, as in *C. aegagrus*; the premaxilla has a long suture with the nasal; and the paroccipital processes are elongated. Males in S ibex have a tendency to develop a boss on the median frontal suture, just below the insertion of the horns; all males have it in *C. walie*, but it only develops in old males in *C. nubiana*.

N ibex vary between species, but all have a shorter, steeper braincase than S species; the nasals slope evenly downward (as in *C. falconeri*); the skull is broader across the orbits, the supraorbital foramina, the zygomatic arches, and the basioccipital region; and the dorsal profile runs fairly straight from the horns to the interorbital region.

The most distinct of the N ibex is *C. pyrenaica*, in which the horns are rounded-rectangular in cross section, but with a well-marked posterointernal keel; their anterointernal ridge is poorly marked, or has rounded knobs that do not extend laterally across the front surfaces; the horns have a homonymous horn curl, like *C. falconeri* and *C. cylindricornis*, and they tend to be lyrate in anterior view; the horn bases are expanded; the horn cores are less straight than in other ibex, curving out laterally; the skull is very broad, its breadth 56% of its length; alone among ibex, the orbits of *C. pyrenaica* do not interrupt the profile line; the nasals are as broad as in *C. aegagrus* (45% of their length in the males); the nasopremaxillary suture is long, about 20 mm; the ethmoid fissure is open, if narrowly, right down to the premaxilla; the median palatal notch is often well in front of the laterals; and the paroccipital processes are long.

Other N ibex have longer, unexpanded nasals, which make a rounded suture with the frontals; the skull is less broad; the nasopremaxillary suture is usually shorter; and the paroccipital processes are short. These ibex divide into a more primitive group (*C. ibex*, *C. sibirica*) and a more derived group (*C. caucasica*, *C. cylindricornis*). The highly derived (Caucasian) group has a shorter braincase than other ibex, except for *C. sibirica*; the skull breadth is 53%–54% of the length (as in *C. sibirica*; in *C. ibex*, it is 52%); the nasals are broad (the breadth is 39% of the length, compared with 36% in the alpine and Siberian ibex), pointed both anteriorly and posteriorly; the median palatal notch is very broad and short, not reaching the back of the toothrow or extending in front of the laterals; the horns are divergent and, above all, the horn bases are rounded in cross section and enormously expanded, the transverse basal breadth of a core being over 90% of the anteroposterior breadth.

The largest and longest-horned species are *C. sibirica* and *C. walie*. The two Caucasian species are equally large, but they have relatively short, massive horns. *C. ibex* and *C. pyrenaica* are medium-sized species, with horns that are usually not spectacular, though occasional specimens of the latter may develop quite extraordinary horns. Finally, *C. nubiana* is the smallest species in the genus, yet it possesses relatively long horns.

Phylogeny of *Capra*

It is extremely difficult to work out a simple morphological cladogram for *Capra* species. Points to be borne in mind are:

—The *C. nubiana* / *C. walie* group have horns similar to *C. ibex* and *C. sibirica*, but their skulls are very different: the narrow skull with the long braincase would be plesiomorphic if *C. aegagrus* is treated as the outgroup, but, as the horn forms could be a heterochronic growth series (see below, under *C. ibex*), it might also be that *C. aegagrus* is morphologically a special derivative from the S ibex group. Possibly supporting the latter hypothesis are the limb markings (which might, however, also be symplesiomorphic—see below) and the forehead convexity, and perhaps the scimitar horn type (but see below).

—The resemblance of the of the horns of *C. cylindricornis* to *Pseudois* has often been noticed and commented upon; but for the fact that the two Caucasian species are remarkably similar cranially, it would be reasonable to see the E Caucasian tur as a sister species to all other *Capra*. A hypothesis to account for this anomaly will be put forward below.

—*C. pyrenaica*, *C. cylindricornis*, and *C. falconeri* all have a homonymous horn curl; this appears to be the correct interpretation, although the direction of the spiral can be extraordinarily difficult to figure out (whereas true goats usually have heteronymous horns, even in most domestic breeds with corkscrew horns). It is likely that extreme spiraling, of either type, is apomorphic. Other Caprini, in the main, also have homonymous horns, so it seems probable that mild homonymy is primitive for *Capra*—but whether it is truly primitive to sickle-shaped horns is moot.

—Probably the strong leg markings (with white pasterns and knees) of *C. aegagrus* and the S ibex are primitive; they occur outside the genus in *Pseudois*.

Thus we simply leave the groupings (as outlined above) as phenetic groups, and, for the present, avoid any attempt to reconstruct their phylogeny.

Curiously, the molecular evidence does little to resolve the phylogeny of *Capra*, it being complex in the extreme. According to the mtDNA sequencing study of Manceau, Després, et al. (1999), *Capra sibirica* forms a separate clade from all other members of the genus, which, in contrast, shows no special clustering between any two species. The placement of *C. sibirica* as a sister species to all others is unexpected and counterintuitive, but Pidancier et al. (2006) resolved this problem—while, at the same time, raising other phylogenetic controversies—by combining mitochondrial cytochrome *b* sequences with Y-chromosome sequences. The authors confirmed previous findings that in the mitochondrial tree, *Capra sibirica* is sister to all other species; but in the Y-chromosome tree, *C. aegagrus* and *C. falconeri* belong to one clade, and all ibex to the other, the maximum-likelihood bootstrap values being 98% in the first case, and 92% in the second. The best explanation of this is that the initial distribution of the species conformed to the mtDNA data, and was then altered by the spread of proto-*sibirica*, which overwhelmed most other species, leaving only *C. aegagrus* and *C. falconeri* intact. When looking at the morphology of the "ibex" species, we can, in fact, see that there could not have been total nuclear swamping; indeed, there was less and less *C. sibirica* influence farther and farther W, where the ibex become progressively less *sibirica*-like in their horns, with more and more retention of the *aegagrus*-like color pattern. We can judge the S ibex—*C. nubiana*, at any rate—to be genetically about 50% proto-*sibirica* and 50% proto-*aegagrus*.

According to the mitochondrial data, the relationships of the non-*sibirica* ibex would be as follows:

((*nubiana* ((*pyrenaica*, *ibex*) (*caucasica*, *cylindricornis*))).

We will consider some of these relationships below.

Capra aegagrus Erxleben, 1777

wild goat, bezoar

The name *Capra aegagrus* Erxleben, 1777, was placed on the Official List of Specific Names in Zoology by ICZN, Opinion 2027 (2003).

Various subspecies have been described for this species, which are listed below, with their type localities.

1777 *Capra aegagrus* Erxleben. Daghestan district, Caucasus.

1875 *Capra blythi* Hume. Sind, Pakistan.

1905 *Capra persica* Matschie. Laristan, SW Iran.

1907 *Capra florstedti* and *Capra cilicica* Matschie. Bulghar Dagh, Turkey.

1918 *Capra hircus neglectus* Zarudny & Bilkevitch. Seistan, E Iran.

1950 *Capra aegagrus turcmenicus* Zalkin. Kopet-Dagh, S Turkmenia.

A further subspecies that has usually been accepted is the Cretan wild goat (*Capra aegagrus cretica*). The origin of this putative taxon was surveyed by Bar-Gal et al. (2002), with the aid of cytochrome *b* and D-loop sequencing. Cretan wild goats cluster quite clearly with domestic goats, especially the Baladi breed of Lebanon, and not with wild bezoar; that is to say, the Cretan taxon is a feral goat, but of very ancient origin, and presumably extremely valuable as indicating something of the morphology of the earliest Middle Eastern domesticated goats.

Zalkin (1950) gave measurements of bezoar goats from the Caucasus and from the Kopet-Dagh, from which it emerges than the only difference between them is the more slender horns of the latter. He gave the following statistics (with the mean and the minimal-maximal range in mm):

1. Skull length (males)
   —Caucasus = 263.4, standard deviation 9.50, 248–278 (*N* = 11)
   —Kopet-Dagh = 261.6, standard deviation 8.45, 250–279 (*N* = 11)
2. Skull length (females)
   —Caucasus = 224.2, standard deviation 6.25, 215–233 (*N* = 6)
   —Kopet-Dagh = 227.5, standard deviation 4.05, 224–234 (*N* = 6)
3. Basal circumference of horn (around the sheath)
   —Caucasus = 224.3, standard deviation 13.2, 200–250 (*N* = 15)
   —Kopet-Dagh = 198.0, standard deviation 13.4, 180–230 (*N* = 37)

On the basis of the horn circumference figures, Zalkin erected the subspecies *Capra aegagrus turcmenicus*.

Our own figures confirm and extend these findings; four male skulls from Sind have a mean skull length of 261.2 mm; a male from Mazanderan (Alborz Mountains) is 260 mm, one from Fars is 261 mm, and two labeled just as "Persia" are 269–286 mm. A skull from the Makran (SE Iran), however, is only 250 mm long—unusually small, at the lower limit of the range given by Zalkin (1950). A female from Shiraz, in SW Iran, has a skull length of 200 mm, and one from Sind is 229 mm.

We measured the horn basal circumference around the core, not the sheath, and found a mean of 186.6 mm for 10 specimens from Turkey and the Caucasus, 166.7 mm for 3 from the Kopet-Dagh, 181.8 mm for 4 from Sind, and 175.0 mm for 2 from S Iran.

Horns from the Kopet-Dagh and Sind average longer (1125 and 1115 mm, respectively) than those from the Caucasus and Turkey (923 mm), and their curve length is 162.7% of their direct base-to-tip distance, compared with 149.8%; but the ranges for the horn characters overlap considerably. Therefore, at present we find no evidence, based on skull or horn characters, for dividing *Capra aegagrus* into subspecies, but it may be that a combination of characters would succeed in separating some populations.

A study by Zeder (2001) of a large collection of skulls from different regions of Iran found that there is a noticeable reduction in size, as represented by metacarpal distal breadth, from N to S, especially in the males. There seems to be a much greater difference in the males between the N Zagros Range and other parts of the Zagros than between those from C and S Zagros, whereas, in so far as there are differences in the females, it is the S Zagros females which stand out. These data seem to conflict somewhat with our own observations on skulls; further study of additional skulls would show either that those from the S Zagros (and presumably the Makran and Pakistan) are smaller than those from farther N, or else that the S populations have more gracile metapodials for the same general skull size.

It is reported in the literature that goats from Sind are lighter in color than those from Iran, on average; that those from Turkey are darker still, brown or red-brown rather than yellow-brown or red-gray, and that their flank- and shoulder-bands are broader. The few localized skins and living specimens we have seen would tend to confirm this, but it is on average only that they differ, and the silvery white coat of old age is found in the males from all three areas.

Capra falconeri (Wagner, 1839)

markhor

There are, again, a large number of subspecific names available in this species.

1839 *Aegoceros (Capra) falconeri* (Wagner). Astor, Kashmir.
1842 *Capra megaceros* Hutton. Kandahar, Afghanistan.
1875 *Capra jerdoni* Hume. Suleiman Range, Punjab, Pakistan.
1898 *Capra falconeri cashmiriensis* Lydekker. Pir-Panjal Range, Kashmir.
1945 *Capra falconeri heptneri* Zalkin. Dashtidjum district, Tadjikistan.
1945 *Capra falconeri ognevi* Zalkin. Kuh-I-Tang Range, Uzbekistan.
1950 *Capra falconeri gilgitensis* Cobb, nom. nud. Gilgit.
1958 *Capra falconeri chitralensis* Cobb. Chitral.

Schaller & Khan (1975) and Schaller (1977) made critical observations on subspeciation in this species. They found, for example, that within Kashmir (including

Gilgit, Chitral, Astor, and the Pir-Panjal Range) a variety of horn types, though always of the open-spiraled type, can be found in any one locality; similarly, the desert ranges of S Pakistan and Afghanistan share a more tightly spiraled type of horn, which also varies greatly. They proposed that there are really only two subspecies in the Indian and Pakistani parts of the range; occasionally, the "wrong" type of spiral can even be found between these two subspecies, though they can be overwhelmingly separated. For the Kashmir type, the name *Capra falconeri falconeri* has priority; for the S type, *Capra falconeri megaceros*. In our sample, and adding measurements from the literature (especially Heptner et al.,1961, who found no reason to separate the two [former] Soviet forms from each other), we find the following data (males only):

From these figures it would appear that the *C. f. heptneri* has remarkably short horns; *C. f. falconeri* has more curled horns and a wider flare; and *C. f. megaceros* has the tight curl and narrow tip-to-tip distance of *C. f. heptneri*, but the very long horns of *C. f. falconeri*. Some of the above ranges do not overlap, but increased sampling might well cause them to do so; however, on the face of it, it does seem likely that we might well have three separate species.

According to Schaller (1977), the Kashmir taxon would have longer hair, especially in winter, and a longer ruff than *C. f. megaceros* (an extremely long ruff also occurs in *C. f. heptneri*, the taxon that is much the commonest in captivity).

Hammer et al. (2008) sequenced the first hypervariable segments of the mitochondrial control region in zoo markhor, and they showed that *C. f. heptneri* is paraphyletic with respect to *C. f. megaceros*. More worryingly, they found that many zoo markhor are introgressed with domestic goat DNA.

Therefore, we provisionally retain all markhor in a single species, but urge a detailed comparison of large samples; if it should prove that three distinct species exist, then this will have great significance for conservation.

Table 61 Horn length for markhor: males

	falconeri	*megaceros*	*heptneri*
Horn length			
curve	708–1355 (20)	755–1130 (12)	750–900
straight	801–1082 (13)	755–782 (12)	580–730
Ratio, straight:curve	66–80 (14)	76–87 (12)	76–93
Tip–tip distance	695–1321 (17)	518–711 (10)	210–600
Ratio, tips:straight length	85–133 (14)	60–85 (10)	36–82?

The Chiltan Goat

Schaller & Khan (1975) maintained that this screw-horned goat (*Capra falconeri chialtanensis* Lydekker, 1913), from the Chiltan and neighboring ranges in Pakistan, is a localized morph of *C. aegagrus*. T. Roberts (1977) and Valdez (1985), on the other hand, argued that it is a hybrid between bezoar and markhor. The only thing that is agreed on is that it is not a form of markhor, as Lydekker thought.

Schaller & Khan (1975) reported that the screw-horned type occurs in the same herds with ordinary bezoar goats on several ranges, and that bezoar and markhor coexist without interbreeding on at least two other ranges. T. Roberts (1977) maintained that there is evidence that markhor formerly occurred on at least one of the ranges where the Chiltan goat is found; according to his observations, the Chiltan goat has some markhor-like features, such as the lack of a dark face pattern. Valdez (1985) depicted three specimens and noted their resemblance to experimental goat-markhor hybrids, one of which he also illustrated.

We have seen two specimens of this form: the type (a frontlet and horns) and a more complete trophy skull (British Museum specimen BM 67.795), lacking only premaxillae. The latter is, overall, markhor-sized, and thus is bigger than any *C. aegagrus*: the biorbital breadth is 146 mm, at the lower end of the markhor range but outside that of the bezoar (in bezoar, the maximum for this measurement is 144 mm, and it is usually below 140 mm). Nasal length is 88 mm; the smallest markhor measures 89 mm while the largest bezoar is 82 mm. For what it is worth the cranial capacity of the Chiltan goat is considerably greater than any *C. aegagrus*, although we have not measured the capacities of any markhor for comparison.

We tend to agree with T. Roberts (1977) and Valdez (1985) that the Chiltan goat is a natural hybrid. Valdez (1985) depicted a Chiltan goat that approaches the bezoar type of horns, and it is presumably a backcross.

Capra pyrenaica Schinz, 1838

Spanish ibex

This distinctive species today lives only in Spain. *C. pyrenaica* formerly occurred in Portugal, but it is now extinct there. Engländer (1986) described its occurrence in SW France in the early Upper Pleistocene and its replacement in the Latest Pleistocene by *C. ibex*.

Five subspecific names are available:

Capra lusitanica Schlegel, 1872. N Portugal (now extinct). On the history and availability of this name, see Almaça (1992).

Capra pyrenaica Schinz, 1838. Pyrenees.

Capra pyrenaica cabrerae Camerano, 1917. Sierra Morena.

Capra pyrenaica nowaki Wyrwoll, 1999. This name was to replace *Capra pyrenaica hispanica* Schimper, 1848, which was supposedly from the Sierra Nevada, but the type series of *hispanica* is, in fact, mixed, and included a specimen from the Pyrenees, according to Wyrwoll (1999).

Capra pyrenaica victoriae Cabrera, 1911. Sierra de Gredos.

Engländer (1986) gave numbers to show that there is a progressive reduction in size from N to S: head plus body length averages 148 cm in nominotypical *C. p. pyrenaica*, 142 cm in *C. p. lusitanica*, 135.5 cm in *C. p. victoriae*, and 121 cm in "*C. p. hispanica*" (or *C. p. nowaki*; see above). The Portuguese form (*C. p. lusitanica*) had the shortest, stoutest horns (mean length 44.5 cm, basal circumference 23.5 cm), the Pyreneean *C. p. pyrenaica* the longest (at 80 cm), and the other two intermediate (73–73.4 cm), while the basal circumference is only 18.9 cm in "*C. p. hispanica*" and 20–22 cm in the other two. The S "*C. p. hispanica*" and Portuguese *C. p. lusitanica* have the narrowest horn spans (29.4 and 25 cm, respectively), with *C. p. pyrenaica* and *C. p. victoriae* much wider (at 42 and 47 cm, respectively). Engländer (1986) noted that a peculiar, somewhat sheep-like horn morph (in which the horns are curled around into a near circle) occurs frequently on some, but not all, of the ranges in the far S; and a horn type that resembles *C. ibex*, except for the absence of knots, can be found on the Sierra Blanca. There seems to be a cline in color, the N ibex being darker, with a more strongly expressed pattern of dark markings; although in the Sierra de Gredos, in C Spain, the dark markings may be often so strong that in summer they seem almost black. Engländer (1986) stated that on the Sierra de Gredos, in C Spain, there are ibex corresponding to all four of Cabrera's described subspecies.

On the basis of mtDNA sequencing, Manceau, Crampe, et al. (1999) found that three specimens from the Pyrenees clustered together, in contrast to specimens from other localities; the support value for the non-Pyreneean clade, however, was only 73%.

Although we have seen very few specimens, from the above evidence we think that all the variation in *C. pyrenaica* can be ascribed to a major N–S cline, with some additional microgeographic variation. Probably no subspecies can be recognized nomenclaturally.

We should comment, however, on the head of a Pyrenean male (the only one we have seen from this population) in the British Museum: at the base, the horns are broader anteroposteriorly than in any other specimen, with prominent round knots along an anteromedian ridge (but not extending across the front of the horns, as in *C. ibex*). If enough localized material becomes available, another look at infraspecific variation would be advisable.

Capra nubiana F. Cuvier, 1825

Nubian ibex

Smallest species of the genus.

We have seen skulls from Israel, the Sinai, and Sudan, as well as an incomplete skull from Aden, S Yemen; and living specimens from Israel and Sudan.

Characteristic measurements of the males are as follows:

There is thus a cline in size from N to S, but a fairly clean difference in tip-to-tip distance between skulls from Israel and the Sinai, on the one hand, and Sudan, on the other: the tips turn markedly inward in the former, less so in the latter. An incomplete skull from Aden appears to be at least as large as the average Sudan specimen.

Adult males in Israel and the Sinai are light brown, with red, yellow, or whitish tones; there is a weak brown shoulder-band, a blackish dorsal stripe, a dark flank band, a white belly, and white wrist marks. The females are lighter in color. Sudanese specimens are dark brown, at least in the adults, with no lighter tones on the body; the shoulder band is dusky; the dorsal stripe is black; the flank-band, on the contrary, is barely visible; the belly is white or off-white; and the wrist is white, as in other populations. The females are barely

Table 62 Measurements of Nubian ibex: males

	Skull length			Horns, tip–tip		
	Mean	Std dev	N	Mean	Std dev	N
Israel	244.5	8.81	4	336.8	171.92	4
Sinai	250.2	10.03	5	345.7	225.50	3
Sudan	258.1	8.81	10	447.7	158.04	11

lighter in color than the males. We have seen no specimens from S Arabia.

Granjon et al. (1990) confirmed that Sudanese specimens are much darker than those from Egypt and the Dead Sea region. A single specimen from Saudi Arabia had different alleles at the MOD (malic-enzyme) locus when compared with 20 specimens from Sudan, Egypt, and the Dead Sea region.

Primarily on the basis of color, we recommend recognition of two subspecies: *Capra nubiana nubiana* F. Cuvier, 1825, from Sudan; and *Capra nubiana sinaitica* Ehrenberg, 1833 (synonym—*Capra arabica* Rüppell, 1835), from Israel and the Sinai. It is possible that *Aegoceros beden* (Wagner, 1835) (synonym—*Capra mengesi* Noack, 1896), from Saudi Arabia, may represent a third subspecies, but evidence is lacking. Given, however, that the color of the adult males appears, on limited evidence, to be quite different between the Sudanese and the Israeli/Sinai specimens, it is possible that we have different species; resolution of this question is also hampered by the lack of material from S Arabia.

Capra walie Rüppell, 1835

walia ibex

This species, from Ethiopia, is the most endangered in the genus. Its taxonomic status is slightly equivocal. *C. walie* shares many characters with *C. nubiana*, as listed earlier, but differs from it characteristically: it is much larger (among the largest of the genus); the nasal breadth is 40% of the length (compared with 38% in *C. nubiana*); the nasofrontal suture is rounded, rather than triangular; and the nasopremaxillary suture is much longer, about 30 mm instead of 18 mm. The horn cores are more expanded, their transverse width 79.5% of their anteroposterior diameter. In *C. walie*, all adult males have a marked convexity on the interfrontal suture; this develops only in old age in *C. nubiana*. In general, *C. walie* may be viewed as a hypermorphic derivative of *C. nubiana*; that is, a logical prolongation of the growth-processes of the latter. Considering the ease of distinguishing the two, *C. walie* has to be retained as a full species.

Capra ibex Linnaeus, 1758

alpine ibex

Medium-sized species, characterized by its horns, with the anteroexternal corner rounded, thus having a "beveled-off" appearance, unlike the squared-off form (with the anteroexternal angle as well developed as the anterointernal angle) of *C. nubiana*, *C. walie*, and *C. sibirica*. Observation of the growth series of *C. nubiana* and *C. sibirica* in captivity shows that the first-year horns of the males resemble the *C. aegagrus* form, that is, narrow, with a prominent anterior keel; the horn then expands, especially posterolaterally, reducing the keel to an anteromedial angle, as in *C. ibex*; then, around sexual maturity, the anterolateral expansion takes place to give the adult a broad-fronted horn. Therefore, either *C. ibex* can be viewed as a neotenous derivative of the more "advanced" ibex, or else the broad-fronted species are seen as hypermorphic. *C. aegagrus* would perhaps be easier to fit in at the base of the latter series, rather than at the top of the former. With age, the horns of *C. ibex* do not become broad-fronted, but the knots become worn down.

Alpine ibex are one of the great success stories of wildlife conservation; as told by Grodinsky & Stuewe (1987), their numbers have recovered from a low of about 50 in 1821 to some 22,000 today.

Capra sibirica (Pallas, 1776)

Siberian ibex

Large. Differing from *C. ibex* in the more tubular orbits; median palatal notch not extending forward to the third molar, or in front of the laterals; ethmoid fissure commonly obliterated with age. Horn bases about equally expanded (the transverse diameter 82%–84% of the anteroposterior diameter), but broad-fronted.

There are, in Ellerman & Morrison-Scott's (1951) list, no fewer than 16 names available for subspecific differentiation in this species; Zalkin (1949) reduced these to three or four, at the same time adding two more. Zalkin's list is as follows:

- *Capra sibirica alaiana* Noack, 1902. Tienshan system, including Transalai and Alai-Pamir. Synonyms—*Capra sibirica merzbacheri* Leisewitz, 1906; *Capra sibirica transalaiana* and *Capra sibirica almasyi* Lorenz, 1907.
- *Capra sibirica dementjevi* Zalkin, 1949. Kuen-Lun Range.
- *Capra sibirica formosovi* Zalkin, 1949. Talas-Alatau and other W Tienshan ranges.
- *Capra sibirica hagenbecki* Noack, 1903. Gobi Altai and Transaltai Gobi.
- *Capra sibirica sakeen* Blyth, 1842. Kashmir. Zalkin had few specimens from this part of the species' range, but he thought this probably represented a distinct subspecies. Synonyms—probably *Capra sibirica skyn* (Wagner, 1844); *Capra sibirica dauvergnii* Sterndale, 1886; *Capra sibirica wardi* Lydekker, 1900; *Capra sibirica*

Table 63 Characteristic measurements for *Capra sibirica sibirica, C. s. alaiana*, and *C. s. sakeen*

	sibirica			*alaiana*			*sakeen*		
	Mean	Std dev	*N*	Mean	Std dev	*N*	Mean	Std dev	*N*
Skull length									
males	270.0	9.25	17	292.0	8.95	20	293.1	9.11	20
females	247.0	6.55	11	254.0	4.45	11	—	—	—
Nasal length, males	86.6	5.10	16	103.9	6.80	18	107.0	4.32	7
Horn (males), basal circumf	20.7	1.36	17	23.0	2.02	43	—	—	—

pedri Lorenz, 1906; *Capra sibirica filipii* Camerano, 1911.

Capra sibirica sibirica (Pallas, 1776). Russian Altai, Sayan, and neighboring ranges in Mongolia. Synonyms—*Capra sibirica lydekkeri* Rothschildi, 1900; *Capra sibirica altaica* and *Capra sibirica fasciata* Noack, 1902; *Capra sibirica lorenzi* Satunin, 1905.

There is also the name *Capra sibirica hemalayanus* Hodgson, 1841, said to be from Nepal, but the species does not occur there.

Heptner et al. (1961) made *C. s. formosovi* a synonym of *C. s. alaiana*, commenting that specimens with and without a light saddle spot (its absence being the distinguishing feature of *C. s. formosovi*) occur in the same populations in the Tienshan system.

Table 63 contains some of the characteristic measurements given by Zalkin (1949) for *C. s. sibirica* and *C. s. alaiana*; we have studied and measured specimens of both (but mainly of *C. s. alaiana*), which tend to confirm his findings, and we include in the table statistics for *C. s. sakeen* (from Kashmir, Ladakh, Pamir, and Afghanistan) taken by ourselves:

Zalkin (1949) also mentioned that the horns of the males (measured along the curve) are normally >1 m long in *C. s. alaiana*, but <1 m in *C. s. sibirica* (though his table shows that actually only 4 out of 6 *alaiana* are >1 m). In our sample of *C. s. alaiana*, however, 10 out of 13 have horns >1 m long (making a total of 14 out of 19), and only 2 in our sample of 8 *C. s. sibirica* are >1 m (one of them only just); in our sample of *C. s. sakeen*, the horns are usually <1 m (only 6 exceptions in 28 specimens). Thus *C. s. sakeen* would seem, in general, to be large like *C. s. alaiana*, but shorter-horned.

G. M. Allen (1940) gave (incomplete) measurements for two specimens from the Gobi Altai: a male with a skull length of 265 mm (thus only slightly below the mean for *C. s. sibirica*), and a female with a skull length of 230 mm (more than two standard deviations below *C. s. sibirica*). The metrical evidence for the validity of *C. s. hagenbecki* would, therefore, seem equivocal.

Zalkin's (1949) evidence for his subspecies *C. s. dementjevi* consisted of two female skulls, much smaller than any other female specimens. We measured a male skull from "near Yarkand," which suggests the Kuen-Lun Range from which *C. s. dementjevi* was described; its measurements fall well within the range of *C. s. alaiana*.

Externally, nominotypical *C. s. sibirica* is described as dirty white or yellowish white, and dark along the flanks, the lumbars, the shoulders, the breast, and the anterior legs. The light area forms a saddle, which is separated by the dark shoulder zone from a light cervical spot. Zoo specimens we have seen, mostly known to be from the Altai, are of this type. *C. s. hagenbecki* is described as uniform gray-brown; that is, lacking the dark/light contrasts of *C. s. sibirica*. If this difference holds true, which would mean that the presence or absence of the saddle is less variable in Siberian/Mongolian ibex than in those from the Tianshan system, then *C. s. hagenbecki* would be a valid subspecies.

C. s. alaiana is described as dark cinnamon-brown, with a light saddle extending to the haunches; again, the saddle is separate from the cervical spot. The relatively few museum skins seen by us are of this type.

As noted above, the absence of the light saddle is said to predominate in some ranges, and to be less frequent in others, so that *C. s. formosovi* is a polymorphic variant, rather than a valid subspecies.

Zalkin (1949) suggested that *C. s. sakeen* is distinguished by the broad saddle-spot being connected to the cervical spot, leaving only the haunches dark. But some specimens we have seen have the cervical and saddle spots separated by the dark shoulder zone.

Capra sibirica thus has two reasonably well-defined subspecies (*C. s. sibirica* and *C. s. alaiana*), a third (*C. s.*

sakeen) that is probably recognizable (note that, if it is combined with *C. s. alaiana*, the name *C. s. sakeen* has priority), and two (*C. s. hagenbecki* and *C. s. dementjevi*) whose validity must remain uncertain for the moment.

Fedosenko & Blank (2001) recognized the following subspecies:

—*Capra sibirica sibirica*—Small; skull length 256–293 mm (males), 239–260 mm (females); horn length 110–117 cm (males; further on it says <100); girth at the horn base 18.5–24 cm; body mass up to 103 kg. Nasals short, their length usually <100 mm, 28%–37% of the skull length (males), or 32%–42% (females). Nasofrontal suture anterior to the anterior edge of the orbit, or at the same level. Males (in winter) relatively light, the back, the sides, and the upper part of the neck mainly dirty white or yellowish white. Dark and light color patches bright, with contrasting borders (Zalkin, 1950; Sobanskiy, 1988).

—*Capra sibirica hagenbecki*—Similar in size and build, but with relatively longer horns (males 74–139 cm, females 16–38 cm); ridges in the middle portion of the horns larger. Color more gray and pale brown, without a light saddle; old males sometimes becoming nearly white (Lydekker, 1913; Bannikov, 1954).

—*Capra sibirica alaiana*—Bigger; skull length 276–310 mm (males), 248–262 mm (females). Nasals long, their length >100 mm, 32%–40% of the skull length for the males, 34%–43% for the females. Nasofrontal suture higher than the anterior edge of the orbit. Horns length usually >100 cm, and up to 147 mm; girth at the horn base 20–28 cm. Body mass of the males up to 130 kg. Adult males (in winter) relatively dark, a dark brown-gray, with less contrast in the dark and light patches. Light saddlelike spot on the dorsum variable in form and size (Heptner et al., 1961).

—*Capra sibirica sakeen*—Like *C. s. alaiana*, but the horns relatively short (males 102–140 cm), more massive at the base, the girth up to 30 cm. Males weighing less, up to 90 kg. Light in color; back and sides pale brown or creamy white; dorsal stripe pale brown; other parts light yellowish brown; beard and tail black-brown (Lydekker 1913; Heptner et al. 1961; T. Roberts 1977; Schaller 1977).

We are uncertain as to what to make of this redescription of geographic variation, which includes features and differences not known to Zalkin (1949), who apparently remains the last author to perform a primary systematic revision.

Caucasian Ibex (*Capra caucasica* and *Capra cylindricornis*)

Capra caucasica and *Capra cylindricornis* are evidently very closely related. They have very short, steeply sloping braincases; extremely broad skulls; broad nasals that are pointed at both ends; and the median palatal notch is exceptionally broad, reaching to the hind surface of the third molars, as do the lateral notches. They are almost equally large in size (skull length about 294–296 mm, equal to *C. sibirica* and *C. walie*), with comparatively short, extremely thick-based horns. The color in both is a medium brown, with little color contrast.

The main difference is in the horns of the males. In *C. caucasica* (the Kuban tur), the horns resemble a shorter, thicker-based, more divergent form of those of *C. ibex*; like the latter, the anterolateral edge is beveled off. The tip-to-tip distance in the males of *C. caucasica* averages 380 mm; of *C. cylindricornis*, 540 mm. In *C. caucasica*, the young males have prominent anterior knobs, but these wear down seemingly earlier than in *C. ibex*. In *C. cylindricornis* (the Daghestan tur) the horns curl inward, then upward again at the tips, in an almost sheep-like fashion, which is, however, somewhat foreshadowed in *C. caucasica*. The transverse diameter of the horn core base is 99% (±5.0%) of the anteroposterior diameter in *C. cylindricornis* (and may exceed it), while in *C. caucasica* this figure is 92% (±10.3%), the second highest in the genus.

There are some differences in the skull; mainly that in *C. caucasica* the ethmoid fissure appears to remain open throughout life, while in *C. cylindricornis* it is often obliterated. Also, the coronal suture is nearly straight in *C. caucasica*, but projects forward at an angle in *C. cylindricornis*; the highest point of the skull is farther to the back of the horn cores in *C cylindricornis*; and the parietal has a depression in the latter (Weinberg, 2002); the length of the premaxillary suture (with the nasals) measures 33.3 ±5.32 mm in *C. caucasica*, but only 21.3 ±8.46 mm in *C. cylindricornis*; the nasals are longer in *C. cylindricornis*, 125.2 ±7.29 mm (41% of the skull length), while in *C. caucasica* they are only 107.3 ±8.08 mm (37% of the skull length). Heptner et al. (1961) state that the Daghestan tur has a shorter braincase, but we do not find this in our data.

In the phylogenetic tree of Manceau, Després, et al. (1999), six samples of *C. caucasica* formed a separate clade from three samples of *C. cylindricornis*, which, curiously, were sister to a clade of two out of four studied specimens of *C. aegagrus*.

C. caucasica inhabits approximately the W third of the Caucasus, *C. cylindricornis* the E two-thirds. Weinberg (2002) specified that the E limit of *C. cylindricornis* is 48° 31' E, in Azerbaijan, whereas the W limits are less well defined. According to Heptner et al. (1961), the ranges of the two species overlap between about 40° E and 43° 30' E, and here, on the S slopes of the Caucasus, they hybridize. Hybrids were said by these authors to be rather infrequent, and Weinberg (2002) even seemed to doubt whether hybridization really occurs; evidently, there is no hybrid population, and all hybrids, it is claimed, conform to a type, implying, if true, that there is little or no backcrossing. Moreover, Heptner et al. (1961) reported no evidence of hybrids from the N slopes. As for the previous descriptions of several taxa in the Caucasus, Heptner et al. put this down to variability not within a hybridizing and backcrossing continuum, but within each of the two species themselves; thus *C. severtzowi*, *C. raddei*, and *C. dinniki* would be synonyms of *C. caucasica*, and *C. pallasii* of *C. cylindricornis*.

The results of our work conform to the implications of Vereschagin's (1959) claim that *C. ibex*, *C. caucasica*, and *C. cylindricornis* represent a kind of transformation series. Moreover, the maturation of horns in *C. caucasica* males is reminiscent of the transition from a triangular cross section and prominent knots, which are *Ovis gmelini* characters, to a rounded cross section and crowded transverse ridges in the ontogeny of argali (the *Ovis ammon* group). So *C. caucasica* could represent a phylogenetic intermediate between *C. ibex* and *C. cylindricornis*.

A quite different interpretation is also possible. First, it is difficult to believe that the hybrids are of lowered fertility when one bears in mind the records of hybridization elsewhere (Vereschagin, 1959; Valdez, 1985). Second, it is not absolutely clear that there are only three morphs (parental forms and hybrids). There seems to be a widespread belief in geographic variation within *C. caucasica*—Vereschagin (1959), for example, treats true *C. caucasica* as transitional between "*C. severtzovi*" and *C. cylindricornis*— but not within the much more widespread *C. cylindricornis*. *C. ibex* and *C. cylindricornis* do not, in general, look like sister species, or like the starting point and end point of a transition series. These two species have much more the appearance, as far as their horns are concerned, of being designed for alternative styles of battering and for taking the force of blows, and thus are not so very closely related. We therefore propose the following scenario.

C. ibex once ranged up to the W Caucasus. The Caucasian population became isolated there by climatic change, perhaps being several times cut off and then rejoined in successive glacial/interglacial cycles, and its genome was periodically swamped by the larger *C. cylindricornis* males dominating the smaller *C. ibex* males. Males formed a fitness series: *cylindricornis* > hybrids > *ibex*. In this scenario, *C. caucasica* represents a stabilized hybrid population, with the W end ("*C. severtzovi*") retaining more *C. ibex* characters than the E end ("*C. caucasica*"). The present situation would be the latest cycle, with *C. cylindricornis* genes again introgressing the W population. We suggest that the two Caucasian forms have to hybridize, because, in occupying much the same altitudinal range, they must compete. Goats, on the other hand, with different altitudinal preferences, will not compete; thus they do not hybridize, even though they surely come within "cruising distance" of each other (*C. aegagrus* with *C. nubiana* or with the turs; *C. falconeri* with *C. sibirica*). The other case within the genus where two species with the same altitudinal preferences meet is that of *C. aegagrus* and *C. falconeri*; and here, too (as we argue above), they hybridize.

There is a clear need for a behavioral field study, preferably accompanied by genetics, to determine the extent of hybridization, the question of backcrossing, and, ultimately, the issue of gene exchange. As far as conservation is concerned, the need is to ensure, as far as possible, the security of both species, as well as the overlap zones, on both the N and S flanks of the Caucasus. Whatever the real genetic situation, we are evidently in the presence of a very interesting and potentially significant biological phenomenon, which must not be allowed to disappear unstudied.

Capra caucasica Güldenstaedt & Pallas, 1783

Kuban tur

1783 *Capra caucasica* Güldenstaedt & Pallas. Between the Malka and Baksan rivers, NE of Mt. Elbrus, N Caucasus.

1888 *Capra severtzowi* Menzbier. W of Elbruz and S of Teberda, Great Caucasus Range.

1901 *Capra raddei* Matschie. Upper course of the Ingur River, S slope of the Great Caucasus Range, Elbruz region.

1905 *Capra dinniki* Satunin. W extremity of the Great Caucasus Range; probably the upper reaches of the Beloe, according to Heptner et al. (1961).

Capra cylindricornis (Blyth, 1841)

Daghestan tur

1841 *Aegoceros pallasii* Rouillier; not Schinz, 1838. Caucasus.
1841 *Ovis cylindricornis* Blyth. Caucasus.

Nilgiritragus Ropiquet & Hassanin, 2005

Nilgiri tahr

2*n* = 58.

Horns parallel, somewhat curved back, transversely wrinkled, the inner surface nearly flat, the outer surface highly convex; low keel on the antero-internal edge; posterior surface rounded. Braincase descending steeply behind the horns, as in *Ovis*. Tympanic tube very short, thick. Frontals flat. Trough between the body of the zygoma and the orbital rim. Nasofrontal suture V-shaped. Anterior nasals very long, pointed. Frontolacrimal suture forming a simple curve. Ethmoid fissure absent, as in *Ovis*. Nasal branch of the premaxilla broad, the ends squarely near the nasal, but not contacting it.

The Nilgiri tahr is sister to *Ovis*, not to the other species of tahr, as was previously assumed (Ropiquet & Hassanin, 2005).

Nilgiritragus hylocrius (Ogilby, 1837)

1837 *Kemas hylocrius* Ogilby. Nilgiri Hills.

Dark yellow-brown above, paler below; dark dorsal stripe. Males becoming much darker with age, developing a grayish saddle. Greatest skull length about 290 mm.

Hill ranges of the W Ghats of India, above the treeline.

Ovis Linnaeus, 1758

sheep

Sheep constitute the culmination of the broad-based-horn method of fighting (Schaffer & Reed, 1972), with, in the males, thick horn bases and horns curled in nearly, or completely, or more than in a full circle, varying from homonymous to heteronymous. Braincase descending steeply behind the horns. Ethmoid fissure absent. Anterior nasals very long, pointed.

An enormous amount has been written about taxonomic variation in *Ovis*; it is very difficult to synthesize this and put it together with our own, much more limited analysis.

When the transverse diameter of the horn is plotted against the longitudinal diameter, the species and species-groups of this genus form a morphocline from *O. gmelini* and *O. vignei* through *O. dalli* and *O. canadensis* to *O. ammon*, but, when plotting the posterior skull breadth against the skull length, the morphocline is from *gmelini*/*vignei* to *ammon*, the *O. canadensis* group falling above the line (i.e., broader skulls), with *O. nivicola* still broader yet (Grubb, 2000a).

Pees & Hemmer (1980) found that specimens of the *O. ammon* group (the precise taxon was not stated) have a larger cranial capacity relative to the other Old World wild sheep.

For our craniometric analyses, we used mainly the dataset of Pfeffer (1967), with some extra specimens measured by ourselves.

Craniometric analyses divide Old World wild sheep into three major groups: (1) *O. gmelini* plus *O. vignei*; (2) *O. ammon*; and (3) *O. nivicola*. In some of our analyses, we utilized this division. There are, on the other hand, four karyogroups: (1) *O. gmelini*, (2) *O. vignei*, (3) *O. ammon*, and (4) *O. nivicola*. We tested these divisions as well. We find that, comparing skulls of the *O. gmelini* and the *O. vignei* groups, the latter have a longer frontal arc compared with the frontal chord, and a higher and somewhat narrower and shorter skull. Comparing skulls of the *O. vignei* / *O. gmelini* group with those of the *O. ammon* group, the latter are much larger and broader-skulled, with shorter foreheads.

Apart from our original analyses, we have made much use of Heptner et al. (1961), Valdez (1982), Geist (1991a) and Vorobeev & van der Ven (2003).

Wild sheep are very habitat specific, favoring hilly, rolling country, though not too steep (at least in the Old World). Thus it is not surprising that there are quite a large number of taxa, and the differences between them, as far as the evidence goes, are quite discrete and consistent. Therefore, we have no option but to recognize a number of species.

Ovis gmelini Group

The name *Ovis orientalis* Gmelin, 1774, was placed on the Official List of Specific Names in Zoology by ICZN, Opinion 2027 (2003), but as it seems to refer to a hybrid (Valdez, 1982), the name nonetheless is unusable (Art. 17.2).

Ovis gmelini Blyth, 1841

red or Armenian sheep

1856 *Ovis anatolica* Valenciennes.
1899 *Ovis ophion urmiana* Günther.
1919 *Ovis ophion armeniana* Nasonov.

2*n* = 54 (NW Iran, Marakan Protected Region, 38° 50' N, 45° 11' E; Koyun-Daghi Island, Lake Urmiah,

38° 28' N, 45° 37' E). Valdez et al. (1978) karyotyped 5 males and 5 females from Mooteh at $2n = 54$, identifying them as *O. isphahanica*, but Valdez (1982) later reidentified them as of this taxon.

Skull length 246–264 mm (males; mean 251 mm), 222–238 mm (females; mean 231mm).

Horns supracervical, with a negative bend (heteronymous), exceptionally almost in one plane (weakly perverted). Horn length up to 670 mm, rarely slightly more; basal circumference 220–270 mm. Frontal surface tapering upward somewhat; outer ridge usually indistinct. Fronto-orbital edge and frontal surface rounded; frontonuchal edge sharp. Hornless females common.

Reddish rusty to rusty cinnamon, darker in winter; light patch of varying size usual on the sides of the males. Muzzle white; lower parts of the limbs white, with a black streak down part of the front. Chest and underparts with a short tress of elongated dark to blackish hair, not reaching up to the throat and the lower jaw.

Transcaucasus, Turkey, NW Iran (vicinity of Tabriz), and W Iran in the Zagros, to at least the C Zagros. Valdez (1982) identified those of Mooteh as belonging to this taxon.

Ovis isphaganica Nasonov, 1910

Esfahan sheep

$2n = 54$ (Murche Khort, 140 km NW of Esfahan, 33° 38' N, 50° 46' E). If Valdez (1982) is correct in stating that the Mooteh sheep are *O. gmelini*, then these karyotype records would refer to that species, and not to *O. isphaganica*.

Horns cervical, with the tips growing toward the neck; fronto-orbital edge rounded; frontonuchal edge sharp. Females at least sometimes horned.

Males (in winter) with a full-length black ruff extending to the brisket; otherwise, resembling *O. gmelini* in being bibless; white saddle-patch; lower parts of the limbs white, with a black streak running down part of the front; muzzle white; chin white. Shorter teeth and shorter, thinner horns than *O. gmelini*.

Valdez (1982) saw this taxon in Kolah-Gazi and Tange-Sayad (SE and SW of Esfahan); evidently they are not the taxon occurring in Mooteh, N of Esfahan.

Ovis laristanica Nasonov, 1909

Laristan sheep

$2n = 54$ (Schmitt & Ulbrich, 1968), confirmed by Valdez et al. (1978) in specimens from Bamou National Park, near Shiraz, and the Hormud Protected Area (27° 35' N, 54° 05' E). Karyotype 0.1, from a zoo.

Horns homonymous. Frontal surface flat, with sharp angles. All females seemingly horned.

Lighter color, with less red than in *O. gmelini* and *O. isphaganica*; straw-brown in summer, brownish in winter. White saddle-patch present, as in *O. gmelini* and *O. isphaganica*. Short black ruff, restricted to the lower neck and the breast. Occasionally a dark shoulder-patch in winter in the males. Like *O. isphaganica*, with small teeth and shorter, thinner horns than *O. gmelini*.

Ovis vignei Group

urial

Ovis vignei Blyth, 1841

Ladakh urial

Skull length 257–272 mm (males); facial part of the skull relatively long (58%–62.5% of the maximum skull length). Anterior profile quite highly flexed; relatively high braincase behind the horns.

Horns almost in one plane (slightly perverted / supracervical). Frontal surface of the horn tapering upward significantly; outer rib variously developed, but usually smooth. Transverse knobs relatively large; temporal and frontal regions relatively convex.

Brown or rusty cinnamon; no light spots on the sides. Tress on the chest and the neck better developed, forming something of a beard on the throat and the angles of the jaw; bib white; ruff black.

Wakhan and Ishkashim in Tajikistan, and across the Afghanistan border; Ladakh, Kashmir.

Ovis punjabiensis Lydekker, 1913

Punjab urial

Cervical horns; redder color than in *O. vignei*.

Salt and Kala Chitta ranges

Our analysis finds that *O. punjabiensis* is quite distinct from *O. vignei*, and evidently has much in common with *O. bochariensis* (perhaps symplesiomorphic: a greater skull height and long nasals, contrasting with a lower skull length and breadth). An ungrouped skull from Faraza, near Tach (39.13° N, 68.48° E, in the mountains N of Dushanbe), identifies with the *O. vignei* sample. The type of *O. blanfordi* Hume, 1877 (Bolan Pass, near Quetta, near the southernmost Afghanistan border with Pakistan), is "ultra-*bochariensis / punjabiensis*," possibly representing a further taxon, but this needs investigation.

Ovis bochariensis Nasonov, 1914
Bukhara urial

For Valdez (1982), this is a synonym of *Ovis cycloceros*, but we are able to differentiate the two cleanly in our analyses.

2*n*=58 (Pyandzh Kara-Tau, 37°–38° N, 69°–70° E; Sar-Sorak Range, 38°–39° N, 69°–70° E).

Skull length 232–255 mm (males), 218–240 (females).

Horns usually in one plane, or close to it (perverted); sometimes positive (homonymous). Outer rib quite faint; transverse folds small. Horn length up to 73 cm, rarely longer; basal circumference 220–270 mm (mean 24.5 mm).

Rusty cinnamon; no light spots on the sides. Tress on the chest and the neck well developed, forming something of a beard on the throat and the face of the lower jaw.

Right bank of the Amu-Darya and lower Pyandzha rivers, E to Darvaza.

Ovis arabica Sopin & Harrison, 1986
Oman wild sheep

Skull length 243 mm (males). Skull relatively broad and short; nasals relatively long (39% of the skull length, compared with 37.6% in *O. punjabiensis*, and 37% in *O. vignei* and *O. bochariensis*), braincase narrow (28%, compared with 31.2% and 30%), toothrow short (25%, compared with 31.2% and 29%).

Horns everted, twisted spirally. Horn length 720 mm, basal girth 200 mm, horn spread 500 mm. Triangular in cross section; frontal surface narrow; nuchal surface concave, bases close together. Mild angulation on the external edge of the horns, and an even poorer one on the internal edge. Horn color grayish brown.

Body rusty reddish, with reddish spots on the cheeks. Beard long, white, and narrow, extending as far as the corners of the lower jaw, where it is bordered on both sides with dark brown streaks. Saudi Arabian specimen (D. L. Harrison & Bates, 1991:figure 273) with a clear dark line on the flanks; white underside; white below the knees and the hocks.

Wadi Kharbora and Hatta, Oman; and near Sharawrah, on the SE edge of the Rub-al-Khali, Saudi Arabia. Uerpmann (1987) discussed the distributional problems raised by the discovery of *O. arabica*. Its skull, in our analysis, is classified very strongly with the *O. vignei* group, even though it seems counterintuitive that it should belong to this species-group, and not to the *O. gmelini* group, which lives on the opposite side of the Persian Gulf.

Ovis cycloceros Hutton, 1842
Turkmenian sheep

1850 *Ovis arkal* Eversmann.
1852 *Ovis arkar* Brandt.
1905 *Ovis vignei varenzovi* Satunin.

2*n*=58 (Kosh Yeilagh Protected Region, 50 km ENE of Shahrud; extreme E Alborz, Mohammad Reza Shah Wildlife Park, 37° 20' N, 56° 07' E; Badkhyz Reserve, Turkmenistan).

Skull length 257–297 mm (males; mean 272 mm), 231–263 (females; mean 249 mm). Facial part elongated, averaging over 60% of the maximum length of the skull. External nares longer and broader than in *O. arabica*; nasals narrow. Profile behind the horns sloping more steeply.

Horns homonymous (definite positive curvature); exceptionally, bent in a single or almost a single plane (highly perverted); markedly larger and heavier than in neighboring species. Horn length up to 920 mm, basal circumference 240–300 mm (mean 260 mm). Outer rib well developed; frontal surface flat or slightly concave, not tapered outward, or only poorly so; temporal surface flat; transfer skulls prominent, well projected.

Yellowish rusty; no light spots on the sides. Tress luxuriant; light-colored beard on the throat and the base of the lower jaw; white bib; black neck ruff.

Mountains of extreme S Turkmenistan (Karabil', Chengurets Mountains E of Kushka, Badkhyz, Gyaez'-Gyadyk), Kopet-Dagh, Great Balkhan, E coast of the Caspian Sea, Mangyshlak, Ustyurt, NE Iran, Afghanistan.

Mangyshlak sheep are often separated from Kopet-Dagh sheep as a subspecies, *Ovis cycloceros arkal* Eversmann, 1850 (e.g., by Zalkin, 1951); they differ primarily in the absence of animals with perverted horns, and by having a more intense beard, both of which are merely extreme developments of *O. cycloceros*.

Our analysis finds that *O. cycloceros*, *O. bochariensis*, and *O. vignei* are, in fact, distinct. The *O. vignei* sample includes specimens from montane Tadjikistan; SW lowland Tadjik specimens are within *O. bochariensis*; and our *O. cycloceros* specimens are from within the range assigned to *O. c. arkal*. Skull length and breadth, and the frontal arc, contrasted with the low frontal chord, distinguish *O. cycloceros*, especially from *O. bochariensis*; greater skull height and nasal length plus the frontal arc, contrasted with the low frontal chord, distinguish *O. vignei* from both the others.

Ovis severtzovi Nasonov, 1914

Kyzylkum sheep

$2n = 56$, as in argali (Lyapunova et al., 1997; Bunch et al., 1998).

Skull length 260–270 mm (males, mean 265 mm). Lacrimal depression poorly developed.

Horns homonymous, resembling *O. cycloceros*. Maximum horn length 730 mm, basal circumference 230–260 mm (mean 240 mm). Fronto-orbital and frontonuchal edges sharp. Vorobeev & van der Ven (2003)—who treated *O. severtzovi* as an argali—stated that the horns are black, whereas in "other argali" they are wax-colored; in fact, it seems that the horns are waxy-colored when clean, but become dark because of ingrained dirt from wear and use (Atanas Tchobanov, pers. comm. to CPG).

Darker; grayish cinnamon to chestnut or grayish brown. Tress weakly developed, short, not reaching the base of the head; beard absent. Rump-patch very pronounced. Slightly paler on the neck; grayish brown on the flanks; belly and rump white. Mane white, tinged with gray. Head darker than the neck, with white face markings; legs dirty white, with dark, reddish brown stripes. Young animals redder.

High altitude locations in the Kyzylkum Desert, especially Aktau and Nuratau. The Nuratau Ridge, in the Kyzylkum, is an extreme NW spur of the W Pamir-Alai Range that is inhabited by *O. bochariensis*; it is separated from the Kara-Tau by the Golodneya Steppe and the Syr Darya River.

Vorobeev & van der Ven (2003) treated *O. severtzovi* as an argali, but they admitted that it is somewhat similar to a urial in several characters: darker than other argali found in Kyrgyzstan, with only the tip of the nose being light; belly noticeably lighter, grayish; virtually no white spots on the face; long hair on the neck light, short; long white ruff (or white interspersed with dark hairs); and white bib. Lyapunova et al. (1997) found that in a 4-year-old ram, the bib was poorly developed, like an argali; whereas in a 6-year-old ram, it was prominent and snow-white, sharply contrasting with the background coat, which was blackish on the anterior side of the forelegs and brown on the shoulders.

The weak growth of the tress resembles argali; the skull and horns resemble the urial species *O. cycloceros*, but, as noted by Lyapunova et al. (1997), the horns extend from the skull at a lower angle than in urial.

O. severtzovi has been shuffled back and forth between the urial and the argali groups. Valdez (1982) stated that "its skull characteristics clearly establish it to be a urial as verified by Sapozhnikov (1976)," and *O. severtzovi* indeed fall in with the *O. vignei* group in our craniometrics. On the other hand, for Vorobeev & van der Ven (2003), it is an argali; this is supported by its karyotype, as well as some other characteristics, such as the poor development of the bib in young animals (developing, however, with age), and the waxy-colored horns. Considering its distribution, between the ranges of these two major species-groups, and the possibility that it may not be entirely isolated from either of them, we suggest that *O. severtzovi* is a (perfectly valid) species of hybrid origin.

Existing Hybrid Populations

Alborz Red Sheep Hybrid (*Ovis gmelini* × *Ovis cycloceros arkal*)

$2n = 54$ or 55 (Imperial Reserve, just E of Teheran, 35° 41' N, 51° 34' N); $2n = 55$–58 (Parvar Protected Region, 50 km N of Semnan, 36° 06' N, 53° 35' E). Valdez et al. (1978) found that in Parvar, one individual had $2n = 55$, eight had $2n = 56$, seven had $2n = 57$, and eight had $2n = 58$; in the Imperial Reserve, five had $2n = 54$ and nine had $2n = 55$—that is to say, there was no evidence of any deficiency of heterozygotes.

The name given to this hybrid population is *Ovis orientalis* Gmelin, 1774, described from the mountains of Gilan and Mazanderan provinces, Persia (restricted by Nasonov [1923] to the E part of the Alborz Mountains). The subspecies *Ovis orientalis erskinei* Lydekker, 1904, was described from the same general area. Valdez (1982:100) illustrated a skull in the Zoological Museum of the Academy of Sciences in St. Petersburg (specimen ZMAS 937), collected by Gmelin, which may be the type, or at any rate from the type series; it has a sharp fronto-orbital edge, yet supracervical horns. Valdez specified that homonymous, supracervical, and intermediate horn types all occur, especially in the vicinity of Teheran; farther W, where the hybrid *O. orientalis* blend with pure Armenian mouflon (*O. gmelini*), supracervical horns predominate; to the E, where they blend with Transcaspian *O. c. arkal*, the horns tend to be homonymous. A saddle-patch, variable in size, may be present or absent; the neck ruff is highly variable, and may be either similar to that of *O. gmelini* or bibless, or with either a full-length black bib or a white one and a long black ruff.

In our craniometric analysis, three skulls of "Alborz hybrids" are classified strongly with the *O. vignei* group, and a further Alborz skull is marginal between the *O. gmelini* and the *O. vignei* groups (*O. c. arkal* belonging to the latter); while, of two skulls from "North

Persia," one identifies with each group. Of two skulls from Shahroud, 36.25° N, 55.00° E (in the far E Alborz, SE of Gorgan), one is more *vignei*, and the other is marginal between the *O. vignei* and the *O. gmelini* groups.

Kerman Hybrid (*O. laristanica* × *O. cycloceros*)

2*n* = 54 and 55 (Khabr-va-Rouchoun Wildlife Refuge, 28° 45' N, 56° 23' E; Kavir National Park). In our analysis, we had only a skull from Panjgur (Mekran), 26.58° N, 64.06° E (just E of the Pakistani border with Iran), which is well within the morphological range of the *O. gmelini* group.

The horns are homonymous, and the frontal surface is flat, with sharp angles. In winter, their color is darker than in *O. laristanica*, with a white saddle-patch. The Kerman hybrids are bibless. A black ruff is present, but it is variable in length: in some it extends the length of the neck, while in others it is limited to the lower half.

Kerman Province in the Khabr-va-Rouchoun Wildlife Refuge and E and S, at least to the Pakistan border.

Ovis ammon Group

argali

A study of the mitochondrial control region in the argali (*O. ammon*) group (Wu C-h. et al., 2003) found that, using domestic sheep as outgroups, *O. nigrimontana* was the first to separate, followed by a four-way split: *O. severtzovi*; *O. darwini*; *O. "adametzik"*(?) plus *O. sairensis* (i.e., *O. collium*; one specimen of each); and *O. hodgsoni* plus *O. dalailamae* (three and two samples respectively, forming somewhat separate clades).

This group of sheep was revised in detail by Geist (1991a), and his revision is broadly followed here, except that we treat all the subspecies as distinct species, since their differences are discrete and consistent.

Ovis nigrimontana Severtzov, 1873

Karatau argali

2*n* = 56 (in zoo specimens from Karatau; Schmitt & Ulbrich, 1968), as in other argali; but the Y chromosome is acrocentric, not metacentric as in other *Ovis*, including other argali (Bunch, 2000).

Skull length 285–315 mm (males, mean 303 mm), 258–274 mm (females, mean 268 mm). Skull profile steeper behind the horns; occipital projects a little beyond the line of the back of the bony horn cores; facial part higher and broader; prealveolar portion of the upper jaw somewhat longer; nares and nasal bones broader.

Horns homonymous, up to 1020 mm long; basal circumference 285–335 mm (mean 310 mm). Geist (1991a) noted that the horns of old rams have a well-developed combat edge (as they do in bighorn sheep).

Darker than *O. cycloceros*, brownish, with a cinnamon or rusty tinge. Tress poorly developed; elongated hair (up to 11 cm) only on the chest and the lower portion of the neck.

Geist (1991a) described *O. nigrimontana* as being dark brown above, lighter on the neck, with the head darker than the neck color; the ruff creamy white (but comparatively short-haired and gradually blending into the darker shoulder and body hair); a distinctly set-off white rump-patch that does not surround the tail; a white belly, set off by a dark flank stripe; a dark rostrum; and a long tail.

Karatau, 43° N, 70° E°.

Based on the mtDNA control region, Wu C-h. et al. (2003) found that *O. nigrimontana* is the sister clade to other argali (but the bootstrap for the other clade is only 73%).

O. nigrimontana is the other species, apart from *O. severtzovi*, whose allocation to a species-group has been regarded as equivocal, though in this case, it is more evidently an argali. To test its affinities, and those of *O. severtzovi*, we performed a special analysis. *O. severtzovi* fell in with *O. vignei*, and *O. nigrimontana* with *O. polii*—but, in both cases, toward the edges of these respective dispersions.

Ovis karelini Severtzov, 1873

Tianshan argali

1873 *Ovis heinsii* Severtsov.
1898 *Ovis sairensis littledale* Lydekker.
1913 *Ovis ammon humei* Lydekker.
1914 *Ovis polii karelini* forma *melanopyga* Nasonov.
1929 *Ovis polii nassonovi* Laptev.

2*n* = 56.

Skull length 305–335 mm (males, mean 324 mm). Facial part longer, higher, and broader, 60%–63% of the maximum skull length. Nares longer and broader. Interpterygoid fossa broad; condyles very large. Skull behind the horns projecting a little behind the line of the back of the bony cores; upper profile steep.

Horn length up to 1290 mm; basal circumference 330–400 mm (mean 360 mm). Terminal portion of the horns relatively weakly bent sideways, rising up almost parallel to the base. Outer ridge not so distinct; base of the horn blunted.

Lighter, either brownish yellow or brownish cinnamon. Tress absent, but the hair on the front of the chest and the base of the neck slightly elongated (up to 10 cm). Geist (1991a) specified that *O. karelini* differs from *O. polii* by having, in winter, a distinct, dark flank stripe segregating the dark body hair from the light belly hair, with the ruff long-haired and white. Vorobeev & van der Ven (2003) described males in winter as mainly gray-colored, light brown on the head, while the neck, the legs, and the belly are beige; in summer, males are bright red to dirty brown.

Tianshan S to Naryn and the Fergana Range, W to the W extremity of the mountains, N to the Chu-Ili Mountains.

Ovis collium Severtzov, 1873
Kazakhstan argali

1898 *Ovis sairensis* Lydekker.
1923 *Ovis polii collium* forma *albula* Nasonov.
1923 *Ovis polii collium* forma *obscura* Nasonov.

Size as in *O. karelini*. Ascending processes of the premaxillae very long, extending posteriorly between the nasals and the maxillae, their ends separated from the lacrimals only by a short distance.

Outer ridge of the horn more faintly developed than in *O. karelini*; temporal surface more convex. Geist (1991a) noted that the horns are similar in shape to those of *O. polii*, but arise from the skull at a steeper angle, tend to be shorter and thicker, with increased rounding of the combat edges, and have a basal circumference of 330–400 mm.

Dark, vivid, brownish cinnamon (light-colored animals also known). Geist (1991a) described *O. collium* as darker than *O. karelini*, with the rostrum gray rather than white.

Kazakhstan W to about 67°–70° E, E to the Irtysh, S to the Saur and Tarbegatai (including in NW Dzungaria), and the N shore of Lake Balkhash.

Geist did not separate this taxon from *O. karelini*, but our craniometric analyses distinguish it clearly by the much shorter horn cores compared with the sheaths, and by the relatively smaller core circumference; in addition, the skull is narrower and lower, compared with *O. karelini*.

Ovis polii Blyth, 1841
Pamir argali, Marco Polo sheep

Skull length 328–355 mm (males, mean 341mm), 283–305 mm (females, mean 295 mm). Skull behind the horns projecting even less. Upper profile steeper. Interpterygoid fossa broader.

Horn length sometimes reaching 1900 mm; basal circumference 350–400 mm (mean 370 mm). Terminal part of the horns highly twisted outward. Geist (1991a) characterized *O. polii* as having well-developed combat ridges that are triangular in cross section; the horns are normally widely flaring, relatively thin, and light in color, and, as life progresses, they break, rather than "broom," as do those of other argali.

Dark grayish reddish brown, sometimes chocolate. Tress on the neck and the chest slightly larger, reaching 18 cm. Geist (1991a) described the color as light reddish brown to chocolate-brown, usually pale grayish brown or buff. Secondary rump-patch present (i.e., a rump-patch extending onto the thighs, the hindlegs, the belly, the chest, and the forelegs, turning these parts white); tail length 10 cm, without hair, and without a median dark stripe. Light flank stripe. Face light. No ruff on the males in summer, nor at any season on the females.

Vorobeev & van der Ven (2003) described the winter coat as white–grayish brown, with the back brown-white to pale brown; old animals have white faces. The legs were grayish in 5- to 7-year-old males. Young animals were more creamy colored.

Pamir, E extremity of Alai, the Aksai Plateau and the Chatyrkel' region, Kokshaal to Khan-Tengri in the N, and Dzhungar Alatau. Also occurs in the Chinese sector of Tianshan and the mountains adjoining the Pamirs.

Very closely related to *O. karelini*. Geist (1991a) stated that "the segregation in pelage and horn characteristics may not be complete: there appear to be enough differences in pelage, horn, and qualitative skull characteristics." We tested this: are *O. karelini* and *O. collium* different from *O. polii*? Not when skulls alone are considered, but both are quite distinct when the horns are taken into consideration. *O. polii* and *O. collium* have a much shorter horn core length—vis-à-vis sheath length, core circumference, and skull breadth—compared with *O. karelini*. *O. collium*—compared with *O. polii* and, to a lesser extent, with *O. karelini*—has a lower, narrower skull vis-à-vis its length.

Ovis hodgsoni Blyth, 1841
Tibetan argali

1841 *Ovis ammonoides* Hodgson.
1873 *Ovis blythi* Severtzov.
1874 *Ovis brookei* Ward.
1888 *Ovis dalai-lamae* Przewalski.
1892 *Ovis henrii* Milne-Edwards.
1913 *Ovis poli* [*sic*] *adametzi* Kowarzik.

Skull length 357–359 mm.

Horns in old rams with a less-developed combat edge than in *O. polii*, less everted, and usually broomed instead of broken.

Uniformly brown (no light saddle-patch between the shoulders), but in spring, the haunches appearing lighter than the torso. Long-haired, light-colored ruff; rump-patch, distinctly set off in the upper parts, surrounding the tail; tail tiny, about 2.5–5.8 cm; dark, partial flank band separating the light belly hair from the dark body hair; distinct segregation into light and dark fields of hair on the hindlegs and the forelegs, with (normally) a dark stripe from the chest to the hoofs.

Whole Tibetan plateau.

***Ovis ammon* Linnaeus, 1758**
Altai argali

1785 *Ovis argali* Boddaert.
1873 *Ovis argali altaica* Severtsov.
1873 *Ovis argali dauricia* Severtsov.
1873 *Ovis argali mongolica* Severtsov.
1923 *Ovis ammon przevalski* Nasonov.

$2n = 56$ (Gorno-Altaisk Autonomous District, W spurs of the Chikhachev Mountains, 50° 03' N, 89° 25' E). Karyotype 1.0 (Mongolian Altai, 46° 30' N, 93° 00' E).

Skull length 345–395 mm (males, mean 368 mm), 305–337 mm (females, mean 318 mm). Skull very similar to *O. polii*, differing in the longer anteronasal profile and the greater width of the brain case.

Horn length up to 1600 mm; basal circumference 400–550 mm (mean 440 mm). Outer rib usually weakly developed, sometimes disappearing in the basal portion. Temporal surface convex, especially at the base, with two weak combat ridges. Horn surface strongly wrinkled.

In summer, dark brown to almost white, with rufous patches; in winter, dirty white. In summer, the rump-patch diffuse, sometimes absent; in winter, the rump-patch distinct, bordered dorsally, lacking a secondary expansion on the thighs. Light underside often blending into the dark body hair. Tail long, with a dark stripe on the top, sometimes surrounded by a rump-patch. Lower limbs light, sometimes with scattered dark hair. Occasionally a light rostral spot; muzzle and lower jaw light.

Russian and Mongolian Altai, Tuva. Formerly found in Trans-Baikalia (this is, in fact, the type locality), but now extinct there.

***Ovis darwini* Przewalski, 1883**
Gobi argali

1873 *Ovis argali mongolica* Severtsov.
1913 *Ovis kozlovi* Nasonov.
1919 *Ovis comosa* Hollister.
1936 *Ovis ammon intermedia* Gromova.

Small in size; skull length 325–357 mm (mean 339 mm). Skull foreshortened; nasals noticeably short. Horns massive, heavily crenellated, broomed, and tight.

For Geist (1991a), "two extremes [of color] are lumped under 'Gobi argali.'" One, in winter, is dark chocolate-brown, with a gray-brown muzzle and face, the muzzle barely lighter than the rest of the face. The light portions (the lower limbs and the underside) are buff to light brown, and blend into the dark body hair. The other, in winter, has dark brown hair, light lower limbs, a large rostral spot and muzzle, and large light areas behind the shoulders and on the withers. The ruff is extensive, but spotted with dark patches.

Tail length, minus hair, 7.5–10 cm. Overall, *O. darwini* is very similar to *O. ammon*, except for the light ruff and smaller size.

Low, isolated, dry desert ranges, such as the Gobi Altai, Yabarai, to about 112° E.

Is *O. darwini* distinct? We performed a craniometric analysis: it is. Compared with both *O. ammon* and *O. hodgsoni*, *O. darwini* has a shorter, narrower, but relatively somewhat higher skull.

***Ovis jubata* Peters, 1876**
Shansi argali

Skull length 315–335 mm.

Horn length up to 320 mm, basal circumference 400–500 mm.

Dark fawn-gray; light to dark ruff, twice as long as the body hair, blending into the dark shoulder hair; well-developed dorsal neck mane. Rump-patch large, white to buff, distinctly bordered; shoulders and back flecked with white in very old rams; belly light, but grading into the dark body hair; lower front legs creamy; distinct stripes down the forelimbs and the hindlimbs. Tail length 7.5 mm, without hair, but with a dark median stripe.

Shansi, Hebei, Inner Mongolia, Shensi.

We have no experience of this taxon, and thus rely on Geist's (1991a) description and analysis.

***Ovis nivicola* Eschscholtz, 1829**
snow sheep

$2n = 52$ (Korobitsyna et al., 1974).

O. nivicola are related to *O. dalli* and *O. canadensis* from North America, but the evidence clearly indicates that they form a strongly distinct species.

Bunch (2006), on the basis of mitochondrial cytochrome *b* sequences, timed the separation of North American sheep (*O. canadensis* and *O. dalli*) from other sheep, including *O. nivicola*, at more than 3 Ma. Interestingly, in the majority-rule tree (his figure 3), *O. nivicola* was on the *O. ammon* / *O. gmelini* branch, not with the North American sheep, although the support values were not high; nonetheless, it is of note that one of the four snow sheep sequences, the one from the Taygonos Peninsula, was further up the same clade, while the other three (from the Maimandzhinskii Range and Cape Yevreinova, farther to the W) were sister to the rest of this clade. The author proposed that the basal *Ovis* had $2n=52$, like *O. nivicola* (and also *Budorcas*), from which, independently, were derived the *O. canadensis* / *O. dalli* number of $2n=54$, the *O. ammon* group's $2n=56$, and the *O. gmelini* / *O. vignei* progression of $2n=54$ and 58.

Chernyavsky (1962) found the following:

1. In *O. canadensis*, the relative length of the nasals is greater than in *O. dalli* and *O. nivicola*.
2. The skull of *O. nivicola* is significantly wider in the orbits than in the two American species, at 49.2% of the basal length (minimum 47%), compared with 44.9% (*O. dalli*) and 44.5% (*O. canadensis*) (maximum in both, 46%).
3. In *O. nivicola*, the rostral and occipital parts are relatively broader than in the North American species.
4. Unlike the American species, the white rump-patch in *O. nivicola* does not extend onto the back above the base of the tail.

In the Late Pleistocene and Early Holocene, snow sheep were found as far SE as the Russian Far East and as far N to the Yana River delta, SW into W Siberia (Boyeskorov, 2001). The shortness of their extremities shows that, in contrast to the *O. ammon* group, snow sheep are adapted to montane rather than rolling or plateau habitats. Nonetheless, in some regions in S Siberia, they were possibly sympatric.

Interestingly, considering the enormous distributional range of the species, the described subspecies, listed below, seem to be rather weak (following Heptner et al., 1961; and other sources):

1. Kamchatka snow sheep, *Ovis nivicola nivicola* Eschscholtz, 1829 (synonym—*O. storcki*)
   —Skull length 265–300 mm (males, mean 280 mm), 245–265 mm (females, mean 255 mm). Horn sheath 170–270 mm (mean 216 mm); basal circumference 250–325 mm (mean 300 mm). Horn core maximum 110 mm; basal circumference 290–360 mm.
   —Color uniform, fairly dark, without significant lightening on the flanks and the head.
   —Kamchatka.
2. Okhotsk snow sheep, *Ovis nivicola alleni* Matschie, 1907 (synonyms—*Aegoceros* (*Ovis*) *montanus*, *O. middendorfi*, *O. n. potanini*)
   —Size, skull, and horns similar to *O. n. nivicola*.
   —Color usually dark, though lighter individuals are known. Head, neck, and chest brown; tip of the muzzle whitish, nose-bridge black. Forehead to occiput white. Body lighter than the neck. Dark band along the spine. White spot on the flanks behind the axilla. Dark band running along the lower flanks from the elbow to the groin invariably distinct in light-colored animals. Rear half of the abdomen white.
   —Stanovoi Range, Dzhugdzhur Range, SW parts of the Kolyma Range, Taygonos Peninsula.
3. Yakutian snow sheep, *Ovis nivicola lydekkeri* Kowarzik, 1913 (synonyms—*O. lenaensis*, *O. n. lydekkeri* forma *albula*, *O. n. lydekkeri* forma *obscura*)
   —Bony horn core shorter (150–210 mm, mean 175 mm), thinner; basal circumference 250–280 mm (mean 260 mm). Horns much shorter (up to 910 mm) and lighter; basal circumference up to 250–330 mm (mean 290 mm).
   —Color very light, sometimes almost white or yellowish white. Dark brown shades on the cheeks, the neck, the chest, and the front of the legs. Dark band along the spine. Axillae brown; brown bands running posteriorly. Underside white, or with some slight darkening in the chest. Front of the nose, the forehead, the top of the head, the tips of the ears, and the upper neck light or whitish. No dark band on the nose-bridge.
   —From the Lena River in the W (region of the Verkhoyansk Range and other ranges), N parts of the Kolyma Range, and the Anadyr region.
4. Putorana snow sheep, *Ovis nivicola borealis* Severtzov, 1873
   —Very little known; dark, fairly uniform, closely resembling *O. n. nivicola*.
   —Isolated in Putorana (Noril').
5. Chukotka snow sheep, *Ovis nivicola tschuktchorum* Zheleznov, 1994
   —According to Železnov (1994:169), "both sexes lack the yellow color and males are without the big white stain which is characteristic of the Okhotsk [*O. n. alleni*] and Yakutsk [*O. n. lydekkeri*] subspecies. The color is generally bright with brownish-grey shades." The

ProOutfitters website (!) says *O. n. tschuktchorum* has a heavier, lighter coat [light in color?] of soft, woolly hair. Železnov (1994) also said the "craniometric measurements" are different, but no measurements appear to be given in the paper, which is mainly in Russian (with an English abstract).

—"It occurs in the Anadyr district on the Chukot Peninsula, without the Koriacky Mountains; from hilly country in an elevation of 500 m to mountains of an elevation of 1300 m" (Železnov, 1994:169).

6. Koryak snow sheep, *Ovis nivicola koriakorum* Chernyavski, 1966
   —Very similar in coloration to *O. n. nivicola*, but about 30% smaller; the white muzzle and rump-patch not as pronounced. Compact in size, with shorter legs in proportion to the body than its S cousin (*O. n. nivicola*). The ProOutfitters website notes that individuals found at the N extremity of the Koryak Range exhibit characteristics of *O. n. tschuktchorum*.
7. Kodar snow sheep, *Ovis nivicola kodarensis* Medvedev, 1994
   —The original description gives the following characters: "The suture between os praesphenoidale and os sphenoidale is absent. Os lacrimale is very narrow and, in adult individuals, this bone is separated from nasal bones by connections of appendixes [shoots?] of maxillar and frontal bones. They have narrow meatus acusticus, long and massive tooth row. Males have a nearly oval transverse section of the processus cornutus in contrast to Yakutian [*O. n. lydekkeri*] and Okhotian [*O. n. alleni*] subspecies which have a rounded-triangular section. Width of skull is maximal for Siberian snow sheep in Asia but the length is comparatively short [?]."
   —The Kodar mountain ridge, E of Lake Baikal, is "a narrow and non-isolated area" (G. G. Boyeskorov, pers. comm. to CPG).

Are the taxa of the *O. nivicola* group distinct? The results of our analyses are equivocal. They seem to show that no, on the characteristics of horns, they are not distinct; and that taking the skull plus the horns together, *O. n. lydekkeri* differs, on average, from the others by its greater horn sheath compared with horn core circumference, and by its shorter, less thick horns.

Table 64 gives the univariate statistics for the skulls and horns of non–North American male *Ovis*.

Ovis canadensis Shaw, 1804

bighorn sheep

Cowan (1940) recognized several subspecies. Ramey (1995), using morphometrics, proteins, and mtDNA, could find no difference between the four supposed subspecies of desert bighorn, and recognized only *O. c. nelsoni*. While bighorn from the Rocky Mountains (*O. c. canadensis*) are certainly larger than those from the desert ranges (*O. c. nelsoni*), the size overlaps greatly, and other differences are not consistent, so we question whether there are any subspecies at all.

Ovis dalli Nelson, 1884

dall sheep

Skull longer and wider than in *O. canadensis*; horns thinner. Color white, or black, or particolored.

S males (the putative subspecies *O. d. stonei*) have a larger skull size than N males; the white N form has a basal length of 235–256 mm, the black S form 258–274 (Cowan, 1940).

A study of the mtDNA control region by Loehr et al. (2006) found that most of the population of *O. dalli* in British Columbia possessed haplotypes evidently derived from the *O. canadensis* population of the Canadian Rockies, although long enough ago (the authors proposed that this would be prior to the last glacial maximum) that the bighorn-derived haplotypes in British Columbian *O. dalli* are somewhat different, forming their own cluster. They also pointed out that the coloration of *O. dalli* changes clinally across the range, from the pure white form in the N to the dark *O. d. stonei* form in the S; that is to say, there is, in fact, no division between the two near-homogeneous populations. Therefore, we are reluctant to recognize any taxonomic distinction, both because the color changes clinally, and the precise relationship of color to size is unclear; and because the molecular evidence, cited above, suggests that the larger size of the S populations may relate to gene flow from *O. canadensis*.

Myotragus Bate, 1909

Balearic sheep

We list this genus here because it survived into the Holocene. General information about the terminal species, *Myotragus balearicus* Bate, 1909, and its evolution, was given by Bover & Alcover (1999). Its extinction is put between 3700 and 2030 BC on Mallorca, between 10,000 and 1930 BC on Menorca, and between 3650 and 300 BC on Cabrera (Bover & Alcover, 2003).

Table 64 Univariate statistics for the skulls and horns of *Ovis*: males

	Skull l	Nas l	Fr chord	Fr arc	Max teeth	Skull ht	Skull br	Horn core l	Horn core circumf	Horn sheath l	Horn sheath circumf
gmelini											
Mean	252.14	93.29	97.14	127.86	74.71	129.00	132.57	310.00	185.00	587.14	238.57
N	7	7	7	7	7	6	7	3	4	7	7
Std dev	5.900	4.716	6.122	4.880	2.360	4.690	3.101	26.458	12.910	64.991	17.491
Min	246	87	86	120	71	123	128	290	170	460	220
Max	264	98	105	135	78	135	137	340	200	670	270
Alborz hybrids											
Mean	263.14	104.29	93.00	128.86	75.33	139.17	140.50	363.33	206.67	795.00	248.33
N	7	7	6	7	6	6	6	3	3	6	6
Std dev	9.494	5.407	6.325	7.883	6.154	6.145	5.541	41.633	5.774	66.858	9.309
Min	252	97	85	118	67	129	134	330	200	690	240
Max	281	113	103	140	84	146	147	410	210	890	260
isphaganica											
Mean	250.33	93.00	93.25	117.50	70.86	131.50	130.88	230.00	165.00	401.40	210.00
N	6	8	8	2	7	8	8	1	1	5	2
Std dev	15.042	2.507	5.007	—	2.911	6.141	4.883	—	—	21.767	.000
Min	225	87	88	110	67	123	125	—	—	370	210
Max	262	95	100	125	74	145	138	—	—	430	210
laristanica											
Mean	238.50	87.00	87.50	114.00	70.50	126.50	127.00	300.00	165.00	480.00	190.00
N	2	2	2	2	2	2	2	2	2	1	1
Min	225	80	83	103	68	110	114	240	145	—	—
Max	252	94	92	125	73	143	140	360	185	—	—
arkal											
Mean	273.39	104.83	94.12	133.00	75.79	148.00	144.47	348.33	199.84	739.72	260.50
N	36	35	24	26	33	29	34	18	19	29	30
Std dev	7.530	6.128	6.707	10.855	3.305	6.897	5.440	50.322	16.863	116.820	18.200
Min	252	95	81	112	70	136	132	275	170	490	208
Max	297	119	108	150	81	162	155	480	225	990	295
bochariensis											
Mean	250.40	93.30	91.13	125.50	73.10	135.67	130.90	310.83	195.38	610.56	248.89
N	10	10	8	8	10	9	10	6	8	9	9
Std dev	12.937	5.078	5.668	9.562	3.348	7.714	9.098	48.828	20.028	91.598	23.289
Min	233	86	83	115	69	124	118	250	173	510	220
Max	272	102	100	140	81	144	151	380	230	730	300
vignei											
Mean	260.90	95.67	103.00	139.50	96.60	127.33	138.70	348.75	288.33	688.33	254.17
N	10	9	8	8	10	9	10	4	6	9	6
Std dev	9.158	5.723	9.472	9.739	43.482	31.468	5.539	76.635	209.396	127.892	15.943
Min	245	87	92	130	72	72	131	285	180	515	235
Max	275	103	117	158	179	163	148	460	715	930	275
blanfordi											
Mean	260.00	109.00	107.00	155.00	72.00	148.00	134.00	—	—	750.00	—
N	1	1	1	1	1	1	1	—	—	1	—
punjabiensis											
Mean	245.00	91.33	89.67	123.17	70.17	136.17	132.17	331.67	182.50	695.00	240.00
N	6	6	6	6	6	6	6	3	4	3	1
Std dev	5.514	5.125	3.559	12.090	3.920	3.869	6.555	106.105	17.059	107.587	—
Min	238	81	83	100	66	132	122	245	163	585	240
Max	252	94	93	135	77	142	142	450	202	800	240
arabica											
Mean	243.00	96.00	—	—	60.00	—	133.00	—	200.00	720.00	—
N	1	1	—	—	1	—	1	—	1	1	—

(*continued*)

Table 64 (continued)

	Skull l	Nas l	Fr chord	Fr arc	Max teeth	Skull ht	Skull br	Horn core l	Horn core circumf	Horn sheath l	Horn sheath circumf
severtzovi											
Mean	265.00	83.40	96.00	—	72.60	143.00	141.40	300.00	190.00	636.00	245.00
N	2	5	5	—	5	1	5	3	3	5	5
Std dev	7.071	10.668	3.742	—	3.362	—	3.647	45.826	5.000	85.615	13.229
Min	260	70	93	—	69	—	138	260	185	550	230
Max	270	93	102	—	76	—	147	350	195	730	265
nigrimontana											
Mean	300.17	116.50	104.00	—	85.00	162.33	161.67	337.50	237.50	908.00	308.00
N	6	6	6	—	5	6	6	4	4	5	5
Std dev	10.128	10.075	7.823	—	2.550	4.274	2.251	23.979	10.408	108.028	19.558
Min	285	102	98	—	83	157	158	315	225	770	285
Max	315	129	119	—	89	168	164	370	250	1020	330
karelini											
Mean	324.74	124.05	117.73	168.11	92.32	175.17	175.53	374.38	278.50	1033.61	351.39
N	19	19	11	9	19	18	19	8	10	18	18
Std dev	10.545	8.727	7.989	11.050	4.230	8.024	8.815	30.640	24.043	124.615	22.932
Min	303	108	102	150	85	153	160	300	240	740	300
Max	338	140	126	190	102	186	191	395	320	1270	400
polii											
Mean	341.45	132.52	128.16	194.06	90.37	177.14	184.58	412.50	311.67	1286.79	383.19
N	31	25	19	17	27	22	31	8	12	28	26
Std dev	11.051	9.042	8.255	17.444	3.553	28.673	5.772	78.921	21.794	224.715	24.516
Min	306	115	116	165	83	120	174	300	275	780	325
Max	355	152	152	235	96	214	198	515	345	1595	430
collium											
Mean	315.00	117.00	107.00	—	86.71	162.83	180.86	319.00	257.00	1037.86	—
N	7	7	1	—	7	6	7	5	5	7	—
Std dev	13.976	12.369	—	—	4.309	20.134	7.925	112.606	73.621	89.063	—
Min	290	97	—	—	79	122	165	125	130	920	—
Max	335	138	—	—	91	175	188	420	320	1150	—
ammon											
Mean	366.42	142.61	—	—	95.59	191.95	197.32	393.33	340.56	1115.56	451.00
N	19	18	—	—	17	19	19	9	9	18	15
Std dev	12.144	10.013	—	—	3.792	12.860	4.900	31.623	21.279	153.392	30.308
Min	345	125	—	—	90	171	190	360	300	820	400
Max	388	158	—	—	101	214	208	440	370	1450	505
darwini											
Mean	298.33	94.33	111.33	—	89.67	138.67	166.00	245.00	193.33	643.33	240.00
N	3	3	3	—	3	3	3	3	3	3	3
Std dev	11.547	15.822	4.509	—	9.292	18.583	5.000	118.216	56.862	218.880	80.467
Min	285	77	107	—	79	118	161	120	130	440	150
Max	305	108	116	—	96	154	171	355	240	875	305
hodgsoni											
Mean	352.00	136.58	140.00	—	89.62	190.83	189.88	365.00	338.75	1068.75	431.91
N	16	12	2	—	13	6	16	4	4	12	11
Std dev	10.602	16.082	—	—	5.026	12.844	11.848	20.817	26.575	64.390	20.413
Min	335	110	136	—	80	173	150	340	310	1010	410
Max	375	162	144	—	97	204	203	390	370	1250	480
nivicola											
Mean	281.78	90.22	115.56	170.00	74.35	151.59	173.09	219.48	302.10	813.87	327.17
N	32	32	9	1	26	29	33	29	31	31	30
Std dev	5.751	4.591	3.941	—	3.019	5.729	7.217	21.016	22.278	123.772	18.555
Min	272	79	112	—	68	140	158	180	250	600	285
Max	296	100	123	—	80	170	187	270	335	1110	360

Table 64 (continued)

	Skull l	Nas l	Fr chord	Fr arc	Max teeth	Skull ht	Skull br	Horn core l	Horn core circumf	Horn sheath l	Horn sheath circumf
lydekkeri											
Mean	277.15	90.25	110.50	145.00	75.42	144.54	169.00	175.56	258.89	708.46	296.54
N	13	12	6	1	12	13	13	9	9	13	13
Std dev	9.608	5.723	9.503	—	3.728	4.875	6.856	18.105	22.469	123.075	22.303
Min	258	81	103	—	70	138	158	150	230	530	265
Max	291	98	127	—	80	153	178	210	290	910	325
alleni											
Mean	274.33	89.67	105.00	—	76.00	144.67	165.33	188.33	268.33	676.67	305.00
N	3	3	1	—	3	3	3	3	3	3	3
Std dev	12.014	6.658	—	—	3.464	7.638	9.292	24.664	33.292	205.994	25.981
Min	262	82	—	—	74	138	155	160	230	460	275
Max	286	94	—	—	80	153	173	205	290	870	320
type of *lenaensis*	281	—	—	—	70	—	—	169	—	865	312
mean of *kodarensis*	288	90	—	—	80	—	167	186	—	1000	320
mean of *koriakorum*	269	81	114	—	82	—	167	186	281	713	306

Almost certainly, it became extinct as a consequence of the earliest human occupation of the islands.

Although at first Lalueza-Fox et al. (2002), using a 338 bp segment of the mitochondrial cytochrome *b* gene, placed *Myotragus* as part of a three-way split with *Ovis* and *Budorcas*, bootstrap support for this placement was below 50%. A later study (Lalueza-Fox et al., 2005) managed to sequence the entire 1143 bp of the cytochrome *b* gene, as well as fragments of 12S and the nuclear 28S rDNA gene, and on this basis achieved higher support for a sister-group relationship just with *Ovis*. They pointed out that the geological date of the isolation of the Balearics—hence, of *Myotragus*—is 5.35 Ma, and on this basis they suggested that the dates of certain other splits could be calculated (see above, under the Caprini).

Rupicapra de Blainville, 1816

chamois

$2n=58$, only one autosomal pair being metacentric (Soma & Kada, 1986).

Relatively slenderly built; short, upright, nearly smooth black horns, characteristically hooked backward. Gazelle-like, black-and-white pattern on the face; body dark (but varying), always with a sharply white underside.

Until quite late in the 20th century, most authors recognized only a single species, *Rupicapra rupicapra*, typified by the chamois of the Alps. Then Lovari (1977) drew attention to the distinctiveness and continued survival of the Abruzzo chamois, which, in a brief review of the genus later (Lovari, 1987), he recognized as a species distinct from the Alpine chamois but conspecific with that of the Pyrenees, thus attributing two species to the genus: *R. pyrenaica* (with subspecies *R. p. parva*, *R. p. pyrenaica*, and *R. p. ornata*) and *R. rupicapra* (with subspecies *R. r. cartusiana*, *R. r. rupicapra*, *R. r. tatrica*, *R. r. carpatica*, *R. r. balcanica*, *R. r. caucasica*, and *R. r. asiatica*). The two species recognized by him are different in cranial features and behavior, as well as genetically (Nascetti et al., 1985). Except for the white face and the white upper throat, with a black line running from the horn through the eye to the snout, *R. rupicapra* is very dark, and nearly black in winter; whereas *R. pyrenaica* has large, light areas on the neck (separated from the white throat by a strong black stripe, running from the ear to the base of the foreleg) and on the haunch. In summer, *R. rupicapra* is lighter, but retains the black legs; whereas *R. pyrenaica* is medium grayish brown all over, except (again) for the black legs. The genetic divergence between *R. rupicapra* and *R. pyrenaica* is considerable; the genetic divergence between the two putative subspecies *R. p. pyrenaica* and *R. p. ornata* is considerably less, although these subspecies appear to have been separated already by the beginning of the last ice age (Masini & Lovari, 1988).

Study of allelic variation in microsatellites (Pérez et al., 2002) confirmed that the deepest divergence

Table 65 Univariate statistics for *Rupicapra*: skulls

	Gt l	Par br	Fr br	Orb vert	Orb horiz	Nas l	Nas br	Max teeth
Males								
cartusiana								
Mean	212.63	61.00	110.67	36.56	37.89	69.00	19.78	62.22
N	8	9	9	9	9	9	9	9
Std dev	3.068	1.323	2.236	1.236	.782	6.325	1.481	2.108
Min	207	60	107	35	37	58	18	59
Max	216	63	114	39	39	77	22	65
rupicapra								
Mean	213.14	63.43	110.14	37.29	39.14	65.83	19.86	58.29
N	7	7	7	7	7	6	7	7
Std dev	4.981	1.512	2.610	.756	.690	6.676	1.345	2.563
Min	205	61	107	36	38	57	18	53
Max	220	65	114	38	40	74	21	61
tatrica								
Mean	213.00	—	110.00	—	—	—	—	—
N	1	—	1	—	—	—	—	—
balcanica								
Mean	217.00	62.00	106.17	37.33	39.33	67.00	20.75	59.29
N	2	6	6	3	3	4	4	7
Std dev	2.828	2.098	3.371	1.528	.577	2.582	1.258	.756
Min	215	60	101	36	39	64	19	58
Max	219	65	111	39	40	70	22	60
asiatica								
Mean	211.00	—	—	—	—	—	28.67	62.67
N	3	—	—	—	—	—	3	3
Std dev	.000	—	—	—	—	—	.577	2.082
Min	211	—	—	—	—	—	28	61
Max	211	—	—	—	—	—	29	65
caucasica								
Mean	209.00	64.00	108.25	38.00	37.00	71.25	—	58.50
N	3	4	4	4	4	4	—	4
Std dev	3.464	1.414	2.217	.816	1.414	3.403	—	3.000
Min	207	62	106	37	36	67	—	55
Max	213	65	111	39	39	74	—	61
carpatica								
Mean	224.00	64.67	115.00	37.00	41.00	69.00	20.67	60.33
N	2	3	3	3	3	3	3	3
Std dev	.000	1.528	2.646	1.000	1.000	6.245	1.528	3.215
Min	224	63	112	36	40	62	19	58
Max	224	66	117	38	42	74	22	64
parva								
Mean	—	—	100.00	—	—	50.00	—	53.00
N	—	—	1	—	—	1	—	1
pyrenaica								
Mean	201.50	59.50	105.25	35.50	37.00	67.25	18.50	54.75
N	4	4	4	4	4	4	4	4
Std dev	3.697	1.291	2.062	2.082	1.414	3.304	.577	2.062
Min	197	58	103	33	35	63	18	52
Max	205	61	108	38	38	71	19	57
ornata								
Mean	204.00	64.50	109.00	35.00	36.00	62.50	—	56.00
N	2	2	1	1	2	2	—	2
Std dev	2.828	.707	—	—	2.828	3.536	—	.000
Min	202	64	—	—	34	60	—	56
Max	206	65	—	—	38	65	—	56

Table 65 (continued)

	Gt l	Par br	Fr br	Orb vert	Orb horiz	Nas l	Nas br	Max teeth
Females								
cartusiana								
Mean	206.80	61.20	104.40	38.00	37.80	64.20	19.40	60.40
N	5	5	5	5	5	5	5	5
Std dev	4.266	1.643	1.342	1.414	.837	4.764	.548	2.702
Min	201	59	103	37	37	59	19	57
Max	211	63	106	40	39	69	20	63
rupicapra								
Mean	208.10	62.64	105.36	37.18	38.55	65.91	19.27	60.00
N	10	11	11	11	11	11	11	11
Std dev	6.641	1.963	3.906	1.328	1.293	4.482	1.272	3.066
Min	200	59	98	35	36	58	17	55
Max	222	65	112	39	40	73	21	66
tatrica								
Mean	211.00	—	106.00	—	—	—	—	—
N	1	—	1	—	—	—	—	—
balcanica								
Mean	208.67	61.67	104.33	36.00	39.00	69.00	20.00	60.20
N	3	3	3	1	2	2	2	5
Std dev	3.055	5.508	6.351	—	.000	16.971	.000	2.168
Min	206	56	97	36	39	57	20	58
Max	212	67	108	36	39	81	20	63
asiatica								
Mean	211.00	—	—	—	—	—	—	63.00
N	1	—	—	—	—	—	—	1
caucasica								
Mean	207.00	65.00	106.00	38.00	37.50	69.50	—	58.00
N	1	2	2	2	2	2	—	2
Std dev	—	4.243	8.485	.000	.707	3.536	—	1.414
Min	207	62	100	38	37	67	—	57
Max	207	68	112	38	38	72	—	59
carpatica								
Mean	216.00	65.00	110.00	35.00	39.00	66.00	19.00	59.00
N	1	1	1	1	1	1	1	1
pyrenaica								
Mean	197.00	60.18	101.82	34.73	36.36	64.00	17.09	55.91
N	10	11	11	11	11	11	11	11
Std dev	3.771	1.079	2.994	1.421	1.120	5.348	.539	1.868
Min	189	58	98	31	34	55	16	51
Max	202	62	108	36	38	72	18	58
ornata								
Mean	202.00	60.67	102.33	35.33	35.67	63.00	—	56.67
N	1	3	3	3	3	3	—	3
Std dev	—	1.155	3.215	.577	.577	1.000	—	2.517
Min	202	60	100	35	35	62	—	54
Max	202	62	106	36	36	64	—	59

was indeed between the *R. rupicapra* group and the *R. pyrenaica* group. Within the *R. rupicapra* group, the diversity within the Alps encompassed that among other localities, although samples from the Carpathians, the Tatra, the Balkans, and the Caucasus all occupied their own distinct clades; within the *R. pyrenaica* group, samples from the Cantabrian Mountains were quite distinct from those from the Pyrenees, although the specimen from the Apennines was more distinct still.

Pemberton et al. (1989) examined the distinctiveness of the Chartreuse chamois (*R. cartusiana*) from the Alpine chamois (*R. rupicapra*) in allelic variation

Table 66 Univariate statistics for *Rupicapra*: horns

	Horn l	Horn br tips	Horn base ap	Horn base diam
Males				
cartusiana				
Mean	236.50	90.13	27.87	26.00
N	8	8	8	8
Std dev	16.000	39.219	2.100	1.195
Min	220	56	25	24
Max	260	155	31	28
rupicapra				
Mean	237.29	96.14	29.36	26.64
N	14	14	14	14
Std dev	10.186	16.129	1.906	1.737
Min	218	75	26	24
Max	259	135	33	30
balcanica				
Mean	235.33	101.50	26.50	23.75
N	3	2	4	4
Std dev	10.970	—	3.109	3.403
Min	223	89	22	19
Max	244	114	29	27
asiatica				
Mean	197.00	59.50	—	—
N	3	2	—	—
Std dev	14.799	—	—	—
Min	180	58	—	—
Max	207	61	—	—
caucasica				
Mean	220.75	86.25	24.00	22.00
N	4	4	4	4
Std dev	12.121	12.764	1.155	1.414
Min	203	72	23	20
Max	229	103	25	23
carpatica				
Mean	232.50	84.00	28.50	26.00
N	2	2	2	2
Std dev	228	76	28	25
Min	237	92	29	27
parva				
Mean	180.00	9.00	—	—
N	1	1	—	—
pyrenaica				
Mean	98.67	45.67	19.67	17.67
N	3	3	3	3
Std dev	3.215	10.970	1.528	2.082
Min	95	33	18	16
Max	101	52	21	20
ornata				
Mean	141.00	—	29.00	24.00
N	1	—	1	1
Females				
cartusiana				
Mean	207.80	93.60	24.20	21.20
N	5	5	5	5
Std dev	9.338	23.104	.837	1.643
Min	197	59	23	19
Max	218	117	25	23
rupicapra				
Mean	199.22	86.00	22.56	20.11
N	18	18	18	18
Std dev	21.945	22.526	1.790	1.875
Min	168	47	19	17
Max	251	131	26	24
balcanica				
Mean	243.00	101.00	24.00	22.00
N	1	1	1	1
asiatica				
Mean	180.00	75.00	—	—
N	1	1	—	—
caucasica				
Mean	185.00	80.00	18.00	15.00
N	2	2	2	2
Std dev	183	48	16	14
Min	187	112	20	16
carpatica				
Mean	250.00	97.00	25.00	25.00
N	1	1	1	1
pyrenaica				
Mean	78.44	36.88	16.44	14.56
N	9	8	9	9
Std dev	11.148	4.224	1.509	1.740
Min	55	30	13	11
Max	93	42	18	17
ornata				
Mean	126.00	—	25.50	22.00
N	2	—	2	2
Min	117	—	22	19
Max	135	—	29	25

across 10 different nuclear loci (a further 45 loci proved to be monomorphic). The Chartreuse chamois was more distinct from the Alpine populations than they were from each other, although there were no fixed differences. Unfortunately, individuals from the Alpine population had already been introduced into the Massif de Chartreuse, and the authors suggested that, if possible, contact between the introduced and indigenous populations should be prevented.

Lovari & Scala (1980) compared the putative taxa of *Rupicapra* by both principal component and discriminant analyses, concentrating on comparing the taxa *R. pyrenaica* and *R. ornata* with each other and with other taxa, using the measurements for all but the nominate *R. r. rupicapra* in the tables in Couturier's (1938) book. Even using simple principal components, *R. pyrenaica* and *R. ornata* were strongly differentiated from other taxa, but differed only on average from each other. Using discriminant analysis, however, *R. pyrenaica, R. ornata, R. carpatica, R. balcanica,* and a *R. caucasica/cartusiana* pairing were all distinct at the one-standard-deviation level on the skulls, but not on the horns, where only *R. pyrenaica* was distinct (there was no sample for the horns of *R. ornata*). Lovari & Scala (1984), using museum samples, some of the measurements in Couturier (1938), and fresh hunter-killed samples, compared horn measurements of *R. asiatica* and found them indistinguishable from other *Rupicapra* taxa.

Information on the described taxon *R. r. tatrica* is difficult to obtain, although a few comparisons have been made by Hrabě & Koubek (1984), who found that there was considerable sexual dimorphism, and that, in general, the Tatra form is larger than that from the Alps, including an introduced population of Alpine chamois in the Jeseniky Mountains of the Czech Republic; on the other hand, in *R. r. tatrica* the nasals are narrower in both sexes, and they are also shorter in the males.

For our analyses, we, like Lovari & Scala (1980), used the datasets of Couturier (1938).

Sexual dimorphism is considerable in all taxa (Turkey and the Caucasus may not be as dimorphic, but the sample size is very small). Although there is only one Carpathian individual of each sex, each is very large (the male is the largest in the entire series). Pyrenees chamois are, sex for sex, smaller than any other except the Abruzzi ones, which average slightly larger.

The differences between Pyrenees and Abruzzi chamois and the others are very clear on the characters of their horns. The single specimen from Santander is much larger than those from the Pyrenees and Abruzzi. Those from Turkey appear rather small, followed by the Caucasus specimens. The difference between Pyrenees chamois and the rest in the relative distance between the horn tips is dramatic; the Santander specimen has very long horns, with the tips very close together. On our discriminant analysis using just horns, the divisions are Pyrenees plus Abruzzi, Carpathians, and the rest.

We propose the arrangement given below, much of which is provisional, although some definite conclusions are possible.

***Rupicapra pyrenaica* Bonaparte, 1845**

Apart from its distinctive pelage, *R. pyrenaica* differs from *R. rupicapra* in its smaller size, the consistent absence of an ethmoid vacuity, the greater angle of the horn cores to the frontals, the presence of a conflict posture, and distinctive courtship patterns. It is not adapted to the cold environments of *R. rupicapra* (Masini & Lovari, 1988).

Pyrenees.

***Rupicapra ornata* Neumann, 1899**

Using simple principal components, Scala & Lovari (1984) were able to distinguish samples of *R. ornata* and *R. pyrenaica* fully (despite their evident close relationship) on the basis of seven skull measurements, but the two overlapped on the basis of seven horn measurements. The authors found no overlap in the length of the frontomaxillary fontanelle: 16–21.6 mm in *R. ornata*, but 3–10 mm in *R. pyrenaica*.

Abruzzi.

***Rupicapra parva* Cabrera, 1911**

R. parva is, as its name suggests, much smaller than the other two SW Mediterranean species. Despite our limited dataset, *R. parva* does seem likely to be distinct, especially given its striking separation from *R. pyrenaica*.

Cantabrian Mountains. The very existence of chamois in this region seems to be poorly known, and more attention should be paid to *R. parva* for conservation purposes, among others.

***Rupicapra rupicapra* Linnaeus, 1758**

1938 *Rupicapra rupicapra cartusiana* Couturier.

The Chartreuse chamois (*R. r. cartusiana*) overlaps in all described character states with chamois from the Alps (*R. rupicapra*), even if they are somewhat different on average. We cannot comment on the distinctiveness of the taxon *Rupicapra rupicapra tatrica* Blahout, 1972, which may or may not be different from the Alpine chamois (the available measurements, however, are very different from those of the Carpathian species, *R. carpatica*).

Alps and neighboring ranges.

***Rupicapra carpatica* Couturier, 1938**

This large, very dark chamois differs absolutely from other taxa of the genus. Again, despite the relative paucity of data, we have no doubt that *R. carpatica* is a thoroughly distinct species.

Carpathians.

?*Rupicapra balcanica* Bolkay, 1925

Evidence as to the distinctiveness of this taxon is lacking, and more specimens need to be studied.

Balkan Mountains.

Rupicapra asiatica Lydekker, 1908

Rupicapra tragus caucasica Lydekker, 1910, almost certainly is a synonym, but the evidence in both cases is poor.

On geographic grounds, it does seem probable that *R. asiatica*, the chamois of Turkey and the Caucasus, is distinct from others, but more evidence is needed.

Nemorhaedus Hamilton Smith, 1827

goral

The spelling of the name *Nemorhaedus* is questionable. Corbet & Hill (1992:270) wrote: "*Naemorhedus* Smith, 1827. . . . Subsequent variation of spelling such as the etymologically correct *Nemorhaedus* are unjustified emendations; there is no evidence of an error in the original publication." Pending a submission to the ICZN (in view of the use of several different spellings in the literature), we will retain the familiar spelling, but it should be noted that it is not technically beyond reproach.

$2n = 55$ for *N. baileyi* (Soma & Kada, 1986; as "*Nemorhaedus goral cranbrooki*"), which, of course, also implies the existence of $2n = 54$ and 56, there being no sex chromosome / autosome translocation; $2n = 55$ for *N. griseus*; and $2n = 56$ for *N. goral*.

Nemorhaedus and *Capricornis*, with which it has been universally linked—even, at times, being placed in the same genus (Groves & Grubb, 1990)—were associated by A. W. Gentry (1992) with *Ovibos*, with which they share some unusual features among the Caprini: notably, the cranial roof is not inclined, and the ethmoid fissure is lost.

Skull generally more apomorphic than any of the serows. Lacrimal fossa small or absent. Lacrimal bone narrowed dorsoventrally. Facial angle steeper than in serows. Zygomatic arch slanting upward along the anteroinferior orbital margin and closely following the lacrimal/malar suture, ending at a masseteric knob on the maxilla. Nuchal surface lower and more horizontal (corresponding to the greater cranial kyphosis) than in serows; frontoparietal angle usually 120°–130°, parieto-occipital angle 135°–150°. Nasal process of the premaxilla narrows posterosuperiorly along the maxillary margin. Nasals relatively long, separated from the maxilla by a hiatus (except in red goral); smaller splint bones (interstitials) sometimes present alongside the nasals, but not filling in the lacuna. No marked lateral nasal processes. Nasofrontal suture very narrow, arch-shaped backward, sometimes with a small median wedge on the frontals (except in red goral, where it resembles that of serows); long frontal processes running lateral to the nasals, reaching forward to the maxillae, but sometimes separated by a long lacrimal tongue. Infraorbital foramen just behind the level of P^2, and halfway up the height of the rostrum. Supraorbital foramina generally single, or with a tiny accessory opening (but commonly double, the two openings equal in size, in Chinese goral). Palatine foramina behind the level of the second molars. Temporal fossa short and high. Distance from the base of the horn cores to the margin of the orbit equal to or greater than the diameter of the orbit, forming short, stout, tapering pedicels. Rostrum more or less parallel-sided; tips of the premaxillae broad.

Horns broad at the base, sometimes hardly ringed at all, but the rings more prominent in Chinese goral; individual and, especially, age variations makes this latter difference rather hard to appreciate.

Preorbital gland very small, not invaginated. Foot gland small, but with a long, back-bent duct.

A discriminant analysis of all goral samples, using just four cranial variables (minimized, in order to be able to include a maximum number of crania), finds that there is overlap between all taxa, but the results are sufficient give an outline of affinities within the genus, with a general differentiation between Himalayan, Russian, Chinese, Burmese, and red goral.

We then separated the Chinese skulls into three groups: W mountains, for which the name *Nemorhaedus griseus* is available; E and C China (prior available name, *Kemas arnouxianus*); and Yunnan, bordering on the region of Burma, whence *Nemorhaedus evansi* was described. Most of the Yunnan skulls are from Tengyueh (= Tengchong, 25° 01' N, 98° 28' E, 1580 m); two are from Huiyao (not found); and one is from Likiang (= Dayan, 26° 48' N, 100° 16' E, 2394 m). Because Likiang is so much farther NE (at least from Tengyueh), we decided to keep this skull separate from the others. *N. griseus* and so-called *Kemas arnouxianus* are larger than the Yunnan goral and *N. evansi*. The Likiang skull is large, like *N. griseus*. Taking this question even further with a discriminant analysis, we found no evidence, at least as far as the cranium is concerned, for differentiation between gorals from W and from E-C China, but *N. evansi* is distinct; there seems to be no justification for separating the Yunnan goral from *N. evansi*, except that the single skull from Likiang is more like *N. griseus*.

The taxonomic differentiation is obviously very marked, but given the (slight) morphometric overlap, the evidence of the external characters is crucial in determining whether the various gorals rank as distinct species.

In discriminant analysis, the Chinese and Russian Far East taxa are totally separate, and thus should be classified as distinct species: *N. griseus* and *N. caudatus*, respectively. The gorals of Shanxi, sometimes separated as a taxon *vidianus*, fall into the range of *N. griseus*. Thus, this brings the number of recognizable goral taxa to three: *N. griseus*, *N. caudatus*, and *N. evansi*.

The Himalayan (*N. goral*, *N. bedfordi*) and red (*N. baileyi*) gorals are of rather small size, below that of (most specimens of) Chinese taxa, but averaging larger than *N. evansi*. The horns of the Himalayan goral(s), however, are as long as those of the Chinese goral (*N. griseus*); only *N. evansi* and *N. baileyi* are short-horned. As far as the lacrimal index is concerned, *N. griseus* and *N. evansi* are alike in having relatively deep, short lacrimals. This offers some support to the separation of a Himalayan taxon and of red goral as separate entities; the number of recognizable taxa of goral therefore now rises to five; whether or not there are two distinct Himalayan taxa will depend on pelage analysis (below).

We find that six taxa show consistent phenotypic differences: the two Himalayan gorals, the Chinese goral, the Burmese goral, the long-tailed goral, and the red goral. Hence, we recognize six species. Using craniometry, they cluster into four groups: (1) the two Himalayan species, *N. goral* and *N. bedfordi*; (2) the Chinese *N. griseus* and the Burmese *N. evansi*; (3) *N. caudatus*; and (4) the single skull representing *N. baileyi*.

Nemorhaedus goral (Hardwicke, 1825)

Himalayan brown goral

1825 *Antilope goral* Hardwicke. Kathmandu.
1908 *Naemorhedus hodgsoni* Pocock. Sikkim.

Skins seen: 21; as well as observations of about 10 living specimens in the Delhi Zoo.

Color either medium brown, with black tips to the hairs, giving a rabbity effect; or slightly grayer, to gray-brown; or pale or dark fawn. Legs browner, to very bright tan, or white on the forelegs only. Underside paler gray. Throat and chin variably white, sometimes interrupted under the jaw; lips white. Dorsal stripe usually weak, sometimes fading behind the withers. Tail short, the distal half black. Short-legged; short-eared.

No nasal interstitials in 25 skulls; no trace of a lacrimal fossa.

Localities of specimens seen by us: Wangdou Potrang, 4500 ft., Bhutan; Ramchu, 11,000 ft., Nepal; Bagaili, E Kumaun; Sikkim; Apoon and Deora, Gorkha, Nepal; Katmandu; near Masuri, Garwhal, Uttar Pradesh; Naini Tal; Molta; Shadra.

Hodgson's and Hardwicke's *Antilope goral*, from "Himalayal range and mts. of Nepaul frontier" was "grey mouse-color (but almost white about the lower part of the neck and throat), and darker . . . inclining to ferrugineous about the legs." The type was in the Barrackpore menagerie; it had been sent from Kathmandu, and, given this information, Lydekker (1906) used the name *Naemorhedus goral* for the E (brown) form and named the W (gray) form *N. bedfordi*. Pocock (1913) argued that, on the basis of the description, Lydekker had used Hardwicke's name wrongly, so Pocock renamed the E (brown) form *N. hodgsoni*. Considering that the type locality of *N. goral* is Kathmandu, where only brown goral occur, and Hardwicke's description of the color could just about fit either brown or gray gorals (though not the extremes), we think that Lydekker's interpretation is more likely.

Ward (1925) placed the boundary between the gray and the brown gorals at roughly the Sardah River between Uttar Pradesh and Nepal; this seems too far E, as specimens from Kumaun, Uttar Pradesh, are all of the present (brown) goral species. The Sutlej River seems a more likely boundary.

Nemorhaedus bedfordi (Lydekker, 1906)

1906 *Urotragus bedfordi* Lydekker. Dharmsala.

Skins seen: 13.

Gray to gray-brown to yellow-gray. Legs lighter, yellower, with a dark brown line down the front, fading on the pasterns. Underside off-white. Chin creamy, this tone reaching the whole way down the throat. Merest trace (if that) of a dorsal stripe. Forehead sometimes slightly darker than the head.

Chamba; Kulu, 7500 ft.; Jagatagukh (or Jagathsukh), Kulu; Kishtwar; Manali, 4 mi. S of the Kulu Valley; Kathai Nullah, Kaj-I-Nag, 7000 ft., Kashmir; Palumpire, Jhelum Valley, Kashmir; N bank of Jhelum River, W of Srinagar; Dharmsala; Wardwan, Kashmir; LohuMang cliffs, Wardwan Valley, Kashmir; Sind Valley, Kashmir.

Pocock (1910a) said that both gray and brown goral occur in Nepal, but there seems to be no evidence for this; likewise, Pocock said they both occur at Chamba, but all of the eight skins sent to the British Museum by Rodon—whom Pocock cited in evidence for this statement—are gray. Tom Roberts (pers. comm. to PG) saw goral in March in C Nepal and thought they looked exactly similar in coat color to the goral in Pakistan, with which he was familiar from his 34 years of experience. He found that the latter tended to look gray in their summer coat and brown in winter, with the males grayer than the females. The evidence of the museum skins, which are of both sexes and from different seasons, does, however, indicate a consistent difference between W (mainly Kashmir) and E Himalayan goral. As noted above (in the morphometrics part of the introductory section to *Nemorhaedus*), there are no evident skull or horn differences, and their taxonomic separation depends on pelage characters alone, which, however, is consistent in our material.

Nemorhaedus griseus Milne-Edwards, 1871

Chinese goral

1871 *Nemorhedus griseus* Milne-Edwards. Moupin, Sichuan.
1874 *Antilope (Nemorhedus) cinerea* Milne-Edwards. Moupin, Sichuan.
1888 *Kemas arnouxianus* Heude. Jixien, Zhejiang.
1890 *Kemas henryanus* Heude. Ichang, Hubei.
1894 *Kemas aldridgeanus* Heude. Yitchang, Hubei.
1894 *Kemas fantozatianus* Heude. Mountains of Kiuntscheou, right bank of the Middle Han River, Hubei.
1894 *Kemas iodinus* Heude. E Sichuan.
1894 *Kemas niger*, *Kemas fargesianus*, *Kemas initialis*, and *Kemas versicolor* Heude. Chenkouting, Sichuan.
1894 *Kemas vidianus* and *Kemas galeanus* Heude. Yu Ho Mountains, S Shaanxi.
1894 *Kemas xanthodeiros* and *Kemas pinchonianus* Heude. W Sichuan.

Skins seen: 27; as well as observations of about 20 living specimens in Chinese zoos.

Dark to light gray, with varying amounts of brown and a varying black overlay (more so in the adults than in the juveniles, and more in the males than in the females); range of colors actually much the same as in *N. goral*. Front and inner side of the shanks very pale, whitish (juvenile) or yellowish (adult), from slightly above the knees or slightly below the hocks, often with a dark line down the front; thigh color sending a wedge down the outer side to halfway down the shank; underside light gray (but the groin and the axillae, as usual, white); throat white, with a golden tone, not extending much to the interramal, but often visible on the lateral surface of the neck; chin dark; dorsal stripe generally clear, thick, dark gray-brown or black, but not always prominent; tail very black, short, but bushy for most of its length; forehead brown or black-brown in the adults (just a brown line in the juveniles); nose zone clearer. Hair fairly long, coarse in all seasons.

Localities of specimens seen by us: Ichang; Sichuan, 6000–7000 ft.; Mupin (= Baoxing), Sichuan, 30° 29' N, 102° 49' E; O-er, near Wei Chow; Wen-chwan Hsien, Si-ho, Sichuan; Wen Chuan, Sichuan, 31° 20' N, 103° 31' E, and vicinity (Tsao Po, 15 mi. to the SW; Chengou Forks and Chen Lliang, both 30 mi. to the W); Loung-Ng'ou-fou, Sichuan; Dun Shih Goh, Sichuan, 30° 25' N, 102° 50' E; Gan Yang Go, Sichuan, 30° 20' N, 102° 30' E; Kuan-Hsien; Ta-tsien-pou (= Tatsienlu, Sichuan, 30° 03' N, 102° 13' E); "Korea"; "E Tibet"; Yenping, Fujian; Patungxian, Yangtze Gorges, Hubei; Wa Shan, Liujiang, and Batang, Sichuan; Beijing; 30 mi. S of Fengxiangfu, Shaanxi, 10,000 ft.; Mt. Hoaysan, 3943 m, Shaanxi; Paochi, Shanxi; Kwei hua cheng, Shanxi; Tung-ling, Hebei; Mai Tai Chao, Shanxi; S of Fengsiangfu, Shaanxi, 34° 24' N, 107° 29' E; Likiang, 27° 04' N, 100° 12' E; Yenping, Fujian, 26° 38' N, 118° 10' E; Tai Pei Shan district, Shaanxi, 33° 56' N, 107° 41' E; Liu-Tsuen, Shaanxi, 34° 02' N, 108° 55' E; Mun San Tsai; Kuei-hua-cheng; Chian Shien; Lianghokou; Hokow, 30° 01' N, 101° 08' E.

Other records: Taukwan, Sichuan (Jacobi, 1923); Gego, 55 mi. SEE of Tatsienlu, 6000 ft. (Orr, 1938); near Ichang, Patung Hsien, and the Yangtze Gorges, Hubei; Wa Shan, Liuyang, and Batang, Sichuan (G. M. Allen, 1940).

The foot gland was bigger in a goral from Shaanxi than in one from the Himalayas (Pocock, 1918).

Interstitials occur in about one-third to one-half of the skulls (Ichang 0/6, Sichuan 7/18, Shaanxi 0/1); even without interstitial bones, the narial notch is not usually as deep as in either species of Himalayan goral (*N. goral* and *N. bedfordi*). The skull, on average, is deeper than in the two Himalayan goral species, and the dorsal outline rises above the orbits, rather than being at about the same level. Usually there is at least a trace of a lacrimal fossa (taking the form of a fine line of minute pits in the bone); the supraorbital foramina is commonly double, the two openings equal in size. The horn rings tend to be more prominent than in the Himalayan gorals; the horn tips are hardly downturned.

Out of the 14 species described by Heude and the two by Milne-Edwards, Sowerby (1917) retained five: *Nemorhedus griseus* (Moupin [= Baoxing], Shaanxi, Hubei); *Kemas niger* (Che-ko-ting), with smaller horns; *N. cinereus* (also from Moupin), with a heavier skull; *K. henryanus* (Ichang), with a wider skull; and *K. arnouxianus* (Zhejiang), isolated from the others.

G. M. Allen (1930) recognized the gorals of Shanxi (Kweihwacheng—actually in Inner Mongolia—and Maitaichao, 45 mi. E of Paotowchen), and also of Tungling (in Hebei), as a distinct form, to which unfortunately he applied the name *Nemorhedus caudatus*. He said that *N. griseus* has a shorter coat than *N. caudatus*, slightly darker, but differing on average only; and that the throat has a whitish patch, with a narrow, ochery border like *N. caudatus*, extending nearly to the lips. He likewise separated *Kemas henryanus* (Ichang, Hubei, Fujian) from *Nemorhedus griseus* (Sichuan, Yunnan), as having a darker neck and a smaller, but bright orange throat-patch.

Later, G. M. Allen (1940) substituted the earlier name *Kemas arnouxianus* for *K. henryanus*, and admitted that *Nemorhedus griseus* is "very little different" from what he called *N. caudatus*. He distinguished the two S races by saying that in *N. griseus*, the throat-patch is whitish, with a narrow, pale ochraceous border, extending nearly to the lips; whereas in *Kemas arnouxianus*, the throat is uniformly pale orange, except in a skin from Fujian.

We think that Allen did not take individual variability into account; in our samples, the full range of throat-patch color, from a small yellow to a large white patch, is seen in Ichang skins.

The thicker pelage in Shanxi goral (Allen's *N. caudatus*) is also variable. It is worth noting that Allen had no specimens of the real *N. caudatus* available to him; he simply assumed that goral from Shanxi are the same as those from the Russian Far East. If Shaanxi goral really are distinct, the name for them would be *N. vidianus* Heude, but there is no available name for those from Shanxi or from the neighborhood of Beijing.

Seventeen Sichuan skulls, with teeth in full wear, ranged from 214 to 225 mm; seven younger skulls, with wear not yet completed on the last molars, from 205 to 221 mm. An adult skull from Shanxi was 209 mm, outside the Sichuan dimensions. This might seem suggestive, but two Shanxi skulls with incomplete tooth wear were 206 and 212 mm—and a British Museum skull from Paochi, Shanxi, measured 226 mm.

As described above (in the morphometrics part of the introductory section for the genus), we wished to test whether the Shanxi and Beijing samples are distinct—perhaps intermediate between *N. griseus* and the true *N. caudatus*—so we performed discriminant analyses on skull measurements. The skulls of *N. caudatus* fell outside the range of *N. griseus*, whereas the Shanxi/Beijing skulls fall well within the latter. Thus, there is no reason to think that Shanxi and Beijing goral are even subspecifically different from *N. griseus*, let alone intermediate between the latter and *N. caudatus*.

Nor is there any reason to separate goral from E China from those from Sichuan. In the discriminant analysis using Sichuan (*Nemorhaedus griseus*), coastal (*Kemas arnouxianus*), and Yunnan and Burmese (*N. evansi*) samples, the *K. arnouxianus* skulls fell within the dispersion of *N. griseus*; on the other hand, Yunnan skulls (except for Likiang) are separate, and the small Burmese sample fell at the edge of the Yunnan one.

Nemorhaedus evansi (Lydekker, 1906)

Burmese goral

1906 *Urotragus evansi* Lydekker. Mt. Victoria, Arakan.

Skins seen: 24.

Very light, fawn or somewhat brown; legs in the Burmese and the Thai specimens mostly golden to creamy gold, with a darker wedge (as in the Sichuan specimens), or browner, with a thick black line down the front; underside light gray-fawn; throat white, with a variable golden tone, extending nearly to the chin, the golden tone fading forward. Dorsal stripe usually fairly prominent, medium dark brown; tail mostly dark brown. Forehead clear medium brown, extending to the midline of the nose. Hair short in available specimens, but there may be seasonal variation.

Much smaller than *N. griseus*.

Localities of specimens seen by us: Byingyi, Kyinmana (Ryinmama?) district, Upper Burma; Dawna Range, Amherst; Doi Angka, 56 km SW of Chiengmai, Thailand; Raheng, 16° 50' N, 99° 05' E; Pakokku district, Upper Burma (W of Mt. Victoria, 21° 17' N, 93° 53' E); Arakan; Thaton district, 16° 50' N, 97° 18' E; Xiaojelu and Tengyueh (25° 00' N, 98° 30' E), Yunnan; Hui-yao; Hsiso-ke-la, Yunnan.

Other localities: Kloss (1923) recorded goral in Thailand at the Me Ping rapids, 17° 30' N.

Lydekker (1906) distinguished *N. evansi* from the gray Himalayan goral (*N. bedfordi*) by its more brownish gray color, with no white on the cheeks; no dark mark on the upper surface of the muzzle; and the legs brown behind, rufous in front. The general color and the leg color are valid differences. Pocock (1913) synonymized *N. evansi* with *N. griseus*, from which, however, the former differs by its (apparently consistent) extremely short hair; average lighter color, and the less contrasting color of the tail and the face; and its much smaller size.

Nemorhaedus caudatus Milne-Edwards, 1867

long-tailed goral

1862 *Antilope (Caprina) crispa* Radde; not Temminck, 1845. Lagar River, a tributary of the Amur River.

1867 *Antilope caudata* Milne-Edwards. Renaming of Radde's *Antilope crispa*.

1894 *Kemas raddeanus* Heude. Amur River.

Skins seen: 8.

Gray-fawn to brown, but paler than *N. griseus*, and with no black overlay. Legs creamy to yellow-gray, with less of a sharp transition to body color than in *N. griseus*; dark line down the front of the foreleg. Underside pale fawn or gray; groin and axillae white; chest darker; throat broadly white, extending forward to the chin, with no golden tone or edging; lips white, this color extending to the mouth angles; forehead and nose usually not very dark; whole head, especially the muzzle, more brownish; often a gazelle-like stripe. Dorsal stripe usually inconspicuous. Tail with a very long brush, usually black nearly to the base, reaching to the hocks or below, the length of this hair giving *N. caudatus* its long-tailed aspect, the tail itself being actually somewhat shorter than in other species (see table 68). Fur longer, softer than in others; underwool ashy brown, differing from the long, pale, buffy, dark-tipped contour hairs. Ears short (see table 68).

Localities of specimens seen by us: Ussuri; I-mien-po, Heilongjiang, 45° 03' N, 128° 04' E; Kentei Mountains, Heilongjiang.

Also from Syrun-Bulyk Ridge, SE Mongolia, *fide* Adlerberg (1932).

For this goral the name *Kemas raddeanus* has often been used, under the misapprehension, which may derive from Heude (via Sowerby), that the type locality of *N. caudatus* was Beijing. This, in turn, has led to speculation on whether the gorals from near Beijing and Shanxi, and from Shaanxi (Heude's *N. vidianus*), might be intermediate between the Amur form and *N. griseus*. Thus Sowerby (1917) said *K. raddeanus* from the Amur region is very large, lighter and grayer than his NE China specimen, which he called *N. caudatus*, and the long tail is white (black in *N. caudatus*); Howell (1930) said *K. raddeanus* has a white tail-tip, more black on the forelegs, and a heavier pelage; and G. M. Allen (1930) described *N. caudatus* as being slightly paler than other Chinese races, with longer underwool in winter, and the gray of the back extending to base of the tail (his sample was from Tungling [E Tombs], Hebei, and Shanxi).

The type locality of *N. caudatus*, however, is the Bureja Mountains, Amurland, and it is quite clearly a renaming of the animal which Radde (1862) described from Amurland and mistakenly called *Antilope crispa* Temminck.

In our small sample, the Ussuri goral are paler than those from Beijing, with a long-haired gray tail, and the shanks are dark down the front, contrasting with the white pasterns. In the Beijing goral, including the ones described by Pocock (1908a) and Milne-Edwards (1867), the pelage is somewhat darker and the tail hair shorter and black, while the shanks and feet are all the same light color in front—in all these features resembling *N. griseus*, to which we assign them.

For Adlerberg (1932), the Ussuri goral was a distinct species, distinguished by its skull characters and by its paler color, whiter throat-patch, very long tail hairs (which are mixed with white), thick underfur, and long fur, even in summer. Heptner et al. (1961) said that in the Russian Far East, three color types occur: gray, foxy, and a very rare white form.

Volf (1976) recorded that crossbreeds between *N. griseus* and *N. caudatus* are sterile, so he supported the view that they are separate species. A pair (identified as *Kemas raddeanus*) sent from the Berlin Tierpark had longer horns than a pair identified as *N. griseus* from Yunnan (although, from the description, it seems more likely that these were *N. evansi*); the Tierpark pair were gray, rather than having a short brown coat, and they had a white tail 20 cm long, as opposed to the short, dark tail of the Yunnan pair. Although there might be other reasons for the hybrid sterility, this is still suggestive.

Nemorhaedus baileyi Pocock, 1914

red goral

1914 *Nemorhaedus baileyi* Pocock. Dre, Yigrong Tso, Po Me, 9000 ft.

1961 *Nemorhaedus cranbrooki* Hayman. 8000 ft., in the Adung Valley, Upper Burma.

Our description is based on the type of *N. baileyi*; the syntypes of *N. cranbrooki*; a mounted skin in the Kunming Institute of Zoology; and about a dozen living specimens in the San Diego and Shanghai zoos.

Warm brown to foxy red, with no black ticking; flanks paler; legs like the body; a mere trace of a dark line on the knees; underside paler; axillae and groin, or the groin only, white; throat and interramal region a paler, yellower version of the body color, sometimes a small white or yellowish patch on the throat. Nose slightly darker than the rest of the head. Dorsal stripe usually clear, but thin, brown. Tail short, the tuft occupying half the length; legs and ears short (see table 68).

Localities of specimens seen by us: Dre, Yigrong Tso, Po Me (30.25° N, 95° E); Gongshan; Adung Valley, 8000 ft., Burma (28.10° N, 97.40° E).

Its distribution extends to Lashio, and perhaps even to Loikaw (19.30° N), according to Hla Aung (1967).

In the single available complete skull, the nasals are not bordered laterally by a lacuna. Instead, they are posteriorly expanded, with a serow-like nasofrontal suture, and form a suture with the maxilla as far forward as the level of the $P^{3/4}$ margin; lateral to them are long splint bones, which fill in the normal nasomaxillary lacuna. A lacrimal fossa is present, though shallow, with a trace of a sharp upper rim.

An incomplete skull in the Bombay Natural History Society, described by Hayman, could not be found during CPG's visit.

The type of *N. baileyi* has a small but clear white patch on the throat; that of *N. cranbrooki* does not. A specimen in San Diego Zoo, from N Burma, had an upper throat that was lighter than the surrounding coat, but not clearly white; the mounted specimen in Kunming, from Gongshan, has a small but noticeable whitish yellow patch. There is only a trace of a dorsal stripe in the *N. baileyi* skin; this is much clearer in the type and paratypes of *N. cranbrooki* and in the San Diego Zoo animal, and strikingly clearly marked in the Kunming specimen from Gongshan.

One of the three paratypes of *N. cranbrooki* is darker and browner than the others, and has a more strongly marked dorsal stripe.

Considering the variability in the zoo specimens, and the improbability of there being two red gorals living so close to each other in the Tibet/Yunnan/Burma borderlands, we think it most likely that the type of *N. baileyi* and the syntypes of *N. cranbrooki* represent one and the same taxon, in its summer and winter coats, respectively. U Tun Yin (1967) cited Francis Kingdom-Ward's report of two "foxy red" gorals in the Tsangpo Gorge, bringing goral of the *N. cranbrooki* description to the vicinity of the *N. baileyi* type locality. U Tun Yin also mentioned seeing them at 14,000 ft. on the Diphuk La Pass, between the Adung Valley and Rima in Tibet. Bailey, the collector of the type specimen, said that people in the valley of the Zayul Chu, a Tibetan tributary of the Tsangpo, wore bright foxy red goral-skin coats.

Capricornis Ogilby, 1837

serow

Preorbital gland very deep, invaginated (in the Himalayan serow; less deep in the Japanese serow). Foot gland bigger than in gorals, but communicating directly to the outside (said to be more goral-like in Japanese serow [Pocock, 1919]). Lacrimal fossa present. Lacrimal bone shorter and higher than in gorals. Forehead swollen and convex. Zygomatic arch continuing below the orbit and onto the face as a sharp crest. Superior nuchal crest following a steep course down the side of the occiput; inferior edges of the occipital condyles above the level of the maxillary toothrow. Frontoparietal angle 141°–155°; parieto-occipital angle 120°–140°. Nasal process of the premaxilla broad, ending bluntly, in front of the level of the anterior premolar. Nasals with short lateral processes, and with the usual posterior expansion; narial notch in front of the level of P^2; nasal forming a suture with the maxilla at this notch, but farther back the two bones merely lying adjacent, with no suture; lacuna commonly occurring at the level of the lacrimomaxillary suture. Infraorbital foramen level with P^2 and low down, on the horizontal plane of the palate. Supraorbital foramina always dividing into at least two (sometimes as many as five). Palatine foramina opposite M^2 or M^1. Temporal fossa long and low. External auditory meatus a vertically elongated oval. Nasal length relatively short; Taiwanese serow have especially short, broad nasal bones. Distance from the base of the horn cores to the margin of the orbit less than the diameter of the orbit. Rostrum swollen in the anterior maxillary region, and tapering toward the narrow free ends of the premaxillae.

Horns narrowing gradually and evenly along their length, closely ribbed with flat rings to more than halfway along.

We divide the species into two informal groups.

Capricornis sumatraensis Group

Preorbital gland very deep. Lacrimal bone relatively long, low. Skull elongated, prognathous. Skull pinched-in under the orbits. Parietals flattened. Skull breadth less than 42% of the length. Cheekteeth 30%–30.6% of the skull length. Horns tend to go back in the plane of the forehead or up at somewhat of an angle. Color black or red, with white facial markings and at least some development of a mane.

Capricornis sumatraensis (Bechstein, 1799)

Sumatran serow

1799 *Antilope sumatraensis* Bechstein. Sumatra.

1814 *Antilope interscapularis* Lichtenstein. Sumatra.

1900 *Nemorhaedus swettenhami* Butler. Larut Hills, Perak.

1908 *Capricornis sumatraensis robinsoni* Pocock. Selangor.

$2n = 46$ (Sumatran specimens)

Skins seen: 21.

Jet-black; no white bases to the hairs, except on the mane and along the midback on either side of the (black) dorsal stripe; mane varies from mostly white, or golden buffy white, to black, with only a few white hairs. Underside not white; short, white jaw streaks quickly becoming red and petering out backward. Legs black, but with distinct (varying) reddish tones toward the hoofs; sometimes with white hairs. Nose not white, reddish above the lips, from the nostrils to the jaw angles; face slightly redder; lips creamy white. Long-legged; ears relatively short (see table 68).

Horns go back in the plane of the face; forehead much more swollen and convex than Indian or Chinese serows. Nasofrontal suture tending to go straight across.

Localities of specimens seen by us: Biserat, Batu, Selangor, S Opis Setul (2° 48' N, 101° 55' E) and Sungei Letue in the Malay Peninsula; "W Siamese Malay States"; Kerinci, at 5600 ft., Padang Besi, Aceh, and Solon in Sumatra; W Sumatra.

We include the Malay Peninsula serow in this species, as the skins we have seen are indistinguishable from other *C. sumatraensis*. Note, however, that the skull of the type of *C. s. robinsoni*, the only Malaysian skull we have been able to find, does not fall into the range of variation for Sumatran skulls, but instead is closer to *C. maritimus*. This may indicate gene flow between the two species, although the skins show no intermediacy.

Butler (1900) distinguished *Nemorhaedus swettenhami* from *Capricornis sumatraensis* by its jet-black legs.

Pocock (1910b) said *C. s. robinsoni* apparently differs from *C. s. sumatraensis* in the mane of the former being more crestlike instead of matlike, and in having mixed black and white hairs; but he recorded *C. s. robinsoni* from Perak, and a specimen resembling *C. s. sumatraensis* from the Siamese frontier—probably, as he suggested, all three names are synonyms. Dolan (1963), too, noted that *C. s. robinsoni* falls in the variation of *Nemorhaedus swettenhami*.

Capricornis thar (Hodgson, 1831)

Himalayan serow

1831 *Antilope thar* Hodgson. Nepal Himalayas.

1832 *Antilope bubalina* Hodgson. Nepal.

1842 *Nemorhaedus (Kemas) proclivus* Hodgson. Alternative for *Antilope thar*.

1908 *Capricornis sumatraensis humei* Pocock. Kashmir.

1908 *Capricornis sumatraensis jamrachi* Pocock. Kalimpong, near Darjeeling.

1908 *Capricornis sumatraensis rodoni* Pocock. Chamba, Punjab.

Skins seen: 20, including the types of *Antilope thar*, *Capricornis sumatraensis humei*, *C. s. jamrachi*, *C. s. rodoni*.

Jet-black, with straw or white hair bases that show through; black tips of the hairs becoming redder lower down the flanks, but sometimes acquiring a buffy tone as the black tips wear off. Long mane mixed black and white, varying, but never with the white predominating. Sharply cream-buff below; inner surfaces of the thighs creamy, continuous with the light underside. Broadly white over the nose, or only on the lip margins; tan patch behind the white. Face sometimes lighter than the body. White extending along the jaw lines, backward in a V, or the interramal region completely white, or occasionally nearly absent. Black tips on the legs becoming red lower down, and the pure white hairs increasingly intermix, with the legs thus becoming creamy white shortly below the knees and the hocks, sometimes with a buffy line down the front. Dorsal stripe sometimes visible, but not usually markedly so.

Horns going up at an angle; forehead flatter than in Sumatran (*C. sumatraensis*) or Burmese (*C. rubidus*) serow.

Localities of specimens seen by us: Kashmir (Dachigam, Pir Panjal; Warapash [Sindh Valley]; Chasma Shahi, 8000 ft.); Rampur, Sutlej Valley; W Garwhal; Kulu; Katnauli, Kumaun, 6500 ft.; Mussoorie; Cheena Peak near Naini Tal, 20° 22' N, 79° 26' E; Nepal; Yodavru, Nepal; Kursiong, Darjeeling; Kalimpong, near Darjeeling; Sikkim; Tong Chu Valley, Bhutan; Lushai Hills, Assam.

Choudhury (1997) gave localities in the following districts for Assam, all on the N bank of the Brahmaputra River: Kokrajhar, Sonitpur, Lakhimpur, Dhemaji. It "usually occurs in the higher hills, above 300 m elevation, but may come down to 100 m in winter. It occurs from tropical rainforest and deciduous forest in Assam to subtropical and temperate forests, both conifers and broadleaf, in Arunachal Pradesh."

Capricornis milneedwardsi David, 1869

white-maned serow

1869 *Capricornis (Antilope) milne-edwardsi* David. Moupin, Sichuan.

1871 *Nemorhedus edwardsii* David.

1888 *Capricornis argyrochaetes* Heude. Tchou-ki, Zhejiang.

1894 *Capricornis maxillaris* Heude. Chaohing, Zhejiang.

1894 *Capricornis platyrhinus, Capricornis cornutus, Capricornis erythropygius, Capricornis microdontus, Capricornis ungulosus, Capricornis nasutus, Capricornis vidianus, Capricornis fargesianus, Capricornis brachyrhinus, Capricornis pugnax, Capricornis longicornis*, and *Capricornis tchrysochaets* Heude. Sichuan.

1899 *Capricornis collasinus* Heude. Kwantung.

1921 *Capricornis osborni* Andrews. Hui-yao, 20 mi. from Tengueh, Yunnan.

1930 *Capricornis sumatraensis montinus* Allen. Likiang Range, Snow Mountains, Yunnan.

Skins seen: 50, including the syntypes of *Nemorhedus edwardsii* and the types of *Capricornis osborni* and *Capricornis sumatraensis montinus*.

Very long, lank hair. Black, tending to reddish, especially on the flanks, the rump, and the tail; much white showing through from the white hair bases. Head browner. Mane varying, but largely silvery. Underside light. Nose not white, but the upper lip white, with a broad white jaw-streak, quickly becoming golden-brown backward, and ending before the gonion. Smudgy light patch on the upper throat. Legs increasingly mingled with foxy red down the upper segments; shanks pure reddish; sometimes a sharp demarcation at the knee. Dorsal stripe present.

Skull sometimes with a sharp upper "shelf" to the preorbital fossa, which is lacking in other species.

Localities of specimens seen by us: NE Peling Mountains (Min Shan, beyond Choni [or Jonê] and Archuen, Gansu, near the Sichuan border, ca. 34° N, 103° E; see Wallace, 1913); Wen-Chwan-Hsien (= Wenqanxian), Si-ho, Sichuan; Loung-Ng'ou-Fou, Sichuan; Towquan, 60 mi. N of Kanshien, Sichuan; Maykeng, near Fatung (or Tatung), 32° N, 117° E; Chengou Creek, Sichuan, 31° 20' N, 103° 00' E; Muli, Mt. Gibbok (or Gibboh), Sichuan; Dan Shi Go, Sichuan, 30° 25' N, 102° 50' E; Moupin (= Baoxing), 30° 29' N, 102° 49' E; Gan Yang Go, Sichuan, 30° 20' N, 102° 30' E; Lianghokou, 29° 07' N, 108° 45' E; Leh; Tatsienlu, 30° 03' N, 102° 13' E; Tao River Valley, Gansu, Tibet border; Kegma, Hui-yao (Huaiyao) and Likiang (Lijiang), Yunnan; May Keng (32° N, 117° E); Mahkawng Ga, 4000 ft. (26.49° N, 98.10° E); Mokanshan, Zhejiang, 30° 26' N, 119° 47' E; Tunglu, Zhejiang, 29° 47' N, 119° 37' E; Chungan Hsien, Fujian, 28° 02' N, 117° 53' E; Yenping Mountains, Fujian, 26° 38' N, 118° 10' E; Fengsiang Fu, 34° 24' N, 107° 29' E; 45 mi. S, Shaanxi.

Other records: "between Ssi-Gu River and Cheicho River, Ssi-Gu district," Gansu (Buchner, 1891); mountains near Taukwan and Kwanhsien, Sichuan (Jacobi, 1923); Tienwan, 50 mi. SEE of Tatsienlu, 5000 ft., Sichuan (Orr, 1938); Wa Shan and Tatsien Lu, Sichuan (G. M. Allen, 1940).

Dolan (1963) described age changes, noting that they account for some of the differences between the described taxa that we have placed as synonyms. He noted that young animals can be black; adults, "strong brown"; and older animals tend to have more white in the mane. He found no consistent differences between Sichuan and C China, so that *C. argyrochaetes* falls in the variation of *C. milneedwardsi*.

Capricornis maritimus Heude, 1888

Indochinese serow

1888 *Capricornis maritimus* Heude. Along bay.

1894 *Capricornis rocherianus* and *Capricornis benetianus* Heude. Along bay.

1897 *Capricornis marcolinus* Heude. Tonkin.

1898 *Capricornis berthetianus* Heude. Tonkin.

1899 *Capricornis gendrelianus* Heude. Tonkin.

1919 *Capricornis sumatraensis annectens* Kloss. Koh Lak.

Skins seen: 19.

Black or dark brown, with long, white hair bases, thus appearing gray or brindled; mane mixed with white or pale yellow-buff hairs, the white bases often making the mane predominately white. Underside not white, often black. Lips white, often back to the level of the eye, with a white moustache mark; short, white or golden-brown jaw streaks; white or golden-brown hairs on the throat, sometimes forming a big patch. Legs jet-black on the upper half, reddish tan or creamy white on the lower half, with a sharp division in between; sometimes a white or red patch on the carpus, and a black mark or mingling or line on the lower shanks. Dorsal stripe black, usually thin, but sometimes thick. Long ears (see table 68).

Localities of specimens seen by us: Trang, Thailand; Ta Chang Tai, W Thailand, 16° 51' N, 99° 03' E; Dawna Range (Amherst and Thaton); SW Siam (collected by Gairdner, so probably the Ratburi district); Koh Lak; Maung Pre, N Thailand; Vinh; Nhatrang; Ninh Binh, Tonkin, 20° 14' N, 106° 00' E; Langson, Tonkin; Baie d'Along (topotype), Tonkin; Than Hoa, N Annam, 19° 49' N, 105° 48' E; Mozok (Mozo, Mogok?), upper Burma; Bernardmyo, Ruby Mines, 5000 ft.; N Shan states, 50 mi. E of Maymyo, 3500 ft.; Hpawshi, ca. 26° 05' N, 98° 35' E; Ngawchaung, Burma, 1800 ft.; near Moulmein (Winyaw River, 16° 19' N, 97° 58' E); Gwongye; Pabya Hills, 600 ft. (76 mi. SSW); Dorngin Hill (N of Moulmein); Zwagaben Mountain, or the Duke of York's Nose (25 mi. N); Tho Toungyen Valley, above Myawadi, 16° 42' N, 98° 30' E; Mt. Mouleyit; Tenasserim; Pegu.

Sowerby (1917) described the Tonkin serow—for which he used Heude's name *C. rocherianus* instead of the earlier *C. maritimus*—as being smaller than the Sichuan serow (*C. milneedwardsi*), with a proportionately deeper skull; horns that are smaller than the Chinese forms; black, inclined to blue-black on the body; the face brown; creamy white legs; and no white or creamy hairs in mane.

Irwin (1914) hunted serow in Thailand, from 15° N (above Lopburi) to 11° 48' N (just below Koh Lak on the peninsula). He stated that there were two kinds of serow in the country, with or without the rufous color, and he never heard of the rufous color occurring N (error for S?) of 12° 40' N, beyond which the serow only have black and whitish hairs in the pelt. He described a skin from Koh Lak as black, grizzled on the back by the whitish hair bases; the mane long and matlike, with hairs 12 in. long, and gray, due to the mixture of black and white hairs; the throat reddish gray; the lower lip and chin white, with some reddish hairs on the head; and the shanks rusty brown. He also mentioned another skin, where the shanks were reddish gray, owing to the mixture of red and white hairs.

Gairdner (1915) shot a serow E of Si-sa-wad, in the Sai Yoke district of Ratburi, Quaa Yai River, at about 14° 30' N. It had a black mane with white hair bases; lips white; rufous markings on the base of the ears, the muzzle, the corners of the mouth, and the throat; and shanks that were rufous, with a blackish median line.

Kloss (1919) named *Capricornis sumatraensis annectens* from Koh Lak, and also received specimens of it from Si-sa-wad, Quaa Yai River, W Siam. He described it as having the lower limbs largely rufous, but this color not extending above the knee and the hock, unlike *C. milneedwardsi*; white hair bases, so the body color looks grizzled; the mane white at the base; a large throat-patch, continuous with the white of the lips; and a black chin. These are precisely the characters of *C. maritimus*. But in the Koh Lak peaks, Kloss said, this form lives in association with animals in which the lower legs are almost entirely black; this suggests that there may be either hybridization or, less likely, sympatry with *C. sumatraensis*.

A specimen collected by Gairdner at Ta Chang Tai (666 ft., E side of the Me Thot [or Me Taw] River, above Raheng, ca. 16° 50' N) was black, with the limbs rufous below the knee or the hock (Chasen & Kloss, 1930). Gyldenstolpe (1916) obtained a serow from Koon Tan, 18° 30' N, 99° 20' E, Thailand, and noted that the shanks were rufous. Hanitsch (1918) saw a serow from Phu'ong Mai, 13.30° N (Annam) with a rufous forehead, shaggy white and black-and-white hairs, and legs that were rusty below the knee. Milner (1921) confirmed that in the Ruby Mines Hills, serow are black with red legs; in the Tharawaddy district, Pegu, he recorded them as being black with red legs and no white anywhere, even on the lips or the muzzle.

The Maung Pre skin (in the Singapore collection) lacks white hair bases on the haunches and the shoulders, so it is blackish red overall ; it becomes red or red-brown on the thighs, gradually growing paler down the leg, to become white on the pasterns.

In addition, these is a skin from the Nam Tamai Valley (27° 42' N, 97° 54' E). One might expect this to be *C. rubidus*, as it is so close to the Adung Valley, but is *C. maritimus*. The skin is gray, with a black dorsal stripe; a prominent white gape streak; a white chest and axillae; whitish buff shanks, with a brown median stripe, becoming reddish where the white meets the black of the upper limb; and the hindlimb reddish below the hock and inside, up to the groin.

Capricornis rubidus Blyth, 1863

Burmese red serow

1863 *Capricornis rubida* Blyth. Arakan Hills.

Skins seen: 6.

Red-brown, with the hair bases black (merely reddish tones on the neck and the flanks), except for the Myitkyina skin, best described as "red-toned black." Mane very short, dark red. Underside white, the throat and the interramal region usually white, but red in Adung, and creamy red in Arakan; white over the nose. Legs red or buffy red, with a brown line in the front. Dorsal stripe black. Hair very short.

Tail very short (see table 68). Horns and forehead somewhat intermediate between *C. sumatraensis* and *C. milneedwardsi*.

Localities of specimens seen by us (all in Burma): Adung Valley, 28.10° N, 97.40° E, 9000 ft.; 150 mi. N of Myitkyina; Mahtun (26.06° N, 95.58° E), 3000 ft.; Arakan, including Paletiva; Mahkawng Ga, Upper Burma, 26° 49' N, 98° 10' E. U Tun Yin (1967) recorded red serow from the following areas in Buma: Upper Chindwin, Pakokku, Maymyo, Mu, Arakan Hill tracts, Laukhaung and Putao, and limestone hills in the Salween Valley. He reported black serow with red legs from Mogok, Mongmit, Maymyo, and Tharrawaddy, and black serow with black legs from

Wetwun, Maymyo, and other localities in the N Shan states.

Blyth (1863) described *"C. rubida"* from the Arakan Hills as "of a red brown color, with black dorsal list; the hair shorter than of the others." The type is reputed to be no longer in existence, but there is a skin (no. 18039) in the Bombay Natural History Society collection from Paletiva, Arakan, which has the typical short red hair with short black bases. Pocock (1910b) quoted "Mr. Bird," who said that red serow are common in the limestone hills of the Salween Valley.

There is evidently some sympatry, with possibly some interbreeding, between *C. rubidus* and *C. maritimus*. Milner (1921) saw red serow in the Maigthon Hills, Mu Forest division; on some hills they tended to be black, with reddish legs, and he did see an all-black one. Peacock (1933) said that in the Chindwin drainage, only red serow appear to be known, though he shot one there that was rather reddish purple; he saw another that was sepia-brown, and still others that were dark brown. He noted that along the Irrawaddy River above Mandalay, the black form is the usual one, while farther down, in the Toungoo and Salween districts, both red and black serows occur; near Mogok the black form is commoner, though a red one was shot there. In Tenasserim, they are all black.

C. rubidus is not closely related to any other species. The shorter mane and the lack of white hair bases distinguish it from *C. thar*, *C. maritimus*, and *C. milneedwardsi*, as well as from the undescribed species listed below.

Capricornis Undescribed Species

Assam red serow

We have seen four skins from the India/Burma border region that cannot be assigned to any known species of serow, and evidently represent an undescribed species.

Strongly red in color, with gray tones; hairs red, with long white bases, not black bases, as in *C. rubidus*. Legs with white knees and fetlocks. Rim of the lower jaw white, continuous with a white patch on the throat. Underside a dirty grayish fawn. Long red mane. Apparently short-tailed (see table 67).

Localities of specimens seen by us: Longpa, 26° 15' N, 95° 25' E, Naga Hills; Mishmi Hills; Minbu, 20° 04' N, 94° 04' E, 2000 ft. (these are all full skins, in the Natural History Museum, London); and Mokokchung (a mounted head in the Oxford University Museum). There are also five skulls without skins, which are from the same general area, and we suppose that they belong to the same species: three from the Garo Hills, 2000 ft., Assam; and two from the Mishmi Hills, one of them specified as being from 4100 to 6000 ft. (i.e., 1230–1830 m.).

Choudhury (1997) stated that this form (which he called *C. rubidus*) is found to the S of the Brahmaputra River, whereas black serow are found to its N. He gave the following districts: Tinsukia, Dibrugarh, Sibsagar, Nagaon, Morigaon, Kamrup, Karbi Anglong, N Cachar Hills, Cachar, Hailakandi, Karimganj. It "occurs from as low as 100 m to the highest parts of Assam (Laika area in Barail Range, 1959 m), from tropical rainforest, deciduous forest, abandoned jhums (shifting cultivation) to subtropical broadleaf forest on the Barails." Choudhury (2003) presented an excellent color photo of a female and young animals of this species.

Pocock (1910b) quoted Annandale and Bentham (in litt.), who said there were skins in Calcutta from Shillong and Chittagong (CPG could not find these skins during a visit to the Zoological Survey of India in Calcutta) with hairs that were white at base, with red tips, thus having no black in them; white knees and fetlocks; a white jaw joining the white throat-patch; and, below, the body a dirty grayish fawn color. This would extend the distribution to the S.

Table 67 Differences between Assam red, Burmese red, and Himalayan serow

	Himalayan (*C. thar*)	Assam red	Burmese red (*C. rubidus*)
Overall color	Jet-black, reddish tinge on the flanks	Red	Red-brown (more red on the neck and the flanks)
Hair bases	Straw-colored or white	White	Black
Legs	Red/white; creamy white below the knees/hocks	Knees and fetlocks whitish	All red or buffy red, with a brown line in the front
Nose, jaw, throat-patch	White	White	White, creamy red, or red
Underside	Sharply creamy buff	Dirty grayish fawn	White
Mane	Long, black/white mixed	Long, red	Very short, dark red

Whereas one might expect the Assam red serow to be either close to Burmese *C. rubidus* or else a series of hybrids between *C. rubidus* and *C. thar*, in craniometrics they lie well away from *C. rubidus* and fully within the range of *C. thar*. The differences between the three in external characters are given in table 67; again, the Assam red serow resembles *C. thar* rather than *C. rubidus*. The Assam red serow is best interpreted as the sister species of *C. thar*, and the erythrism that links it to *C. rubidus* would presumably be convergent (just possibly resulting from ancient gene flow?). In this regard, it is of great interest that the erythristic goral, *Nemorhaedus baileyi*, is found (at higher altitudes) in the same general area.

Other Serow

Capricornis crispus (Temminck, 1845)

Japanese serow

1845 *Antilope crispa* Temminck. Honshu.
1894 *Capricornis pryerianus* Heude. Tokyo.
1898 *Capricornulus saxicola* Heude. Honshu.

2n = 50.

Skins seen: 8; as well as observations of about 10 living specimens in the Beijing and Berlin zoos.

Compared with the *C. sumatraensis* group, the preorbital gland less deep; foot gland said to be more goral-like. Lacrimal fossa occupying more of the lacrimal bone; lacrimal bone shorter and higher, with an anterosuperior angulation; facial angle steeper; crista facialis more steeply downturned; nasal process of the premaxilla ending only slightly in front of the level of the anterior premolar; infraorbital foramen somewhat above the horizontal plane of the palate; nasofrontal suture penetrating the frontals to some extent; short frontal processes running lateral to the nasals, not reaching the maxillae; some orbital tubularity; parietals often strongly vaulted; skull breadth more than 42% of the length; cheekteeth very large, averaging 33% of the skull length. Horns directed more upward, broader at the base, the rings more prominent.

Long white underwool; guard hairs white, with black or red-brown tips, but these wearing off in the course of a season, with the overall tone sometimes becoming whiter or grayer; underside white; dark patch on the withers; thin, wavy, black dorsal stripe. Face blackish brown on the top and the sides of the muzzle and around the eyes; white cheeks, lips, and preorbital glands; white around the rhinarium. Legs blackish brown, with or without a white ring around the hoofs. Long, fluffy hair around the cheeks, the upper neck, and the crown of the head. Short-tailed and short-eared (see table 68).

Locality of a specimen seen by us: 15 mi. E. of Yamagata, Mt. Gannto, Ou Mountains, Honshu. Most specimens are ticketed merely "Japan."

Jass & Mead (2004) have given full information on this species.

Capricornis swinhoii Gray, 1826

Formosan serow

1862 *Capricornis swinhoii* Gray. Taiwan.

Skins seen: 6.

Preorbital and foot glands present, but not described in detail (Dien, 1963).

Compared with the *C. sumatraensis* group, the lacrimal bone shorter and higher, with an anterosuperior angulation; facial angle steeper; crista facialis perhaps more steeply downturned; nasofrontal suture penetrating the frontals to some extent; some orbital tubularity; skull broad. In these characters, *C. swinhoii* resembles *C. crispus*, but it differs strongly in several respects: nasal process of the premaxilla reaching much closer to the nasals; infraorbital foramen on the horizontal plane of the palate; temporal fossa long and low; parietals very strongly flattened; cheekteeth smaller. Like Japanese serow (*C. crispus*), the horns directed more upward, but resembling the *C. sumatraensis* group in form.

Hair short, thick, coarse; much woolly underfur; uniform black-brown, brown, reddish brown, or dark brown, with a grayish or reddish tinge; face gray-brown. Chin, throat, and an area from the posterior jaw up to the ears pale yellowish, russet, or reddish brown; underside of the neck yellow-bay; black zone in the interramal space. Front of the legs, from the knee and the hock to the hoof, brown or black, this area becoming broader near the hoof; reddish brown at the back of the legs. Underside slightly lighter, with a reddish tinge in the middle; hinder part of the belly creamy, sometimes suffused with red, going halfway down the inside of the thighs. Dorsal stripe thin or unclear; slightly lengthened nuchal stripe.

Infants entirely fawn-brown; hair short; white streak behind the corner of the mouth; white patch on the throat; faint dorsal stripe.

Localities of specimens seen by us (all in Taiwan): Nan-Tou Hsien; Wu-Liu Tiung, Shing-I (purchased in Puli); Wu Lai; I-lan I-lan; Puli.

Table 68 gives external measurements for both serow and goral taxa. The short ears of *Capricornis suma-*

Table 68 External measurements for serow and goral taxa

	Tail	Hindfoot	Ear
Serows			
sumatraensis	7.8–11 .0 (4)	26.7 (1)	7.4–11.3 (2)
maritimus	7.1–10.3 (4)	21.4–24.5 (2)	12.3–13.2 (3)
milneedwardsi	4.9–12.3 (7)	22.7–28.9 (8)	11.7–13.3 (6)
rubidus	5.1 (1)	26.9 (1)	11.3 (1)
red/white	7.0 (1)	25.5 (1)	11.4 (1)
swinhoii	6.2–13.7 (7)	22.9–28.0 (3)	10.7–15.2 (7)
crispus	5.7–5.7 (2)	27.0–27.1 (2)	8.9–9.5 (2)
Gorals			
caudatus	9.3 (1)	—	10.9 (1)
griseus	10.1–18.0 (8)	20.9–26.9 (8)	12.0–14.2 (6)
evansi	12.7–14.7 (9)	25.5–28.0 (9)	11.1–14.0 (9)
goral	13.3–13.3 (2)	15.9–23.3 (2)	11.1 (1)
baileyi	7.6–10.3 (2)	20.5–23.4 (2)	9.9–10.3 (2)

traensis, compared with most of the other serows, are very noticeable. Among goral, *Nemorhaedus evansi* has very long distal extremities, as represented by the hindfoot length, while *N. baileyi* has a short tail and short ears.

Ovibos de Blainville, 1816

musk-ox

1816 *Ovibos* de Blainville. Type species: *Bos moschatus* Zimmermann, 1780.

1911 *Bosovis* Kowarzik.

Very broad horn base, covering most of the postorbital region. Horns abruptly curving down in both sexes, curving up again toward the tip; whitish, very rough, fibrous basally, smooth at the tip. Limbs short, stout; tail vestigial; pelage very full, with long, coarse overhair, nearly reaching the ground; very long, thick, woolly underfur. Orbits strongly tubular. Rhinarium narrow. (Following J. A. Allen, 1913.)

Ovibos moschatus (Zimmermann, 1780)

1780 *Bos moschatus* Zimmermann. Hudson Bay.

1900 *Ovibos moschatus wardi* Lydekker.

1905 *Ovibos moschatus niphoecus* Elliot.

1910 *Ovibos moschatus mackenzianus* Kowarzik. Great Slave Lake.

1910 *Ovibos moschatus melvillensis* Kowarzik.

J. A. Allen's (1913) monograph, while thorough and remarkably detailed, is not convincing about geographic variation. For example, in his table of measurements, the range of variation of 16 skulls less than 8 years of age from N Grant Land (on Ellesmere Island), referred to *O. m. wardi*, nearly all completely cover those of the other putative subspecies. The main exceptions are as follows:

—horn breadth at the base—*O. m. wardi* 131–219 mm, *O. m. moschatus* 201–250 mm, *O. m. niphoecus* 192–241 mm

—mastoid breadth as a percentage of the basal length—*O. m. wardi* 38.3%–41.9%, *O. m. moschatus* 37%–40.7%, *O. m. niphoecus* 30.7%–39%

In the photographs, slight differences can be detected in the amount of white on the forehead and between the horns; but these, too, seem to be quite variable. The question needs to be reexamined; in the meantime, we recognize no subspecies in just a single valid species.

Tribe Cephalophini Blyth, 1863

duikers

Grubb (2004a) noted that, strictly speaking, the name Cephalophini Blyth, 1863, is antedated by Sylvicaprini Sundevall, 1845 (as "Sylvicaprina"), but Art. 35.5 stipulates that an older name should not displace a younger but universally used one.

Duikers mostly possess preorbital, foot, and inguinal glands; agouti-banded hairs; relatively short limbs, with the withers lower than the rump; a spatulate I^1, but with the other incisiform teeth narrow and keeled; brachyodont cheek teeth; and relatively poorly developed jaw muscles, especially the masseter, with the areas of origin and insertion on the cranial bones relatively small—in particular, the origin of the masseter does not approach the preorbital fossa, so that the two are not separated by a prominent ridge. The origin of the main head-flexing muscle is not marked by basilar tuberosities; the mastoid is narrow; and the foramen ovale is extremely small (a character shared with *Oreotragus*). Duikers have a long skull, with a relatively stout rostrum, nontubular orbits, and the braincase not strongly flexed back on the facial part of the skull; the nasal bone contributes to the wall of the rostrum and is firmly sutured to the maxilla. The horns are short and straight; the horn cores are keeled, inserted behind the orbits, and much inclined. There is little size difference between the sexes, and quite often females average slightly larger than the males. (Following Grubb & Groves, 2001.)

On an analysis of 12S and 16S sequences, Kuznetsova et al. (2002) favored an association of the Cephalophini with the Reduncini. This association was not strongly supported, however, when a nuclear sequence (b-spectrin) was later analyzed (Kuznetsova & Kholodova, 2003).

In the morphologically based cladogram of Grubb & Groves (2001), the first split separates *Cephalophus* from a branch which combines *Sylvicapra* and *Philantomba*. Within *Cephalophus*, the monophyly of the *C. silvicultor* and the *C. ogilbyi* groups is supported. Within the *C. silvicultor* group, there is a *C. dorsalis* versus *jentinki/silvicultor/spadix* split. *C. adersi* was the sister species to a branch containing both the *C. ogilbyi* and the *C. natalensis* clades in the *C. ogilbyi* group. *C. niger* was the first species to separate from the *C. ogilbyi* clade. *C. rufilatus* was a sister species to the *C. nigrifrons* group, followed by *C. leucogaster*, and then by *C. natalensis* and *C. harveyi*.

The phylogeny of Jansen van Vuuren & Robinson (2001), based (in effect) on the mitochondrial cytochrome *b* and 12S rRNA genes, is almost exactly that of Grubb & Groves (2001), based on morphological characters. The phylogeny of the three duiker genera, however, is slightly different:

(*Philantomba* (*Sylvicapra*, *Cephalophus*)).

Within *Cephalophus* there are three major species-groups, plus two "left over" (*C. zebra* and *C. adersi*); these latter two groups sometimes assort with each other, but more usually appear as separate major lineages of their own.

Our arrangement more or less follows that of Grubb & Groves (2001), except for the recognition of diagnosably distinct "subspecies" as full species.

Sylvicapra Ogilby, 1837

Medium sized; slenderly built, with long, slender legs and particularly elongated distal segments; long neck; broad pelvis, with an elongated and widely spaced ischia; pelage primitively agouti-patterned, usually strongly so; coronal tuft long, narrow, dominated by the forwardly directed hairs. Skull with a straight profile, an enlarged narial opening, somewhat tubular orbits, and an oblique occipital plane; nasals with long median processes; median palatal notch very broad, V-shaped. Horns (present in the males only) long, close together, laterally flattened, rising at a steep angle from the skull, close behind the orbits.

Taxonomic and geographic variation in *Sylvicapra* concerns the following features: overall body tone, degree of speckling, amount of darkening on the mid-dorsal region, tone and development of the face-blaze, amount of black on the tail, color of the patterns, color and development of the stripe on the forelegs, and the degree and tone of whitening on the underparts; overall size, length of the tail, length and shape of the ears; relative skull breadth, elongation of the premaxillae, and development of the nasals. This variation is, however, not at all marked (with a few exceptions); one is as likely to find clinal variation within a single subspecies as to find "stepped" variation between subspecies. Variation in the skull characters is very great within a given population, so that, while it can be broadly stated that the SW races are the largest, and the N ones the smallest, it is in no way possible to distinguish a particular taxon on skull characters alone.

We divide the genus into three species; in effect, this simply splits the two most strikingly different taxa from all the rest. To that extent, it is unsatisfactory. Reexamination of some of the subspecies we have placed in *S. grimmia* may well show that there are a few further species to be recognized.

Sylvicapra grimmia (Linnaeus, 1758)

common bush duiker

The "subspecies" of *S. grimmia*, if that is what they are, can be divided, broadly speaking, into a relatively primitive, unspecialized group (*S. g. caffra*, *S. g. hindei*, *S. g. nyansae*, *S. g. madoqua*, *S. g. campbelliae*, and the Afroalpine forms), and a more derived group. The members of the first group have retained the agouti hair banding: the hair bases are gray (usually light gray, but darker in the Afroalpine races), followed by a brown band (varying from a pale straw color in most to nearly white in *S. g. caffra*, with the band narrow in *S. g. caffra*, and broad in *S. g. madoqua*), and then a dark tip, which is usually quite long. In the second group, this pattern is modified, usually by a lightening of the dark zones and a darkening (or reddening) of the pale zones; by the common elimination of the dark tips; and, therefore, by a lack of overall speckling. Speckling in this group is always seen—variably developed—in the juveniles, and it fades to a greater or lesser extent with maturity. The members of the unspecialized group are really all very much alike; subspecies are difficult to distinguish, and even such geographically distant forms as *S. g. caffra* (SE Africa) and *S. g. madoqua* (Ethiopia) are dismayingly similar in appearance. Very minor differences separate most of the various forms, and the validity of some of them is questionable. The various Afroalpine taxa, some not described, differ in minor (but apparently consistent) ways from one another and from their lowland relatives; the small degree of difference is readily explicable if one calculates a maximum of 10,000 years for the total separation of the E African alpine zones from the lowland savannas. The various taxa of the specialized group differ more from one another, and they are not to be regarded as a monophyletic assemblage.

Sylvicapra grimmia grimmia (Linnaeus, 1758)

1758 *Capra grimmia* Linnaeus. Cape Town (restricted by O. Thomas, 1911).

1811 *Antilope nictitans* Thunberg. Cape of Good Hope.

1816 *Antilope mergens* Desmarest. Cape of Good Hope.

1816 *Cemas cana* Oken; nomenclature in Oken's work has been declared unavailable by the ICZN.

1827 *Antilope burchelli* Hamilton Smith. W side of Caffraria; fixed by A. Roberts (1951) at Zwartwater Poort, Albany district.

1827 *Antilope platous* Hamilton Smith. "Vicinity of the Gareep" (i.e., the Orange River), in Hamilton Smith's contribution to Griffith's Cuvier's Animal Kingdom (4:260), but later (5:345) changed to "mountains on the west side of Caffraria."

1827 *Antilope ptoox* Hamilton Smith. Cape of Good Hope.

On the evidence of eight skulls, this would seem to be the largest subspecies (greatest skull length 186–207 mm). The Leiden skins (4 adult, 4 juvenile) confirm the description given by A. Roberts (1951), who described it as "dull ochraceous tawny," with the development of a medium amount of speckling; facial blaze incomplete, not extending to the intercornual crest; white of the underparts restricted to the throat and the inside of the upper half of the legs; foreleg-stripe well developed. *S. g. grimmia* thus closely resembles *S. g. steinhardti* (below), differing essentially in its darker ground color with more speckling, and probably in its larger size.

S Cape, South Africa.

The W "Caffraria" (Albany district, Algoa Bay) population seem to be essentially intergrades between *S. g. grimmia* and *S. g. steinhardti*. A. Roberts (1951), describing the "Caffraria" population as *S. g. burchelli*, said that they are grayer and less tawny than *S. g. grimmia*, but that they have the same absence of white on the belly.

Sylvicapra grimmia caffra Fitzinger, 1869

1869 *Sylvicapra mergens caffra* Fitzinger. "Kaffirland"; Natal, according to A. Roberts (1951).

1871 *Grimmia irrorata* Gray. Natal.

1926 *Sylvicapra altifrons noomei* Roberts. Maputa River, S Mozambique.

1926 *Sylvicapra grimmia transvaalensis* Roberts. Rustenberg district, Northwest Province, South Africa.

Well-speckled fawn-brown to gray-yellow; mid-dorsal zone darker; facial blaze nearly black, well marked up to the level of the eyes, then fading on the forehead; stripe on the foreleg varying in width, but always well expressed, extending from the hoof to the knee; similar but shorter stripe present on the hindleg. Underside mainly light buff, but the belly white, at least in the midline, this color extending to the groin and the inner sides of the hindlimbs right down to the hoofs, then reappearing on the axillae and the inner sides of the forelimbs, and on the interramal region. Tail nearly all-black on the upper surface, white below. Pasterns black. In general, giving the appearance of a brown animal sprinkled with a whitish color; on individual hairs, the subterminal pale band very narrow, the brown zone below it very broad, the gray bases rather dark.

KwaZulu-Natal, the former Transvaal, S Mozambique (Coguno).

Sylvicapra grimmia steinhardti Zukowsky, 1924

1924 *Sylvicapra grimmia cunenensis* Zukowsky. Otjonganga, SE of the Omuhonga Mountains, N Kaokoveld, Namibia.

1924 *Sylvicapra grimmia omurabae* Zukowsky. Otjomikambo, Grootfontein district.

1924 *Sylvicapra grimmia steinhardti* Zukowsky. Otjikuara, source of the Hoamib River, Namibia.

1924 *Sylvicapra grimmia ugabensis* Zukowsky. Goreis, 45 km W of Outjo, N Damaraland, Namibia.

1926 *Sylvicapra grimmia bradfieldi* Roberts. Quickborn Farm, N of Okahandja, Damaraland.

1942 *Sylvicapra grimmia vernayi* Hill. Kaotwe Pan, Kalahari, Botswana.

Very pale, sandy-colored or fawn; immature animals darker, grayer. Speckling very reduced, with the contrast between the light and the dark bands obliterated, and very few black tips, even in the mid-dorsal region. Facial blaze failing to reach even the level of the eyes; forehead pale ochre. Leg-stripes broad, black, reaching the knees and the hocks; pasterns black. Underside off-white; chest and throat buffy; only the upper parts of the inner limb surfaces white, the rest pale buff. Tail black for the terminal two-thirds.

Namibia; S as far as Port Nolloth, W Cape, South Africa; N to Namburi and Cahama, Angola; E to Kazungula, the Chobe district, and Kuka Pan, Botswana.

A specimen from Bambei, Okovango, represents a gradation toward *S. g. splendidula* (below).

Sylvicapra grimmia splendidula (Gray, 1871)

1871 *Grimmia splendidula* Gray. St. Paul de Loanda, Angola.
1894 *Cephalophus grimmia flavescens* Lorenz. Victoria Falls.
1899 *Cephalophus leucoprosopus* Neumann. St. Paul de Loanda.
1919 *Sylvicapra grimmia uvirensis* Lönnberg. Uvira and Baraka, W of the N end of Lake Tanganyika.

Light, a bright reddish ochre; black tips prominent on the midback, but not elsewhere; face-stripe generally short, as in *S. g. steinhardti*; leg-stripes only sometimes reaching the knee and the hock, but well marked; tail usually black for the terminal three-quarters, with a black line extending forward to the root; underside as in *S. g. steinhardti*, but with the white on the inner surfaces of the limbs extending right to the hoofs, and commonly with a whitish streak from the interramal region to the throat; paler region of the muzzle and the circumocular zone often quite white, clearly marked off. Size less than in *S. g. steinhardti*; females larger than the males.

Angola, from the Mupa district N to Kunungu, the DRC, and across the river to Odzala, Congo Republic; SE to the S end of Lake Tanganyika; NE to Uvira; in Zambia, W of the Luangwa Valley, N as far as Mpika; Zimbabwe.

Sylvicapra grimmia orbicularis (Peters, 1852)

1852 *Antilope altifrons* Peters. Sena.
1852 *Antilope ocularis* Peters. Substitute for *Antilope orbicularis*.
1852 *Antilope (Cephalophus) orbicularis* Peters. Sena, N bank of the Zambezi River, Mozambique (fixed by Ellerman et al., 1953).
1906 *Cephalophus walkeri* Thomas. Tuchila River, Shire highlands, Malawi.
1910 *Cephalophus abyssinicus shirensis* Wroughton. Zomba, Malawi.
1913 *Sylvicapra grimmia deserti* Heller. Voi, Kenya.

Light fawn, without much ochery tinge; black tips in the mid-dorsal region, grayer toward the haunches; facial blaze narrow, usually not reaching above the eyes, but sometimes vaguely continuing to the crest; foreleg-stripe narrow, often reaching above the knee, but the hindleg-stripe not reaching the hock; underparts white to buffy, this tone reaching halfway down the inner side of the limbs; occasionally with a throat streak (commoner in S specimens); pasterns black; tail black for at least the terminal half, sometimes nearly entirely so. Females considerably larger than the males.

N of the Zambezi River in Mozambique and Malawi, Zambia E of the Luangwa Valley, N through Tanzania (except the far NW) to the Kenya coast, inland as far as Voi, and the Juba River, Somalia.

S specimens are more reddish, with a longer, darker facial blaze, and larger; N specimens tend to be very pale, and smaller.

The controversial taxon *Cephalophus walkeri* is a melanistic example of *S. g. orbicularis* (Grubb, 1988); hence, it can be assigned to this subspecies.

Sylvicapra grimmia hindei (Wroughton, 1910)

1910 *Cephalophus abyssinicus hindei* Wroughton. Fort Hall (Murango), Kenya.

An "unspecialized" form, like *S. g. caffra*, with much speckling (especially in the mid-dorsal region) on an ochery ground; facial blaze quite bold, broad and dark, extending to the crest; leg-stripes unclear, often failing to reach the knee and the hock; underside buffy white; throat wholly buff, with no white line; white on the inner sides of the limbs failing to reach the hoofs; pasterns dark brown, not black; tail dark brown above for the terminal three-quarters. Very small (greatest skull length 159–171 mm), broad-skulled.

Kenya highlands, E of the Rift Valley, as far E as Machakos and Sultan Hamud, and N to the latitude of Mt. Kenya; S as far as Moshi, Tanzania.

Sylvicapra grimmia nyansae Neumann, 1910

1910 *Sylvicapra abyssinica nyansae* Neumann. Kavirondo.

Very like *S. g. hindei*, but brighter in color, with less mid-dorsal darkening and whiter underparts; facial blaze deep brown rather than blackish; tail less than half blackened; median line running partially down the throat from the interramal region. White down the inner limbs tending to reach farther toward the hoofs. Size small. Ears shorter.

Kenya W of the Rift Valley, from S Ewaso Ngiro N to Torit in Sudan and Roseires in Ethiopia, on the Blue Nile; into the Rift Valley at Elmenteita; W to the Busoga district, Uganda.

Sylvicapra grimmia altivallis Heller, 1912

1912 *Sylvicapra grimmia altivallis* Heller. Kinangop Peak, Aberdares, 10,500 ft. (= 3200 m).

Grayer than *S. g. hindei*, with heavier speckling, a more marked face-stripe (thick, black, complete), black pasterns, and a drabber tone on the underparts; ear short, blunt; nasal bones relatively short; fur long, thick. *S. g. altivallis* shows its relationship to *S. g. hindei*—which lives at the foot of the mountains whose alpine zones *S. g. altivallis* inhabits, and from which *S. g. altivallis* is evidently derived—in its dull color, the mid-back darkening, the amount of black on the tail, the form of the leg-stripes, and the absence of a white throat-streak.

Afroalpine zone of the Aberdares and also, probably, of Mt. Kenya (per a specimen in the Nairobi Museum). On Mt. Kenya, *S. g. hindei* reaches up to the edge of the forest belt, at 1980 m, and the two taxa are isolated from one another by unsuitable habitat (montane forest).

Sylvicapra grimmia lobeliarum Lönnberg, 1919

1919 *Sylvicapra grimmia lobeliarum* Lönnberg. *Lobelia* zone of Mt. Elgon.

Pinkish gray–toned, with heavier speckling than *S. g. nyansae*; face-stripe thick, chocolate-brown; whole upperside of the tail black; pasterns black; foreleg-stripe usually extending to the knee; no white streak down the throat. Long, thick fur. The only complete skull—the type—is extremely small, smaller than any skull of any race so far (this diminution is equaled only by a few specimens of *S. g. madoqua*), but this and three similar-sized but incomplete skulls in the British Museum are comparatively very broad; no other subspecies shows such an extreme length–breadth relationship. Nasals seem short, but not as shortened as in *S. g. altivallis*. Ear short, but normally pointed.

Afroalpine zone of Mt. Elgon. (On the lower slopes, below the montane forest belt, *S. g. nyansae* reaches elevations of up to 1830 m.).

In most characters, *S. g. lobeliarum* seems the most distinctive of the Afroalpine forms. It has some features recalling *S. g. nyansae*—a less-darkened facial blaze, a rather bright color, a tendency for the white of the underside to extend down the inner side of the limbs—but *S. g. lobeliarum* differs in such features as the pinkish tone of the fur, the blackening of the whole upper surface of the tail, the long leg-stripes, and the lack of a throat-streak.

A Possible Further *Sylvicapra grimmia* Subspecies

This is the duiker of the alpine zone of Mt. Kilimanjaro, described by King (1975). In addition to King's specimens, we have studied a skull in Nairobi and two skins—one immature, and one mature but lacking the head—in London. This still does not amount to material sufficient to characterize a new subspecies, although it is surely distinct.

Like *S. g. altivallis*, this alpine-zone duiker seems fairly clearly derived from *S. g. hindei*, with its gray-brown color; thick, black facial blaze; reduced black on the tail; and off-white underparts not extending down the inner sides of the limbs. It differs from *S. g. altivallis* in its usually heavier speckling, with less black on the tail; the foreleg-stripe that usually reaches the knee; the "squared-off" ears (King 1975); the shorter tail; and even shorter nasals than in *S. g. altivallis*.

Sylvicapra grimmia madoqua (Rüppell, 1835)

1835 *Antilope madoqua* Rüppell; not rendered unavailable by *A. madoka* Hamilton Smith (= *Madoqua saltiana*, a dik-dik [Neotragini]), contra O. Thomas [1892]). Galla, W of Massawa (selected by Grubb & Groves [2001]); and the Kullaf Mountains.

1892 *Cephalophus abyssinicus* Thomas. "Abyssinia."

Generally gray-buff to brownish ochre, with heavy speckling; mid-dorsal zone darkened; face-stripe complete in the highland forms, less so in the lowland specimens (Harrar, Abaya, Gazgay); tail usually nearly all-black above, but the black much reduced—on the tip alone—in the Harrar specimens; foreleg-stripe diffuse, even occasionally absent, but, when present, usually reaching the knee; tending to be buffy-toned in the white areas underneath, but purer white in the Harrar, Lake Abaya, and Lake Tsana specimens; no throat streak; no distal extension of the white color on the inner surfaces of the limbs, except in a few individuals from Shoa and one from Gazgay. Fur generally thick and long, especially in the highland areas (Semyen, Gojjam, Shoa, Arussi). Skull as short as those of the various Kenya taxa, or shorter; skull relatively broad, but not as much as in *S. g. lobeliarum*; short nasals; tail apparently fairly long.

Ethiopian highlands, both E and W of the Rift Valley; from as high as 2990 m in the Sahatu Mountains to as low as 610 m at Hawash; it is not certain, however, that these all belong to one taxon.

S. g. madoqua is a quite variable form, difficult to define, although the various component samples do differ as a whole from the various Kenya taxa and from *S. g. campbelliae*. The Ethiopian taxon (*S. g. madoqua*) differs from the Kenya nonalpine forms in its grayer color, heavier speckling, less developed facial blaze, and longer foreleg-stripe, although all these features are subject to variation. It differs from the Afroalpine taxa of the Aberdares, Kilimanjaro, and Elgon in its grayer tone, less developed facial blaze, and brown (not black) pasterns. It differs from *S. g. campbelliae* in its less developed facial blaze and more vaguely developed leg-stripe, as well as in its (usually) smaller size.

This variability, and the rather poor degree of differentiation, doubtless reflects the fact that this largely Afroalpine subspecies is not isolated by extensive forest belts from its lower-living relatives, unlike the Aberdares, Kilimanjaro and Elgon forms. The Lake Abaya, Hawash, Gazgay, and Harrar populations may be reinvasions of lower habitats, as they bear little relationship to other low-altitude taxa S or W of the Ethiopian highlands—with the possible exception of Gazgay. Detailed descriptions of variations will be found in Grubb & Groves (2001).

As far as the skulls as concerned, the range of values is small; again, nothing clear cut can be said about it. In the males, there seems to be a division in size between those from Shoa and Harrar, and those from Arussi, Semyen, and Gojjam—a nonsensical result geographically, and doubtless the result of parallelism.

Tail length appears greatest in the Shoa specimens, and least in those from Semyen; there may be something in this, as the Semyen Massif is the highest of all, while Shoa is the least high of the various ranges.

Sylvicapra grimmia campbelliae (Gray, 1843)

1843 *Cephalophus campbelliae* Gray. "Sierra Leone"; more likely Nigeria (Pocock, 1910b).

1912 *Sylvicapra grimmia roosevelti* Heller. Rhino camp, NW Uganda.

Dull ochery to gray-buff, heavily speckled, and darker in the mid-dorsal zone; facial blaze dark, usually extending to the crest; tail black, nearly to the root in E specimens, but less and less black toward the W, so that the skin from Ejura, Ashanti, has hardly any black at all. Pasterns brown. Foreleg-stripe well marked, reaching the knee. Buffy white to white below; no throat streak; usually no extension of the white color down the inner surfaces of the limbs. Skull fairly small, with much the same size range as *S. g. madoqua*; averaging large in the SE—Burundi, Ankole—and again in the Shari district, and smaller in size to the NE and the W.

Savanna country, from Burundi and Karagwe in the SE, N to Bahr-el-Ghazal, W via NE Congo, southernmost CAR, S Chad, N Cameroon, and Nigeria to Ghana (Ashanti) and Burkina Faso (Fada-N'Gourma).

Sylvicapra pallidior Schwarz, 1914

Sahel bush duiker

1914 *Sylvicapra grimmia pallidior* Schwarz. Mani, Lower Shari River.

Pale buff, more weakly speckled, but still with mid-dorsal darkening; facial blaze extending to the crest, but more diffuse than in *S. g. campbelliae*; tail dark above for over half of its length; foreleg-stripe long, but diffuse; white always extending down the inside of the limbs to the hoofs. Ears longer. Size very small (greatest skull length 158–165 mm).

Sahel zone, from Mani (furthest W locality) as far E as Gallabat, on the borders of the Ethiopian highlands; N as far as Jebel Marra, Darfur; S as far as Fort Archambault.

Several localities are quite close to the distributional range of *S. g. campbelliae*, but the two taxa are always readily distinguishable, and the savanna-sahel changeover always comes between them. (It is not known whether any form of bush duiker extends into the Sahel W of Lake Chad.)

Sylvicapra coronata (Gray, 1842)

crowned bush duiker

1842 *Cephalophora coronata* Gray. "Western Africa."

Bright orange-yellow in color, with no speckling in the adults (hardly detectable even in the juveniles), but the mid-dorsal region a somewhat darker reddish shade; facial blaze deep red, not extending to the crest; tail black only at the tip; pasterns brown; foreleg-stripe very indistinct, not reaching the knee; hardly ever even a trace of a stripe on the hindlegs; underside yellowish white, with no throat streak; only the upper halves of the limbs whitish on their inner surfaces. Skull small (greatest skull length 161–169 mm), but not with the allometrically determined broadness and short-faced characteristics of small individuals of other races; skull curiously narrow, with long premaxillae.

This very distinctive species has been recorded only from a very small area on the borders of Guinea, Guinea-Bissau, Senegal, and Gambia.

Philantomba Blyth, 1840

1840 *Philantomba* Blyth. Type species: *Antilope philantomba* Hamilton Smith, 1827; = *P. maxwelli* (Hamilton Smith, 1827), by tautonomy.

1852 *Guevi* Gray. Type species: *Antilope maxwelli* Hamilton Smith, 1827.

Very small size; head plus body length <750 mm; thick mat of hair on the forehead, darker and browner than the rest of the face; often a buff or red stripe above each eye; coat identical in color in the adults and the juveniles, with no agouti pattern and no strong contrast between the shaft and the tip; color pattern on the body distinctive, dark on the dorsum, with the flanks usually markedly lighter, sometimes a sharply marked-off white strip on the underside; dorsal tone darkening fairly rapidly on the croup and the tail, with the sharpest transition from the dorsal to the lateral tone taking place on the haunch. Skull small, delicate; constricted rostrum and anterior nasal region, with the anterior width of the nasals only half that at the nasofrontal suture; orbits somewhat tubular; occipital plane oblique; preorbital fossa shallow, but very extensive; horns small, delicate, upturned at the tips, very reduced (or even absent) in the females.

Philantomba maxwelli Group

Differs from the *P. monticola* group in various characters: size much larger; tail longer (106–160 mm); zygomatic breadth greater; orbital borders less protuberant; palate broader; free ends of the nasals broader; no sharp break on the haunches between the dark croup and the light color of the flanks and the lower haunches; cheeks lighter; superciliary streaks more prominent, and much deeper; foot gland larger.

Members of this group are apparently less specialized than the *P. monticola* group, except in their feet and, possibly, their larger size.

Philantomba maxwelli (Hamilton Smith, 1827)

Maxwell's duiker

Color light gray-yellow-ochre to a rather dark gray-brown, slightly browner on the middle of the back; yellow-white below, with the groin white. Hair very rarely reversed on the neck. Females usually hornless; horns of the males comparatively short. Females average larger than the males.

Ghana, Ivory Coast, Liberia, Sierra Leone, Guinea, and Guinea-Bissau to Senegal (type locality: Bignona) and Gambia (type of *Philantomba whitfieldi* Gray, 1850).

Grubb & Groves (2001) recognized the duikers from Yatward and Sherbro islands—the one in the Rokelle River, the other farther down the coast—as a distinct subspecies, *Philantomba maxwelli danei* (Hinton, 1920). But now it seems likely that what the samples from the two separate islands have in common—small size and relatively short nasals—is due to independent insular dwarfing.

Philantomba walteri Colyn et al., 2010

Verheyen's duiker

Color rather light, with the flanks tending to buffy, rather than to gray. Hair on the midline of the neck always reversed. Horns of the males usually longer; females always with horns, although the horns much smaller than in the males. Nasal bones shorter.

From the W bank of the Niger River to Benin and Togo.

Philantomba monticola Group

blue duiker

Smaller size than in the *P. maxwelli* group; color brighter, darkening on the rump and the tail, with a sharply bordered transition to a lighter shade on the haunch; flanks and legs noticeably lighter than the back. Cheeks less lightened than in the *P. maxwelli* group, with less contrast between the midface and the cheeks. Tail shorter, <122 mm, and almost always <100 mm. Skull shorter, narrower, but with somewhat tubular orbits; rostrum sharply constricted, with the palate and the snout narrower, and the nasals narrowing strongly toward the tips.

Unlike *Sylvicapra*, *Philantomba* is quite easy to divide into different species, not only separating *P. maxwelli* (as is universally done), and the newly described *P. walteri*, but separating what has hitherto been regarded as a single species (*P. monticola*) into a number of species.

The various species comprising the *P. monticola* group can be divided into two major groups: a gray-legged one and a red-legged one. The first group contains *P. congicus*, *P. melanorheus*, *P. aequatorialis* (including *P. musculoides*), *P. sundevalli*, and *P. lugens*. In these, the legs are much the same color as the body, even though there is still a well-marked transition of lighter tones on the haunch; also, the hair on the midline of the neck is often reversed. The second group contains *P. simpsoni*, *P. anchietae*, *P. defriesi*, *P. hecki*, *P. bicolor*, and *P. monticola*. In these, the legs and haunches are distinctly red-toned; the fur is thicker and softer; and the neck hair is never reversed. As in *P. maxwelli*, there are populations in which the females are predominantly hornless; as in *P. aequatorialis musculoides*, those in which no female skull examined had horns; and, as in *P. a. aequatorialis*, those in which hornless females predominated in most series.

Gray-Legged Group

Philantomba congicus (Lönnberg, 1908)

west-central African blue duiker

Dorsum bright, strong gray-brown to black; flanks pale grayish, the haunches browner; clear, dark brown, horizontal stripe on the haunches marking the sharp transition from the blackish brown of the croup to the paler haunch. Underside broadly white. Cheeks red-tinged.

Median nuchal hair reversal is usual.

From the E bank of the Niger River SE to the Congo River, and across the Oubangui River as far E as Lisala.

Philantomba melanorheus (Gray, 1846)

Bioko blue duiker

Related to *P. congicus*, but even more contrasty, with a black dorsum and pinkish gray flanks; fur long and thick; very small in size; horns very long.

Median nuchal hair reversal is usual.

Bioko Island.

Philantomba aequatorialis (Matschie, 1892)

East African blue duiker

A drab-brownish species, with very restricted black on the croup; light gray below, with a tolerably sharp transition on the haunches. Nasal bones short.

Philantomba aequatorialis aequatorialis (Matschie, 1892)

Drabber and browner than *P. congicus*; black on the croup more restricted; underside usually light gray; transition on the haunches less sharp than in *P. congicus*, but still very noticeable. Nasal bones, on average, longer. Females usually hornless. Females tend to average slightly larger than the males.

From the Uele district into Uganda, at least to Bunyoro, and probably as far as the Nile; NE to the Imatong Mountains in Sudan; S in the DRC, to 4° S. Duikers in the Parc National de l'Upemba are intermediate between *P. a. aequatorialis* and *P. defriesi*, while in the Kisangani region one finds intermediates with *P. simpsoni*.

Philantomba aequatorialis musculoides (Heller, 1913)

Darker than *P. a. aequatorialis* on the dorsum, the flanks somewhat lighter, with this tone extending farther up the sides; haunch-stripe less well marked. Nasals still shorter. Females apparently always hornless.

Forest regions from E Uganda into Kenya, as far E as the Rift Valley.

Philantomba aequatorialis sundevalli (Fitzinger, 1869)

Light in color, brownish; hardly any color differentiation between the dorsum and the flanks; croup not black, but still with quite a sharp transition between the two colors; haunch somewhat pinkish; white below. Size small; horns fairly long in both sexes.

Pemba, Mafia, and probably Zanzibar; also the E African coast.

Philantomba lugens (Thomas, 1898)

Tanzanian highland blue duiker

Very dark, gray-brown, without much contrast between the flanks and the dorsum; croup black, but the haunches so comparatively dark that the transition line not obvious; legs dark, except for a light streak down the front; underside gray or gray-fawn, with (at most) only a narrow streak of white. Horns very long. Hair generally reversed on the midline of the neck.

Highlands of Tanzania, from Usambaras S to Irangi, and to the borders of Malawi at Tukuyu; also Uhehe, Kigoma district.

Red-Legged Group

Philantomba simpsoni (Thomas, 1910)

Simpson's blue duiker

Mid-dorsal region blackish brown, restricted to a broad band in the middle of the back; flanks red-brown, paling to a broadly white belly; haunches redder than the flanks, with a poorly marked transition stripe; legs brownish, pale, but not white, on the inner surfaces. Stripe above each eye well marked. Size small (greatest skull length 114–135 mm); horns of the males fairly long, but not those of the females (the females occasionally lack horns). Premaxillae often failing to contact the nasals (27 out of 45 specimens), and, in any case, never making more than a point contact with them.

Between the lower Congo and Kasai rivers.

There are specimens apparently intermediate between *P. simpsoni* and *P. anchietae* from the Kwamouth region, S of the lower Kasai; between *P. simpsoni* and *P. defriesi* from the Luluabourg district; and between *P. simpsoni* and *P. aequatorialis* from the Kisangani district.

Philantomba hecki (Matschie, 1897)

Malawi blue duiker

Back light fawn-gray; flanks light red; croup medium to dark brown (not black), but with a fairly sharp transition line to the red-fawn of the haunches; legs reddish tan; underside white, at least in the midline. Skull large (greatest skull length 130–135 mm); nasals long.

Malawi; Zambia E of the Luangwa Valley; N Mozambique.

Intermediates between *P. hecki* and *P. lugens* occur at Kalambo Falls and at Kasanga.

Philantomba bicolor (Gray, 1863)

Zulu blue duiker

Darker than *P. hecki*, but not as dark as *P. lugens*; flanks normally dark rufous-orange; throat with an orange tinge; underside broadly white. Small size (greatest skull length 119–130 mm); nasals short.

KwaZulu-Natal, N to the Zambesi in suitable forested areas.

Philantomba monticola (Thunberg, 1789)

Cape blue duiker

Somewhat light gray or fawn; legs reddish tan; haunches only slightly reddened, with very little transition between the haunches and the croup; tail not very dark. Size as in *P. bicolor* (greatest skull length 115–126 mm); horns longer.

E and W Cape.

Philantomba anchietae (Bocage, 1879)

Angolan blue duiker

Back pale gray-brown; flanks gray, with a hint of red; legs pale red-fawn; haunches red posteriorly, grading into the gray of the flanks anteriorly. Tail black, but this color not extending very far onto the croup; well-marked transition on the haunches. Underside broadly gray-white, this color extending halfway down the legs. Perhaps the largest species overall; greatest skull length 131–138 mm.

N Angola, from about the Cubal River to Dalla Tando, near the DRC border.

Philantomba defriesi (Rothschild, 1904)

Zambian blue duiker

Related to *P. hecki*, but paler and more contrasty; dorsal gray zone dark gray, quite distinct, but restricted to the mid-dorsal region; flanks pale red-fawn; haunch noticeably reddish, as are the legs; croup and tail dark, the haunch transition well marked; wholly white below, with the white extending down the inner aspect of the upper half of the legs, unlike *P. anchietae*. Size very large; greatest skull length 126–140 mm.

Zambia, W of the Muchinga Escarpment; Katanga, N to Lukonzolwa and Kinda.

Specimens from the Parc National de l'Upemba and the surrounding area are evidently hybrids between *P. defriesi* and *P. aequatorialis*.

Cephalophus Hamilton Smith, 1827

Distinguished from *Sylvicapra* and *Philantomba* by its heavy build, convex skull profile, nontubular orbits, vertical occipital plane, and depressed horns. The frontal tuft is shorter and broader-based than in *Sylvicapra*, not forming a thick frontal mat of hair as in *Philantomba*.

Cephalophus zebra Group

Lacking certain derived characters of the other species-groups (their deeper preorbital fossae, greater convexity in the skull profile, a larger tuft, longer horns), and with numerous specializations of its own (a very broad muzzle; dorsal flattening; extremely depressed horns, forming grooves in the parietals for their reception; medially ridged horns; a rounded median palatal notch; zebra-like stripes on the back; a lack of agouti banding, even when young; special hair-tufts on the hocks).

Cephalophus zebra Gray, 1838

banded duiker, zebra duiker

Grubb (2004a) discussed the nomenclature in some depth.

Size small to medium; light reddish yellow, becoming darker and browner on the neck and the middle of the back; 12 or 13 black stripes from behind the shoulders to the tail root, backwardly convex on the midback, but becoming forwardly kinked (and browner) on the flanks, the anterior and posterior 2–3 stripes becoming increasingly faint. Underside a paler version of the upperside. Legs redder, with black tufts on the shanks; pasterns black. Tail with black intermixed above, and with white below. Face, and the very small coronal tuft, maroon; nose with blackish tones. Horns very depressed, forming grooves in the parietals for their reception; midline of the parietals and the posterior frontals convex; horns upturned at the very tips. Skull with a very broad snout; forehead nearly flat; preorbital depressions relatively tiny.

The (extremely restricted) distribution part of Sierra Leone, much of Liberia, and part of Ivory Coast.

Cephalophus silvicultor Group

Distinguished from *C. zebra* by its larger preorbital fossae, somewhat more convex skull profile; lack of parietal grooves; a narrower, deeper median palatal notch; longer and less depressed horns; and other features. Distinguished from the *C. ogilbyi* group by its untufted tail; a lack of sexual dimorphism in the skull form; a less convex skull profile; a deeper median palatal notch, more depressed and cylindrical (not flattened) horns, without much sexual dimorphism; and weakly annulated hair. Distinguished from both *C. zebra* and the *C. ogilbyi* group by its distinctive juvenile pelage, which is dark brown overall, each hair being conspicuously agouti-patterned (with a buffy band), and showing a pronounced color change during maturation. The six species assigned to this species-group show a mosaic of characters; four of them (i.e., except for *C. dorsalis* and *C. castaneus*) have shortened neck hair covering a dermal shield, and there seems little alternative but to assume that this has been independently evolved in *C. jentinki* and in *C. silvicultor* / *C. spadix* (as well as in those species of the *C. ogilbyi* group that sport it), since *C. jentinki* and *C. dorsalis* are otherwise so clearly related (see Kuhn, 1968). In the *C. silvicultor* group, the development of the nuchal dermal shield seems to be size related, or related to the absence of much sexual size difference in the horns, or both.

Cephalophus dorsalis Gray, 1846

western bay duiker

Skull profile evenly convex; supraorbital foramina in grooves; nasal bones ending in median points, without lateral processes; mesopterygoid fossa very narrow; maxillae outbowed anteriorly. Color bright chestnut red in the adults; midfacial blaze and coronal tuft usually black, or black-suffused; size small (head plus body length 760–850 mm, greatest skull length 156–174 mm). Black dorsal stripe from the nape to the tail-tuft, usually broadest on the withers; underside like the upperside, except for a black patch on the chest; legs brown; tail white below. Juveniles colored quite differently (see above). Midfacial hair parting of the juveniles also generally seen in the adults. Inguinal pouches very deep.

W of the Adamawa highlands, W to Sierra Leone and Guinea.

There appears to be a distributional hiatus between Dahomey Gap and Cross River, Nigeria. Unexpectedly, the barrier separating *C. dorsalis* from *C. castaneus* seems to be the Adamawa highlands, not the Niger River, as one might expect.

Cephalophus castaneus Thomas, 1892

Larger than *C. dorsalis* (head plus body length 883–1032 mm, greatest skull length 185–202 mm); darker in color; crest and face usually a deep red, not black; width of the dorsal stripe usually greater.

Cameroon, from Mt. Cameroon S and E; all the forests from the Atlantic coast E through the DRC to the Ituri Forest.

Apparently only sporadically found S of the Congo River. Machado (1969) recorded *C. castaneus* from N Angola.

There is considerable geographic variation within *C. castaneus*, but no clear divisions can be made. The width of the dorsal stripe varies somewhat: specimens from N Cameroon (Yoko, etc.) average only 45.7 mm; those from the Cameroon coast, 60.6 mm; but from all other localities (including Mt. Cameroon), at least 80 mm. The forehead, while usually pure red, may be blackened; the proportion of specimens with black foreheads falls from W to E, from 10 out of 30 in Cameroon (but 3 out of 5 in the Sangha River district) to only 1 out of 9 in E and S DRC.

Note that the type of *Cephalophus castaneus arrhenii* Lönnberg, 1917 (from Beni, the Semliki Valley, Zaire), is not a member of this species, but instead is an example of *C. leucogaster*.

The relative abundance of *C. castaneus* falls markedly from W to E. The apparently localized nature of its range in the DRC leads to much uncertainty as to its exact distribution there.

Cephalophus jentinki Thomas, 1982

Jentink's Duiker

Remarkably particolored. Body medium bluish gray (composed of very short white hairs, mixed in with slightly longer dark brown ones that have light tips and bases); limbs and an extension across the shoulders white; head and neck black-brown. Dark bands down the anterior surfaces of the limbs; hocks the same color; chest black in the midline. Hair coarser on the trunk, more silky on the foreparts. Neck very short-haired, with (apparently) a dermal shield. Lips and muzzle white, with (as in juvenile *C. dorsalis*) dark spots at the bases of the facial vibrissae. Juveniles, and apparently also young adults, with rather long, lateral hoofs. Coronal tuft very short, black. Horns more depressed than in *C. silvicultor*. Ears not furry inside, like *C. dorsalis* and *C. castaneus*.

Known only from Liberia, the neighboring part of Ivory Coast, and Freetown Peninsula, Sierra Leone.

Cephalophus silvicultor (Afzelius, 1815)

western yellow-backed duiker

Largest species of duiker. Skull length >263 mm. General color some shade of gray-brown; hairs with lighter bases. Yellow or creamy white triangle on the back, beginning in a point at about the level of the forelegs, then gradually broadening backward, and ending in a straight line on the apex of the croup; fairly well-defined, sparsely haired area behind this triangle, reaching to the root of the tail; whitish and brown hairs mixed at the base of the tail; lateral to this in some (but not all) individuals, possibly some small, triangular haunch marks, colored like the dorsal triangle, and with their anterior vertices contacting the posterior vertices of the latter. Posterior surface of the rump and the thighs nearly black. Sides of the neck lighter than the general body color; midface dark brown; cheeks and muzzle whitish; groin whitish. Coronal tuft maroon, fairly well developed. No pale vibrissal patches; no midfacial parting at any age; no elongation of the lateral hoofs. Dermal neck shield extremely well developed.

From the far W of Africa E to the Albertine Rift, and S into Zambia.

The skull of *C. silvicultor* differs from those of *C. jentinki*, *C. dorsalis*, and *C. castaneus* in having a more convex frontal region, making the dorsal profile more sinuous; the horns, however, are less depressed. The supraorbital foramina are not sunk into channels. The lateral nasal processes are usually conspicuous, and may articulate with the premaxilla. The choanae and mesopterygoid fossa are broader. The rostrum is narrower, but relatively shorter, than in *C. jentinki*. The horns, equal in size in the two sexes, are often quite noticeably downcurved. Generally, median pillars develop on the buccal sides of the cheekteeth.

Cephalophus silvicultor silvicultor (Afzelius, 1815)

Very large; color medium brown, with grayish tones; dorsal triangle very broad, creamy to ochre-yellow, with no haunch spot.

Sierra Leone, Liberia, Ghana, Togo, and Lagos (W Nigeria).

Cephalophus silvicultor longiceps Gray, 1865

Skins seen: 72.

Somewhat smaller; color tends to be darker, to blackish brown; dorsal triangle considerably narrower, creamy to deep dull gold; haunch spot absent in 44 skins, a trace in 20, well developed in 8.

From the Cross River district to Cameroon, Gabon, the Congo River mouth, and the Congo Republic E to the Uele and Ituri districts and into S Sudan.

Cephalophus silvicultor ruficrista Bocage, 1869

Skins seen: 35.

Size as in *C. s. longiceps*, with merely a slightly shorter face; color lighter (light to dark brown), usually without gray tones; dorsal triangle broader than in most individuals of *C. s. longiceps*, and light in color (white to dull gold); haunch spots always present.

S of the Congo River into Angola and Zambia.

Cephalophus curticeps Grubb & Groves, 2001

eastern yellow-backed duiker

Skins seen: 11.

Much smaller than *C. silvicultor*; greatest skull length <260 mm; color darker; dorsal triangle narrower and darker, gold to very dark golden-brown; haunch spot well developed in nine skins, a trace in the other two, but never actually absent.

Rwanda, Burundi, and W Kenya (Mau Forest, Kericho, Molo, Masi).

Cephalophus spadix True, 1890

Abbott's duiker

Somewhat smaller than *C. curticeps*, and very similar to the latter, but lacking the dorsal triangle. Very dark, blackish brown, with a hint of red; dorsal stripe vague, dark; underside reddish, except for the dark midline; throat and cheeks light gray; groin white; coronal tuft large, maroon. Skull differing from *C. curticeps* in its less convex frontals, shorter rostrum, very reduced auditory bullae, narrower (V-shaped) median palatal notch, and somewhat narrower mesopterygoid fossa. Juvenile pelage lacking the conspicuous dorsal stripe of *C. silvicultor* and *C. curticeps*, despite the fact that what trace of it there is remains in the adults. Occasionally with rudiments, or precursors, of the *C. silvicultor* / *C. curticeps* dorsal markings: traces of white at the base of the tail, or an elongate white spot on either side of the tail base.

Tanzania, from Kilimanjaro and the Usambaras S through the Ulugurus, Uzungwas, and S highlands to the N tip of Lake Malawi (Poroto Mountains, Mt. Rungwe, Mfrika Scarp).

Cephalophus ogilbyi Group

Distinguished from the *C. zebra* and the *C. silvicultor* groups by a greater degree of sexual dimorphism, including longer, thicker horns and a thickening of the bone in the frontal region in the males; the mesopterygoid fossa broader, with the median palatal notch usually at about the same level as the lateral ones, or slightly anterior to them; the horns, and their cores, mediolaterally flattened and conical; the horn sheaths striated, with massive transverse ridges; the adult and the subadult growth stages always clearly distinct. The tail is tufted, sometimes very strongly so; the color pattern of the juveniles is similar to that of the adults, though often darker, and lacks the conspicuous speckling of the *C. silvicultor* group.

The species in the *C. ogilbyi* group are very diverse. Some have a dermal neck shield and shortened neck hair, presumably in parallelism with members of the *C. silvicultor* group; these *C. ogilbyi*–group species also have a deep and sharply margined preorbital fossa, and a heavily developed, bipartite tail-tuft. The other species in the *C. ogilbyi* group lack the above characters; instead, they have developed an interfrontal groove, and the adults have lost the juveniles' agouti pattern of hairs.

***Cephalophus niger* Gray, 1846**
black duiker

Color dark blackish mahogany-brown; individual hairs with a straw-colored base, the rest of the hair black; midback zone darker, due to the dark brown bases of the hairs, this band extending from the withers to the rump, including the upper part of the buttocks. Yellow stripe down the inner side of the upper part of the forelimbs. Forehead, nose, and coronal tuft reddish mahogany; coronal tuft with much black in the center. Sides of the face and the underside of the neck pale gray-brown. Tail black, with a broad, light gray, paintbrush-shaped tuft. Juveniles similar in color, but with less red about the face, and paler below. Neck hair very short in the adults, with a thickened skin; no reversal of the hair tracts. Preorbital pits not very deep, but extending forward to the premaxillae, and downward to the molar alveoli; well-defined groove between the frontals; frontals thickened and convex, but not excessively so. Supraorbital foramina often partly roofed over by the lateral extension of the frontal thickening. Zygomatic arch straight, not flared in the middle. Superior temporal line well marked, extending out (in the adults) into a ridge between the horn core and the postorbital bar.

Sierra Leone E to the Niger River.

Jansen van Vuuren & Robinson (2001) found that *C. niger* is sister to the rest of the *C. ogilbyi* group in their cytochrome *b* tree, but is more closely associated (although with low bootstrap support) with the *C. silvicultor* group in their 12 S tree.

***Cephalophus ogilbyi* Waterhouse, 1838**
Ogilby's duiker

Large; bright ochery color; dorsal stripe dark, narrow, only 11–30 mm broad, extending onto the tail; legs colored like the body; foreleg with a dark line from the carpus to the hoof; neck somewhat darker, grayer, with frequent hair reversal dorsally. Juveniles speckled. Tail-tuft very large. Median nuchal hair-stream (where it occurs) reversed over a wide median zone. Juveniles less strongly speckled. Skull with an extreme inflation of the forehead behind the nasofrontal suture, making a marked frontal boss; frontal boss reduced laterally, so the supraorbital foramina not roofed over; zygomatic arch more curved than in *C. niger*, flaring out in the middle; superior temporal line less developed; rostrum not swollen outward; preorbital fossa smaller; nasals shorter, making a shorter suture with the premaxillae.

Bioko and the opposite mainland on the Nigeria/Cameroon border (Cross River district; Rumpi Hills).

Insular specimens tend to have an even more extreme frontal boss than the mainland ones.

***Cephalophus crusalbum* Grubb, 1978**
white-legged duiker

Smaller; legs white from the knee and the hock to the hoof; dorsal stripe broader (25–60 mm); reversal of the hair-stream on the neck (where it occurs) over a much narrower zone. Juveniles strongly speckled. Skull with a less exaggerated frontal boss.

Gabon, from the Fernan Vaz district S, perhaps even to Mayombe, and inland to the Forêt des Abeilles, C Gabon, where it is sympatric with *C. callipygus*.

Grubb (1978b) described *C. crusalbum* as a subspecies of *C. ogilbyi*, comparing it not only with other reputed taxa of that species, but also with *C. leucogaster*, to which specimens of it had been ascribed in the past.

***Cephalophus brookei* Thomas, 1893**
Brooke's duiker

Very similar to *C. ogilbyi*, but of a paler, dull golden color, becoming red-gold on the dorsum, with a broader (27–66 mm) black line down the back, indistinct in front of the shoulders and narrowing to a thin line on the rump; face and coronal tuft red-gold; underside pale yellow; legs colored like the body, but the inner side of the forelegs with whitish hairs. Neck with extremely short hair, reversed (in most specimens) over a fairly wide zone, starting from a whorl on the withers. Juveniles strongly speckled. Skull with the forehead not so strongly inflated; muzzle bowed outward on either side; preorbital fossa small, shallow (restricted by the bowing of the rostrum); premaxillae widely suturing with the relatively very long nasals.

Known only from a few localities in Ghana, W of the Volta; Liberia; and one locality in Guinea.

Cephalophus callipygus Peters, 1876

Peters's duiker

Color as in *C. ogilbyi*, but becoming bright red on the loins; face and crest rich reddish; underside yellowish, but the chin and the throat white; dorsal stripe distinct, from the withers to the tail-tuft, becoming very wide on the croup and including the buttocks and the hind surface of the hindlimbs, as far down as the hocks. Forelegs diffusely darkened down the front; hindlimbs black below the hocks. Hair in the midline of the neck reversed over a narrow zone in about half the specimens. Midline of the chest and the belly dark. Inguinal pouches less clearly present than in other members of the *C. ogilbyi* group. Skull resembles that of *C. ogilbyi*, with a strongly developed frontal boss.

Forests of Cameroon, S and E of the Adamawa highlands, Gabon (except the coastal district), Rio Muni, the Congo Republic, and the CAR.

Cephalophus weynsi Thomas, 1901

Weyns's duiker

Color chestnut-brown; midback region darker and redder than the flanks, forming an ill-defined dorsal stripe; darkened zone on the withers; underside pale ochery; groin white. Coronal tuft red. Limbs dark brown-gray in color, this color often reaching as high as the shoulders and the stifle. Tail with a black line down the upper surface. Hair-reversal zone on the neck narrow. Inguinal pouches well developed. Skull resembles *C. callipygus*, but without a conspicuous boss.

DRC, E to the C African Rift Valley, and to the Imatong Mountains, Sudan. There is an isolated population on Mt. Kabobo in the SE part of the DRC.

Cephalophus johnstoni Thomas, 1901

Johnston's duiker

Differs from *C. weynsi* in its much smaller size; skull length measurements do not even overlap (Groves & Grubb, 1974).

Forests from the Rutshuru district through Uganda to the W side of the Kenya Rift Valley.

Cephalophus lestradei Groves & Grubb, 1974

Lestrade's duiker

Resembles *C. weynsi* and *C. johnstoni*, but very dark in color, a dark olive-gray-brown, with varying reddish tones; dorsal stripe well marked, but not sharply bordered, ending on the root of the tail; size intermediate between *C. johnstoni* and *C. weynsi*.

Forests of the Nile–Congo divide, Rwanda.

Cephalophus adersi Thomas, 1918

Aders's duiker

Color very light ochery, becoming more reddish on the nape. Underside white; white stripe across the upper thigh, joining the white of the groin; lower segments of the limbs marked with white spots. Coronal tuft dark chestnut, as is the face. Hair reversed on the midline of the nape; neck hair not shortened. Light bases of the hairs very long; hairs very soft. Size very small; nasals broadening distally; nasal notch coinciding laterally with the upper margin of the preorbital fossa. Other skull characters resembling those of *Cephalophus leucogaster, C. arrhenii*, and *C. rufilatus* (see below): deep preorbital fossae, with a fairly sharply defined upper margin (though not as marked as in other three species); no interfrontal groove; no marked frontal convexity.

Zanzibar, and the Arabuko-Sokoke Forest in Kenya.

C. adersi is a species of uncertain and isolated affinities; it may even be distantly related to *C. zebra*, rather than to the present species-group (Jansen van Vuuren & Robinson, 2001).

Cephalophus leucogaster Gray, 1873

western white-bellied duiker

Reddish ochery color, becoming darker and redder on the back; fairly thick, well-defined, black dorsal stripe, usually complete from the occiput to the tail-tuft, the maximum breadth 69 mm, but averaging 36–47 mm in different regions. Face, in the midline, colored like the dorsum (or blacker); crest a mixed black and dark red. Legs dark gray. Underside sharply white, as are the hind surfaces of the buttocks. Tail-tuft white, mixed with black, and very large. Hair short and fine, not shortened on the neck. Skull as in *C. adersi*, but with a more sharply defined upper rim to the preorbital fossa, a shorter rostrum, and a distinct frontal boss. Horns very short in both sexes.

From the Cameroon/Gabon/Congo region, S and W of the Adamawa highlands and W of the Congo-Oubangui river system.

Cephalophus arrhenii Lönnberg, 1917

eastern white-bellied duiker

Color paler, browner. Dorsal stripe broader and more diffuse, its minimum breadth 67 mm, but averaging 109 mm. Females, on average, slightly larger than the males.

NE of the DRC, from Bondo to the upper Ituri region.

Cephalophus rufilatus Gray, 1846

red-flanked duiker

Light reddish, with a striking, broad gray dorsal stripe and facial blaze; legs gray. Underside and the sides of the face ochery. Coronal tuft black. Rostrum long, narrow; frontals evenly convex as far forward as the proximal nasals, and as far back as the bases of the horns; horn bases raised up; supraorbital foramina reduced to a single pair. Very small species. Males larger than the females (unusual among duikers), with a more specialized skull.

Guinea and Gambia, along the forest-edge zone as far E as Yambio in Sudan and Arua in the West Nile district, Uganda.

A forest-edge or bush-and-thicket species, not found in deep forest.

Cephalophus natalensis A. Smith, 1834

Natal red duiker

Pale orange-ochre to reddish chestnut; legs slightly grayer, or not at all darker than the body; midline of the face somewhat darker, grayer. Coronal tuft a deep red, mixed with black. Neck washed with gray, but the hair not shortened. Ears gray-black for most of their length. Tail thin, with a big tassel at the tip; tail red basally, black distally, and white below. Mid-back zone darker, more chestnut-colored than the flanks. Underparts light ochery, tending to white in the groin, the axillae, and the interramal region. Size small; skull length 153–169 mm; skull broad; frontal convexity not strongly developed, and not involving the horn pedicels.

Coastal galley and montane forests, from Durban, via Hluhluwe and SE Transvaal, Mozambique, and the Shire Valley (Malawi), to the coast of S Tanzania, as far N as Kilwa.

Cephalophus harveyi Thomas, 1893

East African red duiker

Brighter, deeper in color than *C. natalensis*, a reddish chestnut to ochery brown; legs dark gray to brownish black; midfacial zone blackish, usually right up to the coronal tuft. Crest with more black in it. Size, on average, larger; skull length 162–184 mm.

From C Tanzania (Dar-es-Salaam, Mpwapwa, Kondoa) N, via the montane forests of Tanzania (Usambaras, Kilimanjaro, Meru) and Kenya (Mt. Kenya, Aberdares, E Rift Scarp to 1800 m) and the coastal forests, as far as the Juba River, Somalia.

On Mt. Kenya and in the Aberdares, it has not been recorded above 1800 m.

Cephalophus nigrifrons Gray, 1871

black-fronted duiker

Very close to *C. harveyi*, but distinguished by several characters: its narrower skull, with the muzzle pinched in, the nasals obscuring the maxillae in dorsal view; the much deeper preorbital fossae; the outbowed premaxillae; the more posteriorly restricted frontal boss; the less frequent nasopremaxillary contact; the elongated hoofs; and the thicker, coarser hair, especially long on the withers. Some populations with noticeable inguinal pouches, lacking in *C. natalensis* or *C. harveyi*. Skull length >169 mm.

Rainforest, including montane forest, from the Cameroon/Gabon region across the DRC to the Great Lakes region and SW Uganda.

An isolated population in the Niger delta, 500 km from any other known occurrence, is indistinguishable from *C. nigrifrons* (Powell & Grubb, 2002).

Cephalophus nigrifrons nigrifrons Gray, 1871

Shining chestnut brown; forelegs black as far up as the elbow; hindlegs variable, but generally black up to the hock. Chest broadly black. Noticeable black sprinkling on the withers, sometimes light or sometimes very heavy. Chin yellow. Forehead generally black, the black continuing onto the facial blaze, but sometimes becoming chestnut.

SE Nigeria S to Mayombe, E across the Congo River to the upper Ituri district, S to the Franz Joseph and Guillaume falls, SE to Kinda (Shaba).

Skins from the Mt. Tshiaberimu area (Butembo, Moera, etc.) are coarser-haired, with the chin tending to be white, rather than yellow, and the forelimbs very black, thus tending toward *C. n. kivuensis*.

Cephalophus nigrifrons kivuensis Lönnberg, 1919

Distinguished by its generally more contrasting coloration: the whole of the limbs gray-black; the tail with much black at the base; the hairs, especially in the dorsal region, with a longer, darker (gray-black) basal zone; and the facial blaze very black and broad. Hair not especially long, but thick and coarse. Chin pale reddish, but varying toward whitish in specimens from the highest altitudes. Smaller than *C. n. nigrifrons*.

Virunga volcanoes and the surrounding area in Rwarda, Burundi (as far S as Bururi), Uganda (Shonga Hill, Rukiga) and the DRC (Masisi, Ngoma, Mt. Kahuzi).

Cephalophus hypoxanthus **Grubb & Groves, 2001**
Itombwe duiker

Distinguished from *C. nigrifrons* by its paler, light yellowish chestnut color; the limbs hardly darkened, gray-brown at most, this color not extending to the elbow on the forelimbs, and, on the hindlimbs, forming (at most) an infusion on the shanks. Chin white. Hair fairly long, but soft, the hair being rather coarse in highland populations of *C. nigrifrons*.

Appears to be restricted to the Itombwe Mountains, W of the N end of Lake Tanganyika.

Cephalophus rubidus **Thomas, 1901**
Rwenzori duiker

A very beautiful duiker. Hair long, thick, deep rufous in color, lacking any black sprinkling; chin white; limbs very black. Tail bushy.

Confined to the upper forested slopes of the Rwenzoris; recorded from 1300 to 4200 m.

Jansen van Vuuren & Robinson (2001) found *C. rubidus* to be part of the *C. ogilbyi* clade; this is such a surprising result that it must surely be questioned as to whether the specimen from which they obtained their sample really was *C. rubidus*.

Cephalophus fosteri **St. Leger, 1934**
Mt. Elgon forest duiker

Very small in size; skull length 153–166 mm. Brownish; face and sides of the neck reddish, not gray; chin white, this color extending to the interramal zone. Hair bases pinkish white. Coronal tuft very short, completely black. Fur thick and coarse; tail very bushy.

Forests of Mt. Elgon, at 2400–3400 m.

Cephalophus hooki **St. Leger, 1934**
Mt. Kenya forest duiker

Very small, though slightly larger than *C. fosteri*; skull length 163 mm ($N=1$). Dull chestnut or reddish gray; face and sides of the neck grayer; chin reddish white, this color often (but not always) extending into the interramal region; hair bases pinkish gray; subterminal bands dark brown, not quite black. Facial blaze bordered by a red stripe. Coronal tuft short, completely black. Fur thick, coarse; tail bushy, with an enormous tuft.

Highland forests of Mt. Kenya and the Aberdares, above 2400 m.

The forehoof is 28–44 mm long (mean 34 mm, $N=8$), resembling that of *C. fosteri* and *C. nigrifrons* (28–37 mm, mean 33 mm, $N=10$); while in *C. harveyi*, a rather larger animal, it measures 21–29 mm (mean 26 mm, $N=18$).

Tribe Oreotragini Pocock, 1910

This tribe is notable for the very upright horn insertions, the absence of lateral prongs on the nasal tips, and the very short metapodials. It was perhaps unexpected that molecular studies should reveal that its closest affinities are with the Cephalophini; the relationship is not close, however, and klipspringers are phylogenetically isolated.

Oreotragus A. Smith, 1834

klipspringers

As might be expected from their ecology, which is restricted to rocky outcrops, the distribution of klipspringers is very spotty, and there are quite distinctive species in different parts of the range. Some klipspringers are distinguished by different pelage features, but in many cases the pelage is hardly different in widely separated species, and the degree and type of sexual dimorphism, as well as skull and tooth sizes, are more diagnostic.

The degree and nature of sexual dimorphism in the different species appear to correlate with aspects of ecology and social behavior, particularly including the size of the pair territories, as listed by S. Craig Roberts (1996).

Oreotragus oreotragus **(Zimmermann, 1783)**
Cape klipspringer

Uniform yellow, speckled with brown; hairs mainly white, but some dark brown; tips yellow. Underside pale (with white-tipped hairs) only on the midline; chin and throat light yellowish. Large, dark brown patch above each hoof. Ears relatively short, whitish, with a thick black line on the rim. Forehead and occiput reddish brown. Sexes the same size.

Horns particularly short. Males the largest among the klipspringers; females (because of the usual reverse sexual dimorphism in the genus) only middling sized.

S coastal regions and W Cape.

Oreotragus transvaalensis **Roberts, 1917**
Transvaal klipspringer

Bright golden-yellow, with strongly contrasting white underparts. Feet above the hoofs broadly brown. Ears short, 80–84 mm.

Horns particularly long. Males medium sized; females tending to be large.

KwaZulu Natal to the E and W Transvaal.

Populations of this species in the Drakensberg, at Gamka and at Springbok, had territory sizes of 10, 15, and 49 ha respectively (S. C. Roberts, 1996). There was a territory size of 21.5 ha in the Sentinel district in Zimbabwe, just N of the Limpopo, for a population that may also belong to *O. transvaalensis*.

Oreotragus stevensoni Roberts, 1946

Stevenson's klipspringer

Much duller in color than *O. transvaalensis*; head, especially, much darker. Extensive white underparts, similar to *O. transvaalensis*. Marks above the hoofs variable.

Horns particularly long. Sexes apparently the same size; one of the smallest forms in both sexes.

W Zimbabwe.

According to S. C. Roberts (1996), a population of this species at Metobo had a mean territory size of 5.5 ha.

Oreotragus tyleri Hinton, 1921

Angolan klipspringer

1924 *Oreotragus oreotragus cunenensis* Zukowsky.
1924 *Oreotragus oreotragus steinhardti* Zukowsky.

Wholly pale sandy ochre (dark yellow, speckled with brown) above; contrast occurring only between the speckled dorsum and the uniform color of the legs (sometimes gray, sometimes not). Underparts conspicuously, broadly, uniformly white. Conspicuously dark brown above the hoofs. Ears yellow, the insides with yellowish white hairs. Differs from the S and E forms in having longer ears and hindfeet.

Females larger than the males; overall, medium sized in both sexes.

The Kaokoveld and S Angola.

A population of *O. tyleri* has the largest known territories: 100 ha (S. C. Roberts, 1996).

Oreotragus centralis Hinton, 1921

Zambian klipspringer

Deep rufous to more yellow-gray, legs contrastingly gray. No black above the forehoofs. Some of the more orange individuals very like *O. aceratos*; others with quite a dazzling yellow-and-black effect.

Females larger than the males; overall, the males medium sized, the females fairly large.

Zambia, and probably extending into neighboring countries.

Oreotragus aceratos Noack, 1899

southern Tanzanian klipspringer

Forequarters ochery, only the hindquarters olive, with the two zones quite well separated; legs gray.

Sexes apparently the same size.

S Tanzania.

O. aceratos is very similar to *O. centralis*, from which it differs in the smaller size of the females, and the (on average) slightly larger teeth.

Oreotragus schillingsi Neumann, 1902

Maasai klipspringer

Generally very similar to *O. saltatrixoides*; very rich orangey color, with contrastingly gray legs, dark pasterns, and a darker midback. A rare variant, occurring in some areas, whitish / lemon yellow, especially on the foreparts. Thighs clear gray or rufous, differing markedly in color from the body.

Uniquely among klipspringers, horns present in the females; as Roosevelt & Heller (1914) state, females are "as well horned as males." Females much larger than the males; overall, the males very small, but the females the largest in the genus, equal to those in *O. centralis*.

S. C. Roberts (1996) described the distribution of this species as being in Kenya, S of Mt. Kenya; in Uganda, as far W as the Dodoth Hills; and in Tanzania, S to Rukwa.

S. C. Roberts also noted that the territories of *O. schillingsi*, studied both in Tsavo and in Gilgil, average less than 2 ha, whereas in all other *Oreotragus* species they are much greater than this. As he argued, this lends support to the female-competition hypothesis.

Kenya Highlands Form

Duller color than the Ankole form; darker, not yellow.

Females apparently smaller than males (but we have seen only one female!). This size feature is perhaps identical to the Ankole form (below).

Oreotragus aureus Heller, 1913

golden klipspringer

General color golden-yellow, lighter than *O. saltatrixoides*. Crown of the head rufous, in marked contrast to the body color. Color difference between the body and the legs very slight, but the rufous crown well marked. Hoof-spot large.

Males larger than the females (unusual in *Oreotragus*); overall, both sexes fairly small.

N Guaso Nyiro.

Ankole Form

More speckled, but gold-orange, not—as in *O. saltatrixoides*—drab gray. Neck, shoulders, and head bay-yellow; snout a contrasting gray. Golden, especially in the midneck and slightly down the midline. Some variability in color (e.g., in the contrast of the yellow parts); one specimen with a rich brown crown.

Table 69 Univariate statistics for *Oreotragus*

	Skull l	Skull br	Teeth	Horn l
Males				
porteousi				
Mean	135.000	78.250	49.500	99.000
N	4	4	4	2
Std dev	4.0825	3.3040	2.6458	—
Min	131.0	74.0	47.0	89.0
Max	139.0	82.0	53.0	109.0
Bogos / N Sudan				
Mean	130.000	78.000	46.000	95.000
N	1	1	1	1
saltatrixoides				
Mean	137.438	82.250	47.500	84.286
N	8	8	8	7
Std dev	1.7816	2.3452	2.6049	9.1645
Min	135.5	78.5	44.0	72.5
Max	141.0	85.0	51.0	100.0
Ankole				
Mean	141.500	79.000	51.500	91.500
N	2	2	2	2
Min	141.0	79.0	51.0	81.0
Max	142.0	79.0	52.0	102.0
NW Kenya				
Mean	137.750	77.500	49.900	93.100
N	4	5	5	5
Std dev	2.2174	4.4721	2.0433	9.8133
Min	135.0	71.0	47.0	85.0
Max	140.0	83.0	52.5	109.0
aureus				
Mean	139.714	79.286	48.643	91.714
N	7	7	7	7
Std dev	2.9560	2.5797	.8997	5.8228
Min	134.5	75.0	47.5	83.0
Max	144.0	82.5	50.0	99.0
Kenya highlands				
Mean	141.265	81.222	49.763	90.136
N	17	18	19	14
Std dev	3.6406	3.1165	2.6372	12.9468
Min	132.5	76.5	45.0	67.0
Max	147.0	86.0	57.0	112.5
schillingsi				
Mean	136.750	77.500	51.000	84.625
N	4	4	6	4
Std dev	1.7078	2.5166	1.6733	1.8875
Min	135.0	75.0	49.0	82.5
Max	139.0	81.0	53.0	87.0
aceratos				
Mean	141.000	82.000	53.000	92.000
N	1	1	1	1
centralis				
Mean	140.875	79.063	50.312	97.875
N	16	16	16	16
Std dev	5.0050	3.1931	2.3301	16.4879
Min	132.0	71.0	46.0	74.0
Max	154.0	83.0	54.0	125.0
W Zambia				
Mean	141.000	81.000	51.000	107.000
N	1	1	1	1
stevensoni				
Mean	135.900	80.700	47.700	104.625
N	5	5	5	4
Std dev	2.5100	1.7176	2.8636	11.8980
Min	134.0	79.0	43.0	88.0
Max	140.0	83.0	50.0	115.0
tyleri				
Mean	139.562	80.591	49.955	77.550
N	8	11	11	10
Std dev	3.8493	4.1402	2.7790	28.4590
Min	133.0	70.0	45.0	—
Max	145.0	85.0	53.0	104.0
transvaalensis				
Mean	142.667	84.000	49.667	109.167
N	3	3	3	3
Std dev	4.5092	4.3589	1.2583	20.0271
Min	138.0	79.0	48.5	87.5
Max	147.0	87.0	51.0	127.0
oreotragus				
Mean	147.000	85.000	50.333	83.667
N	2	3	3	3
Std dev	—	4.5826	2.0817	8.0829
Min	143.0	81.0	48.0	75.0
Max	151.0	90.0	52.0	91.0
Females				
Bogos / N Sudan				
Mean	131.750	78.000	45.000	—
N	2	2	2	—
Min	131.5	76.0	41.0	—
Max	132.0	80.0	49.0	—
somalicus				
Mean	143.333	79.333	48.667	—
N	3	3	3	—
Std dev	2.5166	1.5275	2.5166	—
Min	141.0	78.0	46.0	—
Max	146.0	81.0	51.0	—
saltatrixoides				
Mean	138.300	82.200	48.900	—
N	5	5	5	—
Std dev	2.4393	1.8908	1.2450	—
Min	135.5	80.5	47.0	—
Max	142.0	85.0	50.0	—
Ankole				
Mean	145.333	79.667	50.667	—
N	3	3	3	—
Std dev	3.7859	4.9329	2.8868	—
Min	141.0	74.0	49.0	—
Max	148.0	83.0	54.0	—
NW Kenya				
Mean	134.000	78.000	52.000	—
N	1	1	1	—

(*continued*)

Table 69 (continued)

	Skull l	Skull br	Teeth	Horn l
aureus				
Mean	137.286	78.357	49.000	—
N	7	7	7	—
Std dev	2.2704	1.6511	1.5275	—
Min	134.0	76.0	47.0	—
Max	141.0	80.0	51.0	—
Kenya highlands				
Mean	133.000	82.500	46.000	—
N	1	1	1	—
schillingsi				
Mean	146.667	80.625	52.000	81.00
N	3	4	4	4
Std dev	.5774	3.0923	1.4142	11.460
Min	146.0	76.0	51.0	67
Max	147.0	82.5	54.0	95
aceratos				
Mean	141.667	76.000	49.333	—
N	3	3	3	—
Std dev	2.0817	1.0000	2.0817	—
Min	140.0	75.0	47.0	—
Max	144.0	77.0	51.0	—
centralis				
Mean	146.714	79.000	50.143	—
N	7	7	7	-
Std dev	3.1997	1.7321	4.0591	—
Min	141.0	77.0	46.0	—
Max	151.0	81.0	55.0	—
W Zambia				
Mean	146.000	82.000	52.000	—
N	1	1	1	—
stevensoni				
Mean	136.000	79.000	48.500	—
N	1	1	1	—
tyleri				
Mean	144.313	80.813	50.889	—
N	8	8	9	—
Std dev	3.5349	1.4126	1.9808	—
Min	139.0	79.0	47.0	—
Max	151.0	83.0	54.0	—
transvaalensis				
Mean	149.500	82.500	50.000	—
N	2	2	2	—
Min	142.0	81.0	48.0	—
Max	157.0	84.0	52.0	—
oreotragus				
Mean	143.125	86.875	48.833	—
N	4	4	3	—
Std dev	5.2341	4.2106	1.4434	—
Min	138.0	82.0	48.0	—
Max	149.0	91.5	50.5	—

Females much larger than the males (but specimens of only two males, and three females, available from this region); overall, both sexes fairly large in size. Apart from the size difference, both this and the Kenya highlands form could be synonymous with *O. aureus*.

Oreotragus saltatrixoides (Temminck, 1853)

Ethiopian klipspringer

Golden color, especially deeper on the neck; speckling less extensive on the rump.

Females somewhat larger than the males; overall, both sexes fairly small. Teeth very small.

Ethiopia.

A population of this species at Sankeber had a mean territory size of 8.1 ha (S. C. Roberts, 1996).

Bogos / N Sudan Form

General color uniform grayish olive. Base of the hairs whitish, especially on the back. Underparts white. Black patch on front of the feet, above the hoofs. Thighs only a little lighter than the back. White spot on the outer side of the ear.

Particularly long horns. Both sexes extremely small, the smallest of the genus. Teeth very small.

Animals from Bogos and Sudan are most similar to *O. somalicus*, but they differ in such features as size and horn length.

Oreotragus somalicus Neumann, 1902

Somali klipspringer

An unusually homogeneous species, the coloration in almost all individuals being very much the same: yellowy olive; brown on the crown; legs contrastingly gray in front; underparts white; but individuals vary in having from very little black to almost all black above the hoofs. Hardly any differentiation of the dorsal pelage from the front to the back. Bases of the hairs, especially on the back, reddish gray or reddish brown.

Females medium sized (no male skulls available). Teeth small.

N Somalia.

Oreotragus porteousi Lydekker, 1911

1921 *Oreotragus oreotragus hyatti* Hinton.

Color varying; body color dull yellowy; speckling down the midline; crown and upper muzzle more golden; ears gray; upper forelegs paler; lower forelegs gray. Closely resembling *O. saltatrixoides* and *O. schillingsi* in pelage; variation within each of these species greater than between them.

Distinguished by the particularly long horns. Males rather small in size (no female skulls available). Small teeth.

Only found on the Jos Plateau, N Nigeria.

We have no information about the reputed population in the CAR.

The large size of *Oreotragus transvaalensis*, and, to some extent, of *O. oreotragus*, as well as the small size of *O. porteousi*, are all noteworthy, as is the extreme sexual dimorphism of *O. schillingsi* (despite the small sample sizes of the latter, the figures for the two sexes are so extremely different that this conclusion is quite evidently justified).

Relative horn length is horn length as a percentage of skull length. Thus, particularly long horns characterize *O. porteousi*, apparently the Bogos/Sudan form ($N = 1$), *O. stevensoni*, and *O. transvaalensis*. Particularly short horns characterize *O. oreotragus*.

References

Abril, V.V., J.A. Sarria-Perea, D.S.F. Vargas-Munar & J.M.B. Duarte. 2010. Chromosome evolution. In J.M.B. Duarte & S. González (eds.), Neotropical Cervidology: Biology and Medicine of Latin American Deer. Jaboticabal, Brazil and Gland, Switzerland: Funep, in collaboration with IUCN [International Union for Conservation of Nature], 18–26.

Abril, V.V., A. Vogliotti, D.M. Varela, J.M.B. Duarte & J.L. Cartes. 2010. Brazilian dwarf brocket deer *Mazama nana* (Hensel 1872). In J.M.B. Duarte & S. González (eds.), Neotropical Cervidology: Biology and Medicine of Latin American Deer. Jaboticabal, Brazil and Gland, Switzerland: Funep, in collaboration with IUCN, 160–165.

Ackermann, R.R, J. Rogers & J.M. Cheverud. 2006. Identifying the morphological signatures of hybridization in primate and human evolution. Journal of Human Evolution 51:632–645.

Adlerberg, G. 1932. Critical review of the genera *Nemorhaedus* H. Smith and *Capricornis* Ogilby. Izvestiya Akademii Nauk SSSR, Otdeleniye Matematicheskikh i Estestvennykh Nauk (7) 1932:259–285.

Alados, C.L. 1987. A cladistic approach to the taxonomy of the dorcas gazelles. Israel's Journal of Zoology 34:33–49.

Allard, M.W., M.M. Miyamoto, L. Jarecki, F. Kraus & M.R. Tennant. 1992. DNA systematics and evolution of the artiodactyl family Bovidae. PNAS: Proceedings of the National Academy of Science USA 89:3972–3976.

Allen, G.M. 1930. Bovidae from the Asiatic expeditions. American Museum Novitates 410:1–11.

Allen, G.M. 1939. A checklist of African mammals. Bulletin of the Museum of Comparative Zoology 83:1–763.

Allen, G.M. 1940. The Mammals of China and Mongolia, vol. 2. New York: American Museum of Natural History.

Allen, G.M. & A. Loveridge. 1927. Mammals from the Uluguru and Usambara Mountains, Tanganyika Territory. Proceedings of the Boston Society of Natural History 38:413–441.

Allen, J.A. 1913. Ontogenetic and other variations in musk-oxen, with a systematic review of the musk-ox group, recent and extinct. Memoirs of the American Museum of Natural History (n.s.) 1:103–226.

Allen, J.A. 1915. American deer of the genus *Mazama*. Bulletin of the American Museum of Natural History 34:521–553.

Almaça, C. 1992. Notes on *Capra pyrenaica lusitanica* Schlegel, 1872. Mammalia 56:121–123.

Alpers, D.L., B.J. van Vuuren, P. Arctander & T.J. Robinson. 2004. Population genetics of the roan antelope (*Hippotragus equinus*) with suggestions for conservation. Molecular Ecology 13:1771–1784.

Alves, E., C. Óvilo, M.C. Rodríguez & L. Silió. 2003. Mitochondrial DNA sequence variation and phylogenetic relationships among Iberian pigs and other domestic and wild pig populations. Animal Genetics 34:319–324.

Amato, G., D. Wharton, Z.-Z. Zainuddin & J.R. Powell. 1995. Assessment of conservation units for the Sumatran rhinoceros (*Dicerorhinus sumatrensis*). Zoo Biology 14:395–402.

Amato, G., M.G. Egan & A. Rabinowitz. 1999. A new species of muntjac, *Muntiacus putaoensis* (Artiodactyla: Cervidae) from northern Myanmar. Animal Conservation 2:1–7.

Amato, G., M.G. Egan & G.B. Schaller. 2000. Mitochondrial DNA variation in muntjac: evidence for discovery, rediscovery, and phylogenetic relationships. In E.S. Vrba & G.B. Schaller (eds.), Antelopes, Deer, and Relatives. New Haven: Yale University Press, 285–295.

Amin, R., K. Thomas, R.H. Emslie, T.J. Foose & N. van Strien. 2006. An overview of the conservation status of and threats to rhinoceros species in the wild. International Zoo Yearbook 40:96–117.

Ansell, W.F.H. 1971. Artiodactyla (excluding the genus *Gazella*). Part 15 in J. Meester & H.W. Setzer (eds.), The

Mammals of Africa: An Identification Manual. Washington, DC: Smithsonian Institution Press, 184.

Arctander, P., P.W. Kat, R.A. Aman & H.R. Siegismund. 1996. Extreme genetic differences among populations of *Gazella granti*, Grant's gazelle, in Kenya. Heredity 76:465–475.

Arctander, P., C. Johansen & M.-A. Coutellec-Vreto. 1999. Phylogeography of three closely related African bovids (tribe Alcelaphini). Molecular Biology and Evolution 16:1724–1739.

Asher, R.J. & K.M. Helgen. 2010. Nomenclature and placental mammal phylogeny. BMC [BioMedCentral] Evolutionary Biology 10:102.

Ashley, M.V., D.J. Melnick & D. Western. 1990. Conservation genetics of the black rhinoceros (*Diceros bicornis*), 1: evidence from the mitochondrial DNA of three populations. Conservation Biology 4:71–77.

Ashley, M.V., J.E. Norman & L. Stross. 1996. Phylogenetic analysis of the Perissodactylan family Tapiridae using mitochondrial cytochrome *c* oxidase (COII) sequences. Journal of Mammalian Evolution 3:315–326.

Azzaroli, A. 1985. Taxonomy of Quaternary Alcini (Cervidae, Mammalia). Acta Zoologica Fennica 170:179–180.

Balakrishnan, C.N., S.L. Monfort, A. Gaur, L. Singh & M.D. Sorensen. 2003. Phylogeography and conservation genetics of Eld's deer (*Cervus eldi*). Molecular Ecology 12:1–10.

Banfield, A.W.F. 1961. A revision of the reindeer and caribou, genus *Rangifer*. National Museum of Canada Bulletin 77:1–137.

Bannikov, A.G. 1954. Die Säugetiere der Mongolischen Volksrepublik. Moscow: Verlag Akademie der Wissenschaften der USSR.

Bannikov, A.G. 1963. Die Saiga-Antilope (*Saiga tartarica* L.). Die Neue Brehm-Bücherei No. 320. Wittenberg-Lutherstadt, Germany: A. Ziemsen.

Banwell, D.B. 1997. The Pannonians—*Cervus elaphus pannoniensis*—a race apart. Deer 10:275–277.

Banwell, D.B. 1998. Identification of the Pannonian or Danubian red deer: a maraloid—*Cervus elaphus pannoniensis*. Deer 10:495–497.

Banwell, D.B. 1999. The Sika. Auckland: Halcyon.

Banwell, D.B. 2006. The Rusa, the Sambar, and the Whitetail. Auckland: Halcyon.

Banwell, D.B. 2009. The Red Deer, Part I. Auckland: Halcyon.

Baral, H.S., K.B. Shah & J.W. Duckworth. 2009. A clarification of the status of Indian chevrotain *Moschiola indica* in Nepal. Vertebrate Zoology 59:197–200.

Bard, J.B.L. 1977. A unity underlying the different zebra striping patterns. Journal of Zoology, London 183:527–539.

Bar-Gal, G.K., P. Smith, E. Tchernov, C. Greenblatt, P. Ducos, A. Gardeisen & L.K. Horwitz. 2002. Genetic evidence for the origin of the agrimi goat (*Capra aegagrus cretica*). Journal of Zoology, London 256:369–377.

Barrio, J. 2010. Taruka *Hippocamelus antisensis* (d'Orbigny 1834). In J.M.B. Duarte & S. González (eds.), Neotropical Cervidology: Biology and Medicine of Latin American Deer. Jaboticabal, Brazil and Gland, Switzerland: Funep, in collaboration with IUCN, 77–88.

Baryshnikov, J.F.& A. Tikhonov. 1994. Notes on skulls of Pleistocene saiga of northern Eurasia. Historical Biology 8:209–234.

Basilio, R.P.A. 1952. La vida animal en la Guinea Española. Madrid: Instituto de Estudios Africanos.

Baskevich, M.I. & A.A. Danilkin. 1991. Cytogenetic variability among representatives of the genus *Capreolus* (Artiodactyla, Cervidae). In F. Spitz, G. Janeau, G. Gonzalez, et al. (eds.), Proceedings of the International Symposium "Ongulés/Ungulates 91." Paris: Société Française pour l'Étude et la Protection des Mammifères, 123–127.

Bastos-Silveira, C. & A.M. Lister. 2007. A morphometric assessment of geographical variation and subspecies in impala. Journal of Zoology, London 271:288–301.

Bechstein, J.M. 1799. Thomas Pennant's Allgemeine Übersicht der vierfüssign Thiere. Weimar: Industrie-Comptoirs.

Beja-Pereira, A., P.R. England, N. Ferrand, S. Jordan, A.O. Bakhiet, M.A. Abdalla, M. Mashkour, J. Jordana, P. Taberlet & G. Luikart. 2004. African origins of the domestic donkey. Science 304:178.

Bello-Gutiérrez, J., R. Reyna-Hurtado & W. Jorge. 2010. Central American red brocket deer *Mazama temama* (Kerr 1792). In J.M.B. Duarte & S. González (eds.), Neotropical Cervidology: Biology and Medicine of Latin American Deer. Jaboticabal, Brazil and Gland, Switzerland: Funep, in collaboration with IUCN, 166–171.

van Bemmel, A.C.V. 1948. A further note on *Axis* (*Hyelaphus*) *kuhlii* (Müller & Schlegel) (Ungulata, Cervidae). Treubia 19:403–406.

van Bemmel, A.C.V. 1949. Revision of the rusine deer in the Indo-Australian archipelago. Treubia 20:191–262.

Benirschke, K. & A.T. Kumamoto. 1989. Further studies on the chromosomes of three species of peccary. Advances in Neotropical Mammalogy 1989:309–316.

Benirschke, K., D. Rüedi, H. Müller, A.T. Kumamoto, K.L. Wagner & H.S. Downes. 1980. The unusual karyotype of the lesser kudu, *Tragelaphus imberbis*. Cytogenetics and Cell Genetics 26:85–92.

Benirschke, K., A.T. Kumamoto, G.N. Esra & K.B. Crocker. 1982. The chromosomes of the bongo, *Taurotragus* (*Boocerus*) *eurycerus*. Cytogenetics and Cell Genetics 34:10–18.

Benirschke, K., A.T. Kumamoto, J.H. Olsen, M.M. Williams & J. Oosterhuis. 1984. On the chromosomes of *Gazella soemmeringi* Cretzschmar, 1826. Zeitschrift für Säugetierkunde 49:368–373.

Benirschke, K., A.T. Kumamoto & D.A. Meritt. 1985. Chromosomes of the Chacoan peccary, *Catagonus wagneri* (Rusconi). Journal of Heredity 76:95–98.

Birungi, J. & P. Arctander. 2000. Large sequence divergence of mitochondrial DNA genotypes of the control region within populations of the African antelope, kob (*Kobus kob*). Molecular Ecology 9:1997–2008.

Birungi, J. & P. Arctander. 2001. Molecular systematics and phylogeny of the Reduncini (Artiodactyla: Bovidae) inferred from the analysis of mitochondrial cytochrome *b* gene sequences. Journal of Mammalian Evolution 8:125–147.

Black-Décima, P., R.V. Rossi, A. Vogliotti, J.L. Cartes, L. Maffei, J.M.B. Duarte, S. González & J.P. Juliá. 2010. Brown brocket deer *Mazama gouazoubira* (Fischer 1814). In J.M.B. Duarte & S. González (eds.), Neotropical Cervidology: Biology and Medicine of Latin American Deer. Jaboticabal, Brazil and Gland, Switzerland: Funep, in collaboration with IUCN, 190–201.

Blyth, E. 1842. Report of Curator for April, 1842. Journal of the Asiatic Society of Bengal, 11:444–455.

Blyth, E. 1863. Catalogue of Mammals in the Museum of the Asiatic Society of Bengal. Calcutta: Government Printers.

Bohlken, H. 1958. Zur Nomenklatur der Haustier. Zoologischer Anzeiger 160:167–168.

Bohlken, H. 1967. Beitrag zur Systematik der rezenten Formen der Gattung *Bison* H. Smith, 1827. Zeitschrift für Zoologisches Systematik und Evolutionsforschung 5:54–110.

Boisserie, J.-R. 2005. The phylogeny and taxonomy of Hippopotamidae (Mammalia: Artiodactyla): a review based on morphology and cladistic analysis. Zoological Journal of the Linnaean Society 143:1–26.

Boisserie, J.-R. 2007. Family Hippopotamidae. In D.R. Prothero & S.E. Foss (eds.), The Evolution of Artiodactyls. Baltimore: Johns Hopkins University Press, 106–119.

Bork, A.M., C.M. Strobeck, F.C. Yeh, R.J. Hudson & R.K. Salmon. 1991. Genetic relationship of wood and plains bison based on restriction fragment length polymorphisms. Canadian Journal of Zoology 69:43–48.

Bongso, T.A. & M. Hilmi. 1982. Chromosome banding homologies of a tandem fusion in river, swamp, and crossbred buffaloes (*Bubalus bubalis*). Canadian Journal of Genetics and Cytology 24:667–673.

Bosma, A.A., N.A. de Haan & A.A. Macdonald. 1991. The current status of cytogenetics of the Suidae: a review. Bongo: Zoo-Report, Berlin 18:258–272.

Botezat, E. 1903. Gestaltung und Klassifikation des Geweihe des Edelhirsches, nebst einem Anhange über die Stärke der Karpathenhirsche und die zwei Rassen derselben. Gegenbaurs Morphologisches Jahrbuch 32:104–158.

Bourlière, F. & J. Verschuren. 1960. Introduction à l'Écologie des Ongules du Parc National Albert. Fasc. 1 of Exploration du Parc National Albert, Mission F. Bourlière et J. Verschuren. Brussels: Institut des Parcs Nationaux du Congo Belge.

Bouvrain, G., D. Geraads & Y. Jehenne. 1989. Nouvelles données relatives à la classification des Cervidae (Artiodactyla, Mammalia). Zoologischer Anzeiger 223:82–90.

Bover, P. & J.A. Alcover. 1999. The evolution and ontogeny of the dentition of *Myotragus balearicus* Bate, 1909 (Artiodactyla, Caprinae): evidence from new fossil data. Biological Journal of the Linnaean Society 68:401–428.

Bover, P. & J.A. Alcover. 2003. Understanding Late Quaternary extinctions: the case of *Myotragus balearicus* (Bate, 1909). Journal of Biogeography 30:771–781.

Boyce, W.M. 1999. Population subdivision among desert bighorn sheep (*Ovis canadensis*) ewes revealed by mitochondrial DNA analysis. Molecular Ecology 8:99–106.

Boyeskorov, G. G. 1997. [Chromosomal differences in moose (*Alces alces* L., Artiodactyla, Mammalia)]. Genetika 33:974–978. [In Russian; English translation in Russian Journal of Genetics, 33:825–828.]

Boyeskorov, G.[G.] 1999. New data on moose (*Alces*, Artiodactyla) systematics. Säugetierkundliche Mitteilungen 44:3–13.

Boyeskorov, G.G. 2001. On the systematics and distribution of sheep of the genus *Ovis* (Artiodactyla, Bovidae) in eastern Siberia and the Far East in the Pleistocene and Holocene. Zoologicheskii Zhurnal 80:243–256.

Boyeskorov, G.G., M.V. Shchelchkova & Yu.V. Revin. 1993. Karyotype of moose (*Alces alces* L.) from north-eastern Asia. Doklady Rossiyskoy Akademii Nauk 329:506–508.

Bradley, R.D., F.C. Bryant, L.C. Bradley, M.L. Haynie & R.J. Baker. 2003. Implications of hybridisation between white-tailed deer and mule deer. Southwestern Naturalist 48:654–660.

Bradshaw, C.J.A., Y. Isagi, S. Kaneko, D.M.J.S. Bowman & B.W. Brook. 2006. Conservation value of non-native banteng in northern Australia. Conservation Biology 20:1306–1311.

Braun, A., C.P. Groves, P. Grubb, Q-s. Yang & L. Xia. 2001. Catalogue of the Musée Heude collection of mammal skulls. Acta Zootaxonomica Sinica 26:608–660.

Braun, A., C.P. Groves & P. Grubb. 2002. Rediscovery of the type specimen of *Bubalus mindorensis* Heude, 1888. Mammalian Biology 67:246–249.

Brisson, M.-J. 1762. Regnum Animale in Classes IX distributum, sive Synopsis Methodica, 2nd ed. Lugduni Batavorum [Leiden]: Theodorum Haak.

Brooks, A.C. 1961. A Study of the Thomson's Gazelle (*Gazella thomsonii* Günther) in Tanganyika. London: Her Majesty's Stationery Office.

Brown, D.M., R.A. Brenneman, K.-P. Koepfli, J.P. Pollinger, B. Mila, N.J. Georgiadis, E.E. Louis Jr., G.F.

Grether, D.K. Jacobs & R.K. Wayne. 2007. Extensive population genetic structure in the giraffe. BMC Biology 5:57. doi:10.1186/1741-7007-5-57.

Bubenik, A.B. 1990. Epigenetic, morphological, physiological, and behavioural aspects of evolution of horns, pronghorns, and antlers. In G.A. Bubenik & A.B. Bubenik (eds.), Horns, Pronghorns, and Antlers. New York: Springer-Verlag, 1–113.

Buchner, E. 1891. Die Säugethiere der Gansu-Expedition (1884–87). Mélanges Biologiques Tirés du Bulletin de l'Académie Impériale des Sciences de St. Pétersbourg 13:143–164.

Bunch, T.D. 2000. Cytogenetics, morphology, and evolution of four subspecies of the giant sheep argali (*Ovis ammon*) of Asia. Mammalia 64:199–207.

Bunch, T.D. 2006. Phylogenetic analysis of snow sheep (*Ovis nivicola*) and closely related taxa. Journal of Heredity 97:21–30.

Bunch, T.D. & C.F. Nadler. 1980. Giemsa-band patterns of the tahr and chromosomal evolution of the tribe Caprini. Journal of Heredity 71:110–116.

Bunch, T.D., W.C. Foote & J.J. Spillett. 1976. Translocations of acrocentric chromosomes and their implications in the evolution of sheep. Cytogenetics and Cell Genetics 17:122–136.

Bunch, T.D., S-q. Wang, R.S. Hoffmann, Y-p. Zhang, A-h. Liu, S-y. Lin & W. Wang. 1997. Diploid chromosome number and karyotype of the dalai-lama argali (*Ovis ammon dalailamae* Przewalski, 1888). Encyclia: Journal of the Utah Academy of Sciences, Arts and Letters 74:255–263.

Bunch, T.D., N.N. Vorontsov, E.A. Lyapunova & R.S. Hoffmann. 1998. Chromosome number of Severtzov's sheep (*Ovis ammon severtzovi*): G-banded karyotype comparisons within *Ovis*. Journal of Heredity 89:266–269.

Bunch, T.D., S. Wang, R. Valdez, R.S. Hoffmann, Y. Zhang, A. Liu & S. Lin. 2000. Cytogenetics, morphology, and evolution of four subspecies of the giant sheep argali (*Ovis ammon*) of Asia. Mammalia 64:199–207.

Burney, D.A. & Ramilisonina. 1999. The Kilopilopitsofy, Kidoky, and Bokyboky: accounts of strange animals from Belo-sur-Mer, Madagascar, and the megafaunal "extinction window." American Anthropologist 100:957–966.

Butler, A.L. 1900. On a new serow from the Malay Peninsula. Proceedings of the Zoological Society of London 1900:675–676.

Butynski, T.M. 2000. Taxonomy and distribution of the hirola antelope. Gnusletter 19, 2:11–17.

Butynski, T.M., C.D. Schaaf & G.W. Hearn. 1997. African buffalo *Syncerus caffer* extirpated on Bioko Island, Equatorial Guinea. Journal of African Zoology 111:57–61.

Cabrera, A. 1961. Catalogo de los Mamiferos de America del Sur. Buenos Aires: Editora Coni.

Cansdale, G.S. 1960. Animals of West Africa. Longmans: London.

Cao K-q. 1978. On the time of extinction of the wild mi-deer in China. Acta Zoologica Sinica 24:289–291.

Cap, H., S. Aulagnier & P. Deleporte. 2002. The phylogeny and behaviour of the Cervidae (Ruminantia: Pecora). Ethology, Ecology & Evolution 14:199–216.

Cap, H., P. Deleporte, J. Joachim & D. Reby. 2008. Male vocal behaviour and phylogeny in deer. Cladistics 24:917–931.

Cappellini, I. 2007. Dimorphism in the hartebeest. In D.J. Fairbairn, W.U. Blankenhorn & T. Szekely (eds.), Sex, Size, and Gender Roles: Evolutionary Studies of Sexual Size Dimorphism. New York: Oxford University Press, 124–132.

Cappellini, I. & L.M. Gosling. 2006. The evolution of fighting structures in hartebeest. Evolutionary Ecology Research 8:997–1011.

Carr, S.M., S.W. Ballinger, J.N. Derr, L.H. Blankenship & J.W. Bickham. 1986. Mitochondrial DNA analysis of hybridization between sympatric white-tailed deer and mule deer in West Texas. PNAS: Proceedings of the National Academy of Science USA 83:9576–9580.

Cathey J.C., J.W. Bickham & J.C. Patton. 1998. Introgressive hybridization and non-concordant evolutionary history of maternal and paternal lineages in North American deer. Evolution 52:1224–1229.

Chakraborty, S. 1972. On some cranial features of the living-nation genera of family Rhinocerotidae (Mammalia: Perissodactyla). Proceedings of the Zoological Society, Calcutta 25:123–128.

Chasen, F.N. 1940. A handlist of Malaysian mammals. Bulletin of the Raffles Museum, Singapore, Straits Settlements 15:1–209.

Chasen, F.N. & C.B. Kloss. 1930. On mammals from the Raheng district, western Siam. Journal of the Siam Society, Natural History Supplement 8:61–78.

Chau, B. 1997. Another new discovery in Vietnam. Vietnam Economic News 47:46–47.

Chaveerach, A., W. Kakampuy, A. Tanomtong & W. Sangpakdee. 2007. New Robertsonian translocation chromosomes in captive Thai gaur (*Bos gaurus readei*). Pakistan Journal of Biological Science 10:2185–2191.

Chernyavsky, F.B. 1962. On the systematic relationships and history of the snow sheep of the Old and New Worlds. Bulletin of the Moscow Society of Naturalists, Biology 67:17–26.

Chernyavsky, F.B. 2004. On the taxonomy and history of bighorn sheep (*Pachyceros* subgenus, Artiodactyla). Zoologicheskii Zhurnal 83:1059–1070.

Child, C. & C.R. Savory. 1964. The distribution of large mammal species in Southern Rhodesia. Arnoldia (Rhodesia) 1(14):1–15.

Chorley, J.K. 1956. The distribution of Livingston's suni in Southern Rhodesia. Proceedings and Transactions of the Rhodesia Scientific Association 44:63.

Choudhury, A. 1997. Checklist of the Mammals of Assam. Guawahati, India: Gibbon Books and Assam Science Technology and Environment Council.

Choudhury, A. 2003. Status of serow (*Capricornis sumatraensis*) in Assam. Tigerpaper 30, 2:1–2.

Christofferson, M.L. 1995. Cladistic taxonomy, phylogenetic systematics, and evolutionary ranking. Systematic Biology 44:440–454.

Christy, C. 1924. Big Game and Pygmies. London: MacMillan.

Christy, C. 1929. The African buffaloes. Proceedings of the Zoological Society of London 1929:445–462.

Clifford, A.B. & L.M. Witmer. 2004. Case studies in novel narial anatomy, 3: structure and function of the nasal cavity of saiga (Artiodactyla: Bovidae: *Saiga tatarica*). Journal of Zoology, London 264:217–230.

Cobb, E.H. 1958. The markhor. Oryx 4:381–382.

Collins, W.B. 1956. The tropical forest: an animal and plant association. Nigerian Field 21:427.

Cope, E.D. 1889. On the Mammalia obtained by the Naturalist Exploring Expedition to southern Brazil. American Naturalist 23:128–150.

Corbet, G.B. 1969. The taxonomic status of the pygmy hippopotamus, *Choeropsis liberiensis*, from the Niger Delta. Journal of Zoology, London 158:387–394.

Corbet, G.B. 1978. The Mammals of the Palaearctic Region: A Taxonomic Review. London: British Museum (Natural History).

Corbet, G.B. & J. Clutton-Brock. 1984. Appendix: taxonomy and nomenclature. In I.L. Mason (ed.), Evolution of Domesticated Animals. New York: Longman, 434–438.

Corbet, G.B. & J.E. Hill. 1992. The Mammals of the Indomalayan Region: A Systematic Review. London: Natural History Museum Publications and Oxford University Press.

Corbet, S.W. & T.J. Robinson. 1991. Genetic divergence in South African wildebeest: comparative cytogenetics and analysis of mitochondrial DNA. Journal of Heredity 82:447–452.

Coryndon, S.C. 1978. Hippopotamidae. In V.J. Maglio & H.B.S. Cooke (eds.), Evolution of African Mammals. Cambridge, MA: Harvard University Press, 483–495.

Côté, S.D., J.F. Dallas, F. Marshall, R.J. Irvine, R. Langvatn & S.D. Albon. 2002. Microsatellite DNA evidence for genetic drift and philopatry in Svalbard reindeer. Molecular Ecology 11:1923–1930.

Cotterill, F.P.D. 2003. Insights into the taxonomy of tsessebe antelopes *Damaliscus lunatus* (Bovidae: Alcelaphini) with the description of a new evolutionary species in south-central Africa. Durban Museum Novitates 28:11–30.

Cotterill, F.P.D. 2005. The Upemba lechwe, *Kobus anselli*: an antelope new to science emphasises the conservation importance of Katanga, Democratic Republic of the Congo. Journal of Zoology, London 265:113–132.

Cotterill, F.P.D. & W. Foissner. 2009. A pervasive denigration of natural history misconstrues how biodiversity inventories and taxonomy underpin scientific knowledge. Biodiversity and Conservation 19:291–303. doi:10.1007/s10531-009-9721-4.

Courtois, R., L. Bernatchez, J.-P. Ouellet & L. Breton. 2003. Significance of caribou (*Rangifer tarandus*) ecotypes from a molecular genetics viewpoint. Conservation Genetics 4:393–404.

Couturier, M.A.J. 1938. Le Chamois *Rupicapra rupicapra* (L). Grenoble: B. Arthaud.

Cowan, I. McT. 1940. Distribution and variation in the native sheep of North America. American Midland Naturalist 24:505–580.

Cozens, A.B. 1951. The ranges of the dwarf and royal antelopes. Nigerian Field l6:419.

Cranbrook, Earl of & P.J. Piper. 2009. Borneo records of Malay tapir, *Tapirus indicus* Desmarest: a zooarchaeological and historical review. International Journal of Osteoarchaeology 19:491–507.

Croft, D.A., L.R. Heaney, J.J. Flynn & A.P. Bautista. 2006. Fossil remains of a new, diminutive *Bubalus* (Artiodactyla: Bovidae: Bovini) from Cebu Island, Philippines. Journal of Mammalogy 87:1037–1051.

Cronin, M.A. 1992. Intraspecific mitochondrial DNA variation in North American cervids. Journal of Mammalogy 73:70–82.

Cronin, M.A. 2003. Research on deer taxonomy and its relevance to management. Ecoscience 10:432–442.

Cucchi, T., M. Fujita & K. Dobney. 2009. New insights into pig taxonomy, domestication, and human dispersal in Island East Asia: molar shape analysis of *Sus* remains from Niah Caves, Sarawak. International Journal of Osteoarchaeology 19:508–530.

Custodio, C.C., M.V. Lepiten & L.R. Heaney. 1996. *Bubalus mindorensis*. Mammalian Species 520:1–5.

Dalquest, W.W. 1965. Mammals from the Save River, Mozambique, with descriptions of two new bats. Journal of Mammalogy 46:254–264.

Davis, E.B. 2007. Family Antilocapridae. In D.R. Prothero & S.E. Foss (eds.), The Evolution of Artiodactyls. Baltimore: Johns Hopkins University Press, 227–240.

De Beaux, O. 1911. Über einige Antilopen aus dem Rufijitale. Zoologischer Anzeiger 38:575–582.

Deraniyagala, P.E.P. 1972. The destruction of some existing species of Ceylon's endemic mammals. Loris 2:242–244, 250.

Desmarest, A.G. 1822. Mammalogie, ou Description des Espèces de Mammifères, 2ème Partie. Paris: Veuve Agasse.

Deuve, J. 1972. Les Mammifères du Laos. Vientiane: Ministry of National Education.

Dien Z-m. 1963. The Formosan serow (*Capricornis swinhoii* Gray). Quarterly Journal of the Taiwan Museum 16:97–100.

Ditchkoff, S.S., R.L. Lochmiller, R.E. Masters, S.T. Hoofer & R.A. Van den Bussche. 2001. Major-histocompatibility-complex-associated variation in secondary sexual traits of white-tailed deer (*Odocoileus virginianus*): evidence for good-genes advertisement. Evolution 55:616–625.

Dobroruka, L.J. 1960. Der Karpatenhirsch, *Cervus elaphus montanus* Botezat 1903. Zoologischer Anzeiger 165:481–483.

Dolan, J.M. 1963. Beitrag zur systematischen Gliederung des Tribus Rupricaprini, Simpson 1945. Zeitschrift für Zoologisches Systematik und Evolutionsforschung 1:311–407.

Dolan, J.M. 1966. Notes on *Addax nasomaculatus* (de Blainville, 1816). Zeitschrift für Säugetierkunde 31:23–31.

Dolan, J.M. 1988. A deer of many lands—a guide to the subspecies of the red deer *Cervus elaphus*. ZooNooz 62(10):4–34.

Dollman, J.G. 1927. A new race of Arabian gazelle. Proceedings of the Zoological Society of London 1927:1005.

Dollman, J.G. 1928. A new gazelle shot by HRH the Duke of York, K.G. Natural History Magazine 1, 5:130–131.

Dong, W., Y. Pan & J. Liu. 2004. The earliest *Muntiacus* (Artiodactyla, Mammalia) from the Late Miocene of Yuanmou, southwestern China. Comptes Rendus Palevol 3:379–386.

Dragesco, J., F. Feer & J. Genermont. 1979. Contribution à la connaissance de *Neotragus batesi* de Winton, 1903 (position systématique, données biométriques). Mammalia 43:71–81.

Drüwa, P. 1985. Die Damagazelle (*Gazella dama* ssp. Pallas, 1767), einige Beiträge zur allgemeinen Biologie, Haltung und Zucht im Zoologischen Garten. Zoologischer Garten (n.f.) 55:1–28.

Duarte, J.M.B. 1996. Guia de identificação de cervídeos brasileiros. Jaboticabal, Brazil: Funep.

Duarte, J.M.B. & S. González (eds.). 2010. Neotropical Cervidology: Biology and Medicine of Latin American Deer. Jaboticabal, Brazil and Gland, Switzerland: Funep, in collaboration with IUCN.

Duarte, J.M.B. & W. Jorge. 2003. Morphologic and cytogenetic description of the small red brocket (*Mazama bororo* Duarte, 1996) in Brazil. Mammalia 67:403–410.

Duarte, J.M.B., S. González & J.E. Maldonado. 2008. The surprising evolutionary history of South American deer. Molecular Phylogenetics and Evolution 49:17–22.

Ducos, P. 1968. L' Origine des Animaux Domestiques en Palestine. Publications de l'Institut de Préhistoire de l'Université de Bordeaux No. 6. Bordeaux: Delmas.

Dung, V.V., P. Giao, N. Chinh, T. Tuoc, P. Arctander & J. MacKinnon. 1993. A new species of living bovid from Vietnam. Nature 363:443–445.

Dung, V.V., P. Giao, N. Chinh, D. Tuoc & J. MacKinnon. 1994. Discovery and conservation of the vu quang in Vietnam. Oryx 28:16–21.

Effron, M., M.H. Bogart, A.T. Kumamoto & K. Benirschke. 1976. Chromosome studies in the mammalian subfamily Antilopinae. Genetica 46:419–444.

Eisenberg, J.F. 1989. The Northern Neotropics: Panama, Colombia, Venezuela, Guyana, Suriname, French Guiana. Vol. 1 of Mammals of the Neotropics. Chicago: University of Chicago Press.

Eisenberg, J.F. & K.H. Redford. 1999. The Central Neotropics: Ecuador, Peru, Bolivia, Brazil. Vol. 3 of Mammals of the Neotropics. Chicago: University of Chicago Press.

Eisenberg, J.F., C.P. Groves & K. MacKinnon. 1987. Tapire. In W. Keienberg (ed.), Grzimeks Enzyklopädie Säugetiere. 6 vols. Munich: Kindler, 4:598–608.

Eldredge, N. & J. Cracraft. 1980. Phylogenetic Patterns and the Evolutionary Process. New York: Columbia University Press.

Ellerman, J.R. & T.C.S. Morrison-Scott. 1951. Checklist of Palaearctic and Indian Mammals, 1758 to 1946. London: British Museum (Natural History).

Ellerman, J.R., T.C.S. Morrison-Scott & F.W. Hayman. 1953. Southern African Mammals, 1758 to 1951: A Reclassification. London: British Museum (Natural History).

Ellsworth, D.L., R.L. Honeycutt, N.J. Silvy, J.W. Bickham & W.D. Klimstra. 1994. Historical biogeography and contemporary patterns of mitochondrial DNA variation in white-tailed deer from the southeastern United States. Evolution 48:122–136.

Endo, H., M. Kurohmaru & Y. Hayashi. 1995. An osteometrical study of the cranium and mandible of Ryukyu wild pig in Iriomote Island. Journal of Veterinary Medical Science 56:855–860.

Endo, H., Y. Hayashi, M. Sasaki, Y. Kurosawa, K. Tanaka & K. Yamazaki. 2001. Geographical variation of mandible size and shape in the Japanese wild pig (*Sus scrofa leucomystax*). Journal of Veterinary Medical Science 63:815–820.

Engländer, H. 1986. *Capra pyrenaica* Schinz, 1838—Spanischer Steinbock, Iberiensteinbok. In F. Niethammer & F. Krapp (eds.), Handbuch der Säugetiere Europas, vol. 2. pt. 2. Wiesbaden, Germany: Aula, 405–420.

Escamilo B., L.L., J. Barrio, J. Benavides F. & D.G. Tirira. 2010. Northern pudu *Pudu mephistophiles* (de Winton 1896). In J.M.B. Duarte & S. González (eds.), Neotropical Cervidology: Biology and Medicine of Latin American Deer. Jaboticabal, Brazil and Gland, Switzerland: Funep, in collaboration with IUCN, 133–139.

Estes, R.D. 1969. Behavioural study of East African ungulates, 1963–1965. National Geographic Society Research Reports, 1964 Projects:45–57.

Estes, R.D. 1991. The Behaviour Guide to African Mammals. Berkeley: University of California Press.

Estes, R.D. 2000. Evolution of conspicuous colouration in the Bovidae: female mimicry of male secondary characters as catalyst. In E.S. Vrba & G.B. Schaller (eds.), Antelopes, Deer, and Relatives. New Haven: Yale University Press, 234–246.

Evangelista, P., P. Swartzinski & R. Waltermire. 2007. A profile of the mountain nyala (*Tragelaphus buxtoni*). African Indaba 5:1–46. www.africanindaba.co.za.

Ewer, R.F. 1957. A collection of *Phacochoerus aethiopicus* teeth from the Kalkbank Middle Stone Age site, central Transvaal. Palaeontologia Africana 5:5–20.

Falchetti, E., A. Ceccarelli & C. Mantovani. 1995. Relationship between dominance/subordination and colouring patterns in *Kobus megaceros* (Bovidae, Reduncinae) captive males. Gnusletter 14, 1:5.

Fedosenko, A.K. & D.A. Blank. 2001. *Capra sibirica*. Mammalian Species 675:1–13.

Feer, F. 1979. Observations écologiques sur le néotrague de Bates (*Neotragus batesi* de Winton, 1903, Artiodactyla, Ruminantia, Bovidae) du nordest du Gabon. La Terre et La Vie 33:159–239.

Feng, J., C. Lajia, D.J. Taylor & M.S. Webster. 2001. Genetic distinctiveness of endangered dwarf blue sheep (*Pseudois nayaur schaeferi*): evidence from mitochondrial control region and Y-linked ZFY intron sequences. American Genetic Association 92:9–15.

Flagstad, Ø. & K.H. Røed. 2003. Refugial origins of reindeer (*Rangifer tarandus* L.) inferred from mitochondrial DNA sequences. Evolution 57:658–670.

Flagstad, Ø., P.O. Syvertsen, N.C. Stenseth, J.E. Stacy, I. Olsaker, K.H. Røed & K.S. Jakobsen. 2000. Genetic variability in Swayne's hartebeest, an endangered antelope of Ethiopia. Conservation Biology 14:254–264.

Flagstad, Ø, P.O. Syvertsen, N.C. Stenseth & K.S. Jakobsen. 2001. Environmental change and rates of evolution: the phylogeographic pattern within the hartebeest complex as related to climatic variation. Proceedings of the Royal Society of London, B 268:667–677.

Flamand, J.R.B., D. Vankan, K.P. Gairhe, H. Duong & J.S.F. Barker. 2003. Genetic identification of wild Asian water buffalo in Nepal. Animal Conservation 6:265–270.

Flerov, C.C. 1931. A review of the elks or moose (*Alces* Gray) of the Old World. Comptes Rendus de l'Académie des Sciences de l'URSS 1931:71–74.

Flerov, C.C. 1952 [English translation 1960]. Mammals: Musk Deer and Deer. Vol. 1, no. 2 of Fauna of USSR Mammals. Moscow & Leningrad: Academy of Sciences of the USSR [English translation by Israel Program for Scientific Translations, Washington, DC: National Science Foundation and Smithsonian Institution].

Flerov, C.C. 1965. Comparative craniology of recent representatives of the genus *Bison*. Bulletin of the Moscow Naturalists Society, Biology Division 70:41–54.

Florescu, A., J.A. Davila, C. Scott, P. Fernando, K. Kellner, J.C. Morales, D. Melnick, P.T. Boag & P. van Coeverden de Groot. 2003. Polymorphic microsatellites in white rhinoceros. Molecular Ecology Notes 3:344–345.

Flower, W.H. 1876. On some cranial and dental characters of the existing species of rhinoceroses. Proceedings of the Zoological Society of London 1876:443–457.

Frade, F. & J.A. Silva. 1981. Mamiferos de Mocambique (colecção do Centro de Zoologia). Garcia da Orta, Série de Zoologia 10:1–11.

Frame, G.W. 1982. Wild mammal survey of Empakaai Crater area, Tanganyika Notes Records 88–89:41–55.

Frechkop, S. 1943. Mammifères. Fasc. 1 of Exploration du Parc National Albert, Mission S. Frechkop (1937–1938). Brussels: Institut des Parcs Nationaux du Congo Belge.

Furley, C.W. 1986. Reproductive parameters of African gazelles: gestation, first fertile matings, first parturition, and twinning. African Journal of Ecology 24:121–128.

Furley, C.W., H. Tichy & H.-P. Uerpmann. 1988. Systematics and chromosomes of the Indian gazelle, *Gazella bennetti* (Sykes, 1831). Zeitschrift für Säugetierkunde 53:48–54.

Funk, S.M., S.K. Verma, G. Larson, K. Prasad, L. Singh, G. Narayan & J.E. Fa. 2007. The pygmy hog is a unique genus: 19th century taxonomists got it right first time round. Molecular Phylogenetics and Evolution 45:427–436.

Gairdner, K.G. 1915. Addition to the mammalian fauna of Katburi. Journal of the Natural History Society of Siam 1:252–255.

Galbreath, G.J. & R.A. Melville. 2003. *Pseudonovibos spiralis*: epitaph. Journal of Zoology, London 259:169–170.

Galbreath, G.J., J.C. Mordacq & F.H. Weiler. 2006. Genetically solving a zoological mystery: was the kouprey (*Bos sauveli*) a feral hybrid? Journal of Zoology, London 270:561–564.

Gallina, S., S. Mandujano, J. Bello, H.F. López Arévalo & M. Weber. 2010. White-tailed deer *Odocoileus virginianus* (Zimmermann 1780). In J.M.B. Duarte & S. González (eds.), Neotropical Cervidology: Biology and Medicine of Latin American Deer. Jaboticabal, Brazil and Gland, Switzerland: Funep, in collaboration with IUCN, 101–118.

Garcia, J.E. & E.J.F. de Oliveira. 2010. Biochemical genetics. In J.M.B. Duarte & S. González (eds.), Neotropical Cervidology: Biology and Medicine of Latin American Deer. Jaboticabal, Brazil and Gland, Switzerland: Funep, in collaboration with IUCN, 27–30.

Garfield, B. 2006. The Meinertzhagen Mystery: The Life and Legend of a Colossal Fraud. Washington, DC: Potomac Books.

Gatesy, J. & P. Arctander. 2000. Hidden morphological support for the phylogenetic placement of *Pseudoryx nghetinhensis* with bovine bovids: a combined analysis of gross anatomical evidence and DNA sequences from five genes. Systematic Biology 49:515–538.

Gatesy, J., G. Amato, E. Vrba, G. Schaller & R. DeSalle. 1997. A cladistic analysis of mitochondrial ribosomal DNA from the Bovidae. Molecular Phylogenetics and Evolution 7:303–319.

Gatesy, J., M. Milinkovitch, V. Waddell & M. Stanhope. 1999. Stability of cladistic relationships between Cetacea and high-level Artiodactyla taxa. Systematic Biology 48:6–20.

Geist, V. 1987. On the evolution and adaptations of *Alces*. Swedish Wildlife Research Supplement 1:11–23.

Geist, V. 1989. Environmentally guided phenotype plasticity in mammals and some of its consequences to theoretical and applied biology. In M.N. Bruton (ed.), Alternative Life-History Styles of Animals. Dordrecht: Kluwer Academic, 153–176.

Geist, V. 1991a. On the taxonomy of giant sheep (*Ovis ammon* Linnaeus, 1766). Canadian Journal of Zoology 69:706–723.

Geist, V. 1991b. Phantom subspecies: the wood bison *Bison bison "athabascae"* Rhoads 1897 is not a valid taxon, but an ecotype. Arctic 44:283–300.

Geist, V. 1998. Deer of the World. Mechnicsburg, PA: Stackpole Books.

Gentry, A. 1994. Case 2928: Regnum Animale . . . , ed. 2 (M.-J. Brisson, 1762); proposed rejection, with the conservation of the mammalian generic names *Philander* (Marsupialia), *Pteropus* (Chiroptera), *Glis*, *Cuniculus*, and *Hydrochoerus* (Rodentia), *Meles*, *Lutra*, and *Hyaena* (Carnivora), *Tapirus* (Perissodactyla), *Tragulus* and *Giraffa* (Artiodactyla). Bulletin of Zoological Nomenclature 51:135–146.

Gentry, A., J. Clutton-Brock & C.P. Groves. 2004. The naming of wild animal species and their domestic derivatives. Journal of Archaeological Science 31:645–651.

Gentry, A.W. 1964. Skull characters of African gazelles. Annals and Magazine of Natural History (13) 7:353–382.

Gentry, A.W. 1985. The Bovidae of the Omo group deposits, Ethiopia. In F.C. Howell & Y. Coppens (eds.), Les Périssodactyles, les Artiodactyles (Bovidae). Vol. 1 of Les Faunes Plio-Pleistocènes de la Basse Vallée de l'Omo (Ethiopie). Paris: Éditions du CNRS [Centre National de la Recherche Scientifique], 119–191.

Gentry, A.W. 1990. Evolution and dispersal of African Bovidae. In G.A. Bubenik & A.B. Bubenik (eds.), Horns, Pronghorns, and Antlers. New York: Springer-Verlag, 195–227.

Gentry, A.W. 1992. The subfamilies and tribes of the family Bovidae. Mammal Review 22:1–32.

Gentry, A.W. & J.J. Hooker. 1988. The phylogeny of the Artiodactyla. In M.J. Benton (ed.), Mammals. Vol. 2 of The Phylogeny and Classification of the Tetrapods. Oxford: Clarendon Press, 235–272.

Ghosh, M. 1988. The craniology and dentition in the pigmy hog, with a note on the generic status of *Porcula* Hodgson, 1847. Records of the Zoological Survey of India 85:245–266.

Giao, P.M, D. Tuoc, V.V. Dung, E.D. Wikramanayake, G. Amato, P. Arctander & J.R. Mackinnon. 1998. Description of *Muntiacus truongsonensis*, a new species of muntjac (Artiodactyla: Muntiacidae) from central Vietnam, and implications for conservation. Animal Conservation 1:61–68.

Giere, P., C. Freyer & U. Zeller. 1999. Opening of the mammalian vomeronasal organ with respect to the Glires hypothesis: a cladistic reconstruction of the Therian morphotype. Mitteilungen aus dem Museum für Naturkunde in Berlin, Zoologische Reihe 75:247–255.

Gijzen, A. 1959. Das Okapi: *Okapia johnstoni* (Sclater). Die Neue Brehm-Bücherei No. 231. Wittenberg-Lutherstadt, Germany: A. Ziemsen.

Gilbert, C., A. Ropiquet & A. Hassanin. 2006. Mitochondrial and nuclear phylogenies of Cervidae (Mammalia, Ruminantia): systematics, morphology, and biogeography. Molecular Phylogenetics and Evolution 40:101–117.

Gongora, J. & C. Moran. 2005. Nuclear and mitochondrial evolutionary analyses of collared, white-lipped, and Chacoan peccaries (Tayassuidae). Molecular Phylogenetics and Evolution 34:181–189.

Gongora, J., S. Morales, J.E. Bernal & C. Moran. 2006. Phylogenetic divisions among collared peccaries (*Pecari tajacu*) detected using mitochondrial and nuclear sequences. Molecular Phylogenetics and Evolution 41:1–11.

González, B.A., R.E. Palma, B. Marín & J.C. Marín. 2006. Taxonomic and biogeographical status of guanaco *Lama guanicoe* (Artiodactyla, Camelidae). Mammal Review 36:157–178.

González, S., A. Gravier & N. Brum-Zorrilla. 1991. A systematic subspecifical approach on *Ozotoceros bezoarticus* (Linn. 1758) (pampas deer) from South America. In F. Spitz, G. Janeau, G. Gonzalez, et al. (eds.), Proceedings of the International Symposium "Ongulés/Ungulates 91." Paris: Société Française pour l'Étude et la Protection des Mammifères, 129–132.

González, S., F. Álvarez-Valin & J.E. Maldonado. 2002. Morphometric differentiation of endangered pampas deer (*Ozotocerus bezoarticus*), with description of new subspecies from Uruguay. Journal of Mammalogy 83:1127–1140.

González, S., M. Cosse, F.G. Braga, A.R. Vila, M.L. Merino, C. Dellafiore, J.L. Cartes, L. Maffei & M.G. Dixon. 2010. Pampas deer *Ozotocerus bezoarticus* (Linnaeus 1758). In J.M.B. Duarte & S. González (eds.), Neotropical Cervidology: Biology and Medicine of Latin American Deer. Jaboticabal, Brazil and Gland, Switzerland: Funep, in collaboration with IUCN, 119–132.

González, S., J.M.B. Duarte & J.E. Maldonado. 2010. Molecular phylogenetics and evolution. In J.M.B. Duarte & S. González (eds.), Neotropical Cervidology: Biology and Medicine of Latin American Deer. Jaboticabal, Brazil and Gland, Switzerland: Funep, in collaboration with IUCN, 12–17.

Goodman, S.J., N.H. Barton, G. Swanson, K. Abernethy & J.M. Pemberton. 1999. Introgression through rare hybridisation: a genetic study of a hybrid zone between red and sika deer (genus *Cervus*) in Argyll, Scotland. Genetics 152:355–371.

Granjon, L., M. Vassart & A. Greth. 1990. Genetic variability in the Nubian ibex. Mammalia 54:665–667.

Granjon, L., M. Vassart, A. Greth & E.-P. Cribiu. 1991. Genetic study of sand gazelles (*Gazella subgutturosa marica*) from Saudi Arabia: chromosomal and isozymic data. Zeitschrift für Säugetierkunde 56:169–176.

Graphodatsky, A.S. & S.I. Radjabli. 1985. Chromosomes of three cervid species (Mammalia). Zoologicheskii Zhurnal 64:1275–1279.

Gravlund, P., M. Meldgaard, S. Pääbo & P. Arctander. 1998. Polyphyletic origin of the small-bodied, High-Arctic subspecies of tundra reindeer (*Rangifer tarandus*). Molecular Phylogenetics and Evolution 10:151–159.

Gray, J.E. 1843. List of the Specimens of Mammalia in the Collection of the British Museum. London: British Museum (Natural History).

Gray, J.E. 1871. The Chinese long-tailed goat antelope (*Urotragus caudatus*). Annals and Magazine of Natural History (4) 8:371–372.

Greth, A. 1992. Yemen gazelles. CBSG [Conservation Breeding Specialist Group] News 3, 2:18.

Greth, A., D. Williamson, C. Groves, G. Schwede & M. Vassart. 1993. Bilkis gazelle in Yemen—status and taxonomic relationships. Oryx 27:239–244.

Grobler, J.H. 1980. Body growth and age determination of the sable *Hippotragus niger niger* (Harris, 1838). Koedoe 23:131–156.

Grobler, J.P. & F.H. van der Bank. 1993. Genetic diversity and differentiation of the three extant southern African species of the subfamily Hippotraginae (family: Bovidae). Biochemical Systematics and Ecology 21:591–596.

Grodinsky, C. & M. Stuewe. 1987. With lots of help, alpine ibex return to their mountains. Smithsonian Magazine 18, 9:68–77.

Groves, C.P. 1967a. The rhinoceroses of Southeast Asia. Säugetierkundliche Mitteilungen 15:221–237.

Groves, C.P. 1967b. Geographic variation in the black rhinoceros. Zeitschrift für Säugetierkunde 32:267–276.

Groves, C.P. 1967c. On the gazelles of the genus *Procapra* Hodgson, 1846. Zeitschrift für Säugetierkunde 32:144–149.

Groves, C.P. 1969a. The smaller gazelles of the genus *Gazella* de Blainville, 1816. Zeitschrift für Säugetierkunde 34:38–60.

Groves, C.P. 1969b. Systematics of the anoa (Mammalia, Bovidae). Beaufortia 17:1–12.

Groves, C.P. 1971. Species characters in rhinoceros horns. Zeitschrift für Säugetierkunde 36:238–252.

Groves, C.P. 1974. A note on the systematic position of the muntjac (Artiodactyla, Cervidae). Zeitschrift für Säugetierkunde 39:369–372.

Groves, C.P. 1975a. Notes on the gazelles, 1: *Gazella rufifrons* and the zoogeography of central African Bovidae. Zeitschrift für Säugetierkunde 40:308–319.

Groves, C.P. 1975b. Taxonomic notes on the white rhinoceros *Ceratotherium simum* (Burchell, 1817). Säugetierkundliche Mitteilungen 23:200–212.

Groves, C.P. 1978a. The taxonomic status of the dwarf blue sheep (Artiodactyla: Bovidae). Säugetierkundliche Mitteilungen 26:177–183.

Groves, C.P. 1978b. The extinct Cape rhinoceros, *Diceros bicornis bicornis* (Linnaeus, 1758). Säugetierkundliche Mitteilungen 26:117–128.

Groves, C.P. 1980. Notes on the systematics of *Babyrousa* (Artiodactyla, Suidae). Zoologische Mededelingen 55:29–46.

Groves, C.P. 1981a. Notes on the gazelles, 2: subspecies and clines in the springbok (*Antidorcas*). Zeitschrift für Säugetierkunde 46:189–197.

Groves, C.P. 1981b. Systematic relationships in the Bovini (Artiodactyla, Bovidae). Zeitschrift für Zoologisches Systematik und Evolutionsforschung 19:264–278.

Groves, C.P. 1981c. Notes on the gazelles, 3: The dorcas gazelles of North Africa. Annali del Museo Civico di Storia Naturale di Genova 83:455–471.

Groves, C.P. 1981d. Ancestors for the Pigs: Taxonomy and Phylogeny of the Genus *Sus*. Technical Bulletin No. 3. Canberra: Australian National University, Research School of Pacific Studies, Department of Prehistory.

Groves, C.P. 1982a. Asian rhinoceroses—down but not out. Malayan Naturalist 36:11–22.

Groves, C.P. 1982b. A note on geographic variation in the Indian blackbuck (*Antilope cervicapra* Linnaeus, 1758). Records of the Zoological Survey of India 79:489–503.

Groves, C.P. 1982c. *Bos gaurus* H. Smith, 1827 (Mammalia, Artiodactyla): proposed conservation. Bulletin of Zoological Nomenclature 39:279–280.

Groves, C.P. 1982d. *Antilope depressicornis* H. Smith, 1827, and *Anoa quarlesi* Ouwens, 1910 (Mammalia, Artiodactyla): proposed conservation. Bulletin of Zoological Nomenclature 39:281–282.

Groves, C.P. 1983a. Phylogeny of the living species of rhinoceros. Zeitschrift für Zoologisches Systematik und Evolutionsforschung 21:293–313.

Groves, C.P. 1983b. Notes on the gazelles, 4: the Arabian gazelles collected by Hemprich and Ehrenberg. Zeitschrift für Säugetierkunde 48:371–381.

Groves, C.P. 1983c. Geographic variation in the barasingha or swamp deer (*Cervus duvauceli*). Journal of the Bombay Natural History Society 79:620–629.

Groves, C.P. 1983d. A new subspecies of sable antelope, *Hippotragus niger* (Harris, 1838). Revue Zoologique Africaine 97:821–828.

Groves, C.P. 1985. An introduction to the gazelles. Chinkara 1, 1:4–10.

Groves, C.P. 1986. The taxonomy, distribution, and adaptations of recent equids. In R.H. Meadow & H.P.

Uerpmann (eds.), Equids in the Ancient World. Wiesbaden, Germany: Reichert, 11–65.

Groves, C.P. 1989a. Feral mammals of the Mediterranean islands: documents of early domestication. In J. Clutton-Brock (ed.), The Walking Larder. London: Unwin Hyman, 46–58.

Groves, C.P. 1989b. A catalogue of the genus *Gazella*. In A. Dixon & D. Jones (eds.), Conservation and Biology of Desert Antelopes. London: Christopher Helm, 193–198.

Groves, C.P. 1989c. The gazelles of the Arabian peninsula. In A.H. Abu-Zinada, P.D. Goriup & I.A. Nader (eds.), Wildlife Conservation and Development in Saudi Arabia. Publication No. 3. Riyadh, Saudi Arabia: National Commission for Wildlife Conservation and Development, 237–248.

Groves, C.P. 1993a. The chinkara (*Gazella bennetti*) in Iran, with the description of two new subspecies. Journal of Sciences of the Islamic Republic of Iran 4:166–178.

Groves, C.P. 1993b. Testing rhinoceros subspecies by multivariate analysis. In O.A. Ryder (ed.), Rhinoceros Biology and Conservation. San Diego: Zoological Society of San Diego, 92–100.

Groves, C.P. 1994. Morphology, habitat, and taxonomy. In L. Boyd & K.A. Houpt (eds.), Przewalski's Horse: The History and Biology of an Endangered Species. Albany: State University of New York Press, 39–59.

Groves, C.P. 1995a. On the nomenclature of domestic animals. Bulletin of Zoological Nomenclature 52:137–141.

Groves, C.P. 1995b. Microtaxonomy and its implications for captive breeding. In U. Ganslosser, J.K. Hodges & W. Kaumanns (eds.), Research and Captive Propagation. Fürth, Germany: Filander, 24–28.

Groves, C.P. 1996a. Taxonomic diversity in Arabian gazelles: the state of the art. In A. Greth, C. Magin & M. Ancrenaz (eds.), Conservation of Arabian Gazelles. Riyadh, Saudi Arabia: National Commission for Wildlife Conservation and Development, 8–39.

Groves, C.P. 1996b. The taxonomy of the Asian wild buffalo from the Asian mainland. Zeitschrift für Säugetierkunde 61:327–338.

Groves, C.P. 1997a. Taxonomy of wild pigs (*Sus*) of the Philippines. Zoological Journal of the Linnaean Society 120:163–191.

Groves, C.P. 1997b. The taxonomy of Arabian gazelles. In K. Habibi, A.H. Abuzinada & I.A. Nader (eds.), The Gazelles of Arabia. Publication No. 29. Riyadh, Saudi Arabia: National Commission for Wildlife Conservation and Development, 24–51.

Groves, C.P. 1997c. Die Nashörner—Stammesgeschichte und Verwandtschaft. In Die Nashörner: Begegnung mit urzeitlichen Kolossen. Fürth, Germany: Filander, 14–32.

Groves, C.P. 1999. The advantages and disadvantages of being domesticated (a keynote address). Perspectives in Human Biology 4:1–12.

Groves, C.P. 2000. Phylogenetic relationships within recent Antilopini (Bovidae). In E.S. Vrba & G.B. Schaller (eds.), Antelopes, Deer, and Relatives. New Haven: Yale University Press, 223–233.

Groves, C.P. 2001a. Primate Taxonomy. Washington, DC: Smithsonian Institution Press.

Groves, C.P. 2001b. Why taxonomic stability is a bad idea, or why are there so few species of primates (or are there?). Evolutionary Anthropology 10:192–198.

Groves, C.P. 2002. Taxonomy of living Equidae. In P.D. Moehlman (ed.), Equids: Zebras, Asses, and Horses. Gland, Switzerland: IUCN, 94–107.

Groves, C.P. 2003. Taxonomy of ungulates of the Indian subcontinent. Journal of the Bombay Natural History Society 100:341–362.

Groves, C.P. 2004. The what, why, and how of primate taxonomy. International Journal of Primatology 25:1105–1126.

Groves, C.P. 2005. Domestic and wild mammals: naming and identity. In A. Minelli, G. Ortalli & G. Sangha (eds.), Animal Names. Venice: Istituto Veneto di Scienze, Lettere ed Arti, 151–157.

Groves, C.P. 2006a. The genus *Cervus* in eastern Eurasia. European Journal of Wildlife Research 52:14–22. doi:10.1007/s10344-005-0011-5.

Groves, C.P. 2006b. Tribute to Peter Grubb. Suiform Soundings 6, 2:4.

Groves, C.P. 2007. Current views on the taxonomy and zoogeography of the genus *Sus*. In U. Albarella, K. Dobney, A. Ervynck & P. Rowley-Conwy (eds.), Pigs and Humans: 10,000 Years of Interaction. Oxford: Oxford University Press, 15–29.

Groves, C.P. 2008a. Family Cervidae. In D.R. Prothero & S.E. Foss (eds.), The Evolution of Artiodactyls. Baltimore: Johns Hopkins University Press, 249–256.

Groves, C.P. 2008b. Extended Family: Long Lost Cousins. Arlington, VA: Conservation International.

Groves, C.P. & C. Bell. 2004. New investigations on the taxonomy of the zebras genus *Equus*, subgenus *Hippotigris*. Mammalian Biology 69:182–196.

Groves, C.P. & S. Chakraborty. 1983. The Calcutta collection of Asian rhinoceros. Records of the Zoological Survey of India 80:251–263.

Groves, C.P. & Feng Z-j. 1986. The status of musk deer from Anhui Province, China. Acta Theriologica Sinica 6:101–106.

Groves, C.P. & P. Grubb. 1974. A new duiker from Rwanda. Revue Zoologique Africaine 88:189–196.

Groves, C.P. & P. Grubb. 1982. The species of muntjac (genus *Muntiacus*) in Borneo: unrecognised sympatry in tropical deer. Zoologische Mededelingen 56:203–216.

Groves, C.P. & P. Grubb. 1985. Reclassification of the serows and gorals (*Nemorhaedus*: Bovidae). In S. Lovari (ed.), The Biology and Management of Mountain Ungulates. London: Croom Helm, 45–50.

Groves, C.P. & P. Grubb. 1987. Relationships of living deer. In C. Wemmer (ed.), Biology and Management of the Cervidae. Washington, DC: Smithsonian Institution Press, 21–59.

Groves, C.P. & P. Grubb. 1990. Muntiacidae. In G.A. Bubenik & A.B. Bubenik (eds.) Horns, Pronghorns, and Antlers. New York: Springer-Verlag, 132–68.

Groves, C.P. & P. Grubb. 1993a. The suborder Suiformes. In W.L.R. Oliver (ed.), Pigs, Peccaries, and Hippos: Status Survey and Conservation Action Plan. Gland, Switzerland: IUCN, SSC [Species Survival Commission] Pigs and Peccaries Specialist Group, 1–4.

Groves, C.P. & P. Grubb. 1993b. The Eurasian suids *Sus* and *Babyrousa*: taxonomy and description. In W.L.R. Oliver (ed.), Pigs, Peccaries, and Hippos: Status Survey and Conservation Action Plan. Gland, Switzerland: IUCN, SSC Pigs and Peccaries Specialist Group, 107–111.

Groves, C.P. & C. Guérin. 1980. Le *Rhinoceros sondaicus annamiticus* (Mammalia, Perissodactyla) d'Indochine: distinction taxonomique et anatomique; relations phyletiques. Geobios 13:199–208.

Groves, C.P. & D.L. Harrison. 1967. The taxonomy of the gazelles of Arabia. Journal of Zoology, London 152:381–387.

Groves, C.P. & J. Jayewardene. 2009. The wild buffalo of Sri Lanka. Taprobanica 1:56–62.

Groves, C.P. & M. Karami. 1993. A mammal species new for Iran: *Gazella gazella* Pallas, 1766 (Artiodactyla: Bovidae). Journal of Sciences of the Islamic Republic of Iran 4:81–89.

Groves, C.P. & D.M. Lay. 1985. A new species of the genus *Gazella* from the Arabian Peninsula. Mammalia 49:27–36.

Groves, C.P. & V. Mazák. 1967. Taxonomic problems of Asiatic wild asses: with the description of a new subspecies. Zeitschrift für Säugetierkunde 32:321–355.

Groves, C.P. & E. Meijaard. 2005. Interspecific variation in *Moschiola*, the Indian chevrotain. Raffles Bulletin of Zoology 12:413–421.

Groves, C.P. & O.A. Ryder. 2000. Systematics and phylogeny of the horse. In A.T. Bowling & A. Ruvinsky (eds.), The Genetics of the Horse. New York: CABI, 1–24.

Groves, C.P. & G.B. Schaller. 2000. The phylogeny and biogeography of the newly discovered Annamite artiodactyls. In E.S. Vrba & G.B. Schaller (eds.), Antelopes, Deer, and Relatives. New Haven: Yale University Press, 261–282.

Groves, C.P. & C. Smeenk. 1978. On the type material of *Cervus nippon* Temminck, 1836: with a revision of sika deer from the main Japanese islands. Zoologische Mededelingen 53:11–28.

Groves, C.P. & C. Smeenk. 2007. The nomenclature of the African wild ass. Zoologische Mededelingen 81:121–135.

Groves, C.P. & C.R. Westwood. 1995. Skulls of the blaauwbok, *Hippotragus leucophaeus*. Zeitschrift für Säugetierkunde 60:314–318.

Groves, C.P., Y-x. Wang & P. Grubb. 1995. Taxonomy of musk deer, genus *Moschus* (Moschidae, Mammalia). Acta Theriologica Sinica 15:181–197.

Groves, C.P., G.B. Schaller, G. Amato & K. Khounbouline. 1997. Rediscovery of the wild pig *Sus bucculentus*. Nature 386:335.

Groves, C.P., P. Fernando & J. Robovsky. 2010. The sixth rhino: a taxonomic re-assessment of the critically endangered northern white rhinoceros. PLoS One 5, 4:e9703. doi:10.1371/journal.pone.0009703.

Groves, P. & G.F. Shields. 1996. Phylogenetics of the Caprinae based on cytochrome *b* sequence. Molecular Phylogenetics and Evolution 5:467–476.

Groves, P. & G.F. Shields. 1997. Cytochrome *b* sequences suggest convergent evolution of the Asian takin and Arctic musk-ox. Molecular Phylogenetics and Evolution 8:363–374.

Grubb, P. 1972. Variation and incipient speciation in the African buffalo. Zeitschrift für Säugetierkunde 37:121–144.

Grubb, P. 1977. Notes on a rare deer, *Muntiacus feai*. Annali del Museo Civico de Storia Naturale de Genova 81:202–207.

Grubb, P. 1978a. Patterns of speciation in African mammals. Bulletin of Carnegie Museum of Natural History 6:152–167.

Grubb, P. 1978b. A new antelope from Gabon. Zoological Journal of the Linnaean Society of London 62:373–380.

Grubb, P. 1979. Refuges and dispersal in the speciation of African forest mammals. Paper presented at the Association of Tropical Biologists, 5th International Symposium, 8–13 February 1979, La Guaira, Venezuela.

Grubb, P. 1981. *Equus burchellii*. Mammalian Species 157:1–9.

Grubb, P. 1982a. Refuges and dispersal in the speciation of African forest mammals. In G.T. Prance (ed.), Biological Diversification in the Tropics. New York: Columbia University Press, 537–553.

Grubb, P. 1982b. The systematics of Sino-Himalayan musk deer (*Moschus*), with particular reference to the species described by B.H. Hodgson. Säugetierkundlich Mitteilungen 30:127–137.

Grubb, P. 1985a. Geographical variation in the bushbuck of eastern Africa (*Tragelaphus scriptus*, Bovidae). In K.-L. Schuchmann (ed.), African Vertebrates: Systematics, Phylogeny, and Evolutionary Ecology. Bonn: Zoologisches Forschungsinstitut und Museum Alexander Koenig, 11–27.

Grubb, P. 1985b. The biogeographic significance of forest mammals in eastern Africa. In E. Van der Straeten, V.N. Verheyen & F. De Vree (eds.), Proceedings of the Third International Colloquium on the Ecology and Taxonomy of African Small Mammals. Koninklijk Museum voor Midden-Afrika Zoologische Wetenschappen No. 237. Tervuren, Belgium: Koninklijk Museum voor Midden-Afrika, 75–85.

Grubb, P. 1986. Notes on some antelope subspecies. Gnusletter 1986, Jan:3–4.

Grubb, P. 1988. The status of Walker's duiker, a rare antelope from Malawi. Nyala 12:67–72.

Grubb, P. 1989. The systematic status of the suni (*Neotragus moschatus*) in Malawi. Nyala 13:21–27.

Grubb, P. 1990a. Cervidae of Southeast Asia. In G.A. Bubenik & A.B. Bubenik (eds.), Horns, Pronghorns, and Antlers. New York: Springer-Verlag, 169–179.

Grubb, P. 1990b. Primate geography in the Afrotropical forest biome. In G. Peters & R. Hutterer (eds.), Vertebrates in the Tropics. Bonn: Zoologisches Forschungsinstitut und Museum Alexander Koenig, 187–214.

Grubb, P. 1990c. List of deer species and subspecies. Deer 8:153–155.

Grubb, P. 1993a. The Afrotropical hippopotamuses *Hippopotamus* and *Hexaprotodon*: taxonomy and description. In W.L.R. Oliver (ed.), Pigs, Peccaries, and Hippos: Status Survey and Conservation Action Plan. Gland, Switzerland: IUCN, SSC Pigs and Peccaries Specialist Group, 41–43.

Grubb, P. 1993b. The Afrotropical suids *Phacochoerus*, *Hylochoerus*, and *Potamochoerus*: taxonomy and description. In W.L.R. Oliver (ed.), Pigs, Peccaries, and Hippos: Status Survey and Conservation Action Plan. Gland, Switzerland: IUCN, SSC Pigs and Peccaries Specialist Group, 66–79.

Grubb, P. 1994. Notes on Grant's gazelle. Gnusletter 13, 1–2:4–5.

Grubb, P. 1999a. Types and type localities of ungulates named from southern Africa. Koedoe 42:13–45.

Grubb, P. 1999b. Evolutionary processes implicit in distribution patterns of modern African mammals. In T.G. Bromage & F. Schrenk (eds.), African Biogeography, Climate Change, and Human Evolution. New York: Oxford University Press, 150–164.

Grubb, P. 2000a. Morphoclinal evolution in ungulates. In E.S. Vrba & G.B. Schaller (eds.), Antelopes, Deer, and Relatives. New Haven: Yale University Press, 156–170.

Grubb, P. 2000b. Valid and invalid nomenclature of living and fossil deer, Cervidae. Acta Theriologica 45:289–307.

Grubb, P. 2001a. Review of family-group names of living bovids. Journal of Mammalogy 82:374–318.

Grubb, P., 2001b. Endemism in African rain forest mammals. In W. Weber, L.J.T. White, A. Vedder & L. Jaughtoh Treves (eds.), African Rain Forest Ecology and Conservation: An Interdisciplinary Perspective. New Haven: Yale University Press, 88–100.

Grubb, P., 2002. Types, type locality, and subspecies of the gerenuk *Litocranius walleri* (Artiodactyla: Bovidae). Journal of Zoology, London 257, 4:530–543.

Grubb, P. 2004a. Controversial scientific names of African mammals. African Zoology 39, 1:91–109.

Grubb, P. 2004b. Nomenclature, subspeciation, and affinities among elaphine deer (*Cervus elaphus sensu lato*). In C. Oswald. (ed.), Symposium on Red Deer Taxonomy. Ebersberg, Germany: Christian Oswald-Stiftung, 12–30.

Grubb, P. 2005a. Order Perissodactyla. In D.E. Wilson & D.M. Reeder (eds.), Mammal Species of the World: A Taxonomic and Geographic Reference, 3rd ed., 2 vols. Washington, DC: Smithsonian Institution Press, 1:629–636.

Grubb, P. 2005b. Order Artiodactyla. In D.E. Wilson & D.M. Reeder (eds.), Mammal Species of the World: A Taxonomic and Geographic Reference, 3rd ed., 2 vols. Washington, DC: Smithsonian Institution Press, 1:637–722.

Grubb, P. 2006. Geospecies and superspecies in the African primate fauna. Primate Conservation 20:75–78.

Grubb, P. In press. Evolutionary geography of modern African mammals. In T. Butysnki (ed.), Vegetation, Climate, and Geology of Africa: Evolution of African Mammals; Primates. Vol. 1 of The Mammals of Africa. London: Academic Press.

Grubb, P. & A.L. Gardner, 1998. List of species and subspecies of the families Tragulidae, Moschidae, and Cervidae. In C. Wemmer (ed.), Deer: Status Survey and Conservation Action Plan. Gland Switzerland: IUCN, 6–16.

Grubb, P. & C.P. Groves. 1983. Notes on the taxonomy of the deer (Mammalia, Cervidae) of the Philippines. Zoologische Anzeiger (Jena) 210:119–144.

Grubb, P. & C.P. Groves. 1990. Muntiacidae. In G.A. Bubenik & A.B. Bubenik (eds.), Horns, Pronghorns, and Antlers. New York: Springer-Verlag, 134–168.

Grubb, P. & C.P. Groves. 1993. The neotropical tayassuids *Tayassu* and *Catagonus*: taxonomy and description. In W.L.R. Oliver (ed.), Pigs, Peccaries, and Hippos: Status Survey and Conservation Action Plan. Gland, Switzerland: IUCN, SSC Pigs and Peccaries Specialist Group, 5–7.

Grubb, P. & C.P. Groves. 2001. Appendix C: revision and classification of the Cephalophinae. In V.J. Wilson, Duikers of Africa: Masters of the African Forest Floor. Bulawayo: Chipangali Wildlife Trust, 703–728.

Grubb, P. & W. Oliver. 1991. Pigs and Peccaries Specialist Group report: A forgotten warthog; more bush pigs but fewer forest hogs. Species 17:61–62.

Grubb, P. & C.B. Powell. 2002. Range extension of black-fronted duiker (*Cephalophus nigrifrons* Gray 1871, Artiodactyla, Bovidae): first records from Nigeria. Tropical Zoology 15, 1:89–95.

Grubb, P. & V.J. Wilson. 1990. Classification of antelopes adopted for the antelope survey. In R. East (ed.), West and Central Africa. Part 3 of Antelopes: Global Survey and Regional Action Plans. Gland, Switzerland: IUCN, 3–4.

Grubb, P., R. East & V.J. Wilson. 1989. Classification of antelopes adopted for the antelope survey. In R. East (ed.), Southern and South-Central Africa. Part 2 of

Antelopes: Global Survey and Regional Action Plans. Gland, Switzerland: IUCN, 3–4.

Grubb, P., T.S. Jones, A.G. Davies, E. Edberg, E.D. Starin & J.E. Hill. 1998. Mammals of Ghana, Sierra Leone, and the Gambia. St. Ives, UK: Trendine.

Grubb, P., A.M. Lister & S.R.M. Sumner. 1998. Taxonomy, morphology, and evolution of European roe deer. In R. Andersen, P. Duncan & J.D.C. Linnell (eds.), The European Roe Deer: The Biology of Success. Oslo: Scandinavian University Press, 23–46.

Grubb, P., O. Sandrock, O. Kullmer, T.M. Kaiser & F.J. Schrenk. 1999. Relationships between eastern and southern African mammal faunas. In T.G. Bromage & F. Schrenk (eds.), African Biogeography, Climate Change, and Human Evolution. New York: Oxford University Press, 253–267.

Grubb, P., S.C. Kingswood & D.P. Mallon. 2001. Classification of antelopes adopted for the antelope survey. In D.P. Mallon & S.C. Kingswood (comps.), North Africa, the Middle East, and Asia. Part 4 of Antelopes: Global Survey and Regional Action Plans. Gland, Switzerland: IUCN, 7–9.

Grubb, P., C.P. Groves & C.B. Powell, 2003. Duikers and dwarf antelopes: new or uncertain records. In A. Plowman (ed.), Ecology and Conservation of Small Antelope. Fürth, Germany: Filander, 127–140.

Gu Z-l., X-b. Zhao, N. Li & C-x. Wu. 2007. Complete sequence of the yak (*Bos grunniens*) mitochondrial genome and its evolutionary relationship with other ruminants. Molecular Phylogenetics and Evolution 42:248–255.

Guérin, C. 1965. *Gallogoral* (nov. gen.) *meneghinii* (Rütimeyer, 1878): Un Rupicapriné du Villafranchien d'Europe Occidentale. Documents des Laboratoires de Géologie de Lyon No. 11.

Guo Z-p., E-y. Chen & Y-z. Wang. 1978. A new subspecies of sika deer from Sichuan—*Cervus nippon sichuanicus* subsp. nov. Acta Zoologica Sinica 24:187–192.

Gyldenstolpe, N. 1916. Zoological Results of the Swedish Zoological Expedition to Siam 1911–1912 and 1914–1915: 5, Mammals II. Kungliga Svenska Vetenskapsakademien Handlingar No. 57, 2. Stockholm: Almqvist & Wiksell.

Hall, E.R. & K.R. Kelson. 1959. The Mammals of North America. New York: Ronald Press.

Haltenorth, T. 1963. Klassifikation der Säugetiere: Artiodactyla. Handbuch der Zoologie No. 8(32), pt. 1, 18. Berlin: De Gruyter.

Hamilton Smith, C. 1827. Order VII—Ruminantia (Pecora, Lin.). In vols. 4 and 5 of E. Griffith, W. Pidgeon & C. Hamilton Smith, Cuvier's Animal Kingdom [t.p. of vol. 5: Synopsis of the Species of the Class Mammalia]. London: Whittaker, 33–428.

Hammer, S.E., H.M. Schwammer & F. Suchentrunk. 2008. Evidence for introgressive hybridisation of captive markhor (*Capra falconeri*) with domestic goat: cautions for reintroduction. Biochemical Genetics 46:216–226.

Hammond, J., J.C. Bowman & T.J. Robinson. 1983. Hammond's Farm Animals. London: Edward Arnold.

Hammond, R.L., W. Macasero, B. Flores, O.B. Mohammed, T. Wacher & M.W. Bruford. 2001. Phylogenetic reanalysis of the Saudi gazelle and its implications for conservation. Conservation Biology 15:1123–1133.

Hanitsch, R. 1918. On a serow from Annam. Journal of the Straits Branch of the Royal Asiatic Society 78:59–65.

Hansen-Melander, E. & Y. Melander. 1974. The karyotype of the pig. Hereditas 77:149–158.

Hardwicke, T. 1925. Descriptions of two species of antelope from India. Transactions of the Linnaean Society 14:518–529.

Harley, E.H., I. Baumgarten, J. Cunningham & C. O'Ryan. 2005. Genetic variation and population structure in remnant populations of black rhinoceros, *Diceros bicornis*, in Africa. Molecular Ecology 14:2981–2990.

Harper, F. 1940. The nomenclature and type localities of certain Old World mammals. Journal of Mammalogy 21:191–203, 322–332.

Harris, J.M. 2007. Superfamily Suoidea. In D.R. Prothero & S.E. Foss (eds.), The Evolution of Artiodactyls. Baltimore: Johns Hopkins University Press, 130–150.

Harrison, D.L. 1967. Observations on a wild goat, *Capra aegagrus* (Artiodactyla: Bovidae) from Oman, E. Arabia. Journal of Zoology, London 151:27–30.

Harrison, D.L. & P.J.J. Bates. 1991. The Mammals of Arabia, 2nd ed. Sevenoaks, UK: Harrison Zoological Museum.

Harrison, R.G. 1998. Linking evolutionary patterns and processes: the relevance of species concepts for the study of speciation. In D.J. Howard & S.H. Berlocher (eds.), Endless Forms: Species and Speciation. Oxford: Oxford University Press, 19–31.

Hartl, G.B., R. Göltenboth, M. Grillitsch & R. Willing. 1988. On the biochemical systematics of the Bovini. Biochemical Systematics and Ecology 16:575–579.

Hartl, G.B., H. Burger, R. Willing & F. Suchentrunk. 1990. On the biochemical systematics of the Caprini and the Rupicaprini. Biochemical Systematics and Ecology 18:175–182.

Hartl, G.B., K. Nadlinger, M. Apollonio, G. Markov, F. Klein, G. Lange, S. Findo & J. Markowski. 1995. Extensive mitochondrial DNA differentiation among European red deer (*Cervus elaphus*) populations: implications for conservation and management. Zeitschrift für Säugetierkunde 60:41–52.

Hassanin, A. & E.J.P. Douzery. 1999a. The tribal radiation of the family Bovidae (Artiodactyla) and the evolution of the mitochondrial cytochrome *b* gene. Molecular Phylogenetics and Evolution 13:227–243.

Hassanin, A. & E.J.P. Douzery. 1999b. Evolutionary affinities of the enigmatic saola (*Pseudoryx nghetinhensis*) in the context of the molecular phylogeny of Bovidae. Proceedings of the Royal Society of London, B 266:893–900.

Hassanin, A. & E.J.P. Douzery. 2003. Molecular and morphological phylogenies of Ruminantia and the alternative position of the Moschidae. Systematic Biology 52:206–228.

Hassanin, A. & A. Ropiquet. 2004. Molecular phylogeny of the tribe Bovini (Bovidae, Bovinae) and the taxonomic status of the kouprey, *Bos sauveli* Urbain 1937. Molecular Phylogenetics and Evolution 33:896–897.

Hassanin, A. & A. Ropiquet. 2007. What is the taxonomic status of the Cambodian banteng and does it have close genetic links with the kouprey? Journal of Zoology, London 271:246–252.

Hassanin, A., E. Pasquet & J.-D. Vigne. 1998. Molecular systematics of the subfamily Caprinae (Artiodactyla, Bovidae) as determined from cytochrome *b* sequences. Journal of Mammalian Evolution 5:217–236.

Hassanin, A., A. Seveau, H. Thomas, H. Bocherens, D. Billiou & B.X. Nguyen. 2001. Evidence that the mysterious "linh duong" (*Pseudonovibos spiralis*) is not a new bovid. Comptes Rendus de l'Académie des Sciences (Paris), Sér. 3, Sciences de la Vie 324:71–80.

Hassanin, A., A. Ropiquet, R. Cornette, M. Tranier, P. Pfeffer, P. Candegabe & M. Lemaire. 2006. Has the kouprey (*Bos sauveli* Urbain, 1937) been domesticated in Cambodia? Comptes Rendus de l'Académie des Sciences (Paris), Biologies 329:124–135.

Hassanin, A., A. Ropiquet, A.-L. Gourmand, B. Chardonnet & J. Rigoulet. 2007. Mitochondrial DNA variability in *Giraffa camelopardalis*: consequences for taxonomy, phylogeography, and conservation of giraffes in west and central Africa. Comptes Rendus de l'Académie des Sciences (Paris), Biologies 330:265–274.

Hayman, R.W. 1961. The red goral of the north-east frontier region. Proceedings of the Zoological Society of London 136:317–324.

Helgen, K.M. 2003. Major mammalian clades: a review under consideration of molecular and palaeontological evidence. Mammalian Biology 68:1–15.

Heller, E. 1910. A new sable antelope from British East Africa. Smithsonian Miscellaneous Collections 54, No. 6. Washington, DC: Smithsonian Institution.

Heller, E. 1912. New genera and races of African ungulates. Smithsonian Miscellaneous Collections 60, No. 8. Washington, DC: Smithsonian Institution.

Heller, E. 1913a. New races of antelopes from British East Africa. Smithsonian Miscellaneous Collections 61, No. 7. Washington, DC: Smithsonian Institution.

Heller, E. 1913b. New antelopes and carnivores from British East Africa. Smithsonian Miscellaneous Collections 61, No. 13. Washington, DC: Smithsonian Institution.

Heller, E. 1913c. New races of ungulates and primates from equatorial Africa. Smithsonian Miscellaneous Collections 61, No. 17. Washington, DC: Smithsonian Institution.

Heller, E. 1914. Four new subspecies of large mammals from equatorial Africa. Smithsonian Miscellaneous Collections, 61, No. 22. Washington, DC: Smithsonian Institution.

Hemmer, H. 1983 [English translation 1990]. Domestication: The Decline of Environmental Appreciation. Brunswick, Germany: Vieweg [English translation by Neil Beckhaus, Cambridge: Cambridge University Press].

Heptner, V., A.A. Nasimovich & A.G. Bannikov. 1961 [English translation 1988]. Artiodactyla and Perissodactyla. Vol. 1 of Mammals of the Soviet Union. Moscow: Vysshaya Shkola [English translation, Washington, DC: Smithsonian Institution Libraries and National Science Foundation].

Herring, S.W. 1972. The role of canine morphology in the evolutionary divergence of pigs and peccaries. Journal of Mammalogy 53:500–512.

Herring, S.W. 1974. A biometric study of suture fusion and skull growth in peccaries. Anatomy and Embryology 146:167–180.

Hershkovitz, P. 1957. The type locality of *Bison bison* Linnaeus. Proceedings of the Biological Society of Washington 70:31–32.

Hershkovitz, P. 1963. The nomenclature of South American peccaries. Proceedings of the Biological Society of Washington 76:85–88.

Hershkovitz, P. 1982. Neotropical deer (Cervidae), part 1: pudus, genus *Pudu* Gray. Fieldiana Zoology (n.s.) 11:1–86.

Heude, P.-M. 1894a. Notes sur le genre *Capricornis* (Ogilby, 1836). Mémoires Concernant l'Histoire Naturelle de l'Empire Chinois 2:222–233.

Heude, P.-M. 1894b. Notes sur le genre *Kemas* (Ogilby, 1836). Mémoires Concernant l'Histoire Naturelle de l'Empire Chinois 2:234–245.

Heude, P.-M. 1897a. Notes sur le quatrième capricorne de Tonkin. Mémoires Concernant l'Histoire Naturelle de l'Empire Chinois 3:151–155.

Heude, P.-M. 1897b. Capricornes de Se-tchouan. Mémoires Concernant l'Histoire Naturelle de l'Empire Chinois 3:196–198.

Heude, P.-M. 1898. Capricornes de Moupin, 1. Mémoires Concernant l'Histoire Naturelle de l'Empire Chinois 4:1–9.

Heude, P.-M. 1899. Capricorne de la Chine méridionale. Mémoires Concernant l'Histoire Naturelle de l'Empire Chinois 4:211.

Hillman-Smith, A.K.K. & C.P. Groves. 1994. *Diceros bicornis*. Mammalian Species 455:1–8.

Hla Aung, S. 1967. Observations on the Red goral *Nemorhaedus cranbrooki* and the Burmese takin *Budorcas t. taxicolor* at Rangoon Zoo. International Zoo Yearbook 7: 225–256.

Ho, S.Y.W., M.J. Phillips, A. Cooper & A.J. Drummond. 2005. Time dependency of molecular rate estimates and systematic overestimation of recent divergence times. Molecular Biology and Evolution 22:1561–1568.

Hodgson, B.H. 1831. Contributions in natural history. Gleanings in Science 3:320–324.

Hodgson, B.H. 1832. Mammalia and birds collected in Nepal by B. H. Hodgson. Proceedings of the Zoological Society of London, part 2 (1832):10–16.

Hodgson, B.H. 1834. On the Mammalia of Nepal. Proceedings of the Zoological Society of London, part 2 (1834):95–99.

Holbrook, L.T. 2002. The unusual development of the sagittal crest in the Brazilian tapir (*Tapirus terrestris*). Journal of Zoology, London 256:215–219.

Honey, J.G. 2007. Family Camelidae. In D.R. Prothero & S.E. Foss (eds.), The Evolution of Artiodactyls. Baltimore: Johns Hopkins University Press, 177–188.

van Hooft, W.F., A.F. Groen & H.H.T. Prins. 2002. Phylogeography of the African buffalo based on mitochondrial and Y-chromosomal loci: Pleistocene origin and population expansion of the Cape buffalo subspecies. Molecular Ecology 11:267–279.

Howell, A.B. 1930. Mammals from China in the collection of the United States National Museum. Proceedings of the United States National Museum 75:1–82.

Hrabě, V. & P. Koubek. 1984. Craniometrical characteristics of *Rupicapra rupicapra tatrica* (Mamm., Bovidae). Folia Zoologica 33:73–84.

Hu J.,.S-g. Fang & Q-h. Wan. 2006. Genetic diversity of Chinese water deer (*Hydropotes inermis inermis*): implications for conservation. Biochemical Genetics 44:156–167.

Huang, L., W. Nie, J. Wang, W. Su & F. Yang. 2005. Phylogenomic study of the subfamily Caprinae by cross-species chromosome painting with Chinese muntjac paints. Chromosome Research 13:389–399.

Huang, L., J. Wang, W. Nie, W. Su & F. Yang. 2006. Tandem chromosome fusions in karyotypic evolution of *Muntiacus*: evidence from *M. feae* and *M. gongshanensis*. Chromosome Research 14:637–647.

d'Huart, J.-P. & P. Grubb. 2001. Distribution of the common warthog (*Phacochoerus africanus*) and the desert warthog (*Phacochoerus aethiopicus*) in the Horn of Africa. African Journal of Ecology 39:156–169.

d'Huart, J.-P. & P. Grubb. 2005. A photographic guide to the differences between the common warthog (*Phacochoerus africanus*) and the desert warthog (*Ph. aethiopicus*). Suiform Soundings 5, 2:4–8.

Hundertmark, K.J., G.F. Shields, R.T. Bowyer & C.C. Schwartz. 2002. Genetic relationships deduced from cytochrome *b* sequences among moose. Alces 38:113–122.

Hundertmark, K.J., G.F. Shields, I.G. Udina, R.T. Bowyer, A.A. Danilkin & C.C. Schwartz. 2002. Mitochondrial phylogeography of moose (*Alces alces*): Late Pleistocene divergence and population expansion. Molecular Phylogenetics and Evolution 22:375–387.

Hundertmark, K.J., R.T. Bowyer, G.F. Shields & C.C. Schwartz. 2003. Mitochondrial phylogeography of moose (*Alces alces*) in North America. Journal of Mammalogy 84:718–728.

Imaizumi, Y. 1970. Description of a new species of *Cervus* from the Tsushima Islands, Japan, with a revision of the subgenus *Sika* based on clinal analysis. Bulletin of the National Science Museum (Tokyo) 13:185–194.

Ingles, J.M. 1965. Zambian mammals collected for the British Museum (Natural History) in 1962. Puku 3:75–86.

International Commission on Zoological Nomenclature [ICZN], Opinion 460. 1957. Validation under the plenary powers of the generic name of "*Muntiacus*" Rafinesque, 1815, and designation for the genus so named of a type species in harmony with accustomed usage (Class Mammalia). Opinions and Declarations Rendered by the International Commission on Zoological Nomenclature 15:457–474.

ICZN, Opinion 1348. 1985. *Bos gaurus* H. Smith, 1827 (Mammalia, Artiodactyla): conserved. Bulletin of Zoological Nomenclature 42:279–280.

ICZN, Opinion 1349. 1985. *Antilope depressicornis* H. Smith, 1827 and *Anoa quarlesi* Ouwens, 1910 (Mammalia, Artiodactyla): conserved. Bulletin of Zoological Nomenclature 42:281–282.

ICZN, Opinion 1894. 1998. Regnum Animale . . . , ed. 2 (M.-J. Brisson, 1762): rejected for nomenclatural purposes, with the conservation of the mammalian generic names *Philander* (Marsupialia), *Pteropus* (Chiroptera), *Glis*, *Cuniculus*, and *Hydrochoerus* (Rodentia), *Meles*, *Lutra*, and *Hyaena* (Carnivora), *Tapirus* (Perissodactyla), *Tragulus* and *Giraffa* (Artiodactyla). Bulletin of Zoological Nomenclature 55:64–71.

ICZN, Opinion 2027. 2003. Case 3010: usage of 17 specific name was based on wild species which are pre-dated by or contemporary with those based on domestic animals (Lepidoptera, Osteichthyes, Mammalia); conserved. Bulletin of Zoological Nomenclature 60:81–84.

Irwin, A. J. 1914. Notes on the races of serow, or goat-antelope, found in Siam. Journal of the Natural History Society of Siam 1, 1:19–26.

Iyengar, A., F.M. Diniz, T. Gilbert, D. Woodfine, J. Knowles & N. Maclean. 2006. Structure and evolution of the mitochondrial control region in oryx. Molecular Phylogenetics and Evolution 40:305–314.

Jacobi, A. 1923. Zoologische Ergebnisse der Walter Stötznerschen Expeditionen nach Szetschwan, Osttibet, und Tschili auf Grund der Sammlungen und Beobachtungen Dr. Hugo Weigolds: Mammalia. Abhandlungen und Berichtung des Museen für Tierkunde und Völkerkunde Dresden 16:1–22.

Janeck, L.L., R.L. Honeycutt, R.M. Adkins & S.K. Davis. 1996. Mitochondrial gene sequences and the molecular systematics of the artiodactyl subfamily Bovinae. Molecular Phylogenetics and Evolution 6:107–119.

Janis, C.M. & K.M. Scott. 1987. The interrelationships of higher ruminant families with special emphasis on the members of the Cervoidea. American Museum Novitates 2893:1–85.

Jansen van Vuuren, B. & T.J. Robinson. 2001. Retrieval of four adaptive lineages in duiker antelope: evidence from mitochondrial DNA sequences and fluorescence *in situ* hybridisation. Molecular Phylogenetics and Evolution 20:409–425.

Jarman, P.J. 1972. The development of a dermal shield in impala. Journal of Zoology, London 166:349–356.

Jarman, P.J. 1974. The social organisation of antelope in relation to their ecology. Behaviour 48:215–266.

Jarman, P.J. 2000. Dimorphism in social Artiodactyla: selection upon females. In E.S. Vrba & G.B. Schaller (eds.), Antelopes, Deer, and Relatives New Haven: Yale University Press, 171–179.

Jass, C.M. & J.I. Mead. 2004. *Capricornis crispus*. Mammalian Species 750:1–10.

Jiang Z-g., Z-j. Feng, Z-w. Wang, L-w. Chen, P. Cai & Y-b. Li. 1995. Historical and current distributions of Przewalski's gazelles. Acta Theriologica Sinica 15:241–245.

Jiménez, J.E. 2010. Southern pudu *Pudu puda* (Molina 1782). In J.M.B. Duarte & S. González (eds.), Neotropical Cervidology: Biology and Medicine of Latin American Deer. Jaboticabal, Brazil and Gland, Switzerland: Funep, in collaboration with IUCN, 140–150.

Jorge, W. & K. Benirschke. 1976. Banding and meiotic chromosome studies in a male eland. Genen en Phaenen 19:7–10.

Jorge, W., S. Butler & K. Benirschke. 1976. Studies on a male eland × kudu hybrid. Journal of Reproduction and Fertility 46:13–16.

Karami, M. & C.P. Groves. 1993. A mammal species new for Iran: *Gazella gazella* Pallas, 1766 (Artiodactyla: Bovidae). Journal of Sciences of the Islamic Republic of Iran 4:81–89.

Karami, M., M.R. Hemami & C.P. Groves. 2002. Taxonomic, distributional, and ecological data on gazelles in Iran. Zoology in the Middle East 26:29–36.

Kavar, T. & P. Dovč. 2008. Domestication of the horse: genetic relationships between domestic and wild horses. Livestock Science 116:1–14.

Khaute, L.M. 2010. The Sangai: The Pride of Manipur. Delhi: Kalpaz.

Kikkawa, Y., H. Yonekawa, H. Suzuki & T. Amano. 1997. Analysis of genetic diversity of domestic water buffaloes and anoas based on variations in the mitochondrial gene for cytochrome *b*. Animal Genetics 28:195–201.

Kim, K.S., K. Tanaka, D.B. Ismail, S. Maruyama, H. Matsubayashi, H. Endo, K. Fukuta & J. Kimura. 2004. Cytogenetic comparison of the lesser mouse deer (*Tragulus javanicus*) and the greater mouse deer (*T. napu*). Caryologia 57:229–243.

King, D.G. 1975. The Afro-Alpine grey duiker of Kilimanjaro. Journal of the East Africa Natural History Society and National Museum, 152:1–9.

Kingdon, J. 1979. Large Mammals. Vol. 3B of East African Mammals: An Atlas of Evolution in Africa. London: Academic Press.

Kingdon, J. 1982a. Bovids. Vol. 3C of East African Mammals: An Atlas of Evolution in Africa. London: Academic Press.

Kingdon, J. 1982b. Bovids. Vol. 3D of East African Mammals: An Atlas of Evolution in Africa. London: Academic Press.

Kingdon, J. 1997. The Kingdon Field Guide to African Mammals. San Diego: Academic Press.

Kingswood, S.C. & D.A. Blank. 1996. *Gazella subgutturosa*. Mammalian Species 518:1–10.

Kingswood, S.C. & A.T. Kumamoto. 1988. Research and management of Arabian sand gazelle in the USA. In A. Dixon & D. Jones (eds.), Conservation and Biology of Desert Antelopes. London: Christopher Helm, 212–226.

Kingswood, S.C., A.G. Kumamoto, S.J. Charter, R.A. Aman & O.A. Ryder. 1998. Centric fusion polymorphisms in waterbuck (*Kobus ellipsiprymnus*). Journal of Heredity 89:96–100.

Kingswood, S.C., A.T. Kumamoto, S.J. Charter & M.L. Jones. 1998. Cryptic chromosomal variation in suni *Neotragus moschatus* (Artiodactyla, Bovidae). Animal Conservation 1:95–100.

Klein, D.R., M. Meldgaard & S.G. Fancy. 1987. Factors determining leg length in *Rangifer tarandus*. Journal of Mammalogy 68:642–655.

Kloss, C.B. 1919. On mammals collected in Siam. Journal of the Natural History Society of Siam 3:333–407.

Kloss, C.B. 1923. The goral in Siam. Journal of the Natural History Society of Siam 6, 1:135–137.

Kock, D. 2000. The fallow deer *Dama schaeferi* Hilzheimer, 1926 (Mammalia: Cervidae), enigmatic and forgotten. Zoology in the Middle East 21:9–11.

Koh, H.S. & E. Randi. 2001. Genetic distinction of roedeer (*Capreolus pygargus* Pallas) sampled in Korea. Mammalian Biology 66:371–375.

Korobitsyna, K.V., C.F. Nadler, N.N. Vorontsov & R.S. Hoffmann. 1974. Chromosomes of the Siberian snow sheep, *Ovis nivicola*, and implications concerning the origin of Amphiberingian wild sheep (subgenus *Pachyceros*). Quaternary Research 4:235–245.

Kraus, F. & M.M. Miyamoto. 1991. Rapid cladogenesis among the pecoran ruminants: evidence from mitochondrial DNA sequences. Systematic Zoology 40:117–130.

Kretzoi, M. 1946. On *Bison bonasus hungarorum* n. ssp. Annales Historico-Naturales Musei Nationalis Hungarici 39:105–191.

Krüger, K., C. Gaillard, G. Stranzinger & S. Rieder. 2005. Phylogenetic analysis and species allocation of individual equids using microsatellite data. Journal of Animal Breeding and Genetics 122, Suppl. 1:78–86.

Krumbiegel, I. 1944. Die neuweltlichen Tylopoden. Zoologischer Anzeiger 145:45–70.

Krumbiegel, I. 1952. Lamas. Die Neue Brehm-Bücherei No. 54. Wittenberg-Lutherstadt, Germany: A. Ziemsen.

Krumbiegel, I. 1980. Die unterartliche Trennung des Bisons, *Bison bison* (Linné, 1788), und seine Rückzüchtung. Säugetierkundliche Mitteilungen 28:148–160.

Kruska, D.C.T. 2005. On the evolutionary significance of encephalization in some eutherian mammals: effects of adaptive radiation, domestication, and feralization. Brain Behavior and Evolution 65:73–108.

Kruska, D.C.T. 2007. The effects of domestication on brain size in the evolution of nervous systems. In J.H. Kaas (ed.), Mammals. Vol. 3 of Evolution of Nervous Systems. London: Elsevier, 143–153.

Kuhn, H.-J. 1968. A provisional checklist of the mammals of Liberia. Senckenbergiana Biologica 46:321–340.

Kumamoto, A.T. & M.H. Bogart. 1984. The chromosomes of Cuvier's gazelle. In O.A. Ryder & M.L. Byrd (eds.), One Medicine: Essays in Honour of Kurt Benirschke. Berlin: Springer-Verlag, 101–108.

Kumamoto, A.T., S.C. Kingswood & W. Hugo. 1994. Chromosomal divergence in allopatric populations of Kirk's dik-dik, *Madoqua kirkii* (Artiodactyla, Bovidae). Journal of Mammalogy 75:357–364.

Kumamoto, A.T., S.C. Kingswood, W.E.R. Rebholz & M.L. Houck. 1995. The chromosomes of *Gazella bennetti* and *Gazella saudiya*. Zeitschrift für Säugetierkunde 60:159–169.

Kumamoto, A.T., S.J. Charter, M.L. Houck & M. Frahm. 1996. Chromosomes of *Damaliscus* (Artiodactyla: Bovidae): simple and complex centric fusion rearrangements. Chromosome Research 4:614–621.

Kumamoto, A.T., S.J. Charter, S.C. Kingswood, O.J. Ryder & D.S. Gallagher Jr. 1999. Centric fusion differences among *Oryx dammah*, *O. gazella*, and *O. leucoryx* (Artiodactyla, Bovidae). Cytogenetics and Cell Genetics 86:74–80.

Kumar, S., M. Nagarajan, J.S. Sandhu, N. Kumar, V. Behl & G. Nishanth. 2007. Mitochondrial DNA analyses of Indian water buffalo support a distinct genetic origin of river and swamp buffalo. Animal Genetics 38:227–232.

Kumerloeve, H. 1969. Bemerkungen zum Gazellen-Vorkommen im südöstlichen Kleinasien. Zeitschrift für Säugetierkunde 34:113–120.

Künzel, T. & S. Künzel. 1998. An overlooked population of the beira antelope *Dorcatragus megalotis* in Djibouti. Oryx 32:75–80.

Kuznetsova, M.V. & M.V. Kholodova. 2003. Revision of phylogenetic relationships in the Antilopinae subfamily on the basis of the mitochondrial rRNA and b-spectrin nuclear gene sequences. Doklady Biological Sciences 391:333–336.

Kuznetsova, M.V., M.V. Kholodova & A.A. Luschekina. 2002. Phylogenetic analysis of sequences of the 12S and 16S rRNA mitochondrial genes in the family Bovidae: new evidence. Genetika 38:942–950.

Lalueza-Fox, C., B. Shapiro, P. Bover, J.A. Alcover & J. Bertranpetit. 2002. Molecular phylogeny and evolution of the extinct bovid *Myotragus balearicus*. Molecular Phylogenetics and Evolution 25:501–510.

Lalueza-Fox, C., J. Castresana, L. Sambietro, T. Marquès-Bonet, J.A. Alcover & J. Bertranpetit. 2005. Molecular dating of caprines using ancient DNA sequences of *Myotragus balearicus*, an extinct endemic Balearic mammal. BMC Evolutionary Biology 2005, 5:70. doi: 10.1186/1471-2148-5-70.

Lan H., W. Wang & L-m. Shi. 1995. Phylogeny of *Muntiacus* (based on mitochondrial DNA restriction maps). Biochemical Genetics 33:377–388.

Lange, J. 1970. Ein Beitrag zur phylogenetischen Stellung de Springbockes (*Antidorcas marsupialis* Sundevall, 1847). Zeitschrift für Säugetierkunde 35:65–75.

Lange, J. 1971. Ein Beitrag zur systematischen Stellung der Spiegelgazellen (genus *Gazella* Blainville, 1816 subgenus *Nanger* Lataste, 1885). Zeitschrift für Säugetierkunde 36:1–18.

Lange, J. 1972. Studien an Gazellenschädeln: ein Beitrag zur Systematik der kleineren Gazellen, *Gazella* (de Blainville, 1816). Säugetierkundliche Mitteilungen 20:193–249.

Larson, G., K. Dobney, U. Albarella, M. Fang, E. Matisoo-Smith, J. Robins, S. Lowden, H. Finlayson, T. Brand, E. Willerslev, P. Rowley-Conwy, L. Andersson & A. Cooper. 2007. Worldwide phylogeography of wild boar reveals multiple centers of pig domestication. Science 307:1618–1621.

Larson, G., T. Cucchi, M. Fujita, E. Matisoo-Smith, J. Robins, A. Anderson, E. Rolett, M. Spriggs, G. Dolman, T. Djubiantono, P. Griffin, M. Intoh, E. Keane, P. Kirch, K-t. Li, M. Morwood, L.M. Pedriña, P.J. Piper, R.J. Rabett, P. Shooter, G. Van den Bergh, E. West, S. Wickler, J. Yuan, A. Cooper & K. Dobney. 2007. Phylogeny and ancient DNA of *Sus* provides insights into Neolithic expansion in Island Southeast Asia and Oceania. PNAS: Proceedings of the National Academy of Science USA 104:4834–4839.

Latch, E.K., J.R. Heffelfinger, J.A. Fike & O.E. Rhodes Jr. 2009. Species-wide phylogeography of North American mule deer (*Odocoileus hemionus*). Molecular Ecology 18:1730–1745.

von Lehmann, E. & H. Sägesser. 1981. *Capreolus capreolus* Linnaeus, 1758—Reh. In J. Niethammer & F. Krapp (eds.), Paarhufer, Artiodactyla: Suidae, Cervidae, Bovidae. Vol. 2, pt. 2 of Handbuch der Säugetiere Europas. Weisbaden, Germany: Aula, 233–266.

Lei R-h., Z-g. Jiang, Z. Hu & W-l. Yang. 2003. Phylogenetic relationships of Chinese antelopes (subfamily Antilopinae) based on mitochondrial ribosomal RNA gene sequences. Journal of Zoology, London 261:227–237.

Lei R-h., Z. Hu, Z-g. Jiang & W-l. Yang. 2003. Phylogeography and genetic diversity of the critically endangered Przewalski's gazelle. Animal Conservation 6:361–367.

Leinders, J.J.M. 1979. On the osteology and function of the digits of some ruminants and their bearing on taxonomy. Zeitschrift für Säugetierkunde 44:305–318.

Leinders, J.J.M. & E. Heintz. 1980. The configuration of the lacrimal orifices in pecorans and tragulids (Artiodactyla, Mammalia) and its significance for the distinction between Bovidae and Cervidae. Beaufortia 30:155–162.

Leonard, J.A., N. Rohland, S. Glaberman, R.C. Fleischer, A. Caccone & M. Hofreiter. 2005. A rapid loss of stripes: the evolutionary history of the extinct quagga. Biology Letters 1:291–295.

Leslie, D.M. 2009. *Przewalskium albirostre*. Mammalian Species 849:7–18.

Leslie, D.M. & G.B. Schaller. 2008. *Pantholops hodgsonii*. Mammalian Species 817:1–13.

Leslie, D.M. & K. Sharma. 2009. *Tetracerus quadricornis* (Artiodactyla: Bovidae). Mammalian Species 843:1–11.

Leus, K., G.P. Goodall & A.A. Macdonald. 1999. Anatomy and histology of the babirusa (*Babyrousa babyrussa*) stomach. Comptes Rendus de l'Académie des Sciences (Paris), Sér. 3, Sciences de la Vie 322:1081–1092.

Leuthold, W. 1981. Contact between formerly allopatric subspecies of Grant's gazelle (*Gazella granti* Brooke 1872) owing to vegetation changes in Tsavo National Park, Kenya. Zeitschrift für Säugetierkunde 46:48–55.

Li J-x. & L-h. Xu. 1996. A new subspecies of the Indian muntjac (*Muntiacus muntjak*) in Guangdong, China. Acta Theriologica Sinica 16:25–29.

Li M. & H-l. Sheng. 1998. MtDNA deference and molecular phylogeny among musk deer, Chinese water deer, muntjak, and deer. Acta Theriologica Sinica 18:184–191.

Li M., X-m. Wang, H-l. Sheng, H. Tamate, R. Masuda, J. Nagata & N. Ohtaishi. 1998. Origin and genetic diversity of four subspecies of red deer (*Cervus elaphus*). Zoological Research 19:177–183.

Li M.,Y-g. Li, H-l. Sheng, H. Tamate, R. Masuda, J. Nagata & N. Ohtaishi. 1999. The taxonomic status of *Moschus moschiferus anhuiensis*. Chinese Science Bulletin 44:719–722.

Li M., F-w. Wei, P. Groves, Z-j. Feng & J-c. Hu. 2003. Genetic structure and phylogeography of the takin (*Budorcas taxicolor*) as inferred from mitochondrial DNA sequences. Canadian Journal of Zoology 81:462–468.

Li Z-x. 1981. On a new species of musk deer from China. Zoological Research 2:157–161.

Lichtenstein, H. & W. Peters. 1855. Über *Antilope leucotis* Licht. Pet. In Über Neue Merkwürdige Säugethiere des Königlichen Zoologischen Museums. Berlin: Akademie der Wissenschaften, 14–20.

Lindsey, S.L., M.N. Green & C.L. Bennett. 1999. The Okapi: Mysterious Animal of Congo-Zaire. Austin: University of Texas Press.

Linnaeus (von Linné), C. 1758. Systema Naturae per Regna Tria Naturae, secundum Classes, Ordines, Genera, Species, cum Characteribus, Differentiis, Synonymis, Locis. 10th ed., rev., vol. 1. Stockholm: Laurent Salvius.

Linnaeus (von Linné), C. 1766. Systema Naturae per Regna Tria Naturae, secundum Classes, Ordines, Genera, Species, cum Characteribus, Differentiis, Synonymis, Locis. 12th ed., rev., vol. 1. Stockholm: Laurent Salvius.

Lister, A.M, P. Grubb & S.R.M. Sumner. 1998. Taxonomy, morphology, and evolution of European roe deer. In R. Andersen, P. Duncan & J.D.C. Linnell (eds.), The European Roe Deer: The Biology of Success. Oslo: Scandinavian University Press, 23–46.

Liu X-h., Y-q. Wang, Z-q. Liu & K-y. Zhou. 2003. Phylogenetic relationships of Cervinae based on sequence of mitochondrial cytochrome *b* gene. Zoological Research 24:27–33.

Lizcano, D.J., S.J. Álvarez & C.A. Delgado-V. 2010. Dwarf red brocket deer *Mazama rufina* (Pucheran 1951). In J.M.B. Duarte & S. González (eds.), Neotropical Cervidology: Biology and Medicine of Latin American Deer. Jaboticabal, Brazil and Gland, Switzerland: Funep, in collaboration with IUCN, 177–180.

Lizcano, D.J., E. Yerena, S.J. Álvarez & J.R. Dietrich. 2010. Mérida brocket deer *Mazama bricenii* (Thomas 1908). In J.M.B. Duarte & S. González (eds.), Neotropical Cervidology: Biology and Medicine of Latin American Deer. Jaboticabal, Brazil and Gland, Switzerland: Funep, in collaboration with IUCN, 181–184.

Loehr, J., K. Worley, A. Grapputo, J. Carey, A. Veitch & D.W. Coltman. 2006. Evidence for cryptic glacial refugia from North American mountain sheep mitochondrial DNA. Journal of Evolutionary Biology 19:419–430.

Loftus, R.T., D.E. MacHugh, D.G. Bradley, P.M. Sharp & P. Cunningham. 1994. Evidence for two independent domestications of cattle. PNAS: Proceedings of the National Academy of Science USA 91:2757–2761.

Lönnberg, E. 1917. Mammals Collected in Central Africa by Captain E. Arrhenius. Kungliga Svenska Vetenskapsakademien Handlingar No. 58. Stockholm: Almqvist & Wiksell.

Lorenzen, E.D. & H.R. Siegismund. 2004. No suggestion of hybridisation between the vulnerable black-faced impala (*Aepyceros melampus petersi*) and the common impala (*A. m. melampus*) in Etosha National Park, Namibia. Molecular Ecology 13:3007–3019.

Lorenzen, E.D., B.T. Simonsen, P.W. Kat, P. Arctander & H.R. Siegismund. 2006. Hybridisation between subspecies of waterbuck (*Kobus ellipsiprymnus*) in zones of overlap with limited introgression. Molecular Ecology 15:3787–3799.

Lorenzen, E.D., P. Arctander & H.R. Siegismund. 2006. Regional genetic structuring and evolutionary history of the impala *Aepyceros melampus*. Journal of Heredity 97:119–132.

Lorenzen, E.D., R. de Neergaard, P. Arctander & H.R. Siegismund. 2007. Phylogeography, hybridisation, and

Pleistocene refugia of the kob antelope (*Kobus kob*). Molecular Ecology 16:3241–3252.

Lorenzen, E.D., P. Arctander & H.R. Siegismund. 2008a. Three reciprocally monophyletic mtDNA lineages elucidate the taxonomic status of Grant's gazelles. Conservation Genetics 9:593–601.

Lorenzen, E.D., P. Arctander & H.R. Siegismund. 2008b. High variation and very low differentiation in wide-ranging plains zebra (*Equus quagga*): insights from mtDNA and microsatellites. Molecular Ecology 17:2812–2824.

Loubser, J., J. Brink & G. Laurens. 1990. Paintings of the extinct blue antelope, *Hippotragus leucophaeus*, in the eastern Orange Free State. South African Archaeological Bulletin 45:106–111.

Lovari, S. 1977. The Abruzzo chamois. Oryx 14:47–50.

Lovari, S. 1987. Evolutionary aspects of the biology of chamois, *Rupicapra* spp. (Bovidae, Caprinae). In H. Soma (ed.), The Biology and Management of *Capricornis* and Related Mountain Antelopes. London: Croom Helm, 51–61.

Lovari, S. & C. Scala. 1980. Revision of *Rupicapra* genus, 1: a statistical re-evaluation of Couturier's data on the morphometry of six chamois subspecies. Bolletino Zoologico 47:113–124.

Lovari, S. & C. Scala. 1984. Revision of *Rupicapra* genus, 4: horn biometrics of *Rupicapra rupicapra asiatica* and its relevance to the taxonomic position of *Rupicapra rupicapra caucasica*. Zeitschrift für Säugetierkunde 49:246–253.

Lowe, V.P.W. & A.S. Gardiner. 1975. Hybridisation between red deer (*Cervus elaphus*) and sika deer (*Cervus nippon*) with particular reference to stocks in N.W. England. Journal of Zoology, London 177:553–566.

Ludt, C.J., W. Schroeder, O. Rottmann & R. Kuehn. 2004. Mitochondrial DNA phylogeography of red deer (*Cervus elaphus*). Molecular Phylogenetics and Evolution 31:1064–1083.

Lundrigan, B. 1996. Morphology of horns and fighting behaviour in the family Bovidae. Journal of Mammalogy 77:462–475.

Lyapunova, E.A., T.D. Bunch, N.N. Vorontsov & R.S. Hoffmann. 1997. Chromosome sets and the taxonomy of Severtsov wild sheep (*Ovis ammon severtzovi*). Russian Journal of Zoology 1:387–396.

Lydekker, R. 1900. Great and Small Game of India. London: Rowland Ward.

Lydekker, R. 1905. The gorals of India and Burma. Zoologist 9:81–84.

Lydekker, R. 1906. The white-maned serow. Proceedings of the Zoological Society of London 1905, 2:329–331.

Lydekker, R. 1913. Artiodactyla, Family Bovidae, Subfamilies Bovinae to Ovibovinae (Cattle, Sheep, Goats, Chamois, Serows, Takin, Musk-Oxen, etc.). Vol. 1 of Catalogue of the Ungulate Mammals in the British Museum. London: British Museum (Natural History).

Lydekker, R. 1915a. Artiodactyla, Families Cervidae (Deer), Tragulidae (Chevrotains), Camelidae (Camels and Llamas), Suidae (Pigs and Peccaries), and Hippopotamidae (Hippopotamuses). Vol. 4 of Catalogue of the Ungulate Mammals in the British Museum. London: British Museum (Natural History).

Lydekker, R. 1915b. Perissodactyla (Horses, Tapirs, Rhinoceroses), Hyracoidea (Hyraxes), Proboscidea (Elephants): With Addenda to the Earlier Volumes. Vol. 5 of Catalogue of the Ungulate Mammals in the British Museum. London: British Museum (Natural History).

Lydekker, R. & G. Blaine. 1914a. Artiodactyla, Family Bovidae, Subfamilies Bubalinae to Reduncinae (Hartebeests, Gnus, Duikers, Dik-Diks, Klipspringers, Reedbucks, Waterbucks, etc). Vol. 2 of Catalogue of the Ungulate Mammals in the British Museum. London: British Museum (Natural History).

Lydekker, R. & G. Blaine. 1914b. Artiodactyla, Family Bovidae, Subfamilies Aepycerotinae to Tragelaphinae (Pala, Saiga, Gazelles, Oryx Group, Bushbucks, Kudus, Elands, etc.), Antilocapridae (Prongbuck), and Giraffidae (Giraffes and Okapi). Vol. 3 of Catalogue of the Ungulate Mammals in the British Museum. London: British Museum (Natural History).

Ma S-l., Y-x. Wang & L-m. Shi. 1990. A new species of the genus *Muntiacus* from Yunnan, China. Zoological Research 11:48–53.

Machado A. de Barros. 1969. Mamíferos de Angola Ainda Não Citados ou Pouco Conhecidos. Publicações Culturais da Companhia de Diamantes de Angola, Serviços Culturais, Dundo-Lunda-Angola No. 46. Lisbon: Companhia de Diamantes de Angola, 93–232.

Malbrant, R. & A. Maclatchy. 1949. Mammifères. Vol. 2 of Faune de 1'Équateur Africain Français. Paris: Lechevalier.

Manceau, V., L. Després, J. Bouvet & P. Taberlet. 1999. Systematics of the genus *Capra* inferred from mitochondrial DNA sequence data. Molecular Phylogenetics and Evolution 13:504–510.

Manceau, V., J.-P. Crampe, P. Boursot & P. Taberlet. 1999. Identification of evolutionary significant units in the Spanish wild goat, *Capra pyrenaica* (Mammalia, Artiodactyla). Animal Conservation 2:33–39.

Marcot, J.D. 2007. Molecular phylogeny of terrestrial artiodactyls: conflicts and resolution. In D.R. Prothero & S.E. Foss (eds.), The Evolution of Artiodactyls. Baltimore: Johns Hopkins University Press, 4–18.

Marín, J.C., C.S. Casey, M. Kadwell, K. Yaya, D. Hoces, J. Olazabal, R. Rosadio, J. Rodriguez, A. Spotorno, M.W. Bruford & J.C. Wheeler. 2007. Mitochondrial phylogeography and demographic history of the vicuña: implications for conservation. Heredity 99:70–80.

Marín, J.C., A.E. Spotorno, B.A. González, C. Bonacic, J.C. Wheeler, C.S. Casey, M.W. Bruford, R.E. Palma & E. Poulin. 2008. Mitochondrial DNA variation and systematics of the guanaco (*Lama guanicoe*, Artiodactyla: Camelidae). Journal of Mammalogy 80:269–281.

Marino, M.L. & R.V. Rossi. 2010. Origin, systematics, and morphological radiation. In J.M.B. Duarte & S. González (eds.), Neotropical Cervidology: Biology and Medicine of Latin American Deer. Jaboticabal, Brazil and Gland, Switzerland: Funep, in collaboration with IUCN, 2–11.

Markov, G.G., A.A. Danilkin, C. Gerasimov & K. Nikolov. 1985. A comparative craniometric analysis of *Capreolus capreolus*. Doklady Akademii Nauk SSSR 282:489–493.

Márquez, A., J.E. Maldonado, S. González, M.B. Beccaceci, J.E. Garcia & J.M.B. Duarte. 2006. Phylogeography and Pleistocene demographic history of the endangered marsh deer (*Blastocerus dichotomus*) from the Río de la Plata Basin. Conservation Genetics 7:563–575.

Masembe, C., V.B. Muwanika, S. Nyakaana, P. Arctander & H.R. Siegismund. 2006. Three genetically divergent lineages of the oryx in eastern Africa: evidence for an ancient introgressive hybridisation. Conservation Genetics 7:551–562.

Masini, F. & S. Lovari. 1988. Systematics, phylogenetic relationships, and dispersal of the chamois (*Rupicapra* spp.). Quaternary Research 30:339–349.

Masseti, M., E. Pecchioli & C. Vernesi. 2008. Phylogeography of the last surviving populations of Rhodian and Anatolian fallow deer (*Dama dama dama* L., 1758). Biological Journal of the Linnaean Society 93:835–844.

Matschie, P. 1892. Einige Säugethiere von Deutsch-Ost-Afrika. Sitzungsberichte der Gesellschaft Naturforschender Freunde zu Berlin 1892:130–140.

Matschie, P. 1893. Über die Verbreitung der zur Gattung *Oryx*: Blainv. gehörigen Antilopen. Sitzungsberichte der Gesellschaft Naturforschender Freunde zu Berlin 1893:101–105.

Matschie, P. 1895. Die Säugethiere Deutsch-Ost-Afrikas. Berlin: Reiner.

Matschie, P. 1898a. Einige anscheinend noch nicht beschrieben Säugethiere aus Afrika. Sitzungsberichte der Gesellschaft Naturforschender Freunde zu Berlin 1898:75–81.

Matschie, P. 1898b. Die geographische Verbreitung der Tigerpferde und das Zebra des Kaoko-Feldes in Deutsch-Sudwest Afrika. Sitzungsberichte der Gesellschaft Naturforschender Freunde zu Berlin 1898:169–181.

Matschie, P. 1898c. Eine neue Abart von *Hippotragus bakeri* Heugl. Sitzungsberichte der Gesellschaft Naturforschender Freunde zu Berlin 1898:181–182.

Matschie, P. 1899. Eine anscheinend neue *Adenota* vom Weissen Nil. Sitzungsberichte der Gesellschaft Naturforschender Freunde zu Berlin 1899:15.

Matschie, P. 1900. Ueber *Equus penricei*, Thos. Sitzungsberichte der Gesellschaft Naturforschender Freunde zu Berlin. 1900:231.

Matschie, P. 1901a. Über Kaukasische Steinbocke. Sitzungsberichte der Gesellschaft Naturforschender Freunde zu Berlin 1901:27–33.

Matschie, P. 1901b. Wilde Pferde im Park des Hrn. Falz-Fein in Askania-Nova (Sudrussland). Illustrirte Zeitung No. 3010 (7 Marz 1901):366–367.

Matschie, P. 1903. Giebt es in Mittelasien mehrere Arten von echten Wildpferden? Naturwissenschaftliche Wochenschrift, n.f. 2, 49:581–583.

Matschie, P. 1906. Einige noch nich beschriebene Arten des afrikanischen Büffels. Sitzungsberichte der Gesellschaft Naturforschender Freunde zu Berlin 1906:161–179.

Matschie, P. 1907. Übersicht über die vom verfasser aus seiner Reise gesammelten Tiere. In P. Niedieck (ed.), Kreuzfahrten im Beringmeer. Berlin: Paul Parey, 233–247.

Matschie, P. 1910a. Bemerkungen über die Verbreitung der Tiere in den Deutschen Schutzgebieten. Berlin: published privately.

Matschie, P. 1910b. Die von Herrn Major P. H. G. Powell-Cotton gesammelten Rassen des Wasserbockes (*Kobus*). Sitzungsberichten der Gesellschaft Naturforschender Freunde zu Berlin 1910:411–429.

Matschie, P. 1911a. Ueber einige von Herrn Dr. Holderer in der sudlichen Gobi und in Tibet gesammelte Säugetiere. In K. Futterer (ed.), Zoologie (Nachtrag). Vol. 3, pt. 5 of Durch Asien. Berlin: Reimer, 4–29.

Matschie, P. 1911b. Zoologische Ergebnisse der Expedition des Herrn Hauptmann a. D. Fromm 1908/09 nach Deutsch-Ostaafrika, 4: Mammalia (Gattung *Kobus*). Mitteilungen aus dem Zoologischen Museum in Berlin 5:554–575.

Matschie, P. 1912a. Beuteltücke aus fernen Ländern: die achtzehnte deutsche Gewehausftellung. Deutsche Jäger-Zeitung 59:64–83.

Matschie, P. 1912b. *Gazella (Nanger) soemmerringii sibyllae* subsp. nov. Sitzungsberichte der Gesellschaft Naturforschender Freunde zu Berlin 1912:260–270.

Matschie, P. 1912c. Die von Herrn Major P. H. G. Powell-Cotton gesammelten Rassen der Gattung *Tragelaphus*. Sitzungsberichte der Gesellschaft Naturforschender Freunde zu Berlin 1912:544–567.

Matschie, P. 1913. Eine neue Form der Elenantilope: *Oreas oryx niediecki* nov. subsp. Sitzungsberichte der Gesellschaft Naturforschender Freunde zu Berlin 1913:249–258.

Matschie, P. 1914. Eine neue Art der Kudu-Antilope. Sitzungsberichte der Gesellschaft Naturforschender Freunde zu Berlin 1914:383–393.

Matschie, P. 1916. Die richtige Benennung der Kuhantilope von Baunza. Sitzungsberichte der Gesellschaft Naturforschender Freunde zu Berlin 1916:295.

Matschie, P. 1918a. Die Büffel des Ituri-Urwaldes. Sitzungsberichte der Gesellschaft Naturforschender Freunde au Berlin 1918:1–5.

Matschie, P. 1918b. Ein Schwarzbüffel des Kafue-Gebietes. Sitzungsberichte der Gesellschaft Naturforschender Freunde zu Berlin 1918:133–140.

Matschie, P. 1920. Tierwelt der deutschen Schutzgebiete. In H. Schnee (ed.), Deutsches Kolonial-Lexikon, vol. 3. Leipzig: Quelle & Meyer, 483–494.

Matschie, P. 1922. Bemerkungen uber einige tibetanische Säugetiere. Sitzungsberichte der Gesellschaft Naturforschender Freunde au Berlin 1922:65–75.

Matschie, P. & L. Zukowsky. 1916. Die als *Sigmoceros* bezeichnete Gruppe der Kuhantilopen. Sitzungsberichte der Gesellschaft Naturforschender Freunde zu Berlin 1916:188–211.

Matschie, P. & L. Zukowsky. 1917. Die als *Sigmoceros* bezeichnete Gruppe der Kuhantilopen. Sitzungsberichte der Gesellschaft Naturforschender Freunde zu Berlin 1917:527–550.

Matschie, P. & L. Zukowsky. 1922. Die als *Sigmoceros* bezeichnete Gruppe der Kuhantilopen. Sitzungsberichte der Gesellschaft Naturforschender Freunde zu Berlin 1922:79–144.

Matthee, C.A. & S.K. Davis. 2001. Molecular insights into the evolution of the family Bovidae: a nuclear DNA perspective. Molecular Biology and Evolution 18:1220–1230.

Matthee, C.A. & T.J. Robinson. 1999a. Cytochrome *b* phylogeny of the family Bovidae: resolution within the Alcelaphini, Antilopini, Neotragini, and Tragelaphini. Molecular Phylogenetics and Evolution 12:31–46.

Matthee, C.A. & T.J. Robinson. 1999b. Mitochondrial DNA population structure of roan and sable antelope: implications for the translocation and conservation of the species. Molecular Ecology 8:227–238.

Mattioli, S., R. Fico, R. Lorenzin & G. Nobili. 2003. Messola red deer: physical characteristics, population dynamics, and conservation perspectives. Hystrix: Italian Journal of Mammalogy 14:87–94.

Mayr, E. 1963. Animal Species and Evolution. Harvard: Belknap Press.

Medellín, R.A., A.L. Gardner & J.M. Aranda. 1998. The taxonomic status of the Yucatán brown brocket, *Mazama pandora* (Mammalia: Cervidae). Proceedings of the Biological Society of Washington 111:1–14.

Meijaard, E. & C.[P.] Groves. 2002. Upgrading three subspecies of babirusa (*Babyrousa* sp.) to full species level. Asian Wild Pig News 2, 2:33–39.

Meijaard, E. & C.P. Groves. 2004a. A taxonomic revision of the *Tragulus* mouse-deer (Artiodactyla). Zoological Journal of the Linnaean Society 140:63–102.

Meijaard, E. & C.P. Groves. 2004b. Morphometric relationships between South-east Asian deer (Cervidae, tribe Cervini): evolutionary and biogeographic implications. Journal of Zoology, London 263:179–196.

Mendelssohn, H., Y. Yom-Tov & C.P. Groves. 1995. *Gazella gazella*. Mammalian Species 490:1–7.

Mendelssohn, H., C.P. Groves & B. Shalmon. 1997. A new subspecies of *Gazella gazella* from the southern Negev. Israel Journal of Zoology 43:209–215.

Mentis, M. 1974. Distribution of some wild animals in Natal. Lammergeyer 20:1–68.

Merino, M.L. & R.V. Rossi. 2010. Origin, systematics, and morphological radiation. In J.M.B. Duarte & S. González (eds.), Neotropical Cervidology: Biology and Medicine of Latin American Deer. Jaboticabal, Brazil and Gland, Switzerland: Funep, in collaboration with IUCN, 2–11.

Métais, G. & I. Vislobokova. 2007. Basal Ruminants. In D.R. Prothero & S.E. Foss (eds.), The Evolution of Artiodactyls. Baltimore: Johns Hopkins University Press, 189–212.

Milne-Edwards, A.E. 1867. Observations sur quelques mammifères du nord de la Chine. Annales des Sciences Naturelles, Zoologie (5) 7:375–377.

Milner, C.E. 1921. Distribution of serow in Burma. Journal of the Bombay Natural History Society 28:267–268.

Min, M-s., H. Okumura, D-j. Jo, J-h. An, K-s. Kim, C-b. Kim, N-s. Shin, M-h. Lee, C-h. Han, I.V. Voloshina & H. Lee. 2004. Molecular phylogenetic status of the Korean goral and Japanese serow based on partial sequences of the mitochondrial cytochrome *b* gene. Molecules and Cells 17:365–372.

Miyamoto, M.M., F. Kraus & O.A. Ryder. 1990. Phylogeny and evolution of antlered deer determined from mitochondrial DNA sequences. PNAS: Proceedings of the National Academy of Science USA 87:6127–6131.

Mohr, E. 1952. Der Wisent. Die Neue Brehm-Bücherei No. 74. Wittenberg Lutherstadt, Germany: A. Ziemsen.

Molina, M. & J. Molinari. 1999. Taxonomy of Venezuelan white-tailed deer (*Odocoileus*, Cervidae, Mammalia), based on cranial and mandibular traits. Canadian Journal of Zoology 77:632–645.

Mona, S., E. Randi & M. Tommaseo-Ponzetta. 2007. Evolutionary history of the genus *Sus* inferred from cytochrome *b* sequences. Molecular Phylogenetics and Evolution 45:757–762.

Monfort, A. & N. Monfort 1973. Quelques observations sur les grands mammifères du Parc National de Tai (Côte d'Ivoire). La Terre et La Vie 27:499–506.

Moodley, Y. & M.W. Bruford. 2007. Molecular biogeography: towards an integrated framework for conserving Pan-African biodiversity. PLoS ONE 2 (5):e 454. doi:10.1371/journal.pone.0000454.

Moodley, Y. & E.H. Harley. 2005. Population structuring in mountain zebras (*Equus zebra*): the molecular consequences of divergent demographic histories. Conservation Genetics 6:953–968.

Moodley, Y., M.W. Bruford, C. Bleidorn, T. Wronski, A. Apio & M. Plath. 2009. Analysis of mitochondrial DNA data reveals non-monophyly in the bushbuck (*Tragelaphus scriptus*) complex. Mammalian Biology 74:418–422.

Moscarella, R.A., M. Aguilera & A.A. Escalante. 2003. Phylogeography, population structure, and implications for conservation of white-tailed deer (*Odocoileus virginianus*) in Venezuelan. Journal of Mammalogy 84:1300–1315.

Nadler, C.F., K.V. Korobitsina, R.S. Hoffmann & N.N. Vorontsov. 1973. Cytogenetic differentiation, geographic distribution, and domestication in Palaearctic sheep (*Ovis*). Zeitschrift für Säugetierkunde 38:109–125.

Nagata, J., R. Masuda, H.B. Tamate, S-i. Hamasaki, K. Ochiai, M. Asada, S. Tatsuzawa, K. Suda, H. Tado & M.C. Yoshida. 1999. Two genetically distinct lineages of the sika deer, *Cervus nippon*, in Japanese islands: comparison of mitochondrial D-loop region sequences. Molecular Phylogenetics in Evolution 13:511–519.

Nascetti, G., S. Lovari, P. Lanfranchi, C. Berducou, S. Mattrucci, L. Rossi & L. Bullini. 1985. Revision of *Rupicapra* genus, 3: Electrophoretic studies demonstrating species distinction of chamois populations of the Alps from those of the Apennines and Pyrenees. In S. Lovari (ed.), The Biology and Management of Mountain Ungulates. London: Croom Helm, 56–62.

Nasonov, N. 1923. Geographiceskoe Rasprostranenie Dikih Baranov Starogo Sveta [Distribution Géographique des Mouton Sauvages du Monde Ancien]. St. Petersburg, Russia: Akademii Nauk SSR.

Nchanji, A.C. & F.O. Amubode. 2002. The physical and morphological characteristics of the red-fronted gazelle (*Gazella rufifrons kanuri* Gray 1846) in Waza National Park, Cameroon. Journal of Zoology, London 256:505–509.

Nersting, L.G. & P. Arctander. 2001. Phylogeography and conservation of impala and greater kudu. Molecular Ecology 10:711–719.

Neuhaus, P. & K.E. Ruckstuhl. 2002. The link between sexual dimorphism, activity budgets, and group cohesion: the case of the plains zebra (*Equus burchelli*). Canadian Journal of Zoology 80:1437–1441.

Neumann, O. 1896. Description of a new species of antelope from East Africa. Proceedings of the Zoological Society of London 1896:192–194.

Neumann, O. 1898. On a new antelope of the genus *Hippotragus*. Proceedings of the Zoological Society of London 1898:850–851.

Neumann, O. 1905. Über neue Antilopen-Arten. Sitzungsberichte der Gesellschaft Naturforschender Freunde zu Berlin 1905:88–97.

Neumann, O. 1906. Gazellan und Kuh-Antilopen. Sitzungsberichte der Gesellschaft Naturforschender Freunde zu Berlin 1906:236–248.

Neumann-Denzau, G. & H. Denzau. 1999. Wildesel. Stuttgart: Jan Thorbecke.

Neumann-Denzau, G. & H. Denzau. 2003. The southern kiang *Equus kiang polyodon*. Journal of the Bombay Natural History Society 100:322–340.

Nguyen An Quang Ha. 1997. Another new animal discovered in Vietnam. Vietnam Economic News 38:46–47.

Nijman, I.J., D.C.J. van Boxtel, L.M. van Cann, Y. Marnoch, E. Cuppen & J.A. Lenstra. 2008. Phylogeny of Y chromosomes from bovine species. Cladistics 24:723–726.

Nixon, K.C. & Q.D. Wheeler. 1990. An amplification of the phylogenetic species concept. Cladistics 6:211–223.

O'Gara, B.W. 1990. The pronghorn (*Antilocapra americana*). In G.A. Bubenik & A.B. Bubenik (eds.), Horns, Pronghorns, and Antlers. New York: Springer-Verlag, 231–264.

O'Gara, B.W. 2002. Taxonomy. In D.E. Toweill & J.W. Thomas (eds.), North American Elk: Ecology and Management. Smithsonian Institution Press: Washington, DC, 3–66.

Ohtaishi, N. & Y. Gao. 1990. A review of the distribution of all species of deer (Tragulidae, Moschidae, and Cervidae) in China. Mammal Review 20:125–144.

Okello, J.B.A., S. Nyakaana, C. Masembe, H.R. Siegismund & P. Arctander. 2005. Mitochondrial DNA variation of the common hippopotamus: evidence for any recent population expansion. Heredity 95:206–215.

Orr, R.T. 1938. Mammals from Sikang, China. Proceedings of the California Academy of Sciences 23:307–310.

Oswald, C. 2002. A commentary on the systematics of red deer *Cervus elaphus* L. Deer 12:84–89.

Pallas, P.S. 1766. Miscellanea zoologica, quibus novae imprimis atque obscurae animalum species describuntur et observationibus iconibusque illustrantur. The Hague: Kessinger.

Pallas, P.S. 1767–1777. Spicilegia zoologica, quibus novae imprimis et obscurae animalum species iconibus, descriptionibus atque commentariis illustrantur. Berlin: Gottl. August Lange.

Pallas, P.S. 1811. Zoographia Rosso-Asiatica, 3 vols. St. Petersburg: Academiae Scientarum.

Paterson, H.E.H. 1978. More evidence against speciation by reinforcement. South African Journal of Science 74:369–371.

Paterson, H.E.H. 1980. A comment on "mate recognition" systems. Evolution 34:330–331.

Peacock, E.H. 1933. A Game Book for Burma and Adjoining Territories. London: H.F. & G. Witherby.

Peden, D.G. & G.J. Kraay. 1979. Comparison of blood characteristics in plains bison, wood bison, and their hybrids. Canadian Journal of Zoology 57:1778–1784.

Pees, W. & H. Hemmer. 1980. Hirngrösse und Aktivität bei Wildschafen und Hausschafen (Gattung *Ovis*). Säugetierkundliche Mitteilungen 28:39–45.

Pemberton, J.M., P.W. King, S. Lovari & B. Bauchau. 1989. Genetic variation in the Alpine chamois, with special reference to the subspecies *Rupicapra rupicapra cartusiana*, Couturier, 1938. Zeitschrift für Säugetierkunde 54:243–250.

Perez, M.C. 1984. Revision der Systematik von Gazella (*Nanger*) dama. Zeitschrift des Kölner Zoo 27:103–107.

Pérez, T., J. Albornoz & A. Domínguez. 2002. Phylogeography of chamois (*Rupicapra* spp.) inferred from microsatellites. Molecular Phylogenetics and Evolution 25:524–534.

Perret, J.-L. & V. Aellen. 1956. Mammifères du Cameroun de la collection J.-L. Perret. Revue Suisse de Zoologie 63:395–450.

Peterson, R.L. 1950. A new subspecies of moose from North America. Occasional Papers of the Royal Ontario Museum of Zoology 9:1–7.

Pfeffer, P. 1962. Un cobe de montagne propre au Cameroun: *Redunca fulvorufula adamauae* subspecies nova. Mammalia 26:64–71.

Pfeffer, P. 1967. Le mouflon de Corse (*Ovis ammon musimon* Schreber, 1782): position systématique, écologie et éthologie comparées. Mammalia 31, Suppl.: 1–262.

Pidancier, N., S. Jordan, G. Luikart & P. Taberlet. 2006. Evolutionary history of the genus *Capra* (Mammalia, Artiodactyla): discordance between mitochondrial DNA and Y-chromosome phylogenies. Molecular Phylogenetics and Evolution 40:739–749.

Pienaar, U. de V. 1963. The large mammals of the Kruger National Park—their distribution and present-day status. Koedoe 6:1–37.

Piovezan, U., L.M. Tiepolo, W.M. Tomas, J.M.B. Duarte, D. Varela & J.S. Marinho Filho. 2010. Marsh deer *Blastocerus dichotomus* (Illiger 1815). In J.M.B. Duarte & S. González (eds.), Neotropical Cervidology: Biology and Medicine of Latin American Deer. Jaboticabal, Brazil and Gland, Switzerland: Funep, in collaboration with IUCN, 66–76.

Pitra, C., R. Fürbass & H.-M. Seyfert. 1997. Molecular phylogeny of the tribe Bovini (Mammalia: Artiodactyla): alternative placement of the anoa. Journal of Evolutionary Biology 10:589–600.

Pitra, C., A.J. Hansen, D. Lieckfeldt & P. Arctander. 2002. An exceptional case of historical outbreeding in African sable antelope populations. Molecular Ecology 11:1197–1208.

Pitra, C., J. Fickel, E. Meijaard & C.P. Groves. 2004. Evolution and phylogeny of Old World deer. Molecular Phylogenetics and Evolution 33:880–895.

Pitra, C., P. Vaz Pinto, B.W.J. O'Keefe, S. Williams-Munro, B. Jansen van Vuuren & T.J. Robinson. 2006. DNA-led rediscovery of the giant sable antelope in Angola. European Journal of Wildlife Research 52:145–152.

Pocock, R.I. 1908a. Notes on some species and geographical races of serows (*Capricornis*) and gorals (*Nemorhaedus*), based upon specimens exhibited in the Society's Gardens. Proceedings of the Zoological Society of London 1908:173–202.

Pocock, R.I. 1908b. On the generic names of the rupicaprine ruminants known as serows and gorals. Annals and Magazine of Natural History (8) 1:183–188.

Pocock, R.I. 1910a. The serows, gorals, and takins of British India and the Straits Settlements. Journal of the Bombay Natural History Society 19:807–821.

Pocock, R.I. 1910b. On the specialized cutaneous glands of ruminants. Proceedings of the Zoological Society of London 1910:840–986.

Pocock, R.I. 1913. The serows, gorals, and takins of British India and the Straits Settlements, part 2. Journal of the Bombay Natural History Society 22:296–319.

Pocock, R.I. 1914. Description of a new species of goral (*Nemorhaedus*) shot by Captain F. M. Bailey. Journal of the Bombay Natural History Society 23:32–33.

Pocock, R. I. 1918. On some external characters of ruminant Artiodactyla, part 2: the Antilopinae, Rupicaprinae, and Caprinae, with a note on the penis of the Cephalophinae and Neotraginae. Annals and Magazine of Natural History (9) 2: 125–144.

Pocock, R.I. 1919. On some external characters of ruminant Artiodactyla, part 2. Annals and Magazine of Natural History (9) 11:125–144.

Pocock, R.I. 1945. Some cranial and dental characters of the existing species of Asiatic rhinoceroses. Proceedings of the Zoological Society of London 114:437–450.

Powell, C.P. & P. Grubb. 2002. Range extension of black-fronted duiker (*Cephalophus nigrifrons* Gray 1871, Artiodactyla, Bovidae): first records from Nigeria. Tropical Zoology 15:89–95.

Presidente, P.J.A. & M. Draisma. 1980. Hog deer on Sunday Island, part 1: analyses of body measurements taken in 1978–1979. Australian Deer 5, 3:11–21.

Prothero, D.R. 2007. Family Moschidae. In D.R. Prothero & S.E. Foss (eds.), The Evolution of Artiodactyls. Baltimore: Johns Hopkins University Press, 221–226.

Rabinowitz, A., G. Amato & U Saw Tun Khaing. 1998. Discovery of the black muntjac, *Muntiacus crinifrons* (Artiodactyla, Cervidae), in north Myanmar. Mammalia 62:105–108.

Rabinowitz, A., Than Myint, Saw Tun Khaing & S. Rabinowitz. 1999. Description of the leaf deer (*Muntiacus putaoensis*), a new species of muntjac from northern Myanmar. Journal of Zoology, London 249:427–435.

Radde, G. 1862. Säugetierfauna. Vol. 1 of Reisen in den Süden von Ost-Siberien: im auftrage der Kaiserlichen Russischen Gesellschaft ausgeführt in den Jahren 1855–59. Moscow: Buchdrückerei der Kaiserlichen Universität.

Radinsky, L.B. 1963. The Perissodactyl hallux. American Museum Novitates 2145:1–8.

Rahm, U. 1961. Esquise mammalogiques de basse Côte d'Ivoire. Bulletin de l'Institut Fondamental d'Afrique Noire 23:1229–1265.

Rahm, U. 1966. Les mammifères de la forêt équatoriale de l'est du Congo. Annales de la Musée Royale de l'Afrique Centrale 149:39–121.

Ramey, R.R. 1995. Mitochondrial DNA variation, population structure, and evolution of mountain sheep in the south-western United States and Mexico. Molecular Ecology 4:429–439.

Randi, E., V. Lucchini & C.H. Diong. 1996. Evolutionary genetics of the Suiformes as reconstructed using mtDNA sequencing. Journal of Mammalian Evolution 3:163–194.

Randi, E., N. Mucci, M. Pierpaoli & E. Douzery. 1998. New phylogenetic perspectives on the Cervidae (Artiodactyla) are provided by the mitochondrial cytochrome *b* gene. Proceedings of the Royal Society of London, B 265:793–801.

Randi, E., M. Pierpaoli & A. Danilkin. 1998. Mitochondrial DNA polymorphism in populations of Siberian and European roe deer (*Capreolus pygargus* and *Capreolus capreolus*). Heredity 80:429–437.

Randi, E., N. Mucci, F. Claro-Hergueta, A. Bonnet & E.J.P. Douzery. 2001. A mitochondrial DNA control region phylogeny of the Cervinae: speciation in *Cervus* and implications for conservation. Animal Conservation 4:1–11.

Randi, E., J.-P. d'Huart, V. Lucchini & R. Aman. 2002. Evidence of two genetically deeply divergent species of warthog, *Phacochoerus africanus* and *P. aethiopicus* (Artiodactyla: Suiformes) in East Africa. Mammalian Biology 67:91–96.

Ranjitsinh, M K. 1989. The Indian Blackbuck. Dehradun, India: Natraj.

Rautenbach, I.L. 1982. Mammals of the Transvaal. Pretoria: Ecoplan.

Rautian, G.S., B.A. Kalabushkin & A.S. Nemtsev. 2000. A new subspecies of the European bison, *Bison bonasus montanus* ssp. nov. (Bovidae, Artiodactyla). Doklady Biological Sciences 375:636–640.

Rebholz, W.[E.R.] & E. Harley. 1999. Phylogenetic relationships in the bovid subfamily Antilopinae based on mitochondrial DNA sequences. Molecular Phylogenetics and Evolution 12:87–94.

Rebholz, W.E.R., D. Williamson & F. Rietkerk. 1991. Saudi gazelle (*Gazella saudiya*) is not a subspecies of dorcas gazelle. Zoo Biology 10:485–489.

Redford, K.H. & J.F. Eisenberg. 1992. The Southern Cone: Chile, Argentina, Uruguay, Paraguay. Vol. 2 of Mammals of the Neotropics. Chicago: University of Chicago Press.

Reynolds, S.C. & L.C. Bishop. 2002. Cranio-dental variability in modern and fossil plains zebra (*Equus burchellii* Gray 1824) from east and southern Africa. In M. Mashkour (ed.), Equids in Time and Space: Papers in Honour of Véra Eisenmann. Oxford: Oxbow, 49–60.

Rideout, C.B. & R.S. Hoffmann. 1975. *Oreamnos americanus*. Mammalian Species 63:1–6.

Rieppel, O. 2009. Species as a process. Acta Biotheoretica 57:33–49.

Ritz, L.R., M.-L. Glowatzki-Mullis, D.E. MacHugh & C. Gaillard. 2000. Phylogenetic analysis of the tribe Bovini using microsatellites. Animal Genetics 31:178–185.

Roberts, A. 1913. The collection of mammals in the Transvaal Museum registered up to the 31st March 1913, with descriptions of new species. Annals of the Transvaal Museum 4:65–107.

Roberts, A. 1915. Additions to the collections of mammals in the Transvaal Museum. Annals of the Transvaal Museum 4:116–124.

Roberts, A. 1951. The Mammals of South Africa, ed. R. Bigalke, V. FitzSimons & D.E. Malan. Johannesburg: Trustees of "The Mammals of South Africa" Book Fund.

Roberts, S.C. 1996. The evolution of hornedness in female ruminants. Behaviour 133:399–442.

Roberts, T.J. 1977. The Mammals of Pakistan. London: Ernest Benn.

Robins, J.H., H.A. Ross, M.S. Allen & E. Matisoo-Smith. 2006. *Sus bucculentus* revisited. Nature 440: E7.

Robinson, T.J. 1979. Influence of a nutritional parameter on the size differences of the three springbok subspecies. South African Journal of Zoology 14:13–15.

Robinson, T.J. & J.D. Skinner. 1976. A karyological survey of springbok subspecies. South African Journal of Science 72:147–148.

Robinson, T.J., D.J. Morris & N. Fairall. 1991. Interspecific hybridisation in the Bovidae: sterility of *Alcelaphus buselaphus* × *Damaliscus dorcas* F_1 progeny. Biological Conservation 58:345–356.

Robinson, T.J., A.D. Bastos, K.M. Halanych & B. Heizig. 1996. Mitochondrial DNA sequence relationships of the extinct blue antelope *Hippotragus leucophaeus*. Naturwissenschaften 83:178–182.

Robinson, T.J., V. Trifonov, I. Espie & E.H. Harley. 2005. Interspecific hybridisation in rhinoceroses: confirmation of a black × white rhinoceros hybrid by karyotype, fluorescence *in situ* hybridisation (FISH), and microsatellites analysis. Conservation Genetics 6:141–145.

Roche, J. 1971. Recherches mammalogiques en Guinée forestière. Bulletin du Museum National d'Histoire Naturelle (3) 16:737–781.

Rode, P. 1943. Mammifères Ongulés de l'Afrique Noire. Vol. 2 of Faune de l'Empire Français. Paris: Larose.

Røed, K.H., M.A.D. Ferguson, M. Çrête & T.A. Bergerud. 1991. Genetic variation in transferrin as a predictor for differentiation and evolution of caribou from eastern Canada. Rangifer 11:65–74.

Rookmaaker, L.C [= K.]. 1980. The distribution of the rhinoceros in eastern India, Bangladesh, China, and the Indochinese region. Zoologischer Anzeiger 205:253–268.

Rookmaaker, L.C [= K.]. 1982. The type locality of the Javan rhinoceros (*Rhinoceros sondaicus* Desmarest, 1822). Zeitschrift für Säugetierkunde 47:381–382.

Rookmaaker, L.C [= K.]. 1983. Historical notes on the taxonomy and nomenclature of the recent Rhinocerotidae (Mammalia, Perissodactyla). Beaufortia 33, 4:37–51.

Rookmaaker, L.C [= K.]. 1988. The scientific names of the South African steenbok and grysbok (*Raphicerus campestris* and *R. melanotis*). Mammalia 52:213–217.

Rookmaaker, L.C. [= K.] 1989. The Zoological Exploration of Southern Africa, 1650–1790. Rotterdam: A.A. Balkema.

Rookmaaker, L.C [= K.]. 1991. The scientific name of the bontebok. Zeitschrift für Säugetierkunde 66:190–191.

Rookmaaker, L.C [= K.]. 1995. Subspecies and ecotypes of the black rhinoceros. Pachyderm 20:39–40.

Rookmaaker, L.C. [= K.]. 1997. Records of the Sundarbans rhinoceros (*Rhinoceros sondaicus inermis*) in India and Bangladesh. Pachyderm 24:37–45.

Rookmaaker, L.C [= K.]. 1998. The sources of Linnaeus on the rhinoceros. Svenska Linnésallskapets Årsskrift, Uppsala 1996–1997:61–80.

Rookmaaker, L.C. [= K.]. 2004. *Rhinoceros rugosus*—a name for the Indian rhinoceros. Journal of the Bombay Natural History Society 101:308–310.

Rookmaaker, L.C. [= K.]. 2005. The black rhino needs a taxonomic revision for sound conservation. International Zoo News 52:280–282.

Rookmaaker, L.C [= K.]. & C.P. Groves. 1978. The extinct Cape rhinoceros, *Diceros bicornis bicornis* (Linnaeus, 1758). Säugetierkunde Mitteilungen 26:117–126.

Roosevelt, T. & E. Heller. 1914. Life-Histories of African Game Animals, 2 vols. New York: Charles Scribner's Sons.

van Roosmalen, M.G.M., L. Frenz, P. van Hooft, H.H. de Iongh & H. Leirs. 2006. A new species of living peccary (Mammalia: Tayassuidae) from the Brazilian Amazon. Bonner Zoologische Beiträge 55:105–112.

Ropiquet, A. 2006. Étude des radiations adaptatives au sein des Antilopinae (Mammalia, Bovidae). PhD diss., Université Paris 6.

Ropiquet, A. & A. Hassanin. 2004. Molecular phylogeny of caprines (Bovidae, Antilopinae): the question of their origin and diversification during the Miocene. Journal of Zoological Systematics and Evolutionary Research 43:49–60.

Ropiquet, A. & A. Hassanin. 2005. Molecular evidence for the polyphyly of the genus *Hemitragus* (Mammalia, Bovidae). Molecular Phylogenetics and Evolution 36:154–168.

Ropiquet, A. & A. Hassanin. 2006. Hybrid origin of the Pliocene ancestor of wild goats. Molecular Phylogenetics and Evolution 41:395–414.

Ropiquet, A., M. Gerbault-Seureau, J.L. Deuve, C. Gilbert, E. Pagacova, N. Chai, J. Rubes & A. Hassanin. 2008. Chromosome evolution in the subtribe Bovina (Mammalia, Bovidae): the karyotype of the Cambodian banteng (*Bos javanicus birmanicus*) suggests that Robertsonian translocations are related to interspecific hybridization. Chromosome Research 16:1107–1118.

Rossi, R.V., R. Bodmer, J.M.B. Duarte & R.G. Trovati. 2010. Amazonian brown brocket deer *Mazama nemorivaga* (Cuvier 1817). In J.M.B. Duarte & S. González (eds.), Neotropical Cervidology: Biology and Medicine of Latin American Deer. Jaboticabal, Brazil and Gland, Switzerland: Funep, in collaboration with IUCN, 202–210.

Rössner, G.E. 2007. Family Tragulidae. In D.R. Prothero & S.E. Foss (eds.), The Evolution of Artiodactyls. Baltimore: Johns Hopkins University Press, 213–220.

Roth, T.L., H.L. Bateman, J.L. Kroll, B.G. Steinetz & P.R. Reinhart. 2004. Endocrine and ultrasonographic characterization of a successful pregnancy in a Sumatran rhinoceros (*Dicerorhinus sumatrensis*) supplemented with a synthetic progestin. Zoo Biology 23:219–238.

Ruiz-García, M., C. Vásquez, S. Sandoval, A. Castellanos, F. Kaston, B. de Thoisy & J. Shostell. In press. Phylogeography of the mountain tapir (*Tapirus pinchaque*) and of the Central American tapir (*Tapirus bairdii*) and the origins of the three Latin American tapirs by means of mtCyt-*b* sequences. Molecular Phylogeny and Evolution.

Rumiz, D.I. & E. Pardo. 2010. Peruvian dwarf brocket deer *Mazama chunyi* (Hershkovitz 1959). In J.M.B. Duarte & S. González (eds.), Neotropical Cervidology: Biology and Medicine of Latin American Deer. Jaboticabal, Brazil and Gland, Switzerland: Funep, in collaboration with IUCN, 185–189.

Rushby, C.C. & C.H. Swynnerton. 1946. Notes on some game animals of Tanganyika Territory. Tanganyika Notes and Records 22:14–26.

Ruxton, A.E. & E. Schwarz. 1929. On hybrid hartebeests and on the distribution of *Alcelaphus buselaphus*. Proceedings of the Zoological Society of London 1929:567–583.

Ryder, O.A., A.T. Kumamoto, B.S. Durrant & K. Benirschke. 1989. Chromosomal divergence and reproductive isolation in dik-diks. In D. Otte & J.A. Endler (eds.), Speciation and Its Consequences. Sunderland, MA: Sinauer Associates, 208–225.

Scala, C. & S. Lovari. 1984. Revision of *Rupicapra* genus, 2: A skull and horn statistical comparison of *Rupicapra rupicapra ornata* and *R. rupicapra pyrenaica* chamois. Bolletino Zoologico 51:285–294.

Schaffer, W.M. & C.A. Reed. 1972. The co-evolution of social behaviour and cranial morphology in sheep and goats (Bovidae, Caprini). Fieldiana, Zoology 61:1–62.

Schaller, G.B. 1977. Mountain Monarchs: Wild Sheep and Goats of the Himalaya. Chicago: University of Chicago Press.

Schaller, G.B. 1998. Wildlife of the Tibetan Steppe. Chicago: University of Chicago Press.

Schaller, G.B. & S. Khan. 1975. The status and distribution of markhor (*Capra falconeri*). Biological Conservation 7:185–198.

Schaller, G.B. & A. Rabinowitz. 1995. The saola or spindlehorn bovid *Pseudoryx nghetinhensis* in Laos. Oryx 29:107–114.

Schaller, G.B. & E.S. Vrba. 1996. Description of the giant muntjac (*Megamuntiacus vuquangensis*) in Laos. Journal of Mammalogy 77:675–683.

Schmitt, J. & F. Ulbrich 1968. Die Chromosomen verschiedener Caprini, Simpson, 1945. Zeitschrift für Säugetierkunde 33:180–186.

Schomber, H. 2007. Giraffengazelle und Lamagazelle, 2nd ed. Hohenwarsleben, Germany: Westarp Wissenschaften.

Schonewald, C. 1994. *Cervus canadensis* and *C. elaphus*: North American subspecies and evaluation of clinal extremes. Acta Theriologica 39:431–452.

Schouteden, H. 1944–1946. De zoogdieren van Belgisch-Congo en van Ruanda-Urundi. Annales de la Musée du Congo Belge, C: Zoologie (2) 3:1–576.

Schreiber, A. 1994. A female Bukhara deer *Cervus elaphus bactrianus* Lydekker, 1900 with 68 chromosomes. Acta Theriologica 39:99–102.

Schreiber, A. & G. Nötzold. 1995. One EEP, but how many anoas? In F. Rietkerk, K. Brouwer & S. Smits (eds.), EEP [European Endangered Species Program]Yearbook 1995/96. Amsterdam: EAZA [European Association of Zoos and Aquaria] Executive Office, 419–424.

Schreiber, A., P. Fakler & R. Østerballe. 1997. Blood protein variation in blackbuck (*Antilope cervicapra*), a lekking gazelle. Zeitschrift für Säugetierkunde 62:239–249.

Schreiber, A., I. Seibold, G. Nötzold & M. Wink. 1999. Cytochrome *b* gene haplotypes characterize chromosomal lineages of anoa, the Sulawesi dwarf buffalo (Bovidae: *Bubalus* sp.). Journal of Heredity 90:165–176.

Schwarz, E. 1926. Huftiere aus West und Zentralafrika, Ergebnisse der 2 Deutschen Zentral Afrika Expedition 1910–1911. Zoology 1:832–1044.

Sclater, P.L. & O. Thomas 1895. The Book of Antelopes, vol. 2 [of 4]. London: R.H. Porter.

Scoazec, J.-Y. 1966. An overlooked name of an African ungulate: *Connochaetes taurinus babaulti* Kollman, 1919. Mammalia 60:155–157.

Selous, F.C. 1908. African Nature Notes and Reminiscences. London: MacMillan.

Shackleton, D.M. (ed.). 1997. Wild Sheep and Goats and Their Relatives: Status Survey and Conservation Action Plan for Caprinae. Gland, Switzerland: IUCN.

Sheng, H-l. 1992. Deer in China. Shanghai: East China Normal University.

Shi, L. 1987. Recent trends in mammalian cytogenetics in China. La Kromosomo (2) 45:1458–1467.

Shortridge, G.C. 1934. The Mammals of South West Africa: A Biological Account of the Forms Occurring in that Region, 2 vols. London: William Heinemann.

Simpson, G.G. 1961. Principles of Animal Taxonomy. New York: Columbia University Press.

Skinner, J.D. & C.T. Chimimba. 2005. The Mammals of the Southern African Region, 3rd ed. Cambridge: Cambridge University Press.

Smithers, R.H.N. & J.L.P. Lobão Tello 1976. Check List and Atlas of the Mammals of Moçambique. Salisbury: Trustees of the National Museums and Monuments of Rhodesia.

Sobanskiy, G.G. 1988. [The game animals of the Altai Mountains.] Novosibirsk, Russia: Nauka. [In Russian.]

Sokolov, V.E. & A.A. Lushchekina. 1997. *Procapra gutturosa*. Mammalian Species 571:1–5.

Sokolov, V.E. & V.I. Prikhod'ko. 1997. [Taxonomy of the musk deer *Moschus moschiferus* (Artiodactyla, Mammalia)]. Ivestiia Akademii Nauk, Seriia Biologicheskaia [Biology Bulletin] 24:557–566. [In Russian.]

Sokolov, V.E. & V.I. Prikhod'ko. 1998. [Taxonomy of the musk deer *Moschus moschiferus* (Artiodactyla, Mammalia)]. Ivestiia Akademii Nauk, Seriia Biologicheskaia [Biology Bulletin] 25:28–36. [In Russian.]

Solounias, N. 1999. The remarkable anatomy of the giraffe's neck. Journal of Zoology, London 247:257–268.

Solounias, N. 2007a. Family Giraffidae. In D.R. Prothero & S.E. Foss (eds.), The Evolution of Artiodactyls. Baltimore: Johns Hopkins University Press, 257–277.

Solounias, N. 2007b. Family Bovidae. In D.R. Prothero & S.E. Foss (eds.), The Evolution of Artiodactyls. Baltimore: Johns Hopkins University Press, 278–291.

Soma, H. & H. Kada. 1986. Evolutionary pathway of chromosomes of the Capricornis. In O.A. Ryder & M.L. Byrd (eds.), One Medicine: Essays in Honour of Kurt Benirschke. Berlin: Springer-Verlag, 109–118.

Soma, H., T. Kiyokawa, K. Matayoshi, I. Tarumoto, M. Miyashita & K. Nagase. 1979. The chromosomes of the Mongolian gazelle (*Procapra gutturosa*), a rare species of antelope. Proceedings of the Japan Academy, Ser. B: Physical and Biological Science 55:6–9.

Soma, H., H. Kada, K. Matayoshi, Y. Suzuki, C. Meckvichal, A. Mahannop & B. Vatanaromya. 1983. The chromosomes of *Muntiacus feae*. Cytogenetics and Cell Genetics 35:156–158.

Soma, H., H. Kada & K. Matayoshi. 1987. Evolutionary pathways of karyotypes of the tribe Rupicaprini. In H. Soma (ed.), The Biology and Management of Capricornis and Related Mountain Antelopes. London: Croom Helm, 62–71.

Sopin, L.V. & D.L. Harrison. 1987. On the taxonomic position of the Oman wild sheep. Zoologicheskii Zhurnal 65:952–955.

Sowerby, A. de C. 1917. On Heude's collections of pigs, sika, serows, and gorals in the Sikawei Museum, Shanghai. Proceedings of the Zoological Society of London 1917:7–26.

van Staaden, M.J., M.J. Hamilton & R.K. Chesser. 1995. Genetic variation of woodland caribou (*Rangifer tarandus*) in North America. Zeitschrift für Säugetierkunde 60:150–158.

Stanley, H.F., M. Kadwell & J.C. Wheeler. 1994. Molecular evolution of the family Camelidae: a mitochondrial

DNA study. Proceedings of the Royal Society of London B, 256:1–6.

Stockley, C.H. 1923. On the forms of the Himalayan serow *Capricornis sumatraensis*. Journal of the Bombay Natural History Society 29:824–827.

Stoner, C.J., T.M. Caro & C.M. Graham. 2003. Ecological and behavioural correlates of coloration in artiodactyls: systematic analyses of conventional hypotheses. Behavioural Ecology 14:823–840.

Stott, K.[W]. 1959. Giraffe intergradation in Kenya. Journal of Mammalogy 40:251.

Stott, K.W. & C.J. Selsor. 1981. Further remarks on giraffe intergradation in Kenya and unreported marking variations in reticulated and Maasai giraffes. Mammalia 45:261–263.

Sugiri, N. & N. Hidayet. 1996. The diversity and haematology of anoa from Sulawesi. Paper presented at the Population and Habitat Viability Assessment Workshop for the Anoa (*Bubalua depressicornis*, *Bubalus quarlesi*), 22–26 July 1996, Bogor, Java Barat, Indonesia, IUCN/SSC Conservation Breeding Specialist Group.

Swynnerton, C.H. & R.W. Hayman. 1951. A checklist of the land mammals of the Tanganyika Territory and the Zanzibar Protectorate. Journal of the East Africa Natural History Society 20:274–392.

Tchernov, E., T. Dayan & Y. Yom-Tov. 1987. The paleogeography of *Gazella gazella* and *Gazella dorcas* during the Holocene of the southern Levant. Israel Journal of Zoology 34:51–59.

Thomas, D.C. & P. Everson. 1982. Geographic variation in caribou on the Canadian Arctic islands. Canadian Journal of Zoology 60:2442–2454.

Thomas, H., A. Seveau & A. Hassanin. 2001. The enigmatic new Indochinese bovid, *Pseudonovibos spiralis*, an extraordinary forgery. Comptes Rendus de l'Académie des Sciences (Paris), Sér. 3, Sciences de la Vie 324:81–86.

Thomas, O. 1892. On the Mammalia Collected by Signor Leonardo Fea in Burma and Tenasserim. Viaggio di Leonardo Fea in Birmania e Regioni Vicine, 41. Genoa: Istituto Sordo-Muti.

Thomas, O. 1898. On the Zululand form of Livingstone's antelope (*Nesotragus livingstonianus*). Annals and Magazine of Natural History (7) 2:317.

Thomas, O. 1906. On a new pygmy antelope obtained by Col. J. J. Harrison in the Semliki forest. Annals and Magazine of Natural History (7) 18:148–150.

Thomas, O. 1911. The mammals of the tenth edition of Linnaeus: an attempt to fix the types of the genera and the exact bases and localities of the species. Proceedings of the Zoological Society of London 1911:120–158.

Thomas, O. 1917. Preliminary diagnoses of the new mammals obtained by the Yale National Geographic Society Peruvian expedition. Smithsonian Miscellaneous Collections 63, 4:1–3.

Thouless, C.R. & K. al Bassri. 1991. Taxonomic status of the Farasan Island gazelle. Journal of Zoology, London 223:151–159.

Tun Yin, U. 1967. Wild Animals of Burma. Rangoon: Rangoon Gazette.

Tuoc, D., V.V. Dung, S. Dawson, P. Arctander & J. MacKinnon. 1994. Vù một loài mang lon moi phỗt hiên o Viêt Nam. Thông Tin Khoa Hoc Ky Thuât 3:4–12.

Uerpmann, H.P. 1987. The Ancient Distribution of the Ungulate Mammals in the Middle East. Beihefte zum Tübinger Atlas des Vorderen Orients, Reihe A, Naturwissenschaften, No. 27. Wiesbaden, Germany: Ludwig Reichert.

Urbain, A. & H. Friant. 1942. Recherches anatomiques sur l'antilope royale. Archives du Museum d'Histoire Naturelle, Paris (6) 18:167–179.

Vadhanakul, N., M. Tansatit & W. Tunwattana. 2004. Karyotype of a Thai gaur sire of Khao Kheow Open Zoo. Journal of the Thai Veterinary Medical Association under Royal Patronage 54:35–45.

Valdez, R. 1982. The Wild Sheep of the World. Mesilla, NM: Wild Sheep and Goat International.

Valdez, R. 1985. Lords of the Pinnacles: Wild Goats of the World. Mesilla, NM: Wild Sheep and Goat International.

Valdez, R., C.F. Nadler & T.D. Bunch. 1978. Evolution of wild sheep in Iran. Evolution 32:56–72.

Vanpé, C., J.-M. Gaillard, P. Kjellander, A. Mysterud, P. Magnien, D. Delorme, G. van Laere, F. Klein, O. Liberg & A.J.M. Hewison. 2007. Antler size provides an honest signal of male phenotypic quality in roe deer. American Naturalist 169:481–493.

van Zyll de Jong, C.G. 1986. A Systematic Study of Recent Bison, with Particular Consideration of the Wood Bison (*Bison bison athabascae* Rhoads, 1898). National Museum of Natural Sciences, Publications in Natural Sciences No. 6. Ottawa: National Museums of Canada, National Museum of Natural Sciences.

van Zyll de Jong, C.G., C. Gates, H. Reynolds & W. Olsen. 1995. Phenotypic variation in remnant populations of North American bison. Journal of Mammalogy 76:391–405.

Varela, D.M., R.G. Trovati, K.R. Guzmán, R.V. Rossi & J.M.B. Duarte. 2010. Red brocket deer *Mazama americana* (Erxleben 1777). In J.M.B. Duarte & S. González (eds.), Neotropical Cervidology: Biology and Medicine of Latin American Deer. Jaboticabal, Brazil and Gland, Switzerland: Funep, in collaboration with IUCN, 151–159.

Vasil'ev, V.A., E.P. Steklenev, E.V. Morozova & S.K. Semyenova. 2002. DNA fingerprinting of individual species and intergeneric and interspecific hybrids of the genera *Bos* and *Bison*, subfamily Bovinae. Russian Journal of Genetics 38:415–420.

Vassart, M., L. Granjon & A. Greth. 1991. Genetic variability in the Arabian oryx. Zoo Biology 10:399–408.

Vassart, M., L. Granjon, A. Greth, J.F. Asmodé & E.P. Cribiu. 1992. Genetic research on captive Arabian oryx in Saudi Arabia. In F. Spitz, G. Janeau, G. Gonzalez, et al. (eds.), Proceedings of the International Symposium "Ongulés/Ungulates 91." Paris: Société Française pour l'Étude et la Protection des Mammifères, 595–598.

Vassart, M., L. Granjon, A. Greth & F.M. Catzeflis. 1994. Genetic relationships of some *Gazella* species: an allozyme survey. Zeitschrift für Säugetierkunde 59:236–245.

Vassart, M., A. Pinton, A. Séguéla & C. Dutertre. 1994. New data on chromosomes of peccaries. Mammalia 58:500–507.

Vassart, M., L. Granjon & A. Greth. 1995. Genetic study of *Gazella gazella*: chromosomal and allozyme data. Comptes Rendus de l'Académie des Sciences (Paris), Sér. 3, Sciences de la Vie 318:27–33.

Vassart, M., A. Séguéla & H. Hayes. 1995. Chromosomal evolution in gazelles. Journal of Heredity 86:216–227.

Vereschagin, N.K. 1959 [English translation 1967]. The Mammals of the Caucasus. Leningrad: Izd-vo Akademii Nauk SSSR [English translation by A. Lerman & B. Rabinovich, Jerusalem: Israel Program for Scientific Translations].

Verkaar, E.L.C., I.J. Nijman, M. Beeke, E. Hanekamp & J.A. Lenstra. 2004. Maternal and paternal lineages in cross-breeding bovine species: has wisent a hybrid origin? Molecular Biology and Evolution 21:1165–1170.

Vila, A.R., C.E. Saucedo Galvez, D. Aldridge, E. Ramilo & P.C. González. 2010. South Andean huemul *Hippocamelus bisulcus* (Molina 1782). In J.M.B. Duarte & S. González (eds.), Neotropical Cervidology: Biology and Medicine of Latin American Deer. Jaboticabal, Brazil and Gland, Switzerland: Funep, in collaboration with IUCN, 89–100.

Vogliotti, A. & J.M.B. Duarte. 2010. Small red brocket deer *Mazama bororo* (Duarte 1996). In J.M.B. Duarte & S. González (eds.), Neotropical Cervidology: Biology and Medicine of Latin American Deer. Jaboticabal, Brazil and Gland, Switzerland: Funep, in collaboration with IUCN, 172–176.

Volf, J. 1976. Some remarks on the taxonomy of the genus *Nemorhaedus* H. Smith 1827 (Bovidae: Rupicaprini). Vestnik Ceskoslovenska, Spolia Zoologia 40:75–80.

Von Dueben, C. 1846. In C.J. Sundevall, Ny Antelop-art. Öfversigt af Kongliga Vetenskaps-Akademiens Förhandlingar, Stockholm 3, 7:221–222.

Vorobeev, G.G. & J. van der Ven 2003. Looking at Mammals in Kyrgyzia, Central Asia. Bishkek, Kyrgyzstan: PDC.

Vrba, E.S. 1979. Phylogenetic analysis and classification of fossil and recent Alcelaphini (Mammalia: Bovidae). Biological Journal of the Linnaean Society 11:207–228.

Vrba, E.S. 1980. Evolution, species, and fossils: how does life evolve? South African Journal of Science 76:61–84.

Vrba, E.S. 1997. New fossils of Alcelaphini and Caprinae (Bovidae: Mammalia) from Awash, Ethiopia, and phylogenetic analysis of Alcelaphini. Palaeontologica Africana 34:127–198.

Vrba, E.S. & G.B. Schaller. 2000. Phylogeny of Bovidae based on behaviour, glands, skulls, and postcrania. In E.S. Vrba & G.B. Schaller (eds.), Antelopes, Deer, and Relatives. New Haven: Yale University Press, 203–222.

Vrba, E.S., J.R. Vaismys, J.E. Gatesy, R. DeSalle & K-y. Wei. 1994. Analysis of paedomorphosis using allometric characters: the example of Reduncini antelopes (Bovidae, Mammalia). Systematic Biology 43:92–116.

Waddell, P.J., N. Okada & M. Hasegawa, 1999. Towards resolving the interordinal relationships of placental mammals. Systematic Biology 48:1–5.

Wallace, H.F. 1913. The Big Game of Central and Western China. London: John Murray.

Ward, A.E. 1925. The mammals and birds of Kashmir and the adjacent hill provinces, part 2. Journal of the Bombay Natural History Society 30:118–131.

Webb, S.D. 2000. Evolutionary history of New World Cervidae. In E.S. Vrba & G.B. Schaller (eds.), Antelopes, Deer, and Relatives. New Haven: Yale University Press, 38–64.

Weber, M. & R.A. Medellín. 2010. Yucatán brown brocket deer *Mazama pandora* (Merriam 1901). In J.M.B. Duarte & S. González (eds.), Neotropical Cervidology: Biology and Medicine of Latin American Deer. Jaboticabal, Brazil and Gland, Switzerland: Funep, in collaboration with IUCN, 211–216.

Weinberg, P.J. 2002. *Capra cylindricornis*. Mammalian Species 695:1–9.

Welsch, U., G. van Dyk, D. Moss & F. Feuerhake. 1998. Cutaneous glands of male and female impalas (*Aepyceros melampus*): seasonal activity changes in secretory mechanisms. Cell and Tissue Research 292:377–394.

Wetzel, R.M. 1977. The Chacoan peccary *Catagonus wagneri* (Rusconi). Bulletin of the Carnegie Museum of Natural History 3:1–36.

Wheeler, J.C. 1995. Evolution and present situation of the South American Camelidae. Biological Journal of the Linnaean Society 54:271–295.

Wheeler, J.C. 1998. Evolution and origin of the domestic camelids. Alpaca Registry Journal 3, 1:1–15.

Wilhelmi, E., H.Y. Kaariye & S. Hammer. 2007. Das Al Wabra Wildlife Preservation Dibatag-Projekt Beobachtungen in der Region Ogaden, Südost-Äthiopien. ZGAP [Zoologische Gesellschaft für Arten- und Populationsschutz] Mitteilungen, 23:9–12.

Willerslev, E., M. Gilbert, J. Binladen, S. Ho, P. Campos, A. Ratan, L. Tomsho, R. de Fonseca, A. Sher, T. Kuznetsova, M. Nowak-Kemp, T. Roth, W. Miller & S. Schuster. 2009. Analysis of complete mitochondrial genomes from extinct and extant rhinoceroses reveals lack of phylogenetic resolution. BMC Evolutionary Biology 9:95. doi:10.1986/1471-2148-9-95.

Willows-Munro, S., T.J. Robinson & C.A. Matthee. 2005. Utility of nuclear DNA intron markers at lower taxonomic levels: phylogenetic resolution among nine

Tragelaphus spp. Molecular Phylogenetics and Evolution 35:624–636.

Wilson, D.E. & D.M. Reeder. 2005. Mammal Species of the World, 3rd ed., 2 vols. Baltimore: Johns Hopkins University Press.

Wilson, V.J. (ed.). 2002. Duikers of Africa: Masters of the African Forest Floor. Bulawayo, Zimbabwe: Chipangali Wildlife Trust.

Witmer, L.M., S.D. Sampson & N. Solounias. 1999. The proboscis of tapirs (Mammalia: Perissodactyla): a case study in novel narial anatomy. Journal of Zoology, London 249:249–267.

Woodburne, M O. 1968. The cranial myology and osteology of *Dicotyles tajacu*, the collared peccary, and its bearing on classification. Memoirs of the Southern California Academy of Sciences 7:1–46.

Wu C-h., Y-.p. Zhang, T.D. Bunch, S. Wang & W. Wang. 2003. Mitochondrial control region sequence variation within the argali wild sheep (*Ovis ammon*): evolution and conservation relevance. Mammalia 67:109–118.

Wu H., Q-h. Wan & S-g. Fang. 2004. Two genetically distinct units of the Chinese sika deer (*Cervus nippon*): analyses of mitochondrial DNA variation. Biological Conservation 119:183–190.

Wurster, D.H. 1972. Sex-chromosome translocations and karyotypes in bovid tribes. Cytogenetics 11:197–207.

Wyrwoll, T.W. 1999. Eine Neubeachreibung des Südspanischen Steinbocks. Säugetierkundliche Mitteilungen 44:93–98.

Yalden, D.W. 1978. A revision of the dik-diks of the subgenus *Madoqua* (*Madoqua*). Monitore Zoologico Italiano (n.s.) Supplement 11, 10:245–264.

Yalden, D.W. 2007. Obituary—Peter Grubb (1944–2006). Mammal News 147:4.

Yalden, D.W., M.J. Largen & D. Kock. 1984. Catalogue of the mammals of Ethiopia, 5: Artiodactyla. Monitore Zoologico Italiano (n.s.) Supplement 19, 4:67–221.

Yang F., N.P. Carter, L. Shi & M.A. Ferguson-Smith. 1995. A comparative study of karyotypes of muntjacs by chromosome painting. Chromosoma 103:642–652.

Zainuddin, Z.-Z., M.-T. Abdullah & M.-F.M. Suri. 1990. The husbandry and veterinary care of captive Sumatran rhinoceros at Zoo Melaka, Malaysia. Mammalian Nature Journal 44:1–19.

Zalkin, V.I. 1949. Sistematika sibirskikh gornykh kozlov [Systematics of Siberian ibex]. Biulleten Moskovskogo Obshchestva Ispytatelei Prirody, Otdel Biologicheskii 54:3–21. [In Russian].

Zalkin, V.I. 1950. [Subspecies of wild goat]. Comptes Rendus de l'Académie des Sciences de l'URSS 22:325. [In Russian].

Zeder, M.A. 2001. A metrical analysis of a collection of modern goats (*Capra hircus aegagrus* and *C. h. hircus*) from Iran and Iraq: implications for the study of caprine domestication. Journal of Archaeological Science 28:61–79.

Železnov, N.K. 1994. Obca snežná (*Ovis nivicola* Esch.) v sebrobuýchodnej Sibíri—Snow sheep (*Ovis nivicola* Esch.) in north-eastern Siberia. Folia Venatoria 24:161–169.

Zhang F-f. & Z-g. Jiang. 2006. Mitochondrial phylogeography and genetic diversity of the Tibetan gazelle (*Procapra picticaudata*): implications for conservation. Molecular Phylogenetics and Evolution 41:313–321.

Zukowsky, L. 1910. Halbseitige Bastarde bei geographischen Rassen von Grosswild aus freier Wildbahn. Zoologischer Beobachter der Zoologische Garten: Zeitschrift für Beobachtung, Pflege, und Zucht der Tiere 51:259–272.

Zukowsky, L. 1913a. Ergänzungen zu meinen Arbeiten über *Connochaetes albojubatus* Ths. und *Eudorcas thomsoni* Gthr. Archiv für Naturgeschichte 80:142–146.

Zukowsky, L. 1913b. Ueber einige anscheinend neue Rassen von *Connochaetus albojubatus* Thomas aus Deutsch-Ostafrika. Archiv für Naturgerschichte 79:76–91.

Zukowsky, L. 1924. Ein Wort über die Notwendigkeit der systematischen Bearbeitung der Wisentreste. Pallasia 2, 1:1–11.

Zukowsky, L. 1928a. Nachtrag zu den Bemerkingen über die rassenweise Verschiedenheit der Hirschziegenantilope. Carl Hagenbecks Illustrierte Tier- und Menschenwelt 2:147.

Zukowsky, L. 1928b. Bemerkungen über die rassenweise Verschiedenheit der Hirschziegenantilope. Carl Hagenbecks Illustrierte Tier- und Menschenwelt 2:124–127.

Zukowsky, L. 1961. Über eine verzwergte Inselform des Buschbockes, *Tragelaphus scriptus*, in Tanganjika. Zoologischer Garten (n.f.) 26:54–55.

Zukowsky, L. 1965 [t.p. 1964, published 1965]. Die Systematik der Gattung *Diceros* Gray, 1821. Der Zoologische Garten (n.f.) 30:1–178.

Index